Conceptual Issues in Evolutionary Biology

Second Edition

T185

edited by Elliott Sober

A Bradford Book
The MIT Press
Cambridge, Massachusetts
London, England

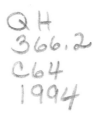

Third printing, 1997

This book was set in Palatino by Asco Trade Typesetting Ltd., Hong Kong and was printed and bound in the United States of America.

Library of Congress Cataloging-in-Publication Data

Conceptual issues in evolutionary biology / edited by Elliott Sober.—2nd ed.
  p.   cm.
"A Bradford book."
Includes bibliographical references and index.
ISBN 0-262-19336-1.—ISBN 0-262-69162-0 (pbk.)
  1. Evolution (Biology) 2. Evolution (Biology)—Philosophy. 3. Biology—Philosophy.
I. Sober, Elliott.
QH366.2.C64   1993
575—dc20   93-8199
                                                                        CIP

Contents

Preface

This anthology brings together a set of essays, written by philosophers and scientists, concerning foundational issues that arise in the theory and practice of evolutionary biology. It supersedes the first edition by the same title, which was published in 1984. About half the chapters in this book did not appear in its immediate ancestor. Philosophy of biology has displayed great vitality since the first edition appeared; the subject has broadened and deepened on several fronts. This second edition reflects many of the important changes that have occurred.

This book should be useful to evolutionary biologists and to philosophers of biology, both in their teaching and in their research. As a teaching tool, this anthology might be used in conjunction with one of the textbooks now available in philosophy of biology: Michael Ruse's *The Philosophy of Biology*, David Hull's *Philosophy of Biological Science*, Alexander Rosenberg's *The Structure of Biological Science*, and my own book, *Philosophy of Biology*.

In the remainder of this Preface, I'll briefly indicate what some of the major issues are that animate the chapters that follow. The first four sections—on Fitness, Function and Teleology, Adaptationism, and Units of Selection—address a connected set of issues having to do with the role played by the concept of natural selection in evolutionary theory. The next four sections—on Essentialism and Population Thinking, Species, Systematic Philosophies, and Phylogenetic Inference—have to do with the nature of species and higher taxa. What are species, how should they be classified, and how should their genealogical relationships to each other be reconstructed? The next section—on the Reduction of Mendelian Genetics to Molecular Biology—examines an important test case for the idea of reductionism in biology. The remaining two sections—on Ethics and Sociobiology and on Cultural Evolution and Evolutionary Epistemology—examine attempts to apply evolutionary theory to problems that are traditionally reserved for philosophy and the human sciences: the nature of morality, the historical changes that cultures experience, and the way human knowledge grows.

I   FITNESS

Darwin accorded a preeminent role to the process of natural selection in his account of how life has evolved. Central to the concept of natural selection is the idea of fitness: if the organisms in a population undergo a selection process, they must differ from each other in terms of their abilities to survive and reproduce.

Herbert Spencer coined the slogan "survival of the fittest" to describe Darwin's theory. Ever since, the theory has been criticized for being an empty truism. Who survives? Those who are fit. And who are the fit? Those who survive. If the theory of natural selection comes to no more than this, then the "theory" is no theory at all. It is a piece of circular reasoning that masquerades as a genuine causal explanation of what we observe.

In chapter 1, "The Propensity Interpretation of Fitness," Susan Mills and John Beatty address this criticism by trying to clarify the probabilistic character of the concept of fitness. Their goal is to describe how the concept of fitness figures in contemporary biology and to show that the charge of circularity is quite misguided.

II   FUNCTION AND TELEOLOGY

Biology, and especially evolutionary biology, is rife with claims concerning what various characteristics are "for." The heart exists for the purpose of pumping blood. Bears have fur in order to ward off the cold.

Functional claims of this sort have quite disappeared from physics. Whereas Aristotle thought that planets, no less than living things, have goals, this teleological conception of the physical world is now a relic of a bygone age. Planets move as they do because of the laws of motion; they do not act as they do for the good of anything.

Darwin is rightly famous for having introduced an important materialist element into the science of life. But rather than banishing functional notions from biology, he showed how they can be domesticated within a materialist framework. Organisms are goal-directed systems because they have evolved. Their behaviors are suited to the tasks of survival and reproduction because natural selection has allowed some traits, but not others, to be passed from ancestors to descendants.

Even if Darwinism legitimates talk of goal and purpose within biology, the question of what such talk means remains to be addressed. The heart does many things. It pumps blood, but it also makes noise and takes up space in our chests. Why are we inclined to say that pumping blood is part of the heart's function, but making noise and taking up space are not? Larry Wright ("Functions") and Robert Cummins ("Functional Analysis") offer two competing accounts of how functional claims should be understood.

Although their views conflict in major ways, both recognize a distinction that is crucial to the practice of evolutionary biology. Biologists sometimes mark this distinction by talking about the difference between adaptations and fortuitous benefits. When sea turtles come out of the ocean to lay eggs, they use their front flippers to dig nests. The flippers are obviously beneficial in this regard, but that is not why flippers evolved. Turtles and their ancestors had front flippers long before any turtle came out of the sea. Flippers provide a fortuitous benefit when they are used to build nests; they are not adaptations for nest building (though they may be adaptations for locomotion).

Used in this way, adaptation is a historical concept. To say that a trait is now an adaptation for some task is not to say that it now helps the organism to perform that task. Rather, the trait is now an adaptation only if its past evolution was due to the fact that it helped perform that task. The fact that our hearts make noise now provides a benefit: physicians can use stethoscopes to diagnosis heart problems. Making noise thus provides a fortuitous benefit; however, the heart is an adaptation for circulating the blood, not for making noise.

III ADAPTATIONISM

Although Darwin accorded a central role to the process of natural selection, he allowed that other processes can influence the course of evolution. How important these other processes are, and how we should endeavor to understand the features of living things that we observe, are the main issues in the controversy concerning adaptationism.

When a biologist studies a complex characteristic of morphology, physiology, or behavior, the first impulse is often to ask, "What is this trait for?" The mind searches for the trait's adaptive significance. We observe the dorsal fins on a dinosaur and immediately set to work thinking about whether the fins are for self-defense, or for thermal regulation, or to attract mates. If one hypothesis of adaptive advantage does not pan out, we discard it and invent another.

In chapter 4, "The Spandrels of San Marco and the Panglossian Paradigm: A Critique of the Adaptationist Programme," Stephen Jay Gould and Richard Lewontin argue that this impulse toward adaptationist thinking has led biologists to neglect the possibility that many traits may be present for reasons quite disjoint from that of adaptive significance. Adaptationism, they suggest, is an assumption that is as pervasive as it is unproven. In chapter 5, "Optimization Theory in Evolution," John Maynard Smith replies to this criticism by outlining the elements that figure in adaptationist explanations; he defends the adaptationist approach by suggesting how particular adaptationist explanations can be tested.

IV UNITS OF SELECTION

Humans are organisms, so it may strike us as entirely natural to think that the parts of organisms exist in order to benefit the organisms that contain them. We have hearts, so we naturally think that the heart exists in order to help the organism in some way. Each of our cells contains genes, so we naturally think that genes exist in order to help the organism in some way.

If we are prepared to think that organs have the function of caring for organisms, why not frameshift this idea up a level and conclude that organisms have the function of caring for the groups in which they live? If an organism's heart is programmed to help the organism survive, maybe the organism is programmed to help its species avoid extinction. And if parts can have the function of helping the wholes in which exist, why can't the opposite relationship also obtain? Why not think of organisms as devices that have the function of guaranteeing the survival and reproduction of the genes they contain?

These questions are central to what is now called the problem of the units of selection. Let us assume that a trait—the opposable thumb, for example—evolved because it was good for the things that possessed it. But what kinds of objects should we regard as the relevant beneficiaries? Did the opposable thumb evolve because it helped individual organisms to survive and reproduce, because it helped the species to avoid extinction, or because it helped some genes make their way into successive generations?

Darwin usually thought of natural selection as a process in which traits evolve because they benefit individual organisms. The two most famous exceptions to this pattern of thinking were his discussion in *The Origin of Species* of sterile castes in the social insects and his discussion in *The Descent of Man* of human morality. In both, instances, Darwin argued that a trait evolved because it benefited the group and in spite of the fact that it was deleterious to individuals possessing it.

Since Darwin's day, the idea of higher-level adaptations has continued to interest evolutionists. For example, in *The Genetical Theory of Natural Selection*, R. A. Fisher suggested that sex may be a group adaptation; perhaps its function is to make the species genetically diverse, so that the species is less likely to go extinct if the environment suddenly changes.

Although group selection thinking was quite common in the heyday of the Modern Synthesis, its fortunes plummeted when George C. Williams published *Adaptation and Natural Selection* in 1966. The most prominent message in Williams's book was that group selectionism is a kind of sloppy thinking. Biologists had uncritically talked about the good of the species, although more parsimonious explanations could be provided at lower levels.

Although Darwin almost always favored individual selection over group selection, Williams's critique of group selection thinking did not conclude that the classical Darwinian picture was the best way to think about adaptation. Rather, he suggested that we descend to a lower level still. The real unit of selection, Williams argued, is not the group, nor even the individual, but the gene. Thus was born the view of natural selection that Richard Dawkins later popularized in his book *The Selfish Gene*.

The first chapter in part IV consists of excerpts from Williams's *Adaptation and Natural Selection*. Williams emphasizes the importance of not assuming that a trait that benefits a group must be a group adaptation. He also argues that hypotheses of group adaptation are less parsimonious than those that posit adaptations at lower levels. In defense of the genic point of view,

Williams contends that genes have a longevity that gene combinations and whole organisms do not. Williams also argues that group selection must be a weak force, compared with individual selection, because groups usually go extinct and found colonies at a slower rate than the rate at which organisms die and reproduce.

In chapter 7, "Levels of Selection: An Alternative to Individualism in Biology and the Human Sciences," David Sloan Wilson argues that the selfish gene theory rests on a fallacy. Wilson does not urge a return to uncritical group selectionism; rather, he defends a pluralistic conception of selection, in which adaptations can evolve for a variety of reasons. According to Wilson, the living world contains selfish genes, but it also contains well-adapted organisms and well-adapted groups.

V ESSENTIALISM AND POPULATION THINKING

"Discerning the essence of things" usually is interpreted as meaning that one has identified what is most important. However, the phrase's familiarity should not lead us to forget that the word "essence" is a vestige of a substantive philosophical doctrine, one that has exercised a considerable influence on metaphysics, both ancient and modern. Essentialism is a doctrine about natural kinds. Gold is a kind of thing; there are many samples of gold, and they are quite different from each other. In the light of this diversity among gold things, how should science endeavor to understand what gold is?

The essentialist replies that science aims to discover properties that are separately necessary and jointly sufficient for being gold. If all and only the specimens of gold have a particular atomic number, then that atomic number may be the essence of gold. Discovery of essences is not an idle exercise but is fundamental to science's search for explanation. We understand what gold is by seeing what all gold things have in common.

Understood in this way, the existence of variation within a kind is a distraction from what is important. Gold things vary from each other, but this variation needs to be set aside. Variation is a veil that we must see through if we are to see what is important.

In "Typological versus Population Thinking" (chapter 8), Ernst Mayr describes how evolutionary biology has rejected essentialism in favor of an alternative doctrine, which he terms population thinking. I try to provide further clarification of this issue in chapter 9, "Evolution, Population Thinking, and Essentialism."

VI SPECIES

When philosophers try to cite examples of natural kinds, they often mention chemical elements and biological species. I just exemplified half this pattern by describing gold as a natural kind whose essence is its atomic number. But is it correct to think of species in the same way? Is the house mouse (*Mus musculus*) a kind of creature in the same way that gold is a kind of stuff?

In "A Matter of Individuality" (chapter 10), David Hull articulates a position that he and Michael Ghiselin have developed. Picking up on Mayr's antiessentialism, they argue that species are individuals, not natural kinds. Species are born and die and have a significant degree of internal cohesiveness while they persist. Hull defends the idea that a species is an integrated gene pool; organisms belong to the same species not because they are similar, but because they reproduce with each other. Individualists such as Hull and Ghiselin have therefore endorsed Ernst Mayr's biological species concept.

Although there are some points of affinity between Hull's individuality thesis and the position advocated by Brent Mishler and Michael Donoghue, the latter authors argue that no single species concept can do justice to the diversity of nature. They defend a pluralistic position, according to which different groups of organisms may require different species concepts.

VII SYSTEMATIC PHILOSOPHIES

Once organisms are grouped into species, the task remains to group those species together into superspecific taxa. The first three chapters in this part present the three main alternatives now in contention; the final chapter, by David Hull, provides an overall assessment of the three approaches.

In "The Continuing Search for Order," Robert Sokal defends the phenetic approach to classification, according to which taxonomic groups are defined by a criterion of overall similarity. Pheneticists have argued that the goals of classification require one to eschew considerations that are specific to a particular theory; the goal is to provide an all-purpose classification, not one that is suited only to the needs of a particular problem.

In stark opposition to the phenetic approach is the cladistic philosophy defended by Willi Hennig in "Phylogenetic Systematics." Hennig argues that genealogical relationship, not overall similarity, should be the criterion of classification. We should place placental wolves and moles together in a taxon that does not include marsupial wolves because the first two are more closely related to each other than either is to the third. Although placental and marsupial wolves may "look alike," Hennig suggests that such impressions cannot form the basis of an objective classification.

The third systematic philosophy is termed evolutionary taxonomy; it is defended by Ernst Mayr in "Biological Classification: Toward a Synthesis of Opposing Methodologies." Evolutionary taxonomy endorses a "synthetic" approach. Sometimes it groups by adaptive similarity and sometimes by genealogical relatedness. Mayr argues that this approach maximizes the information content of the resulting classification.

Historically, pheneticism was a reaction against evolutionary taxonomy, and cladistics became popular among systematists as a reaction against both pheneticism and evolutionary taxonomy. However, the ordering I have imposed on the three chapters is not true to the historical sequence of the ideas they represent. My reason for this arrangement is that it is easier to under-stand what separates these philosophies by envisioning pheneticism and cladism as "extreme" positions and evolutionary taxonomy as a proposed compromise. This does not mean that the intermediate position is in fact the best, a point that emerges clearly from Hull's "Contemporary Systematic Philosophies."

VIII PHYLOGENETIC INFERENCE

The three systematic philosophies presented in part VII can be viewed as reacting differently to the following question: if you knew what genealogical relationships obtained among a set of species, how would you use that infor-mation to construct a classification? Pheneticists would reject the information as irrelevant, evolutionary taxonomists would take the information to be rele-vant but insufficient to forge a taxonomy, and cladists would take the informa-tion to be both relevant and sufficient. In Chapters 16 and 17, respectively, James Farris and Joseph Felsenstein examine the logically prior question of how to figure out the genealogical relationships among a set of taxa.

We do not directly observe that human beings and horses are more closely related to each other than either is to snakes. Rather, we observe various characteristics that the three groups display; what we observe are patterns of similarity and difference. Human beings and horses are warm-blooded; snakes are not. Horses and snakes do not have opposable thumbs; humans do. How are we to use observations such as these to infer phylogenetic relationships?

In "The Logical Basis of Phylogenetic Analysis," James Farris defends a hypothetico-deductive approach to the problem of phylogenetic inference. We should choose genealogical hypotheses on the basis of their explanatory power. The most explanatory hypothesis is the one that is most parsimonious. Farris endorses the cladistic idea that the parsimony of a phylogenetic hypoth-esis is measured by counting the number of independent originations of fea-tures that the hypothesis requires to explain the data. A hypothesis that views robins and sparrows as obtaining their wings from a common ancestor is more parsimonious than one that says that wings evolved separately in the two lineages.

Joseph Felsenstein's approach in "The Detection of Phylogeny" is entirely different. He argues that the relation of hypotheses to data is not hypothetico-deductive in character. Felsenstein maintains that phylogenetic inference is, at base, a statistical problem. Although cladistic parsimony may be a reasonable procedure in some circumstances, it is not so in others.

One of the principal issues that divides Farris and Felsenstein concerns how much one must know about the evolutionary process before one can make inferences concerning genealogical relationships. Farris maintains that rela-tively modest assumptions suffice; Felsenstein argues that the assumptions required for one to use cladistic parsimony, or any other method, are far more substantive.

IX REDUCTION OF MENDELIAN GENETICS TO MOLECULAR BIOLOGY

Philosophers interested in how science changes have devoted a great deal of attention to the issue of theoretical reduction. When one theory replaces another, is this because the latter theory shows that the former theory was false or because the latter theory captures and supplements the insights of the theory it supersedes? The relation of Mendelian genetics to molecular biology has been an important test case for this question. Does modern molecular theory show that Mendel's ideas were false? Or does it show that Mendel was right? Advocates of the latter position tend to say that Mendelian genetics reduces to molecular biology.

This problem has been important to the task of understanding whether, or in what sense, science makes progress. Should we say that Mendel and modern theorists are "talking about the same things," which we can label neutrally by calling them "genes"? Or were the two theories really talking about different things altogether? If the latter, we may be on our way to the conclusion that the two theories are incommensurable, a concept that figured prominently in Thomas Kuhn's *Structure of Scientific Revolutions*.

In chapter 18, "1953 and All That: A Tale of Two Sciences," Philip Kitcher defends the antireductionist position. He argues that the relationship of current molecular theory to Mendel's "laws" is best viewed as one of theory replacement, not theoretical reduction. C. Kenneth Waters takes issue with Kitcher's arguments in the following chapter, "Why the Antireductionist Consensus Won't Survive the Case of Classical Mendelian Genetics."

X ETHICS AND SOCIOBIOLOGY

When E. O. Wilson published *Sociobiology: The New Synthesis* in 1975, he set forth the outlines of a research program that aims to provide evolutionary explanations of social behavior. The book triggered a heated controversy, in large part because of its last chapter, which attempted to apply evolutionary categories to human mind and culture.

Some of the previous parts of this anthology address questions that are pertinent to sociobiology. For example, part III, on adaptationism, is quite germane because sociobiologists usually adopt an adaptationist perspective. Part IV is also relevant. Wilson announced in his book that the fundamental problem of sociobiology is the evolution of altruism; since altruism is often defined as a trait that helps the group in which it occurs though it is deleterious to the organism possessing it, this sociobiological issue immediately engages the problem of the units of selection.

Part X focuses on just one issue in the sociobiology debate: What can evolutionary biology tell us about morality? In *Sociobiology: The New Synthesis*, Wilson suggests that "the time has come for ethics to be removed temporarily from the hands of the philosophers and biologicized." In "Moral Philosophy as Applied Science," Wilson and his coauthor, the philosopher Michael Ruse, set forth their reasons for thinking that this line of inquiry is worthwhile. Philip Kitcher responds skeptically in chapter 21, "Four Ways of 'Biologicizing' Ethics."

XI CULTURAL EVOLUTION AND EVOLUTIONARY EPISTEMOLOGY

Do cultures evolve in a way that is structurally similar to the way that biological populations evolve? And since science is a part of human culture, can one model scientific change by analogizing it with the process of biological evolution?

On the face of it, the analogy is clear-cut. Darwin said that organisms participate in a struggle for existence; evolution occurs because fitter organisms tend to outsurvive their less fit competitors. By the same token, scientific theories compete with each other in a marketplace of ideas; fitter theories are more successful at gaining adherents. By this process, the mix of scientific ideas present in a scientific community evolves.

In chapter 22, "Epistemology from an Evolutionary Point of View," Michael Bradie describes the various specific proposals that this analogy has engendered. Bradie carefully distinguishes this type of evolutionary epistemology from another. Rather than describing a similarity between biological evolution and scientific change, there has been considerable interest in trying to apply models of biological evolution directly to human cognitive capacities. Here we are considering a research program that is tightly connected with the issues that engage sociobiologists. The human mind, like the human heart, is an organ that evolved. What can we learn about our cognitive capacities by taking seriously the fact that we are products of evolution?

In the final chapter of this book, "Models of Cultural Evolution," I draw the same distinction that Bradie emphasizes. My main focus, however, is on models of cultural evolution in general rather than on models of scientific change in particular. The question I pose is whether historians and social scientists have much to gain from models of cultural evolution that treat cultural change as a kind of selection process. Can such models provide a unifying paradigm for the social sciences that plays the same role in the study of human culture that models of biological evolution play in biology as a whole?

ACKNOWLEDGMENT

I am very grateful to Eric Saidel for his meticulous work in proofreading and in preparing the index for this anthology.

REFERENCES

Darwin, C. 1859. *The Origin of Species*. Cambridge: Harvard University Press, 1964 ed.

————. 1872. *The Descent of Man*. Princeton, N.J.: Princeton University Press, 1981.

Dawkins, R. 1976. *The Selfish Gene*. New York: Oxford University Press.

Fisher, R. 1930. *The Genetical Theory of Natural Selection*. New York: Dover, 1958 ed.

Hull, D. 1974. *Philosophy of Biological Science*. Englewood Cliffs, N.J.: Prentice-Hall.

Kuhn, T. 1962. *Structure of Scientific Revolutions*. Chicago: University of Chicago Press.

Rosenberg, A. 1985. *The Structure of Biological Science*. Cambridge: Cambridge University Press.

Ruse, M. 1973. *The Philosophy of Biology*. London: Hutchinson.

Sober, E. 1993. *Philosophy of Biology*. Boulder, Colo.: Westview.

Williams, G. 1966. *Adaptation and Natural Selection*. Princeton, N.J.: Princeton University Press.

Wilson, E. 1975. *Sociobiology: The New Synthesis*. Cambridge: Harvard University Press.

Contributors

John H. Beatty
Department of Ecology, Evolution,
and Behavior
University of Minnesota
Minneapolis, Minnesota 55455

Michael Bradie
Philosophy Department
Bowling Green University
Bowling Green, Ohio 43403

Robert Cummins
Philosophy Department
University of Arizona
Tucson, Arizona 85721

Michael J. Donoghue
Herbarium
Harvard University
Cambridge, Massachusetts 02138

James Farris
Naturhistoriska riksmuseet
Molekylärsystematiska laboratoriet
S-104 05 Stockholm, Sweden

Joseph Felsenstein
Genetics
University of Washington
Seattle, Washington 98195

Stephen Jay Gould
Museum of Comparative Zoology
Harvard University
Cambridge, Massachusetts 02138

Willi Hennig (*Deceased*)

David L. Hull
Philosophy Department
Northwestern University
Evanston, Illinois 60208

Philip Kitcher
Philosophy Department
University of California–San Diego
La Jolla, California 92093

Richard C. Lewontin
Museum of Comparative Zoology
Harvard University
Cambridge, Massachusetts 02138

John Maynard Smith
Biology Department
University of Sussex
Brighton, BN1 9QG, England

Ernst Mayr
Museum of Comparative Zoology
Harvard University
Cambridge, Massachusetts 02138

Susan K. Mills (now Susan M. Finsen)
Philosophy Department
California State University at
San Bernardino
San Bernardino, California 92407

Brent D. Mishler
Botany Department
Duke University
Durham, North Carolina 27708

Michael Ruse
Philosophy Department
University of Guelph
Guelph, Ontario N1G 2W1, Canada

Elliott Sober
Philosophy Department
University of Wisconsin
Madison, Wisconsin, 53706

Robert Sokal
Department of Ecology and
Evolution
State University of New York
Stony Brook, New York 11794-5245

C. Kenneth Waters
Philosophy Department
University of Minnesota
Minneapolis, Minnesota 55455

George C. Williams
Department of Ecology and
Evolution
State University of New York
Stony Brook, New York 11794

David Sloan Wilson
Biology Department
State University of New York
Binghamton, New York 13901

Edward O. Wilson
Museum of Comparative Zoology
Harvard University
Cambridge, Massachusetts 02138

Larry Wright
Philosophy Department
University of California
Riverside, California 92521

I Fitness

1 The Propensity Interpretation of Fitness

Susan K. Mills and John H. Beatty

The concept of "fitness" is a notion of central importance to evolutionary theory. Yet the interpretation of this concept and its role in explanations of evolutionary phenomena have remained obscure. We provide a propensity interpretation of fitness, which we argue captures the intended reference of this term as it is used by evolutionary theorists. Using the propensity interpretation of fitness, we provide a Hempelian reconstruction of explanations of evolutionary phenomena, and we show why charges of circularity which have been levelled against explanations in evolutionary theory are mistaken. Finally, we provide a definition of natural selection which follows from the propensity interpretation of fitness, and which handles all the types of selection discussed by biologists, thus improving on extant definitions.

The testability and logical status of evolutionary theory have been brought into question by numerous authors in recent years (e.g., Manser 1965, Smart 1963, Popper 1974). Many of the claims that evolutionary theory is not testable, that it parades tautologies in the guise of empirical claims, and that its explanations are circular, resulted from misunderstandings which have since been rebuked (e.g., by Ruse 1969, 1973, and Williams 1970, 1973a, 1973b). Yet despite the skilled rejoinders which have been given to most of these charges, the controversy continues to flourish, and has even found its way beyond philosophical and biological circles and into the pages of *Harpers Magazine*. In the spring of 1976, journalist Tom Bethell reported to the unsuspecting public that:

Darwin's theory ... is on the verge of collapse. in his famous book, *On the Origin of Species* ... Darwin made a mistake sufficiently serious to undermine his theory. The machinery of evolution that he supposedly discovered has been challenged, and it is beginning to look as though what he really discovered was nothing more than the Victorian propensity to believe in progress. (1976, p. 72)

Those familiar with the details of evolutionary theory, and with the history of this controversy, will rightfully feel no sympathy with such challenges, and may wonder whether it is worth bothering with them. But the fact is that there is a major problem in the foundations of evolutionary theory which remains unsolved, and which continues to give life to the debate. The definition of fitness remains in dispute, and the role of appeals to fitness in biolo-

From *Philosophy of Science*, 1979, 46:263–286.

gists' explanations is a mystery. This is a problem which ought to concern biologists and philosophers of science, quite independent of the vicissitudes of the controversy which it perpetuates.

Biologists agree on how to *measure* fitness, and they routinely appeal to fitness in their explanations, attributing the relative predominance of certain traits to the relative fitness of those traits. However, these explanations can and have been criticized on the grounds that, given the definitions of fitness offered by most biologists, these explanations are no more than re-descriptions of the phenomena to be explained (e.g., Popper 1974, Manser 1965, Smart 1963). Philosophers have proposed new treatments of fitness designed to avoid these charges of explanatory circularity (e.g., Hull 1974 and Williams 1973a). Unfortunately, none of these interpretations succeeds in avoiding the charges, while providing a definition *useful* to evolutionary theory.

Thus it is high time that an analysis of fitness is provided which reveals the empirical content implicit in evolutionary biologists' explanations. To this end, we propose and defend the *propensity interpretation* of fitness. We argue that the propensity interpretation captures the intended reference of "fitness" as biologists use the term. Further, using this interpretation, we show how references to fitness play a crucial role in explanations in evolutionary theory, and we provide a Hempelian reconstruction of such explanations which reveals the precise nature of this role. We answer the charges of explanatory circularity leveled against evolutionary theory by showing how these charges arise from mistaken interpretations of fitness.

The concepts of fitness and natural selection are closely linked, since it is through the process of natural selection that the fittest gain predominance, according to the theory of evolution. Thus it is not surprising to find misinterpretations of fitness paralleled by misunderstandings of natural selection. The propensity analysis suggests a definition of "selection" which (unlike previously proposed definitions) accords with all the diverse types of selection dealt with by biologists.

But before proceeding with the positive analyses just promised, we consider the charge of explanatory circularity which arises from the lack of a satisfactory interpretation of fitness, and the reasons for the inadequacy of the replies so far offered in answer to the charge.

THE CHARGE OF CIRCULARITY

According to the most frequently cited definitions of "fitness," that term refers to the *actual* number of offspring left by an individual or type relative to the actual contribution of some reference individual or type. For instance, Waddington (1968, p. 19) suggests that the fittest individuals are those which are "most effective in leaving gametes to the next generation." According to Lerner (1958), "the individuals who have more offspring are fitter in the Darwinian sense." Grant (1977, p. 66) construes fitness as "a measure of reproductive success." And Crow and Kimura (1970, p. 5) regard fitness "as

Susan Mills and John Beatty

a measure of both survival and reproduction" (see also Dobzhansky 1970, pp. 101–102; Wilson 1975, p. 585; Mettler and Gregg 1969, p. 93).

These definitions of "fitness" in terms of actual survival and reproductive success are straightforward and initially intuitively satisfying. However, such definitions lead to justifiable charges that certain explanations invoking fitness differences are circular. The explanations in question are those which point to fitness differences between alternate types in a population in order to account for (1) differences in the average offspring contributions of those phenotypes, and (2) changes in the proportions of the types over time (i.e., evolutionary changes). Where fitness is defined in terms of survival and reproductive success, to say that type *A* is fitter than type *B* is just to say that type *A* is leaving a higher average number of offspring than type *B*. Clearly, we cannot say that the difference in fitness of *A* and *B* *explains* the difference in actual average offspring contribution of *A* and *B*, when fitness is defined in terms of actual reproductive success. Yet, evolutionary biologists seem to think that type frequency changes (i.e., evolutionary changes) can be *explained* by invoking the relative fitnesses of the types concerned. For instance, Kettlewell (1955, 1956) hypothesized that fitness differences were the cause of frequency changes of dark- and light-colored pepper moths in industrial areas of England. And he devised experiments to determine whether the frequency changes were correlated with fitness differences. Several philosophers have pointed to the apparent circularity involved in these explanations. Manser (1965) describes Kettlewell's account of the frequency differences in terms of fitness differences as "only a description in slightly theory-laden terms which gives the illusion of an explanation in the full scientific sense" (1965, p. 27).

The whole idea of setting up empirical investigations to determine whether fitness differences are correlated with actual descendant contribution differences seems absurd, given the above definitions of "fitness." If this type of charge is coupled with the assumption that the only testable claims of evolutionary theory are of this variety, (i.e., tests of whether individuals identified as "the fittest" are most reproductively successful), then it appears that evolutionary theory is not testable. As Bethell puts it, "If only there were some way of identifying the fittest before-hand, without always having to wait and see which ones survive, Darwin's theory would be testable rather than tautological" (1976, p. 75).

However, as Ruse (1969) and Williams (1973a) have made clear, this latter charge is mistaken. Evolutionary theory embodies many testable claims. To take but one of many examples cited by Williams, Darwinian evolutionary theory predicted the existence of *transitional forms* intermediate between ancestral and descendant species. The saltationist (creationist) view of the origin of species which was accepted at the time when Darwin wrote on *The Origin of Species* predicted no such plethora of intermediate forms. Ruse has called attention to the predictions concerning distributions of types in populations which can be made on the basis of the Hardy-Weinberg law (1973, p. 36).

While these replies are well taken, they fail to clarify the role of fitness ascriptions in evolutionary theory. We agree with Williams and Ruse that evolutionary theory does make testable claims, and that many of these claims can be seen to be testable without providing an analysis of the role of fitness ascriptions. Nevertheless, some claims of evolutionary theory cannot be shown to be empirical without clarifying the role of "fitness." Moreover, our understanding of other straightforwardly empirical claims of evolutionary theory will be enhanced by an explication of the role of "fitness" in these claims.

WHAT FITNESS IS NOT

There are two questions to be clarified in defining fitness: What sorts of entities does this predicate apply to, and what does it predicate of these entities? Both these questions have received disparate answers from various biologists and philosophers. Fitness has been claimed to apply to types (e.g., Dobzhansky 1970, pp. 101–102; Crow and Kumura 1970) as well as individuals (Lerner 1958, Waddington 1968, p. 19). As will become apparent in the course of the positive analysis, the question of what sorts of entities "fitness" applies to should not be given a univocal answer. Fitness may be predicated of individual organisms, and (in a somewhat different sense) of phenotypes and genotypes. In this section we will only consider the question of what one is predicating of individuals and types in ascribing them a fitness value, according to the various proposals under scrutiny.

Before moving on to alternatives to the definition of "fitness" in terms of actual survival and reproductive success, we need to consider the acceptability of this definition, independent of the criticism that it leads to explanatory circularity. This criticism alone is obviously not sufficient to show that the interpretation is incorrect. For, proponents of this definition can reply that fitness is actual reproductive success, since that is the way biologists use the term, and there is no other feasible definition. The fact that references to fitness lead to explanatory circularity just shows that fitness has no explanatory role to play in evolutionary theory. In fact, Bethell (1976, p. 75) makes this latter claim, and even maintains that biologists have abandoned references to fitness in their accounts of evolutionary phenomena. This is a scandalous claim.[1] A survey of evolutionary journals like *American Naturalist* and *Evolution* reveals that fitness ascriptions still play a major role in explanations of evolutionary phenomena. Indeed, the current literature on evolutionary theory reveals that the notion of fitness is of tremendous concern. Rather than abandoning the notion, modern evolutionary biologists have chosen to refine and extend it. Levins (1968) has raised the problem of fitness in changing environments. Thoday (1953) has pointed to the distinction between short-term and long-term fitness. Analysis of and evidence for "variable fitness" or "frequency dependent fitness" was given by Kojima (1971). The effects of "overdominance with regard to fitness" on the maintenence of polymorphisms continue to be studied. And one very promising model of sociobiological evolution has been developed via an extension of traditional notions of fitness (the new notion is one of "inclusive fitness" [Cf. Hamilton 1964]. As we

Susan Mills and John Beatty

will argue below, biologists are well advised *not* to abandon references to fitness, for such references play a crucial role in explanations of evolutionary phenomena.

Fortunately, we do have grounds quite independent of the issue of explanatory circularity for deeming inadequate definitions of "fitness" in terms of actual survival and reproductive success. For such definitions conflict with biologists' usage of the term, as is demonstrated by the following considerations. Surely two organisms which are genetically and phenotypically identical, and which inhabit the same environment, should be given the same fitness value. Yet where fitness is defined in terms of actual number of offspring left, two such organisms may receive radically different fitness values, if it happens that one of them succeeds in reproducing while the other does not. Scriven (1959) invites us to imagine a case in which two identical twins are standing together in the forest. As it happens, one of them is struck by lightning, and the other is spared. The latter goes on to reproduce while the former leaves no offspring. Surely in this case there is no difference between the two organisms which accounts for their difference in reproductive success. Yet, on the traditional definition of "fitness," the lucky twin is *far* fitter. Most undesirably, such a definition commits us to calling the intuitively less fit of two organisms the fitter, if it happens that this organism leaves the greater number of offspring of the two.[2]

Nor can these counterintuitive results be avoided by shifting the reference of fitness from individual organisms to groups. For, precisely as was the case with individuals, the intuitively less fit subgroup of a population may be chance come to predominate. For example, an earthquake or forest fire may destroy individuals irrespective of any traits they possess. In such a case, we do not wish to be committed to attributing the highest fitness values to whichever subgroup is left.

Since an organism's traits are obviously important in determining its fitness, it is tempting to suggest that fitness be defined entirely independently of survival and reproduction, as some function of traits. Hull (1974) hints at the desirability of such a definition. This suggestion derives *prima facie* support from the fact that given such a definition, explanations of differential offspring contribution which appeal to differences in fitness are noncircular. However, no one has seriously proposed such a definition, and it is easy to see why. The features of organisms which contribute to their survival and reproductive success are endlessly varied and context dependent. What do the fittest germ, the fittest geranium, and the fittest chimpanzee have in common? It cannot be any concretely characterized physical property, given that one and the same physical trait can be helpful in one environment and harmful in another. This is not to say that it is impossible that some as yet unsuspected (no doubt abstractly characterized) feature of organisms may be found which correlates with reproductive success. Rather, it is just to say that we need not, and should not, wait for the discovery of such a feature in order to give the definition of "fitness."

So far, we have seen that we cannot define fitness simply in terms of survival and reproductive success. But we cannot define fitness entirely independently of any reference to survival and reproduction, either. An ingenious alternative to either of these approaches has been offered by Williams (1970, 1973a). She suggests that we regard "fitness" as a primitive term of evolutionary theory, and that we therefore refuse to define it. As she points out, in the formal axiomatization of a theory, it is not possible that all terms be explicitly defined, on pain of circularity. However, the fact that we cannot formally define all the terms of a theory *within* the framework of the theory does not prevent us from stepping outside the theory and explaining the meaning of the term in a broader linguistic framework.[3] Such an explication need not amount to anything as restrictive as an operational definition or an explicit definition making the term eliminable without loss from the theory. Rather, such an explication should allow us to understand what sort of property fitness is, its relation to natural selection, and the role of references to fitness in evolutionary theorists' explanations. Thus, our criticism of Williams is not that she is wrong about fitness but that she does not go far enough. We believe that a more thorough explication is possible, through the *propensity* interpretation of fitness.[4]

PROPENSITY ANALYSIS OF FITNESS

Levins (1968) has remarked that "fitness enters population biology as a vague heuristic notion, rich in metaphor but poor in precision." No doubt this is accurate as a characterization of the unclarity surrounding the role of fitness in evolutionary theory, even among biologists who use the term. But such unclarity is quite compatible with the fact that fitness plays an essential explanatory role in evolutionary theory. It is to the task of increasing the precision of the concept of fitness as well as making explicit this explanatory role that we now turn.

We have already seen that fitness is somehow connected with success at survival and reproduction, although it cannot be defined in terms of actual survival and reproductive success. Why have evolutionary biologists continued to confuse fitness with actual descendant contribution? We believe that the confusion involves a misidentification of the *post facto* survival and reproductive success of an organism with the *ability* of an organism to survive and reproduce. We believe that "fitness" refers to the ability. Actual offspring contribution, on the other hand is a sometimes reliable—sometimes unreliable—indicator of that ability. In the hypothetical cases above, actual descendant contribution is clearly an unreliable indicator of descendant contribution capability. The identical twins are equally *capable* of leaving offspring. And the camouflaged butterfly is more *capable* of leaving offspring than is the noncamouflaged butterfly.

Thus, we suggest that fitness be regarded as a complex *dispositional* property of organisms. Roughly speaking, the fitness of an organism is its *propensity* to survive and reproduce in a particularly specified environment and

Susan Mills and John Beatty

population. A great deal more will have to be added before the substance of this interpretation becomes clear. But before launching into details, let us note a few general features of this proposal.

First, if we take fitness to be a dispositional property of organisms, we can immediately see how references to fitness can be explanatory.[5] The fitness of an organism explains its success at survival and reproduction in a particular environment in the same way that the solubility of a substance explains the fact that it has dissolved in a particular liquid. When we say that an entity has a propensity (disposition, tendency, capability) to behave in a particular way, we mean that certain physical properties of the entity determine, or are causally relevant to, the particular behavior whenever the entity is subjected to appropriate "triggering conditions." For instance, the propensity of salt to dissolve in water (the "water-solubility" of salt) consists in (i.e., "water solubility" *refers to*) its ionic crystalline character, which causes salt to dissolve whenever the appropriate triggering condition—immersion in water—is met. Likewise, the fitness of an organism consists in its having traits which condition its production of offspring in a given environment. For instance, the dark coloration of pepper moths in sooted, industrial areas of England effectively camouflages the moths from predators, enabling them to survive longer and leave more offspring. Thus, melanism is one of many physical properties which constitute the fitness, or reproductive propensity, of pepper moths in polluted areas (in the same sense that the ionic crystalline character of salt constitutes its propensity to dissolve in water).

The appropriate triggering conditions for the realization of offspring contribution dispositions include particular environmental conditions. We do not say that melanic moths are equally fit in polluted and unpolluted environments any more than we claim that salt is as soluble in water as it is in mercury or swiss cheese.[6]

In addition to the triggering conditions which cause a disposition to be manifested, we must, in explaining or predicting the manifestation of a disposition, consider whether any factors other than the relevant triggering conditions were present to interfere with the manifestation. When we say that salt has dissolved in water because it is soluble in water, we assume the absence of disturbing factors, such as the salt's having been coated in plastic before immersion. Likewise, when we explain an organism's (or type's) offspring contribution by referring to its degree of fitness, we assume, for instance, that environmental catastrophes (e.g., atomic holocausts, forest fires, etc.) and human intervention have not interfered with the manifestation offspring contribution dispositions. In general, we want to rule out the occurrence of any environmental conditions which separate successful from unsuccessful reproducers without regard to physical differences between them.

Now let us fill in some of the details of this proposal. First, we must clarify the view of propensities we are presupposing. In our view, propensities are dispositions of *individual objects*. It is each hungry rat which has a tendency or propensity to move in the maze in a certain way, not the class of hungry rats. Classes—abstract objects, in general—do not have dispositions, tendencies, or propensities in any orthodox sense of the term.[7] This aspect of propensities in general is also a feature of the (unexplicated) notion of fitness employed by biologists. Evolutionary biologists often speak of fitness as if it were a phenotypic trait—i.e., a property of individuals. For instance, Wallace (1963, p. 633) remarks, "That instances of overdominance exist, especially in relation to a trait as complex as fitness, is generally conceded."

However, evolutionary biologists also employ a notion of fitness which refers to *types* (e.g., Dobzhansky 1970, pp. 101–102). Fitness cannot be a propensity in this case, although as we will argue, it is a derivative of individual fitness propensities. Thus, we will introduce two definitions of "fitness": Fitness$_1$ of individual organisms and fitness$_2$ of types.

Fitness$_1$: Fitness of Individual Organisms

A paradigm case of a propensity is a subatomic particle's propensity to decay in a certain period of time. Whether a particle decays during some time interval is a qualitative, nonrepeatable property of that particle's event history. It might initially be thought that "propensity to reproduce" is also a qualitative nonrepeatable property of an organism: either it reproduces during its lifetime, or it does not. However, the property of organisms which is of interest to the evolutionary biologist is not the organism's propensity to reproduce or not to reproduce, but rather the *quantity* of offspring which the organism has the propensity to contribute. For the evolutionary biologist is interested in explaining proportions of types in populations, and from this point of view, an organism which leaves one offspring is much more similar to an organism which leaves no offspring than it is to an organism which leaves 100 offspring. Thus, when we speak of "reproductive propensity," this should be understood as a quantitative propensity like that of a lump of radioactive material (considered as an individual) to emit particles over time, rather than as a "yes-no" propensity, like that of an individual particle to decay or not decay during some time interval.

It may have struck the reader that given this quantitative understanding of "propensity to reproduce," there are many such propensities. There is an organism's propensity to leave zero offspring, its propensity to leave 1 offspring, 2 offspring, ..., n offspring (during its lifetime). Determinists might claim that there is a unique number of offspring which an organism is determined to leave (i.e., with propensity 1) in a given environment. For nondeterminists, however, things are more complicated. Organisms may have propensities of different strengths to leave various numbers may have propensities of different strengths to leave various numbers of offspring. The standard dispositions philosophers talk about are tendencies of objects to instantiate certain properties invariably under appropriate circumstances. But besides such "deterministic" dispositions, there are the tendencies of objects to produce one or another of a distribution of outcomes with predetermined frequency. As Coffa (1977) argues, it seems just as legitimate to suppose there are such nondeterministic, "probabilistic" causes as to posit deterministic dispositions.[8]

Susan Mills and John Beatty

If we could assume that there were a unique number of offspring which any organism is determined to produce (i.e., which the organism has propensity 1 to produce), then the fitness$_1$ of an organism could be valued simply as the number of offspring which that organism is disposed to produce. But since it is quite possible that organisms may have a range or distribution of reproductive propensities, as was suggested above, we derive fitness$_1$ values taking these various propensities into consideration.

Unfortunately, we also cannot simply choose the number of offspring which an organism has the *highest* propensity to leave—that is, the mode of the distribution. For in the first place, an organism may not have a *high* propensity to leave any particular number of offspring. In the second place, there may not be one number of offspring which corresponds to the mode of the distribution. For example, an organism might have a .5 propensity to leave 10 offspring and a .5 propensity to leave 20 offspring. And finally, even if there is a number of offspring which an organism has a significantly higher propensity to leave than any other number of offspring, we must take into account the remainder of the distribution of reproductive propensities as well. For example, an organism with a .7 propensity to leave 5 offspring and a .3 propensity to leave 50 offspring is very different from an organism with a .7 propensity to leave 5 offspring and a .3 propensity to leave no offspring, even though each has the propensity to leave 5 offspring as its highest reproductive propensity.[9]

In lieu of these considerations, one might suggest that the fitness$_1$ of an organism be valued in terms of the entire distribution of its reproductive propensities. The simplest way to do this is just to assign distributions as values. For example, the fitness$_1$ of an imaginary organism x might be the following distribution.

number of offspring	1	2	3	4	5	6	7	8	9	10
propensity	.05	.05	.05	.2	.3	.2	.05	.05	.05	

However, our intuitions fail us in regard to the comparison of such distributions. How can we determine whether one organism is fitter than another, on the basis of their distributions alone? For instance, is x fitter or less fit than y and z, whose distributions (below) differ from x's?

number of offspring	1	2	3	4	5	6	7	8	9	10
y					1.0					
z		.5		.3					.2	

In order to avoid the uncertainties inherent in this method of valuation, and still take into account all an organisms' reproductive propensities, we suggest that fitness$_1$ values reflect an organism's *expected number* of offspring. The expected value of an event is the weighted sum of the values of its possible outcomes, where the appropriate weights are the probabilities of the various outcomes. As regards fitness$_1$, the event in question is an individual's total offspring contribution. The possible outcomes $O_1, O_2, \ldots, O_n$ are contributions of different numbers of offspring. Values $(1, 2, \ldots, n)$ of the outcomes correspond to the number of offspring left. And the weighting probability for each outcome 0_i is just the organism's propensity to contribute i offspring. Thus the imaginary organisms x, y, and z above all have the same expected number of offspring, or fitness value, of 5.

We propose, then, that "individual fitness" or "fitness$_1$" be defined as follows:

The *fitness*$_1$ of an organism x in environment E equals $n =_{df} n$ is the expected number of descendants which x will leave in E.[10]

It may have occurred to the reader that the fitness values assigned to organisms are not literally propensity values, since they do not range from 0 to 1. But this does not militate against our saying that the fitness of an organism is a complex of its various reproductive propensities. Consider for comparison another dispositional property of organisms: their intelligence. If everyone could agree that a particular intelligence test really measured intelligence, then an organism's intelligence could be defined as the expected score on this test. (We would not value intelligence as the score actually obtained in a particular taking of the test, for reasons precisely analogous to those which militate against definitions of fitness in terms of actual numbers of organisms left. Intelligence is a competence or capacity of organisms, rather than simply a measure concept.) Obviously, intelligence would not be valued as the strength of the propensity to obtain a *particular* score. Similarly, it is the expected number of offspring which determines an organism's fitness values, not the strength of the propensity to leave a particular number of offspring.

Fitness$_2$: Fitness of Types

Having defined fitness$_1$, we are in a position to define the fitness$_2$ of types. As will become apparent in what follows, it is the fitness of types which figures primarily in explanations of microevolutionary change.

Intuitively, the fitness of a type (genotype of phenotype) reflects the contribution of a particular gene or trait to the expected descendant contribution (i.e., the fitness$_1$) of possessors of the gene or trait. Differences in the contributions of alternate genes or traits would be easy to detect in populations of individuals which were phenotypically identical except in regard to the trait or gene in question. In reality, though, individuals differ with regard to many traits, so that the contribution of one or another trait to fitness$_1$ is not so straightforward. In fact, the notion of any simple, absolute contribution is quite meaningless. For a trait acts in conjunction with many other traits in influencing the survival and reproductive success of its possessors. Thus, its contribution to different organisms will depend upon the different traits it is associated with in those organisms.

Yet, in order to explain the evolution and/or persistence of a gene or its phenotypic manifestation in a temporally extended population, we would like to show that possessors of the gene or trait were generally better able to

Susan Mills and John Beatty

survive and reproduce than possessors of alternate traits or genes. (By "alternate genes" we mean alternate alleles, or alternate genes at the same locus of the chromosome. "Alternate traits" are phenotypic manifestations of alternate genes.) In other words, we want to invoke the *average* fitness$_1$ of the members of each of the types under consideration. Let us refer to average fitness$_1$ as "fitness$_2$." Given some information about the fitness$_2$ of each of a set of alternate types in a population, and given some information about the mechanisms of inheritance involved, we can predict and explain the evolutionary fate of the genes or traits which correspond to the alternate types. For instance, if we knew that possessors of a homozygous-based trait were able to contribute a higher average number of offspring than possessors of any of the alternate traits present in the population, we would have good grounds for predicting the eventual predominance of the trait in the population.

As the above discussion suggests, we actually invoke *relative* fitness$_2$ values in predictions and explanations of the evolutionary fate of genes and traits. That is, we need to know whether members of a particular type have a *higher* or *lower* average fitness$_1$ in order to predict the fate of the type. In order to capture this notion, and to accommodate biologists' extensive references to "relative fitness" or "Darwinian fitness," we introduce "relative fitness$_2$." Given a set of specified alternate types, there will be a type which is fittest in the fitness$_2$ sense (i.e., has highest average fitness$_1$, designated "Max Fitness$_2$"). Using this notion of Max Fitness$_2$, we define relative fitness$_2$ as follows:

The relative fitness$_2$ of type X in E
$=_{df}$ the fitness$_2$ of X in E/Max fitness$_2$ in E

The role of relative fitness$_2$ ascriptions in evolutionary explanations has been acknowledged (for instance by Williams's "condition 3" in her analysis of functional explanations [1976]). Yet very little attention has been paid to the establishment of these ascriptions. Perhaps we should say a few words about these claims. For it might be supposed that the *only* way in which fitness$_2$ ascriptions can be derived is through measurements of actual average offspring contributions of types. If this were the case, even though "fitness$_2$" is not *defined* in terms of such measures (so that explanations employing fitness$_2$ ascriptions to explain actual offspring contribution differences would not be formally circular), claims concerning the influence of fitness$_2$ differences upon offspring contribution could not be *tested*. This would obviously be disastrous for our analysis.

Evolutionary biologists frequently *derive* relative fitness claims from optimality models (e.g., Cody 1966); this is basically an engineering design problem. It involves determining, solely on the basis of design considerations, which of a set of specified alternate phenotypes maximizes expected descendant contribution. The solution to such a problem is only optimal relative to the other *specified* alternatives (there may be an unspecified, more optimal solution). Thus, optimality models provide some insight into the relative fitness of members of alternate types.

The theorems derived from optimality models can be confirmed by measurements of actual descendant contribution. Such measures can also be used to generate fitness$_2$ ascriptions. Given evidence that descendant contribution was affected primarily or solely by individual propensities for descendant contribution, we can infer that descendant contribution measurements are indicative of individual or type fitness.

Explaining Microevolutionary Phenomena

Having elaborated the notions of fitness$_1$ and relative fitness$_2$, we hope to show how these concepts function in explanations of evolutionary phenomena. Perhaps the clearest means of showing this is to work through an example of such an explanation. The example we are going to consider involves a change in the proportion of the two alleles at a single chromosomal locus, and a change in the frequency of genotypes associated with this locus, in a large population of organisms. In this population, at the locus in question, there are two alleles, A and a. A is fully dominant over a, so that AA and Aa individuals are phenotypically indistinguishable with respect to the trait determined by this locus. This trait is the "natural gun" trait. All individuals which are either homozygous (AA) or heterozygous (Aa) at this locus have a natural gun, whereas the unfortunate individuals of genotype aa have no gun. Let us suppose that for many generations this population has lived in peace in an environment E, in which no ammunition is available. (Were the terminology not in question, we would say that there had been no "selective pressure" for or against the natural gun trait.) However, at generation n, environment E changes to environment E', by the introduction of ammunition usable by the individuals with natural guns. At generation n, the proportion of A alleles is .5 and the proportion of a alleles is .5, with the genotypes distributed as follows:

AA: .25 Aa: .50 aa: .25.

What we want to explain is that in generation $n + 1$, the new frequency of genotypes is as follows:

AA: .29 Aa: .57 aa: .14.

Let us suppose that the large size of this population makes such a change in frequency extremely improbable ($p = .001$) on the basis of chance.

We need two pieces of information concerning this population in order to explain the change in frequency. We need to know (1) the relative fitness$_2$ of the natural gun and non-natural gun types, and (2) whether any conditions obtain which would interfere with the actualization of the descendant contribution propensities which the relative fitness$_2$ valuations reflect. As was noted above, the fact that an organism does not survive and reproduce in an environment in which periodic cataclysms occur is no indication of its fitness (any more than the failure of salt to dissolve in water when coated with plastic would count against its solubility).

Susan Mills and John Beatty

The latter qualification, stating that no factors other than fitness$_2$ differences were responsible for descendant contribution, corresponds to the "extremal clause," which, as Coffa (1977, p. 194) has made clear, is a component in the specification of most scientific laws. Such clauses state that no physical properties or events relevant to the occurrence of the outcome described in the law (other than those specified in the initial conditions) are present to interfere with that outcome. In stating scientific laws, the assumption is often tacitly made that no such disturbing factors are present. But as Coffa has pointed out, it is important to make this assumption explicit in an extremal clause. For, no scientific law can be falsified by an instance in which the event predicted by the law fails to occur, unless the extremal clause is satisfied. Thus, our ability to fill in the details of the extremal clause will determine our ability to distinguish between contexts which count as genuine falsifications of a law and contexts which do not. The fact that evolutionary theorists are fairly specific about the types of conditions which interfere with selection is an indication in favor of the testability of claims about fitness. As noted above, the influence of fitness upon offspring contribution is disturbed by any factors which separate successful from unsuccessful reproducers without regard to physical differences between them. In addition, certain other evolutionary factors such as mutation, migration, and departures from panmixia may disturb the systematic influence of fitness differences between types upon proportions of those types in subsequent generations.

Let us suppose that we do know the relative fitnesses$_2$ of the natural gun and non-natural gun types, and let us suppose the natural selection conditions are present (i.e., nothing is interfering with the manifestation of the fitness propensities). This information together with the relevant laws of inheritance will allow us to predict (and explain) the frequencies of types in generation $n + 1$. We need not detail the principles of inheritance which allow this computation here (since they are available in any genetics text) other than to note that the Hardy-Weinberg Law allows us to compute the relative frequencies of types in a population, given information about the heritability of the types in question together with hypotheses about fitness$_2$ differences.

In light of these considerations, we construct the promised schema as follows:

1. In E', in generation n, the distribution of genotypes is:

AA: .25 Aa: .50 aa: .25.

2. $(x)(AAx \supset tx)$ & $(x)(Aax \supset tx)$ & $(x)(aax \supset -tx)$

3. In E', the relative fitness$_2$ of type t is 1.0.

4. In E', the relative fitness$_2$ of type not-t is 0.5.

5. For any three distinct genotypes X, Y, Z (generated from a single locus), if the proportions of X, Y, Z in generation n are P, Q, and R, respectively, and if the relative fitnesses$_2$ of genotypes X, Y, and Z are $F(Y)$ and $F(Z)$, respectively, then the proportion of X in generation $n + 1$ is:

$P \cdot F(X)/P \cdot F(X) + Q \cdot F(Y) + R \cdot F(Z)$.

6. $EC(E)$.

7. Given the size of population P, the probability that the obtained frequencies were due to chance is less than .001.

In E' at generation $n + 1$ the frequency of genotypes is:

AA: .29 Aa: .57 aa: .14.

This explanation is of the inductive-statistical variety, with the strength of the connection between explanans and explanadum determined, as indicated in premise (7), by the size of the population. Premise (1) is, obviously, a statement of the initial conditions. Premise (2) allows us to determine which genotypes determine each phenotype: all individuals with genotype AA or Aa have trait t, and all individuals of genotype aa lack trait t. Premises (3) and (4) indicate the relative fitness$_2$ of types t and not-t in environment E. Premise (5) is the above-mentioned consequence of the Hardy-Weinberg Law which allows computation of the expected frequencies in generation $n + 1$, given information about reproductive rates at generation n, together with information about initial frequences of individuals of each genotype at generation n. Premise (6) affirms that the extremal clause (EC) was satisfied—that is, that the "natural selection conditions" were present for the environment (E) in question. Thus we can infer that propensities to contribute descendants will be reflected in actual reproductive rates. Each genotype receives the relative fitness$_2$ associated with the phenotype it determines, as indicated in premise (2). Thus by substitution of the values provided in premises (3) and (4) in formula (5) (i.e., $X = AA$, $F(X) = 1.0$, $P = .25$; $Y = Aa$, $F(Y) = 1.0$, $Q = .50, \ldots$, etc.) we can obtain the values which appear in the explanandum.

To summarize, from knowledge of (1) initial frequencies of genotypes in generation n, (2) the relative fitness$_2$ of those genotypes, and (3) the fact that the extremal clause was satisfied, we can infer what the frequencies of genotypes will be in generation $n + 1$.

Of course, in this admittedly artificial example, it was presumed that the appropriate relative fitness$_2$ values were known. This suggests that we somehow investigated reproductive *capabilities*, and not just reproductive differences. We must emphasize, however, that actual reproductive differences may be regarded as measures of differences in reproductive capability, as long *as the measured differences are statistically significant*. This is the means of fitness determination in many, if not most, evolutionary investigations. But this must not mislead the reader into identifying fitness with actual reproductive contribution. For *statistically significant* differences would not be required to establish fitness differences in this case. Rather, statistically significant differences are required to establish that certain variables (fitness differences, in this case) are causally connected with other variables (in this case, differences in offspring contribution). Statistically significant differences are thus quite appropriate measures for fitness differences, given the propensity interpretation of fitness.

Having explained the role of statistical significance in measuring fitness differences, we can now consider a more realistic example of the role of fitness in population biology. Certainly one of greatest controversies in the history of population genetics concerns the differences in fitness of heterozygotes and homozygotes. The importance of the controversy lies in the fact that if heterozygotes are generally fitter than homozygotes, then breeding groups will retain a greater amount of genetic variation then if homozygotes were generally superior in fitness. And the amount of variation present in a population is of considerable importance to the evolutionary fate of the population. (For instance, greater variation provides some "flexibility" in the sense that a genetically variable population has more alternatives for adapting to changing environmental conditions.) Theodosius Dobzhansky, a principal protagonist in this controversy, maintained that heterozygotes at many loci were fitter than homozygotes at the same loci, and he and his collaborators gathered a good deal of statistically significant data to support this contention.

For instance, in one article, it was reported that members of the species *Drosophila pseudoobscura* which were heterozygous in regard to the structure of their third chromosome were more viable than the flies which were homozygous. Dobzhansky et al. correlated viability differences (note that *viability* differences are dispositional property differences) with fitness differences, and they performed a statistical analysis on their data, in order to conclude that "heterosis [heterozygote superiority in fitness] has ... developed during the experiment, as indicated by the attainment of equilibrium and by a study of the viability of the flies derived from the cage. Both tests gave statistically significant results" (1951, p. 263). Again, statistical significance would be of no concern if fitness were identified straightforwardly with offspring contribution. Statistical significance is important, however, if fitness is identified with phenotypic properties causally connected with offspring contribution.

As these examples demonstrate, fitness ascriptions play not only a legitimate, but a crucial role in explanations of evolutionary change. While biologists have not been able to justify their usage of the concept of "fitness," their usage of that concept has nevertheless been consistent and appropriate. Philosophers have accused biologists of giving circular explanations of evolutionary phenomena because they have only taken into account the definitions of fitness biologists explicitly cite, and they have not looked for the interpretation implicit in biologists' usage.

A Propensity Analysis of Natural Selection

One consequence of our propensity interpretation of fitness is that the analysis also points to an improved definition of "natural selection." As was noted earlier, the concepts of fitness and natural selection are inextricably bound— so much so that misinterpretations of fitness are reflected in misinterpretations of natural selection.

Thus, according to one of the more popular interpretations of natural selection, that process occurs whenever two or more individuals leave different numbers of offspring, or whenever two or more types leave different average numbers of offspring. For example, Crow and Kimura (1970) stipulate that "selection occurs when one genotype leaves a different number of progeny than another" (p. 173). Insofar as it is correct to say that the *fittest* are *selected*, this definition of "selection" clearly reflects a definition of "fitness" in terms of actual descendant contribution.

But surely these definitions (see also Wallace 1963, p. 160; Wilson 1975, p. 489) do not adequately delimit the reference of "natural selection." For evolutionary biologists do not refer to just any case of differential offspring contribution as "natural selection." For instance, if predatory birds were to kill light- and dark-colored moths indiscriminately, and yet by chance killed more light than dark ones, we would not attribute the differential offspring contribution of light and dark moths to natural selection. But if the dark coloration acted as camouflage, enabling the dark moths to escape predation and leave more offspring, we would attribute the resulting differential offspring contribution to the action of natural selection. For only in the latter case are differences in offspring contribution due to differences in offspring contribution dispositions.

Thus, Kettlewell (1955, 1956) did not presume to have demonstrated the occurrence of natural selection simply by pointing out the dramatic increase in frequency of dark-colored pepper moths within industrial areas of England. In order to demonstrate that selection (vs. chance fluctuations, migration, etc.) had accounted for the change, Kettlewell had to provide evidence that the dark-colored moths were better able to survive and reproduce in the sooted forests of these regions. Nor did Cain and Sheppard (1950, 1954) and Ford (1964) consider differential contribution to be a sufficient demonstration of natural selection in their celebrated accounts of the influence of selection on geographical distribution. In order to support the hypothesis that natural selection had affected the geographic distribution of various color and banding-pattern traits of snails of the species *Cepaea nemoralis*, these men argued that the colors and band-patterns peculiar to an area were correlated with the background color and uniformity of that area. More precisely, yellow snails were predominant in green areas; red and brown snails were predominant in beechwoods ("with their red litter and numerous exposures of blackish soil" [Ford 1964, p. 153]); and unbanded shells were predominant in more uniform environments. These traits effectively camouflaged their possessors from the sight of predators (Ford 1964, p. 155), thus *enabling* suitably marked snails to contribute more offspring than the unsuitably marked snails.

In each of these cases, selection is construed as involving more than just differential perpetuation. Rather, selection involves differential perpetuation caused by differential reproductive capabilities. So, just as we amended traditional definitions of "fitness" to take into account descendant contribution propensities, we must also amend traditional definitions of "selection" so as to emphasize the role of differential descendant contribution propensities. Selection, properly speaking, involves not just the differential contribution of descendants, but a differential contribution *caused* by differential propensities to contribute. On the basis of these considerations, let us define "individual selection" and "type selection" as follows:

Natural selection is occurring in population P in environment E with regard to organisms x, y, z (members of P) $=_{df} x$, y, z differ in their descendant contribution dispositions in E, and these differences are manifested in E in P.

Natural selection is occurring in population P in environment E with regard to types X, Y, Z (included in P) $=_{df}$ members of X, Y, Z types differ in their average descendant contribution dispositions in E, and these differences are manifested in E in P.

We know from our previous analysis that when organisms leave numbers of offspring which reflect their reproductive propensities (i.e., when reproductive propensities are manifesting themselves) in a particular environment, this implies that no factors are interfering with the manifestation of these propensities (cf. our remarks on extremal clauses above). Put more positively, we have grounds for believing that, for example, no cataclysms, cases of human intervention, and so forth are occurring. Of course, the occurrence of natural selection is not precluded by the incidence of such factors. Fitter individuals might leave more offspring than less fit individuals (on account of their fitness differences), even though non-discriminating factors are operating to minimize the reproductive effects of fitness differences. In other words, the incidence of non-discriminating factors will not necessarily override the effects of fitness differences. Thus, we do not have to rule out the occurrence of non-selective factors in our definition of "natural selection." But in explanations (such as our Hempelian schema above) of the precise evolutionary effects of selection, we must take these non-selective factors into account.

CONCLUSION

A science may well progress even though its practitioners are unable to account for aspects of its foundations in any illuminating way. We believe that this has been the case with evolutionary theory, but that the *propensity* analysis of fitness which we have described captures the implicit content in biologists' usage of the term. The propensity interpretation allows us to reconstruct explanations of microevolutionary phenomena in such a way that these explanations appear to be entirely respectable and noncircular. By their form, and by inspection of the premises and conclusion, such explanations appear to satisfy Hempelian adequacy requirements for explanations, and even appear to incorporate recent modifications of the Hempelian model for inductive explanations (Coffa 1974). We chose an example of microevolutionary change, since we wanted the least complicated instance possible in order to illuminate the form of explanations utilizing fitness ascriptions. We know of no reason to believe that a similar reconstruction could not be given for the case of macroevolutionary change.[11]

NOTES

We wish to thank Professor Michael Ruse, for initially drawing our attention to the problems of the logical status of evolutionary theory, and for insightful criticisms of an early draft of this chapter. We are heavily indebted to Alberto Coffa, for providing us with explications both of propensities and of the nature of explanation, and for innumerable criticisms and suggestions. Ron Giere also suggested that the propensity interpretation was a little more complex than we originally suspected. However, we claim complete originality for our mistakes.

1. Bethell may have been misled by the fact that evolutionary biologists recognize mechanisms of evolutionary change other than fitness differences (e.g., drift). Nevertheless, there is no question that fitness differences have been and still are considered effective in producing evolutionary changes.

2. The counter-intuitiveness of the traditional definition is also suggested by the following hypothetical case. Imagine two butterflies of the same species, which are phenotypically identical except that one (C) has color markings which camouflage it from its species' chief predator, while the second (N) does not have such markings and is hence more conspicuous. If N nevertheless happens to leave more offspring than C, we are committed on the definition of fitness under consideration to conclude that (1) both butterflies had the same degree of fitness before reaching maturity (i.e., zero fitness) and (2) in the end, N is fitter, since it left more offspring than C.

3. Gary Hardegree suggested this to us in conversation.

4. As we recently learned, Mary Williams supports the propensity interpretation and has, independently, worked toward an application of this interpretation.

5. Where fitness is defined as a propensity we can also squeeze the empirical content out of the phrase "survival of the fittest" (i.e., the claim that the fittest survive), which has frequently been claimed to be tautological (e.g., by Bethell 1976, Popper 1974, and Smart 1963. Just as the claim that "the soluble (substance) is dissolving" is an empirical claim, so the claim that those which could gain predominance in a particular environment are in fact gaining predominance, is an empirical claim. In short, to claim that a dispositional property is manifesting itself is to make an empirical claim. Such a claim suggests that the conditions usually known to trigger the manifestation are present, and no factors are present to override this manifestation. It seems plausible to interpret "the survival of the fittest" as a loose way of claiming that the organisms which are leaving most offspring are also the most fit. That this is a plausible interpretation of Darwin's use of the phrase is also suggested by Darwin's concern (in *The Origin of Species*) to demonstrate that conditions favoring natural selection are widely in effect. But it should be emphasized that nothing hinges on providing such an interpretation for "the survival of the fittest." This catch-phrase is not an important feature of evolutionary theory, in spite of the controversy its alleged tautological status has generated.

6. As this discussion suggests, an organism's fitness is not only a function of the organism's traits, but also of characteristics of the organism's environment. Actually, this function may be even more complicated. For evolutionary biologists have also noted that the fitness of an individual may depend upon the characteristics of the population to which it belongs. For instance, there is evidence of "frequency dependent selection" in several species of Drosophila (Kojima 1971). This kind of selection is said to occur whenever the fitness of a type depends upon the frequency of the type. Some types appear to be fitter, and are selected for, when they are rare. Thus, fitness is relative to environmental and population characteristics. And consequently, the appropriate triggering conditions for the realization of descendant contribution dispositions include environmental and population structure conditions.

7. Given propensities apply to individual objects, (rather than chance set-ups or sequences of trials) we also take them to be ontologically real—not merely epistemic properties. Our view

Susan Mills and John Beatty

is similar to Mellor's (for a good review of the views on propensities, cf. Kyburg 1974), but it most closely follows Coffa's analysis (1977, and his unpublished dissertation, *Foundations of Inductive Explanation*).

8. While an organism has a number of different propensities to leave n offspring, for different values of n, we do not have the additional complication that an organism has a number of different propensities to leave a particular number of offspring, n. An object has many different *relative probabilities* to manifest a given property, depending on the reference class in which it is placed. (In practice, choice of reference classes is dependent on our knowledge of the statistically relevant features of the situation.) But an object's *propensity* to manifest a certain property is a function of all of the causally relevant features of the situation, independent of our knowledge or ignorance of these factors. The totality of causally relevant features detemines the unique correct reference class, and thus the unique strength of the propensity to manifest the property in question. (Thus it cannot be the case that an object has more than one propensity to manifest a particular property in a particular situation.)

9. It might initially be thought that these examples are highly artificial, since there are no such "bimodal" organisms. But organisms tend to have offspring in litters and swarms. For such organisms, their offspring contribution propensities will cluster around multiples of numbers typical of the litter or hatching size.

10. A note of clarification is in order concerning our definition of "fitness$_1$." It is not clear whether "expected descendant contribution" refers to expected offspring contribution, or expected second-generation descendant contribution, or expected 100th generation descendant contribution. The problem can be illustrated as follows. One kind of individual may contribute a large number of offspring which are all very well adapted to the environment into which they are born, but cannot adapt to environmental changes. As a result, an individual of this type contributes a large number of offspring at time t, but due to an environmental change at $t + \Delta t$, these offspring in turn leave very few offspring, so that the original individual actually has very few second- or third-generation descendants. On the other hand, individuals of an alternate type may leave fewer offspring, yet these offspring may be very adaptable to environmental changes. Thus, although an individual of the latter type contributes a lower average number of offspring at time t, that individual may have a greater descendant contribution at $t + \Delta t$. Which individual is fitter? We suggest differentiating between long-term fitness and short-term fitness—or between first-generation fitness, second-generation fitness, ..., n-generation fitness. Thus, the latter type is fitter in the long term, while the former is fitter in the short term.

11. A great deal more needs to be done by way of clarifying the concepts of fitness and natural selection, given the many uses biologists make of these concepts. But we believe that the broad analyses we have given provide an adequate framework within which further distinctions and clarifications can be made. For example, within the categories of fitness$_1$ and relative fitness$_2$, distinctions can be drawn between short- and long-term fitness, by distinguishing between propensities to leave descendants in the short run (in the next few generations) *vs.* propensities to leave descendants in the long run (cf. note 10).

The propensity interpretation also lends itself to the much-discussed notion of "frequency dependent fitness," wherein the fitness of a type differs according to the frequency of the type. Certain cases of mimicry have been explained via reference to frequency dependent fitness. For instance, it has been suggested that the mimetic resemblance of a prey species to a distasteful model may enhance the survival of the mimics so long as they are rare, because individual predators most readily learn to avoid the distasteful type (and hence the mimic) when the model is more common than the mimic. Surely the survival *ability* of the mimics, and not just their survival rates, are enhanced by the scarcity of their type.

The sociobiological notion of "inclusive fitness" also seems susceptible to a propensity analysis. Biologists have invoked this notion in order to explain the evolution of certain altruistic traits. The idea (very simply) is that some of the organisms benefiting from an altruistic action may be genetically related to the altruistic actor, and may therefore share the behavioral trait which led to the action (if the trait is genetically based). Thus, although an altruistic action may decrease the fitness$_1$ of the actor, it may increase the fitness$_2$ of the altruistic trait. As a result, the trait may come to predominate within the population. "Inclusive fitness" values have been proposed as appropriate indicators of the evolutionary fate of altruistic traits. These values take into account not only the effect of altruistic actions upon the fitness of the actors, but also the probability that the action will benefit genetic relatives, and the extent of the benefit to relatives (cf. Hamilton 1964). Our colleague Greg Robischon is currently considering a propensity interpretation of inclusive fitness.

REFERENCES

Bethell, T. (1976). "Darwin's Mistake." *Harper's Magazine*, 70–75.

Cain, A. J., and P. M. Sheppard (1950). "Selection in the Polymorphic Land Snail *Cepala Nemoralis*." vol. 4, *Heredity*: 275–294.

Cain, A. J., and P. M. Sheppard (1954). "Natural Selection in Cepaea." *Genetics* 39:89–116.

Cody, M. (1966). "A General Theory of Clutch Size." *Evolution*, 20:174–184.

Coffa, J. A. (1974). "Hempel's Ambiguity." *Synthese*, 28:141–163.

———. (1977). "Probabilities: Reasonable or True?" *Philosophy of Science*, 43:186–198.

Crow, J. F., and Kimura, M. (1970). *An Introduction to Population Genetics Theory*. New York: Harper and Row.

Dobzhansky, T. (1970). *Genetics of the Evolutionary Process*. New York: Columbia University Press.

Dobzhansky, T., and Levene, H. (1951). "Development of Heterosis Through Natural Selection in Experimental Populations of Drosophila Pseudoobscura." *American Naturalist* 85:246–264.

Ford, E. B. (1964). *Ecological Genetics*. New York: Wiley.

Grant, V. (1977). *Organismic Evolution*. San Francisco: W. H. Freeman.

Hamilton, W. D. (1964). "The Genetical Evolution of Social Behavior. I." *Journal of Theoretical Biology*, 7:1–16.

Hull, D. (1974). *Philosophy of Biological Theory*. Englewood Cliffs. New Jersey: Prentice-Hall.

Kettlewell, H. B. D. (1955). "Selection Experiments on Industrial Melanism in the Lepidoptera." *Heredity*, 9:323–342.

———. (1956). "Further Selection Experiments on Industrial Melanism in the Lepidoptera." *Heredity*, 10:287–301.

Kojima, K. (1971). "Is There a Constant Fitness Value for a Given Genotype?" *Evolution*, 25:281–285.

Kyburg, H. (1974). "Propensities and Probabilities." *British Journal for the Philosophy of Science*, 25, no. 4:358–374.

Lerner, I. M. (1958). *The Genetic Basis of Selection*. New York: Wiley.

Levins, R. (1970). "Fitness and Optimization." *Mathematical Topics in Population Genetics*. New York: Springer Verlag.

———. (1968). *Evolution in Changing Environments*. Princeton: Princeton University Press.

Manser, A. R. (1965). "The Concept of Evolution." *Philosophy*, 40:18–34.

Mettler, L. E., and Gregg, T. G. (1969) *Population Genetics and Evolution*. Englewood Cliffs, New Jersey: Prentice-Hall.

Popper, K. (1974). "Intellectual Autobiography." *The Philosophy of Karl Popper* (Shilpp, ed.). LaSalle, Illinois: Open Court.

Ruse, M. (1969). "Confirmation and Falsification of Theories of Evolution." *Scientia*, 1–29.

———. (1973). *The Philosophy of Biology*. London: Hutchinson.

Scriven, M. (1959). "Explanation and Prediction in Evolutionary Theory." *Science*, 130:477–482.

Smart, J. J. C. (1963). *Philosophy and Scientific Realism*. London: Routledge and Kegan Paul.

Thoday, J. M. (1953). "Components of Fitness." *Symposia of the Society for Experimental Biology*, 7:96–113.

Waddington, C. H. (1968). "The Basic Ideas of Biology." *Towards a Theoretical Biology*, vol. 1. Chicago: Aldine.

Wallace, B. (1968). *Topics in Population Genetics*. New York: W. W. Norton.

———. (1963). "Further Data on the Over-dominance of Induced Mutations." *Genetics* 48:633–651.

Williams, M. B. (1970). "Deducing the Consequences of Evolution: A Mathematical Model." *Journal of Theoretical Biology*, 29:343–385.

———. (1973a). "The Logical Status of Natural Selection and Other Evolutionary Controversies: Resolution by Axiomatization." *Methodological Unity of Science* (Bunge, ed.). Dordrecht, Holland: Reidel.

———. (1973b). "Falsifiable Predictions of Evolutionary Theory." *Philosophy of Science*, 40:518–537.

———. (1976). "The Logical Structure of Functional Explanations in Biology." *Proceedings of the Philosophy of Science Association 1976*: East Lansing: Philosophy of Science Association, 37–46.

Wilson, E. O. (1975). *Sociobiology*. Cambridge, Massachusetts: Harvard University Press.

Wright, S. (1955). "Classification of the Factors of Evolution." *Cold Spring Harbor Symposia on Quantitative Biology*, 20:16–24.

II   Function and Teleology

2 Functions

Larry Wright

The notion of function is not all there is to teleology, although it is sometimes treated as though it were. Function is not even the central, or paradigm, teleological concept. But it *is* interesting *and* important; and it is still not as well understood as it should be, considering the amount of serious scholarship devoted to it during the last decade or two. Let us hope this justifies my excursion into these murky waters.

Like nearly every other word in English, "function" is multilaterally ambiguous. Consider:

1. $y = f(x)$/The pressure of a gas is a function of its temperature.

2. The Apollonaut's banquet was a major state function.

3. I simply can't function when I've got a cold.

4. The heart functions in this way ... [something about serial muscular contractions].

5. The function of the heart is pumping blood.

6. The function of the sweep-second hand on a watch is to make seconds easier to read.

7. Letting in light is one function of the windows of a house.

8. The wood box next to the fireplace currently functions as a dog's sleeping quarters.

It is interesting to notice that the word "function" has a spectrum of meanings even within the last six illustrations, which are the only ones at all relevant to a teleologically oriented study. Numbers 3, 4, and 8 are substantially different from one another, but they are each, from a teleological point of view, peripheral cases by comparison with 5, 6, and 7, which are the usual paradigms. And even these latter three are individually distinct in some respects, but much less profoundly than the others.

Quite obviously, making some systematic sense of the logical differentiation implicit in categorizing these cases as peripheral and paradigmatic is a major task of this chapter. But a clue that we are on the right track here can

From *Philosophical Review*, 1973, 82:139–168.

be found in a symptomatic grammatical distinction present in the last six illustrations: in the peripheral cases the word "function" is itself the verb, whereas in the more central cases "function" is a noun, used with the verb "to be." And since the controversy revolves around what *the function* of something *is*, the grammatical role of "function" in 5, 6, and 7 makes them heavy favorites for the logical place of honor in this discussion.

SOME RUDIMENTARY DISTINCTIONS

1. *Functions versus goals.* There seems to be a strong temptation to treat functions as representative of the set of central teleological concepts which cluster around goal-directedness. However, even a cursory examination of the usual sorts of examples reveals a very important distinction. Goal-directedness is a behavioral predicate. The *direction* is the direction of behavior. When we do speak of objects (homing torpedoes) or individuals (General MacArthur) as being goal-directed, we are speaking indirectly of their behavior. We would argue against the claim that they are goal-directed by appeal to their behavior (for example, the torpedo, or the General, did not *change course* at the appropriate time, and so forth). On the other hand, many things have *functions* (for example, chairs and windpipes) which do not behave *at all*, much less goal-directedly. And behavior can have a function without being goal-directed— for example, pacing the floor or blinking your eye. But even when goal-directed behavior has a function, very often its function is quite different from the achievement of its *goal*. For example, some fresh-water plankton diurnally vary their distance below the surface. The *goal* of this behavior is to keep light intensity in their environment relatively constant. This can be determined by experimenting with artificial light sources. The *function* of this behavior, on the other hand, is keeping constant the oxygen supply, which normally varies with sunlight intensity. There are many instances to be found in the study of organisms in which the function of a certain goal-directed activity is not some further goal of that activity, as it usually is in human behavior, but rather some natural concomitant or consequence of the immediate goal. Other examples are food-gathering, nest-making, and copulation. Clearly function and goal-directedness are not congruent concepts. There is an important sense in which they are wholly distinct. In any case, the relationship between functions and goals is a complicated and tenuous one; and becoming clearer about the nature of that relationship is one aim of this essay.

2. *A function versus the function.* Recent analyses of function, including all those treated here, have tended to focus on *a* function of something, by contrast with *the* function of something. This tendency is understandable; for any analysis of this sort aims at generality, and "a function" would seem intrinsically more general than "the function" because it avoids one obvious restriction. This generality, however, is superficial: the notion of *a* function is derivable from the notion of *the* function (more than one thing meets the criteria) just as easily as the reverse (only one thing meets the criteria). Furthermore, the notion of *a* function is much more easily confused with certain

peripheral, quasi-functional ascriptions which are examined below. In short, the discussion of this chapter is concerned with *a* function of X only insofar as it is the sort of thing which would be *the* function of X if X had no others. Accordingly, I take the definite-article formulation as paradigmatic and will deal primarily with it, adding comments in terms of the indefinite-article formulation parenthetically, where appropriate.

3. *Function versus accident.* Very likely the central distinction of this analysis is that between the *function* of something and other things it does which are *not* its function (or one of its functions). The function of a telephone is effecting rapid, convenient communication. But there are many other things telephones do: take up space on my desk, disturb me at night, absorb and reflect light, and so forth. The function of the heart is pumping blood, not producing a thumping noise or making wiggly lines on electrocardiograms, which are also things it does. This is sometimes put as the distinction between a function, and something done merely "by accident." Explaining the propriety of this way of speaking—that is, making sense of the function/accident distinction—is another aim, perhaps the *primary* aim of the following analysis.

4. *Conscious versus natural functions.* The notion of accident will raise some interesting and important questions across another rudimentary distinction: the distinction between natural functions and consciously designed ones. Natural functions are the common organismic ones such as the function of the heart, mentioned above. Other examples are the function of the kidneys to remove metabolic wastes from the bloodstream, and the function of the lens of the human eye to focus an image on the retina. Consciously designed functions commonly (though not necessarily) involve artifacts, such as the telephone and the watch's sweep hand mentioned previously. Other examples of this type would be the function of a door knob, a headlight dimmer switch, the circumferential grooves in a pneumatic tire tread, or a police force. Richard Sorabji has argued[1] that "designed" is too strong as a description of this category, and that less elaborate conscious effort would be adequate to give something a function of this sort. I think he is right. I have used the stronger version only to overdraw the distinction hyperbolically. In deference to his point I will drop the term "designed" and talk of the distinction as between natural and conscious functions.

Of the two, natural functions are philosophically the more problematic. Several schools of thought, for different reasons, want to deny that there are natural functions, as opposed to conscious ones. Or, what comes to the same thing, they want to deny that natural functions are functions in anything like the same sense that conscious functions are. Some theologians want to say that the organs of organisms get their functions through God's conscious design, and hence these things *have* functions, but not natural functions *as opposed to* conscious ones. Some scientists, like B. F. Skinner, would *deny* that organs and organismic activity have functions *because* there is no conscious effort or design involved.

Now it seems to me that the notion of an organ's having a function—both in everyday conversation and in biology—has no strong theological commitments. Specifically, it seems to me consistent, appropriate, and even common for an atheist to say that the function of the kidney is elimination of metabolic wastes. Furthermore, it seems clear that conscious and natural functions are functions in the same sense, despite their obvious differences. Functional ascriptions of either sort have a profoundly similar ring. Compare "the function of that cover is to keep the distributor dry" with "the function of the epiglottis is to keep food out of the windpipe." It is even more difficult to detect a difference in what is being requested: "What is the function of the human windpipe?" versus "What is the function of a car's exhaust pipe?" Certainly no analysis should begin by supposing that the two sorts are wildly different, or that only one is really legitimate. That is a possible *conclusion* of an analysis, not a reasonable presupposition. Accordingly, the final major aim of this analysis will be to make sense of natural functions, both as functions in the same sense as consciously contrived ones, and as functions independent of any theological presuppositions—that is, independent of conscious purpose. It follows that this analysis is committed to finding a way of stating what it is to be a function—even in the conscious cases—that does not rely on an appeal to consciousness. If no formulation of this kind can be found despite an honest search, only then should we begin to take seriously the view that we actually mean something quite different by "function" in these two contexts.

SOME ANALYSES OF FUNCTION

The analysis of function for which I wish to argue grew out of a detailed critical examination of several recent attempts in the literature to produce such an analysis, and it is best understood in that context. For this reason, and because it will help clarify the aims I have sketched above, I will begin by presenting the kernel of that critical examination.

The first analysis I want to consider is an early one by Morton Beckner.[2] Here Beckner contends that to say something s has function F' in system s' is to say, "There is a set of circumstances in which: F' occurs when s' has s, AND F' does not occur when s' does not have s" (p. 113).[3] For example, "The human heart has the function of circulating blood" means that there is a set of circumstances in which circulation occurs in humans when they have a heart, and does not when they do not. Translated into the familiar jargon, s has function F' in s' if and only if there is a set of circumstances containing s which are sufficient for the occurrence of F' and which also require s in order to be sufficient for F'. Now it is not clear whether the "requirement" here is necessity or merely nonredundancy. If it is necessity, then under the most natural interpretation of "circumstances" (environment), it is simply mistaken. There are *no* circumstances in which, for example, the heart is absolutely irreplaceable: we could always pump blood in some other way. On the other hand, if the requirement here is only nonredundancy, the mistake is more subtle.

In this case Beckner's formula would hold for cases in which s merely *does* F', but in which F' is not the function of s. For example, the heart is a nonredundant member of a set of conditions or circumstances which are sufficient for a throbbing noise. But making a throbbing noise is not a function of the heart; it is just something it does—accidentally. In fact, there are even dysfunctional cases which fit the formula: in some circumstances, livers are non-redundant for cirrhosis, but cirrhotic debilitation could not conceivable be the (or a) function of the liver. So this analysis fails on the functional/accidental distinction: it includes too much.

After first considering a view essentially similar to this one, John Canfield has offered a more elaborate analysis.[4] According to Canfield: "A function of I (in S) is to do C *means* I does C and that C is done is useful to S. For example, '(In vertebrates) a function of the liver is to secrete bile' means 'the liver secretes bile, and that bile is secreted in vertebrates is useful to them'" (p. 290). Canfield recognizes that natural functions are the problematic ones, but he devotes his attention solely to those cases. He treats only the organs and parts of organisms studied by biology, to the exclusion of the consciously designed functions of artifacts. As a result of this emphasis, his analysis is, without modification, almost impossible to apply to conscious functions. But even with appropriate modifications, it turns out to be inadequate to the characterization of either conscious or natural function.

In the conscious cases, there is an enormous problem in identifying the system S, *in* which I is functioning, and *to* which it must be useful. The function of the sweep-second hand of a watch is to make seconds easier to read. It would be most natural to say that the system *in which* the sweep hand is functioning—by analogy with the organismic cases—is the watch itself; but it is hard to make sense of the easier reading's being useful to the mechanism. On the other hand, the best candidate for the system *to which* the easier reading is useful is the person wearing the watch; but this does not seem to make sense as the system *in which* the sweep hand is functioning.

The crucial difficulty of Canfield's analysis begins to appear at this point: no matter what modifications we make in his formula to avoid the problem of identifying the system S, we must retain the requirement that C be useful. This is really the major contribution of his analysis, and to abandon it is to abandon the analysis. The difficulty with this is that, for example in the watch case, it is clearly not necessary that easily read seconds be useful to the watch-wearer—or anyone else—in order that making seconds easier to read be the function of the sweep hand of that wearer's watch. My watch has a sweep-second hand, and I occasionally use it to time things to the degree of accuracy it allows: it is useful to me. Now suppose I were to lose interest in reading time to that degree of accuracy. Suppose my life changed radically so that nothing I ever did could require that sort of chronological precision. Would that mean the sweep hand on my particular watch no longer has the function of making seconds easier to read? Clearly not. If someone were to ask what the sweep hand's function was ("What's it do?" "What's it there for?") I would still have to say it made seconds easier to read, although I might yawningly append an autobiographical note about my utter lack of interest in that feature. Similarly, the function of that button on my dashboard is to activate the windshield washer, even if all it does is make the mess on the windshield worse, and hence is not useful at all. That would be its *function* even if I never took my car out of the garage—or if I broke the windshield.

It is natural at this point to attempt to patch up the analysis by reducing the requirement that C be useful to the requirement that C *usually* be useful. But this will not do either, because it is easy to think of cases in which we would talk of something's having a function even though doing that thing was quite *generally* of no use to anybody—for example, a machine whose function was to count Pepsi Cola bottle caps at the city dump; or MIT's ultimate machine of a few years back, whose only function was to turn itself off. The source of the difficulty in all of these cases is that what the thing in question (watch, washer button, counting machine) was *designed* to do has been left out of the calculation. And, of course, in these cases, if something is designed to do X, then doing X is its function even if doing X is generally useless, silly, or even harmful. In fact, intention is so central here that it allows us to say the function of I is to do C, even when I cannot even *do* C. If the windshield washer switch comes from the factory defective, comes from the factory defective, and is never repaired, we would still say that its *function* is to activate the washer system; which is to say: that is what it was *designed* to do.

It might appear that this commits us to the view that natural and consciously contrived functions cannot possibly be the same sort of function. If conscious intent is what *determines* the function an artifact has got, there is no parallel in natural functions. I take this to be mistaken, and will show why later. For now it is only important to show, from this unique vantage, the nature of the most formidable obstacle to be overcome in unifying natural and conscious functions.

The argument thus far has shown that meeting Canfield's criteria is not necessary for something to be a function. It can easily be shown that meeting them is also not sufficient. We are always hearing stories about the belt buckles of the Old West or on foreign battlefields which save their wearers' lives by deflecting bullets. From several points of view that is a very useful thing for them to do. But that does not make bullet deflection the function— or even *a* function—of belt buckles. The list of such cases is endless. Artifacts do all kinds of useful things which are not their functions. Blowouts cause you to miss flights that crash. Noisy wheel bearings cause you to have the front end checked over when you are normally too lazy. The sweep hand of a watch might brush the dust off the numbers, and so forth.

All this results from the inability of Canfield's analysis to handle what we took to be one of the fundamental distinctions of function talk: accidental versus nonaccidental. Something can do something useful purely by accident, but it cannot have, as its function, something it does only by accident. Something that I does by accident cannot be the function of I. The cases above allow us to begin to make some fairly clear sense of this notion of accident, at least for artifacts. Buckles stop bullets only by accident. Blowouts only accidentally keep us off doomed airplanes. Sweep hands only accidentally brush dust, if they do it at all. And this brings us back to the grammatical distinction

I made at the outset when I divided the list of illustrations into "central" and "peripheral" ones. When something does something useful by accident rather than design, as in these examples, we signal the difference by a standard sort of "let's pretend" talk. Instead of using the verb "to be" or the verb "to have," and saying the thing in question *has* such and such a function, or saying that *is* its function, we use the expression "functioning as." We might say the belt buckle *functioned as* a bullet shield, or the blowout *functioned as* divine intervention, or the sweep hand *functions as* a dust brush. Canfield's analysis does not embrace this distinction at all.

So far I have shown only that Canfield's formula fails to handle conscious functions. This means it is incapable of showing natural functions to be functions in the same full-blooded sense as conscious ones, which is indeed serious; but that, it might be argued, really misses the point of his analysis. Canfield is not interested in conscious functions. He would be happy just to handle natural functions. For the reasons set down above, however, I am looking for an analysis which will *unify* conscious and natural functions, and it is important to see why Canfield's analysis cannot produce that unification. Furthermore, Canfield's analysis has difficulties in handling natural functions that closely parallel the difficulties it has with conscious functions; which is just what we should expect if the two are functions in the same sense.

For example, it is absurd to say with Pangloss that the function of the human nose is to support eyeglasses. It is absurd to suggest that the support of eyeglasses is even one of its functions. The function of the nose has something to do with keeping the air we breathe (and smell) warm and dry. But supporting a pince-nez, just as displaying rings and warpaint, is something the human nose does, and is useful to the system having the nose: so it fits Canfield's formula. Even the heart throb, our paradigm of nonfunction, fits the formula: the sound made by the heart is an enormously useful diagnostic aid, not only as to the condition of the heart, but also for certain respiratory and neurological conditions. More bizarre instances are conceivable. If surgeons began attaching cardiac pacemakers to the sixth rib of heart patients, or implanting microphones in the wrists of CIA agents, we could then say that these were useful things for the sixth rib and the wrist (respectively) to do. But that would not make pacemaker-hanging a function of the sixth rib, or microphone concealment a function of the human wrist.

There seems to be the same distinction here that we saw in conscious functions. It makes perfectly good sense to say the nose *functions as* an eyeglass support; the heart, through its thump, *functions as* a diagnostic aid; the sixth rib *functions as* a pacemaker hook in the circumstances described above. This, it seems to me, is precisely the distinction we make when we say, for example, that the sweep-second hand *functions as* a dust brush, while denying that brushing dust is one of the sweep hand's functions. And it is here that we can make sense of the notion of accident in the case of natural functions: it is merely fortuitous that the nose supports eyeglasses; it is happy chance that the heart throb is diagnostically significant; it would be the merest serendipity if the sixth rib were to be a particularly good pacemaker hook. It is (would be) only *accidental* that (if) these things turned out to be useful in these ways.

Accordingly, we have already drawn a much stronger parallel between natural functions and conscious functions than Canfield's analysis will allow.

Thus far I have ignored Canfield's analysis of usefulness: "[In plants and animals other than man, that C is done is useful to S means] if, *ceteris paribus*, C were not done in S, then the probability of that S surviving or having descendants would be smaller than the probability of an S in which C is done surviving or having descendants" (p. 292). I have ignored it because its explicit and implicit restrictions make it even more difficult to work this analysis into the unifying one I am trying to produce. Even within its restrictions (natural functions in plants and animals other than man), however, the extended analysis fails for reasons very like the ones we have already examined. Hanging a pacemaker on the sixth rib of a cardiovascularly inept lynx would be useful to that cat in precisely Canfield's sense of "useful": it would make it more likely that the cat would survive and/or have descendants. Obviously the same can be said for the diagnostic value of an animal's heart sounds. So usefulness—even in this very restricted sense—does not make the right function/accident distinction: some things do useful things which are not their functions, or even one of their functions.

The third analysis I wish to examine is a more recent one by Morton Beckner.[5] This analysis is particularly interesting for two reasons. First, Beckner is openly (p. 160) trying to accommodate both natural and conscious functions under one description. Second, he wants to avoid saying things like (to use his examples), "A function of the heart is to make heart sounds" and "A function of the Earth is to intercept passing meteorites." So his aims are very like the ones I have argued for: to produce a unifying analysis, and one which distinguishes between functions and things done by accident. And since the heart sound is useful, and intercepting metoeorites could be (perhaps already is), Beckner would probably agree in principle with the above criticism of Canfield.

Beckner's formulation is quite elaborate, so I will present it in eight distinct parts, clarify the individual parts, and then offer an illustration before going on to raise difficulties with them collectively as an analysis of the concept of function. That formulation is:

P has function F in S if and only if[6]

1. P is a part of S (in the normal sense of "part").

2. P contributes to F. (P's being part of S makes the occurrence of F more likely.)

3. F is an activity in or of the system S.

4. S is structured in such a way that a significant number of its parts contribute to the activities of other parts, and of the system itself.

5. The parts of S and their mutual contributions are identified by the same conceptual scheme which is employed in the statement that P has function F in system S.

6. A significant number of critical parts (of *S*) and their activities definitionally contribute to one or more activities of the whole system *S*.

7. *F* is or contributes to an activity *A* of the whole system *S*.[7]

8. *A* is one of those activities of *S* to which a significant number of critical parts and their activities definitionally contribute.

Two points of clarification must be made at once. First, the notion of "the same conceptual scheme" in number 5 is obscure in some respects, and the considerable attention devoted to it by Beckner does not help very much. In general all one can say is that *P*, *F*, and the other parts and activities of *S* must be *systematically* related to one another. But in practice the point is easier to make. For example, if we wish to speak of removing metabolic wastes as the function of the human kidney, the relevant conceptual scheme contains other human organs, life, and perhaps ecology in general, but not atoms, molecular bonds, and force fields. The second point concerns the "definitional contribution" in number 6. A part (or activity) makes a definitional contribution to an activity if that contribution is part of what we mean by the word which refers to that part (or activity). For example, part of what we mean by "heart" in a biological or medical context is "something which pumps blood": we would allow considerable variation in structure or appearance and still call something a heart if it served that function. Beckner illustrates how all these steps work together, once again using the heart.

It is true that a function of the heart is to pump blood. The heart does pump blood; the body is a complex system of parts that by definition aid in certain activities of the whole body, such as locomotion, self-maintenance, copulation; the concepts "heart" and "blood" are recognizably components of the scheme we employ in describing this complex system; and blood-pumping does contribute to activities of the whole organism to which many of its organs, tissues and other parts definitionally contribute. (p. 160)

There are several difficulties with this analysis. They appear below, roughly in order of increasing severity.

First, Beckner's problems with the system *S* are in some ways worse than Canfield's; for Beckner explicitly wants to include artifacts, and in addition he says much more definite things about the relationship among *P*, *F*, and *S*. So in this case, when we say the function of a watch's sweep hand is making seconds easier to read, we must not only find a system *of* which the sweep hand is a part, and *in* or *of* which "making seconds easier to read" is an activity, but this activity must be or contribute to one to which a number of the system's critical parts definitionally contribute. In the case of natural functions of the organs and other parts of organisms, the system *S* is typically a natural unit, easy to subdivide from the environment: the organism itself. But for the conscious functions of artifacts, such systems, if they can be found at all, must be hacked out of the environment rather arbitrarily. With no more of a guide than Beckner has given us, there is nothing like a guarantee that we can always find such a system. Accordingly, when our minds boggle—as I take it they do in trying to conceive of "making seconds easier to read" being an activity at all, much less one meeting all of the other conditions of this analysis—we have to say that the analysis is at best too obscure to be applicable to such cases, and is perhaps just mistaken.

A second difficulty stems directly from the first. It is not at all clear that functions—even natural functions—have to be activities at all, let alone activities of the sort required by Beckner. Making seconds easier to read is an example, but there are many others: preventing skids in wet weather, keeping your pants up, or propping open my office door. All of these things are legitimate functions (of tire treads, belts, and doorstops, respectively); none is an activity in any recognizable sense.

Third, we noticed in our discussion of Canfield that something could do a useful thing by accident, in the appropriate sense of "accident." Similarly, a part of a system meeting all of Beckner's criteria might easily make a contribution to an activity of that system also quite by accident. For example, an internal-combustion engine is a system satisfying Beckner's criteria for S. If a small nut were to work itself loose and fall under the valve-adjustment screw in such a way as to adjust properly a poorly adjusted valve, it would make an accidental contribution to the smooth running of that engine. We would never call the maintenance of proper valve adjustment the *function* of the nut. If it got the adjustment right it was just an accident. But on Beckner's formulation, we would have to call that its function. The nut does keep the valve adjusted; the engine is a complex system of parts that by definition aid in certain activities of the whole body, such as generation of torque and self-maintenance (lubrication, heat dissipation); the concepts "nut", "valve," and "valve adjustment" are components of the scheme we employ in describing this complex system; and proper valve adjustment does contribute to the smooth running of the (whole) engine, which is an activity to which many of the other parts of the engine definitionally contribute (flywheel, connecting rod, exhaust ports).

The final difficulty is also related to one we raised for Canfield's analysis. There we noticed that if an artifact was explicitly designed to do something, *that* usually *determines* its function, irrespective of how well or badly it does the thing it was supposed to do. An analogous point can be made here. If X was designed to do Y, then Y is X's function regardless of what contributions X does in fact make or fail to make. For example, the *function* of the federal automotive safety regulations is to make driving and riding in a car safer. And this is so even if they actually have just the opposite effect, through some psychodynamic or automotive quirk.

So in spite of their enormous differences, this analysis and Canfield's fail for very similar reasons: problems with the notion of system S, failure to rule out some accidental cases, and general inability to account for the obvious role of design.

There have been several other interesting attempts in the recent literature to provide an analysis of function. Most notable are those by Carl Hempel,[8] Hugh Lehman,[9] Richard Sorabji,[10] Francisco Ayala,[11] and Michael Ruse.[12] The last two of these do a somewhat better job on the function/accident distinction than the ones we have examined. But other than that, a discussion of these analyses would be largely redundant on the discussions of Beckner and Canfield. So I think we have gone far enough in clarifying the issues to begin constructing an alternative analysis.

AN ALTERNATIVE VIEW

The treatments we have so far considered have overlooked, ignored, or at any rate failed to make one important observation: that functional ascriptions are—intrinsically, if you will—explanatory. Merely saying of something, X, that it has a certain function, is to offer an important kind of explanation of X. The failure to consider this, or at least take it seriously, is, I think, responsible for the systematic failure of these analyses to provide an accurate account of functions.

There are two related considerations which urge this observation upon us. First, the "in order to" in functional ascriptions is a teleological "in order to." Its role in functional ascriptions (the heart beats in order to circulate blood) is quite parallel to the role of "in order to" in goal ascriptions (the rabbit is running in order to escape from the dog). Accordingly, we should expect functional ascriptions to be explanatory in something like the same way as goal ascriptions.[13] When we say that the rabbit is running in order to escape from the dog, we are explaining *why* the rabbit is running. If we say that John got up early in order to study, we are offering an explanation of his getting up early. Similarly in the functional cases. When we say that the distributor has that cover in order to keep the rain out, we are explaining *why* the distributor has that cover. And when we say the heart beats in order to pump blood, we are ordinarily taken to be offering an explanation of why the heart beats. This last sort of case represents the most troublesome problem in the logic of function, but it must be faced squarely, and, once faced, I think its solution is fairly straightforward.

The second consideration which recommends holding out for the explanatory status of functional ascriptions is the contextual equivalence of several sorts of requests. Consider:

1. What is the function of X?

2. Why do C's have X's?

3. Why do X's do Y?

In the appropriate context, each of these is asking for the function of X, "What is the function of the heart?" "Why do humans have a heart?" "Why does the heart beat?" All are answered by saying, "To pump blood," in the context we are considering. Questions of the second and third sort, being "Why?" questions, are undisguised requests for explanations. So in this context functional attributions are presumed to be explanatory. And why-form function requests are by no means bizarre or esoteric ways of asking for a function. Consider:

Why do porcupines have sharp quills?

Why do (some) watches have a sweep-second hand?

Why do ducks have webbed feet?

Why do headlight bulbs have two filaments?

These are rather ordinary ways of asking for a function. And if that is so, then it is ordinarily supposed that a function explains why each of these things is the case. The function of the quills is why porcupines *have* them, and so forth.

Moreover, the kind of explanatory role suggested by both of these considerations is not the anemic, "What's it good for?" sort of thing often imputed to functional explanations. It is rather something more substantial than that. If to specify the function of quills is to explain why porcupines *have* them, then the function must be the reason they *have* them. That is, the ascription of a function must be explanatory in a rather strong sense. To choose the weaker interpretation, as Canfield does,[14] is once again to run afoul of the function-accident distinction. For, to use his example, if "Why do animals have livers?" is a request for a function, it cannot be rendered, "What is the liver good for?" Livers are good for many things which are not their functions, just like anything else. Noses are good for supporting eyeglasses, fountain pens are good for cleaning your fingernails, and livers are good for dinner with onions. No, the *function* of the liver is that *particular* thing it is good for which explains why animals have them.

Putting the matter in this way suggests that functional ascription-explanations are in some sense etiological, concern the causal background of the phenomenon under consideration. And this is indeed what I wish to argue: functional explanations, although plainly not causal in the usual, restricted sense, do concern how the thing with the function *got there*. Hence they *are* etiological, which is to say "causal" in an extended sense. But this is still a very contentious view. Functional and teleological explanations are usually *contrasted with* causal ones, and we should not abandon that contrast lightly: we should be driven to it.

What drives us to this position is the specific difficulty the best-looking alternative accounts have in making the function/accident distinction. We have seen that no matter how useful it is for X to do Z, or what contribution X's doing Z makes within a complex system,[15] these sorts of consideration are never sufficient for saying that the function of X is Z. It could still turn out that X did Z only by accident. But all of the accident counterexamples can be avoided if we include as part of the analysis something about how X came to be there (where-ever): namely, that it is there *because it does* Z—with an etiological "because." The buckle, the heart, the nose, the engine nut, and so forth were not there *because* they stop bullets, throb, support glasses, adjust the valve, and all the other things which were falsely attributed as functions, respectively. Those pseudofunctions could not be called upon to explain how those things *got* there. This seems to be what was missing in each of those cases.

In other words, saying that the function of X is Z is saying at least that

(1) X is there *because* it does Z.
     or
     Doing Z is the *reason* X is there.
     or
     That X does Z is *why* X is there.

where "because," "reason," and "why" have an etiological force. And it turns out that "X is there because it does Z,"[16] with the proper understanding of "because," "does", and "is there" provides us with not only a necessary condition for the standard cases of functions, but also the kernel of an adequate analysis. Let us look briefly at those key terms.

"Because" is, of course, to be understood in its explanatory rather than evidential sense. It is not the "because" in, "It is hot because it is red." More important, "because" is to be taken (as it ordinarily is anyway) to be indifferent to the philosophical reasons/causes distinction. The "because" in, "He did not go to class because he wanted to study" and in, "It exploded because it got too hot" are both etiological in the appropriate way.[17] And finally, it is worth pointing out here that in this sense "A because B" does not require that B be either necessary or sufficient for A. Racing cars have airfoils because they generate a downforce (negative lift) which augments traction. But their generation of negative lift is neither necessary nor sufficient for racing cars to have wings: they could be there merely for aesthetic reasons, or they could be forbidden by the rules. Nevertheless, if you want to know why they are there, it is because they produce negative lift. All of this comes to saying "because" here is to be taken in its ordinary, conversational, causal-explanatory sense.

Complications arise with respect to "does" primarily because on the above condition "Z is the function of X" is reasonably taken to entail "X does Z." Although in most cases there is no question at all about what it is for X to do Z, the matter is highly context-dependent and so perhaps I should mention an extreme case, if only as notice that we should include it. In some contexts we will allow that X does Z even though Z never occurs. For example, the button on the dashboard activates the windshield washer system (that is what it does, I can tell by the circuit diagram) even though it never has and never will. An unused organic or organismic emergency reaction might have the same status. All that seems to be required is that X be *able* to do Z under the appropriate conditions; for example, when the button is pushed or in the presence of a threat to safety.

The vagueness of "is there" is probably what Beckner and Canfield were trying to avoid by introducing the system S into their formulations. It is much more difficult, however, to avoid the difficulties with the system S than to clarify adequately this more general placemarker. "Is there" is straightforward and unproblematic in most contexts, but some illustrations of importantly different ways in which it can be rendered might be helpful. It can mean something like "is where it is," as in, "Keeping food out of the windpipe is the reason the epiglottis is where it is." It can mean "*C*'s have them," as in "animals have hearts because they pump blood." Or it can mean merely "exists (at all)," as in, "Keeping snow from drifting across roads (and so forth) is why there are snow fences."

Now, saying that (1), understood in this way, should be construed as a necessary condition for taking Z to be the function of X, is merely to put in precise terms the moral of our examination of the function/accident distinction. We saw above that the accident counterexamples could not meet this requirement. On the other hand, this condition *is* met in all of the center-of-the-page cases. This is quite easy to show in the conscious cases. When we say the function of X is Z in these cases, we are saying that at least some effort was made to get X (sweep hand, button on dashboard) where it is precisely because it does Z (whatever). Doing Z is the reason X is there. *That* is why the effort was made. The reason the sweep-second hand is there is that it makes seconds easier to read. it is there *because* it does that. Similarly, rifles have safeties because they prevent accidental discharge.

It is only slightly less obvious how natural functions can satisfy (1): We can say that the natural function of something—say, an organ in an organism—is the reason the organ is there by invoking natural selection. If an organ has been naturally differentially selected for by virtue of something it does, we can say that the reason the organ is there is that it does that something. Hence we can say animals have kidneys *because* they eliminate metabolic wastes from the blood-stream; porcupines have quills *because* they protect them from predatory enemies; plants have chlorophyll *because* chlorophyll enables plants to accomplish photosynthesis; the heart beats *because* its beating pumps blood. And each of these can be rather mechanically put in the "reason that" form. The reason porcupines have quills is that they protect them from predatory enemies, and so forth.

It is easy to show that this formula does not represent a sufficient condition for being a function, which is to say there is something more to be said about precisely what it is to be a function. The most easily generable set of cases to be excluded is of this kind: oxygen combines readily with hemoglobin, and that is the (etiological) reason it is found in human bloodstreams. But there is something colossally fatuous in maintaining that the function of that oxygen is to combine with hemoglobin, even though it is there because it does that. The function of the oxygen in human bloodstreams is providing energy in oxidation reactions, not combining with hemoglobin. Combining with hemoglobin is only a means to that end. This is a useful example. It points to a contrast in the notion of "because" employed here which is easy to overlook and crucial to an elucidation of functions.

As I pointed out above, if producing energy is the function of the oxygen, then oxygen must be there (in the blood) because it produces energy. But the "because" in, "It is there because it produces energy," is importantly different from the "because" in, "It is there because it combines with hemoglobin."

They suggest different *sorts* of etiologies. If carbon monoxide (CO), which we know to combine readily with hemoglobin, were suddenly to become able to produce energy by appropriate (nonlethal) reactions in our cells and further, the atmosphere were suddenly to become filled with CO, we could properly say that the reason CO was in our bloodstreams was that it combines readily with hemoglobin. We could not properly say, however, that CO was there because it produces *energy*. And that is precisely what we could say about oxygen, on purely evolutionary-etiological grounds.

All of this indicates that it is the nature of the etiology itself which determines the propriety of a functional explanation; there must be specifically functional etiologies. When we say the function of X is Z (to do Z) we are saying that X is there because it does Z, but with a further qualification. We are explaining how X came to be there, but only certain kinds of explanations of how X came to be there will do. The causal/functional distinction is a distinction *among* etiologies; it is not a contrast between etiologies and something else.

This distinction can be displayed using the notion of a causal consequence.[18] When we give a functional explanation of X by appeal to Z ("X does Z"), Z is always a consequence or result of X's being there (in the sense of "is there" sketched above).[19] So when we say that Z is the function of X, we are not only saying that X is there because it does Z, we are also saying that Z is (or happens as) a result or consequence of X's being there. Not only is chlorophyll in plants *because* it allows them to perform photosynthesis, photosynthesis is a *consequence* of the chlorophyll's being there. Not only is the valve-adjusting screw there *because* it allows the clearance to be easily adjusted, the possibility of easy adjustment is a *consequence* of the screw's being there. Quite obviously, "consequence of" here does not mean "guaranteed by." "Z is a consequence of X," very much like "X does Z" earlier, must be consistent with Z's not occurring. When we say that photosynthesis is a consequence of chlorophyll, we allow that some green plants may never be exposed to light, and that all green plants may at some time or other not be exposed to light. Furthermore, this consequence relationship does not mean that whenever Z *does* occur, happen, obtain, exist, and so forth, it is as a consequence of X. There is room for a multiplicity of sufficient conditions, overdetermined or otherwise. Other things besides the adjusting screw may provide easy adjustment of the clearance. This (the inferential) aspect of consequence, as that notion is used here, can be roughly captured by saying that there are circumstances (of recognizable propriety) in which X is nonredundant for Z. The aspect of "consequence" of central importance here, however, is its asymmetry. "A is a consequence of B" is in virtually every context incompatible with "B is a consequence of A." The source of this asymmetry is difficult to specify, and I shall not try.[20] It is enough that it be clearly present in the specific cases.

Accordingly, if we understand the key terms as they have been explicated here, we can conveniently summarize this analysis as follows:

The function of X is Z means

(2)   (a) X is there because it does Z,

    (b) Z is a consequence (or result) of X's being there.

The first part, (a), displays the etiological form of functional ascription-explanations, and the second part, (b), describes the convolution which distinguishes functional etiologies from the rest. It is the second part, of course, which distinguishes the combining with hemoglobin from the producing of energy in the oxygen-respiration example. Its combining with hemoglobin is emphatically not a consequence of oxygen's being in our blood; just the reverse is true. On the other hand, its producing energy *is* a result of its being there.

The very best evidence that this analysis is on the right track is that it seems to include the entire array of standard cases we have been considering, while at the same time avoiding several very persistent classes of counter-examples. In addition to this, however, there are some more general considerations which urge this position upon us.[21] First, and perhaps most impressive, this analysis shows what it is about functions that is teleological. It provides an etiological rationale for the functional "in order to," just as recent discussions have for other teleological concepts. The role of the consequences of X in its own etiology provides functional ascription-explanations with a convoluted forward orientation which precisely parallels that found by recent analyses in ascription-explanations employing the concepts goal and intention.[22] In a functional explanation, the consequences of X's being there (where it is, and so forth) must be invoked to explain why X is there (exists, and so forth). Functional characterizations, by their very nature, license these explanatory appeals. Furthermore, as I hinted earlier, (b) is often simply implicit in the "because" of (a). When this is so, the "because" is the specifically teleological one sometimes identified as peculiarly appropriate in functional contexts. The peculiarly functional "because" is the normal etiological one, except that it is limited to consequences in this way. The request for an explanation as well will very often contain this implicit restriction, hence limiting the appropriate replies to something in terms of this "because"—that is, to functional explanations. "Why is it there?" in some contexts, and "What does it do?" in most, unpack into, "What consequences does it have that account for its being there?"

The second general consideration which recommends this analysis is that it both accounts for the propriety of, and at the same time elucidates the notion of, natural selection. To make this clear, it is important first to say something about the unqualified notion of selection, from which natural selection is derived. According to the standard view, which I will accept for expository purposes, the paradigm cases of selection involve conscious choice, perhaps even deliberation. We can then understand other uses of "select" and "selection" as extensions of this use: drawing attention to specific individual *features* of the paradigm which occur in subconscious or nonconscious cases. Of course, the range of extensions arrays itself into a spectrum from more or less literal to openly metaphorical. Now, there is an important distinction within the paradigmatic, conscious cases. I can say I selected something, X, even though I cannot give a reason for having chosen it: I am asked to select a ball from among those on the table in front of me. I choose the blue one and am asked why I did. I may say something like, "I don't know; it just struck me, I guess." Alternately, I could without adding much give something which has the form of a reason: "Because it is blue. Yes, I'm sure it was the color." In both of these cases I want to refer to the selection as "mere discrimination," for reasons which will become apparent below. On the other hand, there are a number of contexts in which another, more elaborate reply is possible and natural. I could say something of the form, "I selected X because it does Z." where Z would be some possibility opened by, some advantage that would accrue from, or some other result of having (using, and so forth) X. "I chose American Airlines because its five-across seating allows me to stretch out." Or "They selected DuPont Nomex because of the superior protection it affords in a fire."[23] Let me refer to selection by virtue of resultant advantage of this sort as "consequence-selection." Plainly, it is this kind of selection, as opposed to mere discrimination, that lies behind conscious functions: the consequence *is* the function. Equally plainly, it is specifically this kind of selection of which *natural* selection represents an extension.

But the parallel between natural selection and conscious consequence-selection is much more striking than is sometimes thought. True, the presence or absence of volition is an important difference, at least in some contexts. We might want to say that *natural* selection is really *self*-selection, nothing is *doing* the selecting; give the nature of X, Z, and the environment, X will *automatically* be selected. Quite so. But here the above distinction between kinds of conscious selection becomes crucial. For consequence-selection, by contrast with mere discrimination, deemphasizes volition in just such a way as to blur its distinction from natural selection on precisely this point. Given our criteria, we might well say that X *does* select itself in conscious consequence-selection. By the very nature of X, Z, and our criteria (the implementation of which may be considered the environment), X will automatically be selected.[24] The cases are very close indeed.

Let us now see how this analysis squares with the desiderata we have developed. First, it is quite clearly a unifying analysis: the formula applies to natural and conscious functions indifferently. Both natural and conscious functions are functions by virtue of their being the reason the thing with the function "is there," subject to the above restrictions. The differentiating feature is merely the *sort* of reason appropriate in either case: specifically, whether a conscious agent was involved or not. But in the functional-explanatory context which we are examining, the difference is minimal. When we explain the presence or existence of X by appeal to a consequence Z, the overriding consideration is that Z must be or create conditions conducive to the survival or maintenance of X. The exact *nature* of the conditions is inessential to the possibility of this form of explanation: it can be looked upon as a matter of mere etiological detail, nothing in the essential form of the explanation. In any given case something could conceivably get a function through either sort of consideration. Accordingly, this analysis begs no theological questions. The organs of organisms could logically possibly get their functions through God's conscious design; but we can also make perfectly good sense of their functions in the absence of divine intervention. And in either case they would be functions in precisely the same sense. This of course was accomplished only by disallowing explicit mention of intent or purpose in accounting for conscious functions. Nevertheless, the above formula can account for the very close relationship between design and function which the previous analyses could not. For, excepting bizarre circumstances, in virtually all of the usual contexts, X was designed to do Z simply entails that X is there because it results in Z.

Second, this analysis makes a clear and cogent distinction between function and accident. The things X can be said to do by accident are the things it results in which cannot explain how it came to be there. And we have seen that this circumvents the accident counterexamples brought to bear on the other analyses. It is merely accidental that the chlorophyll in plants freshens breath. But what it does for plants when the sun shines is no accident—that is why it is there. Furthermore, in this sense, "X did Z accidentally" is obviously consistent with X's doing Z having well-defined causal antecedents, just like the normal cases of other sorts of accident (automobile accidents, accidental meetings, and so forth). Given enough data it could even have been predictable that the belt buckle would deflect the bullet. But such deflection was still in the appropriate sense accidental: that is not why the buckle was there.

Furthermore, it is worth noting that something can get a function—either conscious or natural—*as the result of* an accident of this sort. Organismic mutations are paradigmatically accidental in this sense. But that only disqualifies an organ from functionhood for the first—or the first few—generations. If it survives by dint of its doing something, then that something becomes its function on this analysis. Similarly for artifacts. For example, if an earthquake shifted the rollers of a transistor production-line conveyor belt, causing the belt to ripple in just such a way that defective transistors would not pass over the ripple, while good transistors would, we could say that the ripple was *functioning as* a quality control sorter. But it would be incorrect to say that the ripple *had* the function of quality control sorting. It does not *have* a function at all. It is there only be accident. Sorting can, however, *become* its function if its sorting ability ever becomes a reason for preserving the ripple: if, for example, the company decides against repairing the conveyor belt *for that reason*. This accords nicely with Richard Sorabji's comment that in conscious cases, saying the function of X is Z requires at least "that some efforts are or would if necessary be made" to obtain Z from X.[25]

Third, the notion of something having more than one function is derivative. It is obtained by substituting something like "partly because"[26] for "because" in the formula. Brushing dust off the numbers is one of the functions of the watch's sweep-second hand if that feature is *one* of the (restricted, etiological) reasons the sweep hand is there. Similarly in the case of natural functions. If two or three things that livers do all contribute to the survival of organisms which have livers, we must appeal to all three in an evolutionary account of why those organisms have livers. Hence the liver would have more than one function in such organisms: we would have to say that each one was *a* function of the liver.

Happily, the analysis I am here proposing also accounts for the undoubted attractiveness of the other analyses we have examined. Beckner's first analysis is virtually included in this one under the rubric, "X does Z." The rest of the formula can be thought of as a qualification to avoid some rather straightforward counter-examples which Beckner himself is concerned to circumvent in his more recent attempt. Canfield's "usefulness" is even easier to accommodate: the usefulness of something, Z, which X does is *very usually* an informative way of characterizing why X has survived in an evolutionary process, or the reason X was consciously constructed. The important point to notice is that this is only *usually* the case, not necessarily: not all useful Z's can explain survival and some things are constructed to do wholly useless things. As for Beckner's most recent analysis, the complex, mutually contributory relationship among parts central to it is precisely the sort of thing often responsible for the survival and reproduction of organisms on one hand, and for the construction of complex mechanisms on the other. Again the valuable features of that analysis are incorporated in this one.

There is still one sort of case in which we clearly want to be able to speak of a function, but which offends the letter of this analysis as it stands. In several contexts, some of which we have already examined, we want to be able to say that X has the function Z, even though X cannot be said to do Z. X is not even *able* to do Z under the requisite conditions. In the cases of this sort I have already mentioned (the defective washer switch and ineffective governmental safety regulations), it has seemed necessary to italicize (emphasize, underline) the word "function" in order to make its use plausible and appropriate. This is a logical flag: it signals that a special or peculiar contrast is being made, that the case departs from the paradigms in a systematic but intelligible way. Accordingly, an analysis has to make sense of such a case as a variant.

On the present analysis, the italic type signals the dropping of the (usually presumed) second condition. X does *not* result in Z, although, paradoxically, doing Z *is* the reason X is there. Of course, in the abstract, this sounds fatuous. But we have already seen cases in which it is natural and appropriate. That *is* the reason X (switch, safety regulations) is there. And a slightly more defensive formulation of (2) will include them directly: a functional ascription-explanation accounts for X's being there by appeal to X's resulting in Z. These cases *do* appeal to X's resulting in Z to explain the occurrence of X, even though X does *not* result in Z. So the form of the explanation is functional even in these peculiar cases.

Interestingly, this account even handles the exotic fact that these italicized functions of X can cease being even italicized functions without dispensing with or directly altering X. (Something that X did not do can stop being its function!) For example, if the ineffective safety regulations were superseded by another set, and were merely left on the books through legislative sloth or expediency, we would no longer even say that had the (italicized) *function* of making driving less dangerous. But, of course, that would no longer be the reason they were there. The explanation would then have to appeal to legislative sloth or expediency. This is usually done with verb tenses: that *was* its function, but is not any longer; that was why it was there at one time, but is not why it is still there. A similar treatment can be given vestigial organs, such as the vermiform appendix in humans.

NOTES

1. Richard Sorabji, "Function," *Philosophical Quarterly*, 14 (1964), 290.

2. Morton Beckner, *The Biological Way of Thought* (New York, 1959), chap. 6.

3. Beckner gives an alternative formulation in which we can speak of *activities* as having functions, instead of *things*. I have abbreviated it here for convenience and clarity. The logical points are the same.

4. John Canfield, "Teleological Explanations in Biology," *British Journal for the Philosophy of Science*, 14 (1964).

5. Morton Beckner, "Function and Teleology," *Journal of the History of Biology*, 2 (1969).

6. As before, Beckner gives an alternative formulation so that we can speak either of a thing or of an activity having a function. My treatment will be limited to things, but again the logical points are the same.

7. Beckner seems to suggest (p. 160, top) that F must *be* an activity of the whole system S, which, of course, would conflict with part of 3. But his illustration, reproduced below, suggests the phrasing I have used here.

8. Carl Hempel, "The Logic of Functional Analyses," in L. Gross, ed., *Symposium on Sociological Theory* (New York, 1959).

9. Hugh Lehman, "Functional Explanations in Biology," *Philosophy of Science*, vol. 32 (1965).

10. Sorabji, *op. cit.*

11. Francisco J. Ayala, "Teleological Explanation in Evolutionary Biology," *Philosophy of Science*, vol. 37 (1970).

12. Michael E. Ruse, "Function Statements in Biology," *Philosophy of Science*, vol. 38 (1971).

13. This is not to abandon, or even modify, the previous distinction between functions and goals: the point can be made in this form only *given* the distinction. Nevertheless, support is provided for the analysis I am presenting here by the fact that the "in order to" of goal-directedness can be afforded a parallel treatment. For that parallel treatment see my paper "Explanation and Teleology," in the June 1972 issue of *Philosophy of Science*.

14. Canfield, p. 295.

15. It is sometimes urged that this sort of thing is all a teleological explanation is asserting; this is all "why?" asks in these contexts.

16. I take the other forms to be essentially equivalent and subject, *mutatis mutandis*, to the same explication.

17. Of course, it follows that the notion of a *reason* offered in one of the alternative formulations is the standard conversational one as well: the reason it exploded was that it got too hot.

18. The qualification "causal" here serves merely to indicate that this is not the purely inferential sense of "consequence." I am not talking about the result or consequence of an argument—e.g., necessary conditions for the truth of a set of premises. The precise construction of "consequence" appropriate here will become clear below.

19. It is worth recalling here that "is there" can only sometimes, but not usually, be rendered "exists (at all)." So, contrary to many accounts, what is being explained, and what Z is the result of, can very often *not* be characterized as "that X exists" *simpliciter*.

20. It is often claimed that the asymmetry is temporal, but there are many difficulties with this view. Douglas Gasking, in "Causation and Recipes," *Mind* (October 1955), attempts to account for it in terms of manipulability, with some success. But manipulability is even less generally applicable than time order, so, as far as I know, the problem remains.

21. The following considerations are intended primarily as support for the entire analysis considered as a whole. Since (*a*) has already been examined extensively, however, I have biased the argument slightly to emphasize (*b*).

22. The primary discussions of this sort I have in mind are those in Charles Taylor's *Explanation of Behavior* (London: Routledge & Kegan Paul, 1964) and the literature to which it has given rise.

23. Of course the advantage is not always stated explicitly: "I chose American because of its five-across seating." But for it to be selection of the sort described here, as opposed to mere discrimination, something like an advantage must be at least implicit.

24. This is a version of the old problem about the tension between rationality and freedom.

25. Sorabji, *op. cit.*, p. 290.

26. Again, it is worth pointing out that "partly" here does not indicate that "because," when *not* so qualified, represents a sufficient condition relationship. It merely serves to indicate that more than one thing plays an explanatorily relevant role in this particular case. More than one thing must be mentioned to answer adequately the functional "why?" question in this context. But that answer, as usual, need not provide a sufficient condition for the occurrence of X.

3   Functional Analysis

Robert Cummins

I

A survey of the recent philosophical literature on the nature of functional analysis and explanation, beginning with the classic essays of Hempel in 1959 and Nagel in 1961, reveals that philosophical research on this topic has almost without exception proceeded under the following assumption.[1]

(A)   The point of functional characterization in science is to explain the presence of the item (organ, mechanism, process, or whatever) that is functionally characterized.

(B)   For something to perform its function is for it to have certain effects on a containing system, which effects contribute to the performance of some activity of, or the maintenance of some condition in, that containing system.

Putting these two assumptions together, we have: a function-ascribing statement explains the presence of the functionally characterized item i in a system s by pointing out that i is present in s because it has certain effects on s. Give or take a nicety, this fusion of (A) and (B) constitutes the core of almost every recent attempt to give an account of functional analysis and explanation. Yet these assumptions are just that: assumptions. They have never been systematically defended; generally they are not defended at all. I think there are reasons to suspect that adherence to (A) and (B) has crippled the most serious attempts to analyze functional statements and explanation, as I will argue in sections II and III below. In section IV, I will briefly develop an alternative approach to the problem. This alternative is recommended largely by the fact that it emerges as the obvious approach once we take care to understand why accounts involving (A) and (B) go wrong.

II

I begin this section with a critique of Hempel and Nagel. The objections are familiar for the most part, but it will be well to have them fresh in our minds as they form the backdrop against which I stage my attack on (A) and (B).

From *Journal of Philosophy*, 1975, 72:741–764.

Hempel's treatment of functional analysis and explanation is a classic example of the fusion of (A) and (B). He begins by considering the following singular function-ascribing statement.

(1)   The heartbeat in vertebrates has the function of circulating the blood through the organism.

He rejects the suggestion that "function" can *simply* be replaced by "effect" on the grounds that, although the heartbeat has the effect of producing heartsounds, this is not its function. Presuming (B) from the start, Hempel takes the problem to be how one effect—the having of which is the function of the heartbeat (circulation)—is to be distinguished from other effects of the heartbeat (e.g., heartsounds). His answer is that circulation, but not heartsounds, ensures a necessary condition for the "proper working of the organism." Thus Hempel proposes (2) as an analysis of (1).

(2)   The heartbeat in vertebrates has the effect of circulating the blood, and this ensures the satisfaction of certain conditions (supply of nutriment and removal of waste) which are necessary for the proper working of the organism.

As Hempel sees the matter, the main problem with this analysis is that functional statements so construed appear to have no explanatory force. Since he assumes (A), the problem for Hempel is to see whether (2) can be construed as a deductive nomological explanans for the presence of the heartbeat in vertebrates and, in general, to see whether statements having the form of (2) can be construed as deductive nomological explananda for the presence in a system of some trait or item that is functionally characterized.

Suppose, then, that we are interested in explaining the occurrence of a trait i in a system s (at a certain time t), and that the following functional analysis is offered:

(a) At t, s functions adequately in a setting of kind c (characterized by specific internal and external conditions).

(b) s functions adequately in a setting of kind c only if a certain necessary condition, n, is satisfied.

(c) If trait i were present in s then, as an effect, condition n would be satisfied.

(d) Hence, at t, trait i is present is s.[2]

(d), of course, does not follow from (a)–(c), since some trait i' different from i might well suffice for the satisfaction of condition n. The argument can be patched up by changing (c) to (c'): "condition n would be satisfied in s only if trait i were present in s," but Hempel rightly rejects this avenue on the grounds that instances of the resulting schema would typically be false. It is false, for example, that the heart is a necessary condition for circulation in vertebrates, since artificial pumps can be, and are, used to maintain the flow of blood. We are thus left with a dilemma. If the original schema is correct, then functional explanation is invalid. If the schema is revised so as to ensure the validity of the explanation, the explanation will typically be unsound, having a false third premise.

Ernest Nagel offers a defense of what is substantially Hempel's schema with (c) replaced by (c').

A teleological statement of the form, "The function of A in a system S with organization C is to enable S in the environment E to engage in process P," can be formulated more explicitly by: every system S with organization C and in environment E engages in process P; if S with organization C and in environment E does not have A, then S does not engage in P; hence, S with organization C must have A.[3]

Thus he suggests that (3) is to be rendered as (4):

(3)   The function of chlorophyll in plants is to enable them to perform photosynthesis.

(4)   A necessary condition of the occurrence of photosynthesis in plants is the presence of chlorophyll.

So Nagel must face the second horn of Hempel's dilemma: (3) is presumably true, while (4) may well be false. Nagel is, of course, aware of this objection. His rather curious response is that, as far as we know, chlorophyll *is* necessary for photosynthesis in the green plants.[4] This may be so, but the response will not survive a change of example, Hearts are *not* necessary for circulation, artificial pumps having actually been incorporated into the circulatory systems of vertebrates in such a way as to preserve circulation and life.

A more promising defense of Nagel might run as follows. While it is true that the presence of a working heart is not a necessary condition of circulation in vertebrates under all circumstances, still, under *normal* circumstances— most circumstances, in fact—a working heart is necessary for circulation. Thus it is perhaps true that, at the present stage of evolution, a vertebrate that has not been tampered with surgically would exhibit circulation only if it were to contain a heart. If these circumstances are specifically included in the explanans, perhaps we can avoid Hempel's dilemma. Thus instead of (4) we should have:

(4')   At the present stage of evolution, a necessary condition for circulation in vertebrates that have not been surgically tampered with is the operation of a heart (properly incorporated into the circulatory system).

(4'), in conjuction with statements asserting that a given vertebrate exhibits circulation and has not been surgically tampered with and is at the present stage of evolution, will logically imply that that vertebrate has a heart. It seems, then, that the Hempelian objection could be overcome if it were possible, given a true function-as-cribing statement like (1) or (3), to specify "normal circumstances" in such a way as to make it true that, in those circumstances, the presence of the item in question is a necessary condition for the performance of the function ascribed to it.

This defense has some plausibility as long as we stick to the usual examples drawn from biology. But if we widen our view a bit, even within biology, I think it can be shown that this defense of Nagel's position will not suffice. Consider the kidneys. The function of the kidneys is to eliminate wastes from the blood. In particular, the function of my blood. Yet the presence of my left kidney is not, in normal circumstances, a necessary condition for the removal of the relevant wastes. Only if something seriously abnormal should befall my right kidney would the operation of my left kidney become necessary, and this only on the assumption that I am not hooked up to a kidney machine.[5]

A less obvious counterexample derives from the well-attested fact of hemispherical redundancy in the brain. No doubt it is in principle possible to specify conditions under which a particular duplicated mechanism would be necessary for normal functioning of the organism, but (a) in most cases we are not in a position to actually do this, though we are in a position to make well-confirmed statements about the functions of some of these mechanisms, and (b) these circumstances are by no means the normal circumstances. Indeed, given the fact that each individual nervous system develops somewhat differently owing to differing environmental factors, the circumstances in question might well be different for each individual, or for the same individual at different times.

Apparently Nagel was pursuing the wrong strategy in attempting to analyze functional ascriptions in terms of necessary conditions. Indeed, we are still faced with the dilemma noticed by Hempel: an analysis in terms of necessary conditions yields a valid but unsound explanatory schema; analysis in terms of sufficient conditions along the lines proposed by Hempel yields a schema with true premises, but validity is sacrificed.

Something has gone wrong, and it is not too difficult to locate the problem. An attempt to explain the presence of something by appeal to what it does—its function—is bound to leave unexplained why something else that does the same thing—a functional equivalent—isn't there instead. In itself, this is not a serious matter. But the accounts we have been considering assume that explanation is a species of deductive inference, and one cannot deduce hearts from circulation. This is what underlies the dilemma we have been considering. At best, one can deduce circulators from circulation. If we make this amendment, however, we are left with a functionally tainted analysis; "the function of the heart is to circulate the blood" is rendered "a blood circulator is a (necessary/sufficient) condition of circulation, and *the heart is a blood circulator.*" The expression in italics is surely as much in need of analysis as the analyzed expression. The problem, however, runs much deeper than the fact that the performance of a certain function does not determine how that function is performed. The problem is rather that to "explain" the presence of the heart in vertebrates by appeal to what the heart *does* is to "explain" its presence by appeal to factors which are causally irrelevant to its presence. Even if it were possible, as Nagel claimed, to *deduce* the presence of chlorophyll from the occurrence of photosynthesis, this would fail to explain the presence of chlorophyll in green plants in just the way deducing the presence and height of a building from the existence and length of its shadow would fail to explain why the building is there and has the height it does. This is not because all explanation is causal explanation: it is not. But to explain the presence of a naturally occurring structure or physical process—to explain why it is there, why such a thing exists in the place (system, context) it does—this does require specifying factors which causally determine the appearance of that structure or process.[6]

There is, of course, a sense in which the question "Why is x there?" is answered by giving x's function. Consider the following exchange. X asks Y, "Why is that thing there [pointing to the gnomon of a sundial]?" Y answers, "Because it casts a shadow on the dial beneath, thereby indicating the time of day." It is exchanges of this sort that most philosophers have had in mind when they speak of functional explanation. But it seems to me that, although such exchanges do represent genuine explanations, the use of functional language in this sort of explanation is quite distinct from its explanatory use in science. In section IV below I will sketch what I think *is* the central explanatory use of functional language in science. Meanwhile, if I am right, the evident propriety of exchanges like that imagined between X and Y has led to premature acceptance of (A), hence to concentration on what is, from the point of view of scientific explanation, an irrelevant use of functional language. For it seems to me that the question, "Why is x there?" can be answered by specifying x's function only if x is or is part of an artifact. Y's answer, I think, explains the presence of the gnomon because it rationalizes the action of the agent who put it there by supplying a *reason* for putting it there. In general, when we are dealing with the result of a deliberate action, we may explain the result by explaining the action, and we may explain a deliberate action by supplying the agent's reason for doing it. Thus when we look at a sundial, we assume we *know* in a general way how the gnomon came to be there: someone deliberately put it there. But we may wish to know *why* it was put there. Specifying the gnomon's function allows us to formulate what we suppose to be the unknown agent's reason for putting it there, viz., a belief that it would cast a shadow such that . . . , and so on. When we do this, we are elaborating on what we assume is the crucial causal factor in determining the gnomon's presence, namely a certain deliberate action.

If this is on the right track, then the viability of the sort of explanation in question should depend on the assumption that the thing functionally characterized is there as the result of deliberate action. If that assumption is evidently false, specifying the thing's function will not answer the question. Suppose it emerges that the sundial is not, as such, an artifact. When the ancient building was ruined, a large stone fragment fell on a kind of zodiac mosaic and embedded itself there. Since no sign of the roof remains, Y has mistakenly supposed the thing was designed as a sundial. As it happens, the local people have been using the thing to tell time for centuries, so Y is right about the function of the thing X pointed to.[7] But it is simply false that the thing is there because it casts a shadow, for there is no agent who put it there "because it casts a shadow." Again, the function of a bowl-like depression in a huge stone may be to hold holy water, but we cannot explain why it is there by appeal to its function if we know it was left there by prehistoric glacial activity.

If this is right, then (A) will lead us to focus on a type of explanation which will not apply to natural systems: chlorophyll and hearts are not "there" as the result of any deliberate action; hence the essential presupposition of the explanatory move in question is missing. Once this becomes clear, to continue to insist that there *must* be *some* sense in which specifying the function of chlorophyll explains its presence is an act of desperation born of thinking there is no other explanatory use of functional characterization in science.

Why have philosophers identified functional explanation exclusively with the appeal to something's function in explaining why it is there? One reason, I suspect, is a failure to distinguish teleological explanation from functional explanation, perhaps because functional concepts do loom large in "explanations" having a teleological form. Someone who fails to make this distinction, but who senses that there is an important and legitimate use of functional characterization in scientific explanation, will see the problem as one of finding a legitimate explanatory role for functional characterization within the teleological form. Once we leave artifacts and go to natural systems, however, this approach is doomed to failure, as critics of teleology have seen for some time.

This mistake probably would have sorted itself out in time were it not the case that we do reason from the performance of a function to the presence of certain specific processes and structures, e.g., from photosynthesis to chlorophyll, or from coordinated activity to nerve tissue. This is perfectly legitimate reasoning: it is a species of inference to the best explanation. Our best (only) explanation of photosynthesis requires chlorophyll, and our best explanation of coordinated activity requires nerve tissue. But once we see what makes this reasoning legitimate, we see immediately that inference *to* an explanation has been mistaken for an explanation itself. Once this becomes clear, it becomes equally clear that (A) has matters reversed: given that photosynthesis is occurring in a particular plant, we may legitimately infer that chlorophyll is present in that plant precisely because chlorophyll enters into our best (only) explanation of photosynthesis, and given coordinated activity on the part of some animal, we may legitimately infer that nerve tissue is present precisely because nerve tissue enters into our best explanation of coordinated activity in animals.

To attempt to explain the heart's presence in vertebrates by appealing to its function in vertebrates is to attempt to explain the occurrence of hearts in vertebrates by appealing to factors which are causally irrelevant to its presence in vertebrates. This fact has given "functional explanation" a bad name. But it is (A) that deserves the blame. Once we see (A) as an undefended philosophical hypothesis about how to construe functional explanations rather than as a statement of the philosophical problem, the correct alternative is obvious: what we can and do explain by appeal to what something does is the behavior of a containing system.[8]

A much more promising suggestion in the light of these considerations is that (1) is appealed to in explaining *circulation*. If we reject (A) and adopt this suggestion, a simple deductive-nomological explanation with circulation as the explicandum turns out to be a sound argument.

(5) a. Vertebrates incorporating a beating heart in the usual way (in the way s does) exhibit circulation.

b. Vertebrate s incorporates a beating heart in the usual way.

c. Hence, s exhibits circulation.

Though by no means flawless, (5) has several virtues, not the least of which is that it does not have biologists passing by an obvious application of evolution or genetics in favor of an invalid or unsound "functional" explanation of the presence of hearts. Also, the redundancy examples are easily handled, e.g., the removal of wastes is deduced in the kidney case.

The implausibility of (A) is obscured in examples taken from biology by the fact that there are two distinct uses of function statements in biology. Consider the following statements.

(a) The function of the contractile vacuole in protozoans is elimination of excess water from the the organism.

(b) The function of the neurofibrils in the ciliates is coordination of the activity of the cilia.

These statements can be understood in either of two ways. (1) They are generally used in explaining how the organism in question comes to exhibit certain characteristics or behavior. Thus (a) explains how excess water, accumulated in the organism by osmosis, is eliminated from the organism; (b) explains how it happens that the activity of the cilia in paramecium, for instance, is coordinated. (2) They may be used in explaining the continued survival of certain organisms incorporating structures of the sort in question by indicating the survival value which would accrue to such organisms in virtue of having structures of that sort. Thus, (a) allows us to infer that incorporation of a contractile vacuole makes it possible for the organism to be surrounded by a semi-permeable membrane, allowing the passage of oxygen into, and the passage of wastes out of, the organism. Relatively free osmosis of this sort is obviously advantageous, and this is made possible by a structure which solves the excess water problem. Similarly, ciliates incorporating neurofibrils will be capable of fairly efficient locomotion, the survival value of which is obvious.[9]

The second sort of use occurs as part of an account which, if we are not careful, can easily be mistaken for an explanation of the presence of the sort of item functionally characterized, and this has perhaps encouraged philosophers to accept (A). For it might seem that natural selection provides the missing causal link between what something does in a certain type of organism and its presence in that type of organism. By performing their respective functions the contractile vacuole and the neurofibrils help species incorporating them to survive, and thereby contribute to their own continued presence in organisms of those species, and this might seem to explain the presence of those structures in the organisms incorporating them.

Plausible as this sounds, it involves a subtle yet fundamental misunderstanding of evolutionary theory. A clue to the mistake is found in the fact that the contractile vacuole occurs in marine protozoans which have no excess water problem but the reverse problem. Thus the function and effect on survival of this structure is not the same in all protozoans. Yet the explanation of its presence in marine and fresh-water species is almost certainly the same. This fact reminds us that the processes actually responsible for the occurrence of contractile vacuoles in protozoans are totally insensitive to what that structure does. Failure to appreciate this point not only lends spurious plausibility to (A) as applied to biological examples, but seriously distorts our understanding of evolutionary theory. Whether an organism o incorporates s depends on whether s is "specified" by the genetic "plan" which o inherits and which, at a certain level of abstraction, is characteristic of o's species. Alterations in the plan are not the effects of the presence or exercise of the structures the plan specifies. This is most obvious when the genetic change is the result of random mutation. Though not all genetic change is due to random mutation, some certainly is, and that fact is enough to show that specifying the function of a biological structure cannot, in general, explain the presence of that structure. If a plan is altered so that it specifies s' rather than s, then the organisms inheriting this plan will incorporate s' regardless of the function or survival value of s' in those organisms. If the alteration is advantageous, the number of organisms inheriting that plan may increase, and, if it is disadvantageous, their number may decrease. But this typically has no effect on the plan, and therefore no effect on the occurrence of s' in the organisms in question.

One sometimes hears it said that natural selection is an instance of negative feedback. If this is meant to imply that the relative success or failure of organisms of a certain type can affect their inherited characteristics, it is simply a mistake: the characteristics of organisms which determine their relative success are determined by their genetic plan, and the characteristics of these plans are typically independent of the relative success of organisms having them. Of course, if s is very disadvantageous to organisms having a plan specifying s, then organisms having such plans may disappear altogether, and s will no longer occur. We could, therefore, think of natural selection as reacting on the *set* of plans generated by weeding out the bad plans: natural selection cannot alter a plan, but it can trim the set. Thus we may be able to explain why a given plan is not a failure by appeal to the functions of the structures it specifies. Perhaps this is what some writers have had in mind. But this is not to explain why, e.g., contractile vacuoles occur in certain protozoans; it is to explain why the sort of protozoan incorporating contractile vacuoles occurs. Since we cannot appeal to the relative success or failure of these organisms to explain why their genetic plan specifies contractile vacuoles, we cannot appeal to the relative success or failure of these organisms to explain why they incorporate contractile vacuoles.

Once we are clear about the explanatory role of functions in evolutionary theory, it emerges that the function of an organ or process (or whatever) is appealed to in order to explain the biological capacities of the organism containing it, and from these capacities conclusions are drawn concerning the chances of survival for organisms of that type. For instance, appeal to the function of the contractile vacuole in certain protozoans explains how these organisms are able to keep from exploding in fresh-water. Thus evolutionary biology does not provide support for (A), but for the idea instanced in (5): identifying the function of something helps to explain the capacities of a containing system.[10]

(A) misconstrues functional explanation by misidentifying what is explained. Let us abandon (A), then, in favor of the view that functions are appealed to in explaining the capacities of containing systems, and turn our attention to (B).

Whereas (A) is a thesis about functional explanation, (B) is a thesis about the analysis of function-ascribing statements. Perhaps when divorced from (A), as it is in (5), it will fare better than it does in the accounts of Hempel and Nagel.

III

In spite of the evident virtues of (5), (5a) has serious shortcomings as an analysis of (1). In fact it is subject to the same objection Hempel brings to the analysis which simply replaces "function" by "effect": vertebrates incorporating a working heart in the usual way exhibit the production of heartsounds, yet the production of heartsounds is not a function of hearts in vertebrates. The problem is that whereas the production of certain effects is essential to the heart's performing its function, there are some effects the production of which is irrelevant to the functioning of the heart. This problem is bound to infect any "selected effects" theory, i.e., any theory built on (B).

What is needed to establish a selected effects theory is a general formula which identifies the appropriate effects.[11] Both Hempel and Nagel attempt to solve this problem by identifying the function of something with just those effects which contribute to the maintenance of some special condition of, or the performance of some special activity of, some containing system. If this sort of solution is to be viable, there must be some principled way of selecting the relevant activities or conditions of containing systems. For no matter which effects of something you happen to name, there will be some activity of the containing system to which just those effects contribute, or some condition of the containing system which is maintained with the help of just those effects. Heart activity, for example, keeps the circulatory system from being entirely quiet, and the appendix keeps people vulnerable to appendicitis.[12]

Hempel suggests that, in general, the crucial feature of a containing system, contribution to which is to count as the functioning of a contained part, is that the system be maintained in "adequate, or effective, or proper working order."[13] Hempel explicitly declines to discuss what constitutes proper working order, presumably because he rightly thinks that there are more serious problems with the analysis he is discussing than those introduced by this phrase. But it seems clear that for something to be in working order is just for it to be capable of performing its functions, and for it to be in adequate or effective or proper working order is just for it to be capable of performing its functions adequately or effectively or properly. Hempel seems to realize this himself, for in setting forth a deductive schema for functional explanation, he glosses the phrase in question as "functions adequately."[14] More generally, if we identify the function of something x with those effects of x which contribute to the performance of some activity a or to the maintenance of some condition c of a containing system s, then we must be prepared to say as well that a function of s is to perform a or to maintain c. This suggests the following formulation of "selected effects" theories.

(6)   The function of an F in a G is f just in case (the capacity for) f is an effect of an F incorporated in a G in the usual way (or: in the way *this* F is incorporated in this G), and that effect contributes to the performance of a function of the containing G.

It seems that any theory based on (B)—what I have been calling "selected effects" theories—must ultimately amount to something like (6).[15] Yet (6) cannot be the whole story about functional ascriptions.

Suppose we follow (6) in rendering, "The function of the contractile vacuole in protozoans is elimination of excess water from the organism." The result is (7).

(7)   Elimination of excess water from the organism is an effect of a contractile vacuole incorporated in the usual way in a protozoan, and that effect contributes to the performance of a function of a protozoan.

In order to test (7) we should have to know a statement of the form "f is a function of a protozoan." Perhaps protozoans have no functions. If not, (7) is just a mistake. If they do, then presumably we shall have to appeal to (6) for an analysis of the statement attributing such a function and this will leave us with another unanalyzed functional ascription. Either we are launched on a regress, or the analysis breaks down at some level for lack of functions, or perhaps for lack of a plausible candidate for containing system. If we do not wish to simply acquiesce in the autonomy of functional ascriptions, it must be possible to analyze at least some functional ascriptions without appealing to functions of containing systems. If (6) can be shown to be the only plausible formulation of thories based on (B), then no such theory can be the whole story.

Our question, then, is whether a thing's function can plausibly be identified with those of its effects contributing to production of some activity of, or maintenance of some condition of, a containing system, where performance of the activity in question is not a function of the containing system. Let us begin by considering Hempel's suggestion that functions are to be identified with the production of effects contributing to the proper working order of a containing system. I claimed earlier that to say something is in proper working order is just to say that it properly performs its functions. This is fairly obvious in cases of artifacts or tools. To make a decision about which sort of behavior counts as working amounts to deciding about the thing's function. To say something is working, though not behaving or disposed to behave in a way having anything to do with its function, is to be open, at the very least, to the charge of arbitrariness.

When we are dealing with a living organism, or a society of living organisms, the situation is less clear. If we say, "The function of the contractile vacuole in protozoans is elimination of excess water from the organism," we do make reference to a containing organism, but not, apparently, to its function (if any). However, since contractile vacuoles do a number of things having nothing to do with their function, there must be some implicit principle of selection at work. Hempel's suggestion is that, in this context, to be in "proper working order" is simply to be alive and healthy. This works reasonably well for certain standard examples, e.g. (1) and (3): circulation does contribute to health and survival in vertebrates, and photo-synthesis does contribute to health and survival in green plants.[16] But once again, the principle will not stand a change of example, even within the life sciences. First, there are cases in which proper functioning is actually inimical to health and life: functioning of the sex organs results in the death of individuals of many species (e.g., certain salmon). Second, a certain process in an organism may have effects which contribute to health and survival but which are not to be confused with the function of that process: secretion of adrenalin speeds metabolism and thereby contributes to elimination of harmful fat deposits in overweight humans, but this is not a function of adrenalin secretion in overweight humans.

A more plausible suggestion along these lines in the special context of evolutionary biology is this:

(8)   The functions of a part or process in an organism are to be identified with those of its effects contributing to activities or conditions of the organism which sustain or increase the organism's capacity to contribute to survival of the species.

Give or take a nicety, (8) doubtless does capture a great many uses of functional language in biology. For instance, it correctly picks out elimination of excess water as the function of the contractile vacuole in fresh water protozoans only, and correctly identifies the function of sexual organs in species in which the exercise of these organs results in the death of the individual.[17]

In spite of these virtues, however, (8) is seriously misleading and extremely limited in applicability even within biology. Evidently, what contributes to an organism's capacity to maintain its species in one sort of environment may undermine that capacity in another. When this happens, we might say that the organ (or whatever) has lost its function. This is probably what we would say about the contractile vacuole if fresh-water protozoans were successfully introduced into salt water, for in this case the capacity explained would no longer be exercised. But if the capacity explained by appeal to the function of a certain structure continued to be exercised in the new environment, though now to the individual's detriment, we would not say that that structure had lost its function. If, for some reason, flying ceased to contribute to the capacity of pigeons to maintain their species, or even undermined that capacity to some extent,[18] we would still say that a function of the wings in pigeons is to enable them to fly. Only if the wings ceased to function as wings, as in the penguins or ostriches, would we cease to analyze skeletal structure and the like functionally with an eye to explaining flight. Flight is a capacity which cries out for explanation in terms of anatomical functions regardless of its contribution to the capacity to maintain the species.

What this example shows is that functional analysis can properly be carried on in biology quite independently of evolutionary considerations: a complex capacity of an organism (or one of its parts or systems) may be explained by appeal to a functional analysis regardless of how it relates to the organism's capacity to maintain the species. At best, then, (8) picks out those effects which will be called functions when what is in the offing is an application of evolutionary theory. As we shall see in the next section, (8) is misleading as well in that it is not *which* effects are explained but the style of explanation that makes it appropriate to speak of functions. (8) simply identifies effects which, as it happens, are typically explained in that style.

We have not quite exhausted the lessons to be learned from (8). The plausibility of (8) rests on the plausibility of the claim that, for certain purposes, we may assume that a function of an organism is to contribute to the survival of its species. What (8) does, in effect, is identify a function of an important class of (uncontained) containing systems without providing an analysis of the claim that a function of an organism is to contribute to the survival of its species.

Of course, an advocate of (8) might insist that it is no part of his theory to claim that maintenance of the species is a function of an organism. But then the defense of (8) would have to be simply that it describes actual usage, i.e., that it is in fact effects contributing to an organism's capacity to maintain its species which evolutionary biologists single out as functions. Construed in this way, (8) would, at most, tell us *which* effects are picked out as functions; it would provide no hint as to *why* these effects are picked out *as functions*. We know why evolutionary biologists are interested in effects contributing to an organism's capacity to maintain its species, but why call them functions? This is precisely the sort of question a philosophical account of function-ascribing statements should answer. Either (8) is defended as an instance of (6)—maintenance of the species is declared a function of organism—or it is defended as descriptive of usage. In neither case is any philosophical analysis provided. For in the first case (8) relies on an unanalyzed (and undefended) function-ascribing statement, and in the second it fails to give any hint as to the point of identifying certain effects as functions.

The failings of (8) are I think bound to cripple any theory which identifies a thing's functions with effects contributing to some antecedently specified type of condition or behavior of a containing system. If the theory is an instance of (6), it launches a regress or terminates in an unanalyzed functional ascription; if it is not an instance of (6), then it is bound to leave open the very question at issue, viz., why are the selected effects seen as functions?

IV

In this section I will sketch briefly an account of functional explanation which takes seriously the intuition that it is a genuinely distinctive style of explanation. The assumptions (A) and (B) form the core of approaches which seek to minimize the differences between functional explanations and explanations not formulated in functional terms. Such approaches have not given much attention to the characterization of the special explanatory strategy science employs in using functional language, for the problem as it was conceived in such approaches was to show that functional explanation is not really different in essentials from other kinds of scientific explanation. Once the problem is conceived in this way, one is almost certain to miss the distinctive features of functional explanation, and hence to miss the point of functional description. The account of this section reverses this tendency by placing primary emphasis on the kind of problem which is solved by appeal to functions.

Functions and Dispositions

Something may be capable of pumping even though it does not function as a pump (ever) and even though pumping is not its function. On the other hand, if something functions as a pump in a system s, or if the function of something in a system s is to pump, then it must be capable of pumping in s.[19] Thus function-ascribing statements imply disposition statements; to attribute a function to something is, in part, to attribute a disposition to it. If the function of x in s is to ϕ, then x has a disposition to ϕ in s. For instance, if the function of the contractile vacuole in fresh-water protozoans is to eliminate excess water from the organism, then there must be circumstances under which the contractile vacuole would actually manifest a disposition to eliminate excess water from the protozoan which incorporates it.

To attribute a disposition d to an object a is to assert that the behavior of a is subject to (exhibits or would exhibit) a certain law-like regularity: to say a has d is to say that a would manifest d (shatter, dissolve) were any of a certain range of events to occur (a is put in water, a is struck sharply). The regularity associated with a disposition—call it the dispositional regularity— is a regularity which is special to the behavior of a certain kind of object and obtains in virtue of some special facts(s) about that kind of object. Not everything is water-soluble: such things behave in a special way in virtue of certain (structural) features special to water-soluble things. Thus it is that dispositions require explanation: if x has d, then x is subject to a regularity in behavior special to things having d, and such a fact needs to be explained.

To explain a dispositional regularity is to explain how manifestations of the disposition are brought about given the requisite precipitating conditions. In what follows I will describe two distinct strategies for accomplishing this. It is my contention that the appropriateness of function-ascribing statements corresponds to the appropriateness of the second of these two strategies. This, I think, explains the intuition that functional explanation is a special *kind* of explanation.

Two Explanatory Strategies[20]

The Instantiation Strategy Since dispositions are properties, not events, to explain a disposition requires explaining how it is instantiated. To explain an event, we cite its cause, and to explain an event type requires a recipe (law) for constructing causal explanations of its tokens. But dispositions, being properties, not events, are not explicable as effects. The *acquisition* of a property is an event, but explaining the acquisition of a property is quite distinct from explaining the property itself. One can explain why/how a thing became fragile without thereby explaining fragility, and one can explain why/how something changed properties—e.g., why something changed temperature—without thereby explaining the property that changed. To explain a property one must show how that property is instantiated in the things that have it.

Simple dispositions are explained by exhibiting their instantiations: water solubility is instantiated as a certain kind of molecular structure, temperature as (average) kinetic energy of molecules, flammability as a kind of subatomic structure (allowing for bonding with oxygen at relatively low temperatures). When we understand how a disposition is instantiated, we are in a position to understand why the dispositional regularity holds of the disposed objects.

Brian O'Shaughnessy has provided an example that allows a particularly simple illustration of this strategy.[21] Consider the disposition he calls elevancy: the tendency of an object to rise in water of its own accord. To explain elevancy, we must explain why freeing a submerged elevant object causes it to rise.[22] This we may do as follows. In every case, the ratio of an elevant object's mass to its nonpermeable volume is less than the density (mass per unit volume) of water: that is how elevancy is instantiated. Once we know this, we may apply Archimedes' Principle, which tells us that water exerts an upward force on a submerged object equal to the weight of the water displaced. In the case of an elevant object, this force evidently exceeds the weight of the object by some amount f. Freeing the object changes the net force on it from zero to a net force of magnitude f in the direction of the surface, and the object rises accordingly. Here we subsume the connection between freeings and risings under a general law connecting changes in net force with changes in motion by citing a feature of elevant objects which allows us (via Archimedes' Principle) to represent freeing them under water as an instance of introducing a net force in the direction of the surface.

The Analytical Strategy Rather than deriving the dispositional regularity that specifies d (in a) from the facts of d's instantiation (in a), the analytical strategy proceeds by analyzing a disposition d of a into a number of other

Robert Cummins

dispositions $d_1, \ldots, d_n$ had by a or components of a such that programmed manifestation of the d_i results in or amounts to a manifestation of d.[23] The two strategies will fit together into a unified account if the analyzing dispositions (the d_i) can be made to yield to the instantiation strategy.

When the analytical strategy is in the offing one is apt to speak of capacities (or abilities) rather than of dispositions. This shift in terminology will put a more familiar face on the analytical strategy,[24] for we often explain capacities by analyzing them. Assembly-line production provides a transparent example of what I mean. Production is broken down into a number of distinct tasks. Each point on the line is responsible for a certain task, and it is the function of the components at that point to complete that task. If the line has the capacity to produce the product, it has it in virtue of the fact that the components have the capacities to perform their designated tasks, and in virtue of the fact that when these tasks are performed in a certain organized way—according to a certain program—the finished product results. Here we can explain the line's capacity to produce the product—i.e., explain how it is able to produce the product—by appeal to certain capacities of the components and their organization into an assembly line. Against this background we may pick out a certain capacity of an individual component the exercise of which is its function on the line. Of the many things it does and can do, its function on the line is doing whatever it is that we appeal to in explaining the capacity of the line as a whole. If the line produces several products—i.e., if it has several capacities—then, although a certain capacity c of a component is irrelevant to one capacity of the line, exercise of c by that component may be its function with respect to another capacity of the line as a whole.

Schematic diagrams in electronics provide another obvious illustration. Since each symbol represents any physical object whatever having a certain capacity, a schematic diagram of a complex device constitutes an analysis of the electronic capacities of the device as a whole into the capacities of its components. Such an analysis allows us to explain how the device as a whole exercises the analyzed capacity, for it allows us to see exercises of the analyzed capacity as programmed exercise of the analyzing capacities. In this case the "program" is given by the lines indicating how the components are hooked up. (Of course, the lines are themselves function-symbols.)

Functional analysis in biology is essentially similar. The biologically significant capacities of an entire organism are explained by analyzing the organism into a number of "systems"—the circulatory system, the digestive system, the nervous system, etc. each of which has its characteristic capacities.[25] These capacities are in turn analyzed into capacities of component organs and structures. Ideally, this strategy is pressed until physiology takes over—i.e., until the analyzing capacities are amenable to the instantiation strategy. We can easily imagine biologists expressing their analyses in a form analogous to the schematic diagrams of electrical engineering, with special symbols for pumps, pipes, filters, and so on. Indeed, analyses of even simple cognitive capacities are typically expressed in flow-charts or programs, forms designed specifically to represent analyses of information-processing capabilities generally.

Perhaps the most extensive use of the analytical strategy in science occurs in psychology, for a large part of the psychologist's job is to explain how the complex behavioral capacities of organisms are acquired and how they are exercised. Both goals are greatly facilitated by analysis of the capacities in question, for then acquisition of the analyzed capacity resolves itself into acquisition of the analyzing capacities and the requisite organization, and the problem of performance resolves itself into the problem of how the analyzing capacities are exercised. This sort of strategy has dominated psychology ever since Watson attempted to explain such complex capacities as the ability to run a maze by analyzing the performance into a series of conditioned responses, the stimulus for each response being the previous response, or something encountered as the result of the previous response.[26] Acquisition of the complex capacity is resolved into a number of distinct cases of simple conditioning—i.e., the ability to learn the maze is resolved into the capacity for stimulus substitution, and the capacity to run the maze is resolved into abilities to respond in certain simple ways to simple stimuli. Watson's analysis proved to be of limited value, but the analytic strategy remains the dominant mode of explanation in behavioral psychology.[27]

Functions and Functional Analysis

In the context of an application of the analytical strategy, exercise of an analyzing capacity emerges as a function: it will be appropriate to say that x functions as a ϕ in s, or that the function of x in s is ϕ-ing, when we are speaking against the background of an analytical explanation of some capacity of s which appeals to the fact that x has a capacity to ϕ is s. It is appropriate to say that the heart functions as a pump against the background of an analysis of the circulatory system's capacity to transport food, oxygen, wastes, and so on, which appeals to the fact that the heart is capable of pumping. Since this is the usual background, it goes without saying, and this accounts for the fact that "the heart functions as a pump" sounds right, and "the heart functions as a noise-maker" sounds wrong, in some context-free sense. This effect is strengthened by the absence of any actual application of the analytical strategy which makes use of the fact that the heart makes noise.[28]

We can capture this implicit dependence on an analytical context by entering an explicit relativization in our regimented reconstruction of function-ascribing statements.

(9) x functions as a ϕ in s (or: the function of x in s is to ϕ) relative to an analytical account A of s's capacity to ψ just in case x is capable of ϕ-ing in s and A appropriately and adequately accounts for s's capacity to ψ by, in part, appealing to the capacity of x to ϕ in s.

Sometimes we explain a capacity of s by analyzing it into other capacities of s, as when we explain how someone ignorant of cookery is able to bake cakes by pointing out that he or she followed a recipe each instruction of which requires no special capacities for its execution. Here we don't speak of, e.g., stirring as a function of the cook, but rather of the function of stirring. Since stirring has different functions in different recipes, and at different points in the same recipe, a statement like, "The function of stirring the mixture is to keep it from burning to the bottom of the pan," is implicitly relativized to a certain (perhaps somewhat vague) recipe. To take account of this sort of case, we need a slightly different schema: where e is an activity or behavior of a system s (as a whole), the function of e in s is to ϕ relative to an analytical account A of s's capacity to ψ just in case A appropriately and adequately accounts for s's capacity to ψ by, in part, appealing to s's capacity to engage in e.

(9) explains the intuition behind the regress-ridden (6): functional ascriptions do require relativization to a "functional fact" about a containing system—i.e., to the fact that a certain capacity of a containing system is appropriately explained by appeal to a certain functional analysis. And, like (6), (9) makes no provision for speaking of the function of an organism except against a background analysis of a containing system (the hive, the corporation, the eco-system). Once we see that functions are appealed to in explaining the capacities of containing systems, and indeed that it is the applicability of a certain strategy for explaining these capacities that makes talk of functions appropriate, we see immediately why we do not speak of the functions of uncontained containers. What (6) fails to capture is the fact that uncontained containers can be functionally analyzed, and the way in which function-analytical explanation mediates the connection between functional ascriptions (x functions as a ϕ, the function of x is to ϕ) and the capacities of the containers.

Function-Analytical Explanation

If the account I have been sketching is to draw any distinctions, the availability and appropriateness of analytical explanations must be a nontrivial matter.[29] So let us examine an obviously trivial application of the analytical strategy with an eye to determining whether it can be dismissed on principled grounds.

(10)   Each part of the mammalian circulatory system makes its own distinctive sound, and makes it continuously. These sounds combine to form the "circulatory noise" characteristic of all mammals. The mammalian circulatory system is capable of producing this sound at various volumes and various tempos. The heartbeat is responsible for the throbbing character of the sound, and it is the capacity of the heart to beat at various rates that explains the capacity of the circulatory system to produce a variously tempoed sound.

Everything in (10) is, presumably, true. The question is whether it allows us to say that the function of the heart is to produce a variously tempoed throbbing sound.[30] To answer this question we must, I think, get clear about the motivation for applying the analytical strategy. For my contention will be that the analytical strategy is most significantly applied in cases very unlike that envisaged in (10).

The explanatory interest of an analytical account is roughly proportional to (i) the extent to which the analyzing capacities are less sophisticated than the analyzed capacities, (ii) the extent to which the analyzing capacities are different in type from the analyzed capacities, and (iii) the relative sophistication of the program appealed to, i.e., the relative complexity of the organization of component parts/processes which is attributed to the system. (iii) is correlative with (i) and (ii): the greater the gap in sophistication and type between analyzing capacities and analyzed capacities, the more sophisticated the program must be to close the gap.

It is precisely the width of these gaps which, for instance, makes automata theory so interesting in its application to psychology. Automata theory supplies us with extremely powerful techniques for constructing diverse analyses of very sophisticated tasks into very unsophisticated tasks. This allows us to see how, in principle, a mechanism such as the brain, consisting of physiologically unsophisticated components (relatively speaking), can acquire very sophisticated capacities. It is the prospect of promoting the capacity to store ones and zeros into the capacity to solve problems of logic and recognize patterns that makes the analytical strategy so appealing in cognitive psychology.

As the program absorbs more and more of the explanatory burden, the physical facts underlying the analyzing capacities become less and less special to the analyzed system. This is why it is plausible to suppose that the capacity of a person and a machine to solve a certain problem might have substantially the same explanation, while it is not plausible to suppose that the capacities of a synthesizer and a bell to make similar sounds have substantially similar explanations. There is no work for a sophisticated hypothesis about the organization of various capacities to do in the case of the bell. Conversely, the less weight borne by the program, the less point to analysis. At this end of the scale we have cases like (10) in which the analyzed and analyzing capacities differ little if at all in type and sophistication. Here we could apply the instantiation strategy without significant loss, and thus talk of functions is comparatively strained and pointless. It must be admitted, however, that there is no black-white distinction here, but a case of more-or-less. As the role of organization becomes less and less significant, the analytical strategy becomes less and less appropriate, and talk of functions makes less and less sense. This may be philosophically disappointing, but there is no help for it.

CONCLUSION

Almost without exception, philosophical accounts of function-ascribing statements and of functional explanation have been crippled by adoption of assumptions (A) and (B). Though there has been widespread agreement that extant accounts are not satisfactory, (A) and (B) have escaped critical scrutiny, perhaps because they were thought of as somehow setting the problem rather than as part of proffered solutions. Once the problem is properly diagnosed, however, it becomes possible to give a more satisfactory and more illuminating account in terms of the explanatory strategy which provides the motivation and forms the context of function-ascribing statements. To ascribe a function to something is to ascribe a capacity to it which is singled out by its role in an analysis of some capacity of a containing system. When a capacity of a containing system is appropriately explained by analyzing it into a number of other capacities whose programmed exercise yields a manifestation of the analyzed capacity, the analyzing capacities emerge as functions. Since the appropriateness of this sort of explanatory strategy is a matter of degree, so is the appropriateness of function-ascribing statements.

NOTES

1. Cf. Carl Hempel, The logic of functional analysis, in *Aspects of Scientific Explanation*, New York, Free Press, 1965, reprinted from Llewellyn Gross, ed., *Symposium on Sociological Theory*, New York, Harper and Row, 1959; and Ernest Nagel, *The Structure of Science*, New York, Harcourt, Brace and World, 1961, chapter 12, section I. The assumptions, of course, predate Hempel's 1959 essay. See, for instance, Richard Braithwaite, *Scientific Explanation*, Cambridge, Cambridge University Press, 1955, chapter X; and Israel Scheffler, Thoughts on teleology, *British Journal for the Philosophy of Science*, 11 (1958). More recent examples include Francisco Ayala, Teleological explanations in evolutionary biology, *Philosophy of Science*, 37 (1970); Hugh Lehman, Functional explanations in biology, *Philosophy of Science*, 32 (1965); Richard Sorabji, Function, *Philosophical Quarterly*, 14 (1964); and Larry Wright, Functions, *Philosophical Review*, 82 (1973).

2. Hempel, p. 310.

3. Nagel, p. 403.

4. Ibid., p. 404.

5. It might be objected here that although it is the function of the kidneys to eliminate waste, that is not the function of a particular kidney unless operation of that kidney *is* necessary for removal of wates. But suppose scientists had initially been aware of the existence of the left kidney only. Then, on the account being considered, anything they had said about the function of that organ would have been false, since, on that account, *it has no function in organisms having two kidneys!*

6. Even in the case of a designed artifact, it is at most the designer's *belief* that x will perform f in s which is causally relevant to x's presence in s, not x's actually performing f in s. The nearest I can come to describing a situation in which x performing f in s is causally relevant to x's presence in s is this: the designer of s notices a thing like x performing f in a system like s, and this leads to belief that x will perform f in s, and this in turn leads the designer to put x in s.

7. *Is* casting a shadow the function of this fragment? Standard use may confer a function on something: if I standardly use a certain stone to sharpen knives, then that is its function, or if I standardly use a certain block of wood as a door stop, then the function of that block is to hold my door open. If nonartifacts *ever* have functions, appeals to those functions cannot explain their presence. The things functionally characterized in science are typically not artifacts.

8. A confused perception of this fact no doubt underlies (B), but the fact that (B) is nearly inseparable from (A) in the literature shows how confused this perception is.

9. Notice that the second use is parasitic on the first. It is only because the neurofibrils explain the coordinated activity of the cilia that we can assign a survival value to neurofibrils: the survival value of a structure s hangs on what capacities of the organism, if any, are explicable by appeal to the functioning of s.

10. In addition to the misunderstanding about evolutionary theory just discussed, biological examples have probably suggested (A) because biology was the *locus classicus* of teleological explanation. This has perhaps encouraged a confusion between the teleological *form* of explanation, incorporated in (A), with the explanatory role of functional ascriptions. Function-ascribing statements do occur in explanations having a teleological form, and when they do, their interest is vitiated by the incoherence of that form of explanation. It is the legitimate use of function-ascribing statements that needs examination, i.e., their contribution to nonteleological theories such as the theory of evolution.

11. Larry Wright (op. cit.) is aware of this problem but does not, to my mind, make much progress with it. Wright's analysis rules out "the function of the heart is to produce heart-sounds," on the ground that the heart is not there because it produces heartsounds. I agree. But neither is it there because it pumps blood. Or if, as Wright maintains, there is a sense of "because" in which the heart *is* there because it pumps blood and not because it produces heartsounds, then this sense of "because" is as much in need of analysis as "function." Wright does not attempt to provide such an analysis, but depends on the fact that, in many cases, we are able to use the word in the required way. But we are also able to use "function" correctly in a variety of cases. Indeed, if Wright is right, the words are simply interchangeable with a little grammatical maneuvering. The problem is to make the conditions of correct use explicit. Failure to do this means that Wright's analysis provides no insight into the problem of how functional theories are confirmed, or whence they derive their explanatory force.

12. Surprisingly, when Nagel comes to formulate his general schema of functional attribution, he simply ignores this problem and thus leaves himself open to the trivialization just suggested. Cf. Nagel, p. 403.

13. Hempel, p. 306.

14. Ibid., p. 310.

15. Hugh Lehman (op. cit.) gives an analysis that appears to be essentially like (6).

16. Even these applications have their problems. Frankfurt and Poole, Functional explanations in biology, *British Journal for the Philosophy of Science*, 17 (1966), point out that heartsounds contribute to health and survival via their usefulness in diagnosis.

17. Michael Ruse has argued for a formulation like (8). See his Function statements in biology, *Philosophy of Science*, 38 (1971), and *The Philosophy of Biology*, London, Hutchinson, 1973.

18. Perhaps, in the absence of serious predators, with a readily available food supply, and with no need to migrate, flying simply wastes energy.

19. Throughout this section I am discounting appeals to the intentions of designers or users. *x* may be intended to prevent accidents without actually being capable of doing so. With reference to this intention, it *would* be proper in certain contexts to say, "*x*'s function is to prevent accidents, though it is not actually capable of doing so."

There can be no doubt that a thing's function is often identified with what it is typically or "standardly" used to do, or with what it was designed to do. But the sorts of things for which it is an important scientific problem to provide functional analyses—brains, organisms, societies, social institutions—either do not have designers or standard or regular uses at all, or it would be inappropriate to appeal to these in constructing and defending a scientific theory because the designer or use is not known—brains, devices dug up by archaeologists—or because there is some likelihood that real and intended functions diverge—social institutions, complex computers. Functional talk may have originated in contexts in which reference to intentions and purposes loomed large, but reference to intentions and purposes does not figure at all in the sort of functional analysis favored by contemporary natural scientists.

20. For a detailed discussion of the two explanatory strategies sketched here, see Cummins, *The Nature of Psychological Explanation*, Bradford Books/MIT Press, Cambridge, 1983. In the original

version of this paper, I called the two strategies the Subsumption Strategy and the Analytical Strategy. I have retained the latter term, but the former I have replaced. What I was calling the subsumption strategy in 1975 was simply a confusion, a conflation of causal subsumption of events, and the nomic derivation of a property via the facts of its instantiation. Since functions are dispositions and dispositions are properties, only the latter is relevant here.

21. Brian O'Shaughnessy, The powerlessness of dispositions, *Analysis*, October (1970). See also my discussion of this example in Dispositions, states and causes, *Analysis*, June (1974).

22. Also, we must explain why submerging a free elevant object causes it to rise, and why a free submerged object's becoming elevant causes it to rise. One of the convenient features of elevancy is that the same considerations dispose of all these cases. This does not hold generally: gentle rubbing, a sharp blow, or a sudden change in temperature may each cause a glass to manifest a disposition to shatter, but the explanations in each case are significantly different.

23. By "programmed" I simply mean organized in a way that could be specified in a program or flow chart: each instruction (box) specifies manifestation of one of the d_i such that if the program is executed (the chart followed), a manifests d.

24. Some might want to distinguish between dispositions and capacities, and argue that to ascribe a function to x is in part to ascribe a *capacity* to x, not a disposition as I have claimed. Certainly (1) is strained in a way (2) is not.

(1)   Hearts are disposed to pump. Hearts have a disposition to pump. Sugar is capable of dissolving. Sugar has a capacity to dissolve.

(2)   Hearts are capable of pumping. Hearts have a capacity to pump. Sugar is disposed to dissolve. Sugar has a disposition to dissolve.

25. Indeed, what makes something part of, e.g., the nervous system is that its capacities figure in an analysis of the capacity to respond to external stimuli, coordinate movement, etc. Thus there is no question that the glial cells are part of the brain, but there is some question whether they are part of the nervous system or merely auxiliary to it.

26. John B. Watson, *Behaviorism*, New York, W. W. Norton, 1930, chapters IX and XI.

27. Writers on the philosophy of psychology, especially Jerry Fodor, have grasped the connection between functional characterization and the analytical strategy in psychological theorizing but have not applied the lesson to the problem of functional explanation generally. The clearest statement occurs in J. A. Fodor, The appeal to tacit knowledge in psychological explanation, *Journal of Philosophy*, 65 (1968), 627–640.

28. It is sometimes suggested that heartsounds do have a psychological function. In the context of an analysis of a psychological disposition appealing to the heart's noise-making capacity, "The heart functions as a noise-maker" (e.g., as a producer of regular thumps) would not even *sound* odd.

29. Of course, it might be that only arbitrary distinctions are to be drawn. Perhaps (9) describes usage, and usage is arbitrary, but I am unable to take this possibility seriously.

30. The issue is not whether (10) forces us, via (9), to say something false. Relative to *some* analytical explanation, it may be true that the function of the heart is to produce a variously tempoed throbbing. But the availability of (10) should not support such a claim.

III   Adaptationism

4

The Spandrels of San Marco and the Panglossian Paradigm: A Critique of the Adaptationist Programme

Stephen Jay Gould and Richard C. Lewontin

An adaptationist program has dominated evolutionary thought in England and the United States during the past forty years. It is based on faith in the power of natural selection as an optimizing agent. It proceeds by breaking an organism into unitary "traits" and proposing an adaptive story for each considered separately. Trade-offs among competing selective demands exert the only brake upon perfection; nonoptimality is thereby rendered as a result of adaptation as well. We criticize this approach and attempt to reassert a competing notion (long popular in continental Europe) that organisms must be analyzed as integrated wholes, with *Baupläne* so constrained by phyletic heritage, pathways of development, and general architecture that the constraints themselves become more interesting and more important in delimiting pathways of change than the selective force that may mediate change when it occurs. We fault the adaptationist program for its failure to distinguish current utility from reasons for origin (male tyrannosaurs may have used their diminutive front legs to titillate female partners, but this will not explain why they got so small); for its unwillingness to consider alternatives to adaptive stories; for its reliance upon plausibility alone as a criterion for accepting speculative tales; and for its failure to consider adequately such competing themes as random fixation of alleles, production of nonadaptive structures by developmental correlation with selected features (allometry, pleiotropy, material compensation, mechanically forced correlation), the separability of adaptation and selection, multiple adaptive peaks, and current utility as an epiphenomenon of nonadaptive structures. We support Darwin's own pluralistic approach to identifying the agents of evolutionary change.

INTRODUCTION

The great central dome of St. Mark's Cathedral in Venice presents in its mosaic design a detailed iconography expressing the mainstays of Christian faith. Three circles of figures radiate out from a central image of Christ: angels, disciples, and virtues. Each circles is divided into quadrants, even though the dome itself is radially symmetrical in structure. Each quadrant meets one of the four spandrels in the arches below the dome. Spandrels—the tapering triangular spaces formed by the intersection of two rounded arches at right angles (figure 4.1)—are necessary architectural by-products of mounting a dome on rounded arches. Each spandrel contains a design admirably fitted into its tapering space. An evangelist sits in the upper part flanked by the heavenly cities. Below, a man representing one of the four biblical

From *Proc. R. Soc. London*, 1978, 205:581–598.

Figure 4.1 One of the four spandrels of St. Mark's; seated evangelist above, personification of river below.

rivers (Tigris, Euphrates, Indus, and Nile) pours water from a pitcher in the narrowing space below his feet.

The design is so elaborate, harmonious, and purposeful that we are tempted to view it as the starting point of any analysis, as the cause in some sense of the surrounding architecture. But this would invert the proper path of analysis. The system begins with an architectural constraint: the necessary four spandrels and their tapering triangular form. They provide a space in which the mosaicists worked; they set the quadripartite symmetry of the dome above.

Such architectural constraints abound, and we find them easy to understand because we do not impose our biological biases upon them. Every fan-vaulted ceiling must have a series of open spaces along the midline of the vault, where the sides of the fans intersect between the pillars (figure 4.2). Since the spaces must exist, they are often used for ingenious ornamental effect. In King's College Chapel in Cambridge, for example, the spaces contain bosses alternately embellished with the Tudor rose and portcullis. In a sense, this design represents an "adaptation," but the architectural constraint is clearly primary.

Stephen Jay Gould and Richard Lewontin

Figure 4.2 The ceiling of King's College Chapel.

The spaces arise as a necessary by-product of fan vaulting; their appropriate use is a secondary effect. Anyone who tried to argue that the structure exists because the alternation of rose and portcullis makes so much sense in a Tudor chapel would be inviting the same ridicule that Voltaire heaped on Dr. Pangloss: "Things cannot be other than they are.... Everything is made for the best purpose. Our noses were made to carry spectacles, so we have spectacles. Legs were clearly intended for breeches, and we wear them." Yet evolutionary biologists, in their tendency to focus exclusively on immediate adaptation to local conditions, do tend to ignore architectural constraints and perform just such an inversion of explanation.

As a closer example, recently featured in some important biological literature on adaptation, anthropologist Michael Harner has proposed (1977) that Aztec human sacrifice arose as a solution to chronic shortage of meat (limbs of victims were often consumed, but only by people of high status). E. O. Wilson (1978) has used this explanation as a primary illustration of an adaptive, genetic predisposition for carnivory in humans. Harner and Wilson ask us to view an elaborate social system and a complex set of explicit justifications involving myth, symbol, and tradition as mere epiphenomena generated by the Aztecs as an unconscious rationalization masking the "real" reason for it all: need for protein. But Sahlins (1978) has argued that human sacrifice represented just one part of an elaborate cultural fabric that, in its entirety, not only represented the material expression of Aztec cosmology, but also performed such utilitarian functions as the maintenance of social ranks and systems of tribute among cities.

We strongly suspect that Aztec cannibalism was an "adaptation" much like evangelists and rivers in spandrels, or ornamented bosses in ceiling spaces: a secondary epiphnomenon representing a fruitful use of available parts, not a cause of the entire system. To put it crudely: a system developed for other reasons generated an increasing number of fresh bodies; use might as well be made of them. Why invert the whole system in such a curious fashion and view an entire culture as the epiphenomenon of an unusual way to beef up the meat supply? Spandrels do not exist to house the evangelists. Moreover, as Sahlins argues, it is not even clear that human sacrifice was an adaptation at all. Human cultural practices can be orthogenetic and drive toward extinction in ways that Darwinian processes, based on genetic selection, cannot. Since each new monarch had to outdo his predecessor in even more elaborate and copious sacrifice, the practice was beginning to stretch resources to the breaking point. It would not have been the first time that a human culture did itself in. And, finally, many experts doubt Harner's premise in the first place (Ortiz de Montellano 1978). They argue that other sources of protein were not in short supply, and that a practice awarding meat only to privileged people who had enough anyway, and who used bodies so inefficiently (only the limbs were consumed, and partially at that), represents a mighty poor way to run a butchery.

We deliberately chose nonbiological examples in a sequence running from remote to more familiar: architecture to anthropology. We did this because the primacy of architectural constraint and the epiphenomenal nature of adaptation are not obscured by our biological prejudices in these examples. But we trust that the message for biologists will not go unheeded: if these had been biological systems, would we not, by force of habit, have regarded the epiphenomenal adaptation as primary and tried to build the whole structural system from it?

THE ADAPTATIONIST PROGRAM

We wish to question a deeply engrained habit of thinking among students of evolution. We call it the adaptationist program, or the Panglossian paradigm. It is rooted in a notion popularized by A. R. Wallace and A. Weismann, (but not, as we shall see, by Darwin) toward the end of the nineteenth century: the near omnipotence of natural selection in forging organic design and fashioning the best among possible worlds. This program regards natural selection as so powerful and the constraints upon it so few that direct production of adaptation through its operation becomes the primary cause of nearly all organic form, function, and behavior, Constraints upon the pervasive power of natural selection are recognized, of course (phyletic inertia primarily among them, although immediate architectural constraints, as discussed in the last section, are rarely acknowledged). But they are usually dismissed as unimportant or else, and more frustratingly, simply acknowledged and then not taken to heart and invoked.

Studies under the adaptationist program generally proceed in two steps:

1. An organism is atomized into "traits" and these traits are explained as structures optimally designed by natural selection for their functions. For lack of space, we must omit an extended discussion of the vital issue, "What is a trait?" Some evolutionists may regard this as a trivial, or merely a semantic problem, It is not. Organisms are integrated entities, not collections of discrete objects. Evolutionists have often been led astray by inappropriate atomization, as D'Arcy Thompson (1942) loved to point out. Our favorite example involves the human chin (Gould 1977, pp. 381–382; Lewontin 1978). If we regard the chin as a "thing," rather than as a product of interaction between two growth fields (alveolar and mandibular), then we are led to an interpretation of its origin (recapitulatory) exactly opposite to the one now generally favored (neotenic).

2. After the failure of part-by-part optimization, interaction is acknowledged via the dictum that an organism cannot optimize each part without imposing expenses on others. The notion of "trade-off" is introduced, and organisms are interpreted as best compromises among competing demands. Thus interaction among parts is retained completely within the adaptationist program. Any suboptimality of a part is explained as its contribution to the best possible design for the whole. The notion that suboptimality might represent anything other than the immediate work of natural selection is usually not entertained. As Dr. Pangloss said in explaining to Candide why he suffered from venereal disease: "It is indispensable in this best of worlds. For if Columbus, when visiting the West Indies, had not caught this disease, which poisons the source of generation, which frequently even hinders generation, and is clearly opposed to the great end of Nature, we should have neither chocolate nor cochineal." The adaptationist program is truly Panglossian. Our world may not be good in an abstract sense, but it is the very best we could have. Each trait plays its part and must be as it is.

At this point, some evolutionists will protest that we are caricaturing their view of adaptation. After all, do they not admit genetic drift, allometry, and a variety of reasons for nonadaptive evolution? They do, to be sure, but we make a different point. In natural history, all possible things happen sometimes; you generally do not support your favored phenomenon by declaring rivals impossible in theory. Rather, you acknowledge the rival but circumscribe its domain of action so narrowly that it cannot have any importance in the affairs of nature. Then, you often congratulate yourself for being such an undogmatic and ecumenical chap. We maintain that alternatives to selection for best overall design have generally been relegated to unimportance by this mode of argument. Have we not all heard the catechism about genetic drift: it can only be important in populations so small that they are likely to become extinct before playing any sustained evolutionary role (but see Lande 1976).

The admission of alternatives in principle does not imply their serious consideration in daily practice. We all say that not everything is adaptive; yet, faced with an organism, we tend to break it into parts and tell adaptive stories as if trade-offs among competing, well-designed parts were the only constraint upon perfection for each trait. It is an old habit. As Romanes complained about A. R. Wallace in 1900: "Mr. Wallace does not expressly maintain the abstract impossibility of laws and causes other than those of utility and natural selection.... Nevertheless, as he nowhere recognizes any other law or cause ... he practically concludes that, on inductive or empirical grounds, there is *no* such other law or cause to be entertained."

The adaptationist program can be traced through common styles of argument. We illustrate just a few; we trust they will be recognized by all:

1. If one adaptive argument fails, try another. Zig-zag commissures of clams and brachiopods, once widely regarded as devices for strengthening the shell, become sieves for restricting particles above a given size (Rudwick 1964). A suite of external structures (horns, antlers, tusks), once viewed as weapons against predators, become symbols of intraspecific competition among males (Davitashvili 1961). The Eskimo face, once depicted as "cold engineered" (Coon et al. 1950), becomes an adaptation to generate and withstand large masticatory forces (Shea 1977). We do not attack these newer interpretations; they may all be right. We do wonder, though, whether the failure of one adaptive explanation should always simply inspire a search for another of the same general form, rather than a consideration of alternatives to the proposition that each part is "for" some specific purpose.

2. If one adaptive argument fails, assume that another must exist; a weaker version of the first argument. Costa and Bisol (1978), for example, hoped to find a correlation between genetic polymorphism and stability of environment in the deep sea, but they failed. They conclude (1978, pp. 132, 133): "The degree of genetic polymorphism found would seem to indicate absence of correlation with the particular environmental factors which characterize the sampled area. The results suggest that the adaptive strategies of organisms belonging to different phyla are different."

3. In the absence of a good adaptive argument in the first place, attribute failure to imperfect understanding of where an organism lives and what it does. This is again an old argument. Consider Wallace on why all details of color and form in land snails must be adaptive, even if different animals seem to inhabit the same environment (1899, p. 148): "The exact proportions of the various species of plants, the numbers of each kind of insect or of bird, the peculiarities of more or less exposure to sunshine or to wind at certain critical epochs, and other slight differences which to us are absolutely immaterial and unrecognizable, may be of the highest significance to these humble creatures, and be quite sufficient to require some slight adjustments of size, form, or color, which natural selection will bring about."

4. Emphasize immediate utility and exclude other attributes of form. Fully half the explanatory information accompanying the full-scale Fiberglass *Tyrannosaurus* at Boston's Museum of Science reads: "Front legs a puzzle:

Stephen Jay Gould and Richard Lewontin

how *Tyrannosaurus* used its tiny front legs is a scientific puzzle; they were too short even to reach the mount. They may have been used to help the animal rise from a lying position." (We purposely choose an example based on public impact of science to show how widely habits of the adaptationist program extend. We are not using glass beasts as straw men; similar arguments and relative emphases, framed in different words, appear regularly in the professional literature.) We don't doubt that *Tyrannosaurus* used its diminutive front legs for something. If they had arisen *de novo*, we would encourage the search for some immediate adaptive reason. But they are, after all, the reduced product of conventionally functional homologues in ancestors (longer limbs of allosaurs, for example). As such, we do not need an explicitly adaptive explanation for the reduction itself. It is likely to be a developmental correlate of allometric fields for relative increase in head and hindlimb size. This nonadaptive hypothesis can be tested by conventional allometric methods (Gould 1974, in general; Lande 1978, on limb reduction) and seems to us both more interesting and fruitful than untestable speculations based on secondary utility in the best of possible worlds. One must not confuse the fact that a structure is used in some way (consider again the spandrels, ceiling spaces, and Aztec bodies) with the primary evolutionary reason for its existence and conformation.

TELLING STORIES

All this is a manifestation of the rightness of things, since if there is a volcano at Lisbon it could not be anywhere else. For it is impossible for things not to be where they are, because everything is for the best.
—Dr. Pangloss on the great Lisbon earthquake of 1755, in which up to 50,000 people lost their lives

We would not object so strenuously to the adaptationist program if its invocation, in any particular case, could lead in principle to its rejection for want of evidence. We might still view it as restrictive and object to its status as an argument of first choice. But if it could be dismissed after failing some explicit test, then alternatives would get their chance. Unfortunately, a common procedure among evolutionists does not allow such definable rejection for two reasons. First, the rejection of one adaptive story usually leads to its replacement by another, rather than to a suspicion that a different kind of explanation might be required. Since the range of adaptive stories is as wide as our minds are fertile, new stories can always be postulated. And if a story is not immediately available, one can always plead temporary ignorance and trust that it will be forthcoming, as did Costa and Bisol (1978), cited above. Second, the criteria for acceptance of a story are so loose that many pass without proper confirmation. Often, evolutionists use *consistency* with natural selection as the sole criterion and consider their work done when they concoct a plausible story. But plausible stories can always be told. The key to historical research lies in devising criteria to identify proper explanations among the substantial set of plausible pathways to any modern result.

We have, for example (Gould 1978) criticized Barash's (1976) work on aggression in mountain bluebirds for this reason. Barash mounted a stuffed male near the nests of two pairs of bluebirds while the male was out foraging. He did this at the same nests on three occasions at ten-day intervals: the first before eggs were laid, the last two afterward. He then counted aggressive approaches of the returning male toward both the model and the female. At time one, aggression was high toward the model and lower toward females but substantial in both nests. Aggression toward the model declined steadily for times two and three and plummeted to near zero toward females. Barash reasoned that this made evolutionary sense, since males would be more sensitive to intruders before eggs were laid than afterward (when they can have some confidence that their genes are inside). Having devised this plausible story, he considered his work as completed (1976, pp. 1099, 1100):

The results are consistent with the expectations of evolutionary theory. Thus aggression toward an intruding male (the model) would clearly be especially advantageous early in the breeding season, when territories and nests are normally defended.... The initial aggressive response to the mated female is also adaptive in that, given a situation suggesting a high probability of adultery (i.e., the presence of the model near the female) and assuming that replacement females are available, obtaining a new mate would enhance the fitness of males.... The decline in male-female aggressiveness during incubation and fledgling stages could be attributed to the impossibility of being cuckolded after the eggs have been laid.... The results are consistent with an evolutionary interpretation.

They are indeed consistent, but what about an obvious alternative, dismissed without test by Barash? Male returns at times two and three, approaches the model, tests it a bit, recognizes it as the same phoney he saw before, and doesn't bother his female. Why not at least perform the obvious test for this alternative to a conventional adaptive story: expose a male to the model for the *first* time after the eggs are laid?

After we criticized Barash's work, Morton et al. (1978) repeated it, with some variations (including the introduction of a female model), in the closely related eastern bluebird *Sialia sialis.* "We hoped to confirm," they wrote, that Barash's conclusions represent "a widespread evolutionary reality, at least within the genus *Sialia.* Unfortunately, we were unable to do so." They found no "anticuckoldry" behavior at all: males never approached their females aggressively after testing the model at any nesting stage. Instead, females often approached the male model and, in any case, attacked female models more than males attacked male models. "This violent response resulted in the near destruction of the female model after presentations and its complete demise on the third, as a female flew off with the model's head early in the experiment to lose it for us in the brush" (1978, p. 969). Yet, instead of calling Barash's selected story into question, they merely devise one of their own to render both results in the adaptationist mode. Perhaps, they conjecture, replacement females are scarce in their species and abundant in Barash's. Since Barash's males can replace a potentially "unfaithful" female, they can afford to be choosy and possessive. Eastern bluebird males are stuck with uncommon mates and had best be respectful. They conclude: "If we did not support Barash's suggestion that male bluebirds show anticuckoldry adaptations, we suggest that both studies still had 'results that are consistent with the expectations of evolutionary theory' (Barash 1976, p. 1099), as we presume any careful study would." But what good is a theory that cannot fail in careful study (since by "evolutionary theory," they clearly mean the action of natural selection applied to particular cases, rather than the fact of transmutation itself)?

THE MASTER'S VOICE REEXAMINED

Since Darwin has attained sainthood (if not divinity) among evolutionary biologists, and since all sides invoke God's allegiance, Darwin has often been depicted as a radical selectionist at heart who invoked other mechanisms only in retreat, and only as a result of his age's own lamented ignorance about the mechanisms of heredity. This view is false. Although Darwin regarded selection as the most important of evolutionary mechanisms (as do we), no argument from opponents angered him more than the common attempt to caricature and trivialize his theory by stating that it relied exclusively upon natural selection. In the last edition of the *Origin*, he wrote (1872, p. 395):

As my conclusions have lately been much misrepresented, and it has been stated that I attribute the modification of species exclusively to natural selection, I may be permitted to remark that in the first edition of this work, and subsequently, I placed in a most conspicuous position—namely at the close of the Introduction—the following words: "I am convinced that natural selection has been the main, but not the exclusive means of modification." This has been of no avail. Great is the power of steady misinterpretation.

Romanes, whose once famous essay (1900) on Darwin's pluralism versus the panselectionism of Wallace and Weismann deserves a resurrection, noted of this passage (1900, p. 5): "In the whole range of Darwin's writings there cannot be found a passage so strongly worded as this: it presents the only note of bitterness in all the thousands of pages which he has published." Apparently, Romanes did not know the letter Darwin wrote to *Nature* in 1880, in which he castigated Sir Wyville Thomson for caricaturing his theory as panselectionist (1880, p. 32):

I am sorry to find that Sir Wyville Thomson does not understand the principle of natural selection.... If he had done so, he could not have written the following sentence in the Introduction to the Voyage of the Challenger: "The character of the abyssal fauna refuses to give the least support to the theory which refers the evolution of species to extreme variation guided only by natural selection." This is a standard of criticism not uncommonly reached by theologians and metaphysicians when they write on scientific subjects, but is something new as coming from a naturalist.... Can Sir Wyville Thomson name any one who has said that the evolution of species depends only on natural selection? As far as concerns myself, I believe that no one has brought forward so many observations on the effects of the use and disuse of parts, as

I have done in my "Variation of Animals and Plants under Domestication"; and these observations were made for this special object. I have likewise there adduced a considerable body of facts, showing the direct action of external conditions on organisms.

We do not now regard all of Darwin's subsidiary mechanisms as significant or even valid, though many, including direct modification and correlation of growth, are very important. But we should cherish his consistent attitude of pluralism in attempting to explain Nature's complexity.

A PARTIAL TYPOLOGY OF ALTERNATIVES TO THE ADAPTATIONIST PROGRAM

In Darwin's pluralistic spirit, we present an incomplete hierarchy of alternatives to immediate adaptation for the explanation of form, function, and behavior.

1. No adaptation and no selection at all. At present, population geneticists are sharply divided on the question of how much genetic polymorphism within populations and how much of the genetic differences between species is, in fact, the result of natural selection as opposed to purely random factors. Populations are finite in size, and the isolated populations that form the first step in the speciation process are often founded by a very small number of individuals. As a result of this restriction in population size, frequencies of alleles change by *genetic drift*, a kind of random genetic sampling error. The stochastic process of change in gene frequency by random genetic drift, including the very strong sampling process that goes on when a new isolated population is formed from a few immigrants, has several important consequences. First, populations and species will become genetically differentiated, and even fixed for different alleles at a locus in the complete absence of any selective force at all.

Second, alleles can become fixed in a population *in spite of natural selection*. Even if an allele is favored by natural selection, some proportion of populations, depending upon the product of population size N and selection intensity s, will become homozygous for the less fit allele because of genetic drift. If Ns is large, this random fixation for unfavorable alleles is a rare phenomenon, but if selection coefficients are on the order of the reciprocal of population size ($Ns = 1$) or smaller, fixation for deleterious alleles is common. If many genes are involved in influencing a metric character like shape, metabolism, or behavior, then the intensity of selection on each locus will be small and Ns per locus may be small. As a result, many of the loci may be fixed for nonoptimal alleles.

Third, new mutations have a small chance of being incorporated into a population, even when selectively favored. Genetic drift causes the immediate loss of most new mutations after their introduction. With a selection intensity s, a new favorable mutation has a probability of only $2s$ of ever being incorporated. Thus one cannot claim that, eventually, a new mutation of just the right sort for some adaptive argument will occur and spread. "Eventually"

Stephen Jay Gould and Richard Lewontin

becomes a very long time if only one in 1,000 or one in 10,000 of the "right" mutations that do occur ever get incorporated in a population.

2. No adaptation and no selection on the part at issue; form of the part is a correlated consequence of selection directed elsewhere. Under this important category, Darwin ranked his "mysterious" laws of the "correlaton of growth." Today, we speak of pleiotropy, allometry, "material compensation" (Rensch 1959, pp. 179–187) and mechanically forced correlations in D'Arcy Thompson's sense (1942; Gould 1971). Here we come face to face with organisms as integrated wholes, fundamentally not decomposable into independent and separately optimized parts.

Although allometric patterns are as subject to selection as static morphology itself (Gould 1966), some regularities in relative growth are probably not under immediate adaptive control. For example, we do not doubt that the famous 0.66 interspecific allometry of brain size in all major vertebrate groups represents a selected "design criterion," though its significance remains elusive (Jerison 1973). It is too repeatable across too wide a taxonomic range to represent much else than a series of creatures similarly well designed for their different sizes. But another common allometry, the 0.2 to 0.4 intraspecific scaling among homeothermic adults differing in body size, or among races within a species, probably does not require a selectionist story, though many, including one of us, have tried to provide one (Gould 1974). R. Lande (personal communication) has used the experiments of Falconer (1973) to show that selection upon *body size alone* yields a brain-body slope across generations of 0.35 in mice.

More compelling examples abound in the literature on selection for altering the timing of maturation (Gould 1977). At least three times in the evolution of arthropods (mites, flies, and beetles), the same complex adaptation has evolved, apparently for rapid turnover of generations in strongly r-selected feeders on superabundant but ephemeral fungal resources: females reproduce as larvae and grow the next generation within their bodies. Offspring eat their mother from inside and emerge from her hollow shell, only to be devoured a few days later by their own progeny. It would be foolish to seek adaptive significance in paedomorphic morphology per se; it is primarily a by-product of selection for rapid cycling of generations. In more interesting cases, selection for small size (as in animals of the interstitial fauna) or rapid maturation (dwarf males of many crustaceans) has occurred by progenesis (Gould 1977, pp. 324–336), and descendant adults contain a mixture of ancestral juvenile and adult features. Many biologists have been tempted to find primary adaptive meaning for the mixture, but it probably arises as a by-product of truncated maturation, leaving some features "behind" in the larval state, while allowing others, more strongly correlated with sexual maturation, to retain the adult configuration of ancestors.

3. The decoupling of selection and adaptation.

(i) Selection without adaptation. Lewontin (1979) has presented the following hypothetical example: "A mutation which doubles the fecundity of individuals will sweep through a population rapidly. If there has been no change in efficiency of resource utilization, the individuals will leave no more off-spring than before, but simply lay twice as many eggs, the excess dying because of resource limitation. In what sense are the individuals or the population as a whole better adapted than before? Indeed, if a predator on immature stages is led to switch to the species now that immatures are more plentiful, the population size may actually decrease as a consequence, yet natural selection at all times will favour individuals with higher fecundity."

(ii) Adaptation without selection. Many sedentary marine organisms, sponges and corals in particular, are well adapted to the flow régimes in which they live. A wide spectrum of "good design" may be purely phenotypic in origin, largely induced by the current itself. (We may be sure of this in numerous cases, when genetically identical individuals of a colony assume different shapes in different microhabitats.) Larger patterns of geographic variation are often adaptive and purely phenotypic as well. Sweeney and Vannote (1978), for example, showed that many hemimetabolous aquatic insects reach smaller adult size with reduced fecundity when they grow at temperatures above and below their optima. Coherent, climatically correlated patterns in geographic distribution for these insects—so often taken as a priori signs of genetic adaptation—may simply reflect this phenotypic plasticity.

"Adaptation"—the good fit of organisms to their environment—can occur at three hierarchical levels with different causes. It is unfortunate that our language has focused on the common result and called all three phenomena "adaptation": the differences in process have been obscured, and evolutionists have often been misled to extend the Darwinian mode to the other two levels as well. First, we have what physiologists call "adaptation": the phenotypic plasticity that permits organisms to mold their form to prevailing circumstances during ontogeny. Human "adaptations" to high altitude fall into this category (while others, like resistance of sickling heterozygotes to malaria, are genetic, and Darwinian). Physiological adaptations are not heritable, though the capacity to develop them presumably is. Second, we have a "heritable" form of non-Darwinian adaptation in humans (and, in rudimentary ways, in a few other advanced social species): cultural adaptation (with heritability imposed by learning). Much confused thinking in human sociobiology arises from a failure to distinguish this mode from Darwinian adaptation based on genetic variation. Finally, we have adaptation arising from the conventional Darwinian mechanism of selection upon genetic variation. The mere existence of a good fit between organism and environment is insufficient for inferring the action of natural selection.

4. Adaptation and selection but no selective basis for differences among adaptations. Species of related organisms, or subpopulations within a species, often develop different adaptations as solutions to the same problem. When "multiple adaptive peaks" are occupied, we usually have no basis for asserting that one solution is better than another. The solution followed in any spot is a result of history; the first steps went in one direction, though others would have led to adequate prosperity as well. Every naturalist has his favorite illustration. In the West Indian land snail *Cerion*, for example, populations

Stephen Jay Gould and Richard Lewontin

living on rocky and windy coasts almost always develop white, thick, and relatively squat shells for conventional adaptive reasons. We can identify at least two different developmental pathways to whiteness from the mottling of early whorls in all *Cerion*, two paths of thickened shells and three styles of allometry leading to squat shells. All twelve combinations can be identified in Bahamian populations, but would it be fruitful to ask why—in the sense of optimal design rather than historical contingency—*Cerion* from eastern Long Island evolved one solution, and *Cerion* from Acklins Island another?

5. Adaptation and selection, but the adaptation is a secondary utilization of parts present for reasons of architecture, development, or history. We have already discussed this neglected subject in the first section on spandrels, spaces, and cannibalism. If blushing turns out to be an adaptation affected by sexual selection in humans, it will not help us to understand why blood is red. The immediate utility of an organic structure often says nothing at all about the reason for its being.

ANOTHER, AND UNFAIRLY MALIGNED, APPROACH TO EVOLUTION

In continental Europe, evolutionists have never been much attracted to the Anglo-American penchant for atomizing organisms into parts and trying to explain each as a direct adaptation. Their general alternative exists in both a strong and a weak form. In the strong form, as advocated by such major theorists as Schindewolf (1950), Remane (1971), and Grassé (1977), natural selection under the adaptationist program can explain superficial modifications of the *Bauplan* that fit structure to environment: why moles are blind, giraffes have long necks, and ducks webbed feet, for example. But the important steps of evolution, the construction of the *Bauplan* itself and the transition between *Baupläne*, must involve some other unknown, and perhaps "internal," mechanism. We believe that English biologists have been right in rejecting this strong form as close to an appeal to mysticism.

But the argument has a weaker—and paradoxically powerful—form that has not been appreciated, but deserves to be. It also acknowledges conventional selection for superficial modifications of the *Bauplan*. It also denies that the adaptationist program (atomization plus optimizing selection on parts) can do much to explain *Baupläne* and the transitions between them. But it does not therefore resort to a fundamentally unknown process. It holds instead that the basic body plans of organisms are so integrated and so replete with constraints upon adaptation (categories 2 and 5 of our typology) that conventional styles of selective arguments can explain little of interest about them. It does not deny that change, when it occurs, may be mediated by natural selection, but it holds that constraints restrict possible paths and modes of change so strongly that the constraints themselves become much the most interesting aspect of evolution.

Rupert Riedl, the Austrian zoologist who has tried to develop this thesis for English audiences (1977 and 1975, translated into English by R. Jeffries in 1978) writes:

The living world happens to be crowded by universal patterns of organization which, most obviously, find no direct explanation through environmental conditions or adaptive radiation, but exist primarily through universal require- ments which can only be expected under the systems conditions of complex organization itself.... This is not self-evident, for the whole of the huge and profound thought collected in the field of morphology, from Goethe to Remane, has virtually been cut off from modern biology. It is not taught in most American universities. Even the teachers who could teach it have disappeared.

Constraints upon evolutionary change may be ordered into at least two categories. All evolutionists are familiar with *phyletic* constraints, as embodied in Gregory's classic distinction (1936) between habitus and heritage. We ac- knowledge a kind of phyletic inertia in recognizing, for example, that humans are not optimally designed for upright posture because so much of our *Bauplan* evolved for quadrupedal life. We also invoke phyletic constraint in explaining why no molluscs fly in air and no insects are as large as elephants.

Developmental constraints, a subcategory of phyletic restrictions, may hold the most powerful rein of all over possible evolutionary pathways. In complex organisms, early stages of ontogeny are remarkably refractory to evolution- ary change, presumably because the differentiation of organ systems and their integration into a functioning body is such a delicate process so easily derailed by early errors with accumulating effects. Von Baer's fundamental embryolog- ical laws (1828) represent little more than a recognition that early stages are both highly conservative and strongly restrictive of later development. Haeckel's biology law, the primary subject of late nineteenth-century evolu- tionary biology, rested upon a misreading of the same data (Gould 1977). If development occurs in integrated packages and cannot be pulled apart piece by piece in evolution, then the adaptationist program cannot explain the alteration of developmental programs underlying nearly all changes of *Bauplan.*

The German palaeontologist A. Seilacher, whose work deserves far more attention than it has received, has emphasized what he calls *"bautechnischer, or architectural,* constraints" (Seilacher 1970). These arise not from former adapta- tions retained in a new ecological setting (phyletic constraints as usually understood), but as architectural restrictions that never were adaptations but rather were the necessary consequences of materials and designs selected to build basic *Baupläne.* We devoted the first section of this chapter to nonbio- logical examples in this category. Spandrels must exist once a blueprint spe- cifies that a dome shall rest on rounded arches. Architectural constraints can exert a far-ranging influence upon organisms as well. The subject is full of potential insight because it has rarely been acknowledged at all.

In a fascinating example, Seilacher (1972) has shown that the divaricate form of architecture (figure 4.3) occurs again and again in all groups of mol-

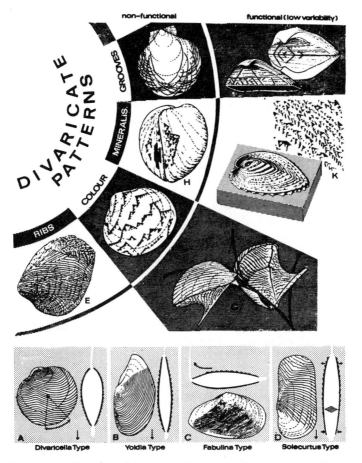

Figure 4.3 The range of divaricate patterns in molluscs. E, F, H, and L are nonfunctional in Seilacher's judgment. A–D are functional ribs (but these are far less common than nonfunctional ribs of the form E). G is the mimetic *Arca zebra*. K is *Corculum*. See text for details.

luscs, and in brachiopods as well. This basic form expresses itself in a wide variety of structures: raised ornamental lines (not growth lines because they do not conform to the mantle margin at any time), patterns of coloration, internal structures in the mineralization of calcite and incised grooves. He does not know what generates this pattern and feels that traditional and nearly exclusive focus on the adaptive value of each manifestation has diverted attention from questions of its genesis in growth and also prevented its recognition as a general phenomenon. It must arise from some characteristic pattern of inhomogeneity in the growing mantle, probably from the generation of interference patterns around regularly spaced centers; simple computer simulations can generate the form in this manner (Waddington and Cowe 1969). The general pattern may not be a direct adaptation at all.

Seilacher then argues that most manifestations of the pattern are probably nonadaptive. His reasons vary but seem generally sound to us. Some are based on field observations: color patterns that remain invisible because clams possessing them either live buried in sediments or remain covered with a periostracum so thick that the colors cannot be seen. Others rely on more general principles: presence only in odd and pathological individuals, rarity as a developmental anomaly, excessive variability compared with much reduced variability when the same general structure assumes a form judged functional on engineering grounds.

In a distinct minority of cases, the divaricate pattern becomes functional in each of the four categories (figure 4.3). Divaricate ribs may act as scoops and anchors in burrowing (Stanley 1970), but they are not properly arranged for such function in most clams. The color chevrons are mimetic in one species (*Pteria zebra*) that lives on hydrozoan branches; here the variability is strongly reduced. The mineralization chevrons are probably adaptive in only one remarkable creature, the peculiar bivalve *Corculum cardissa* (in other species they either appear in odd specimens or only as postmortem products of shell erosion). This clam is uniquely flattened in an anterioposterior direction. It lies on the substrate, posterior up. Distributed over its rear end are divaricate triangles of mineralization. They are translucent, while the rest of the shell is opaque. Under these windows dwell endosymbiotic algae!

All previous literature on divaricate structure has focused on its adaptive significance (and failed to find any in most cases). But Seilacher is probably right in representing this case as the spandrels, ceiling holes, and sacrificed bodies of our first section. The divaricate pattern is a fundamental architectural constraint. Occasionally, since it is there, it is used to beneficial effect. But we cannot understand the pattern or its evolutionary meaning by viewing these infrequent and secondary adaptations as a reason for the pattern itself.

Galton (1909, p. 257) contrasted the adaptationist program with a focus on constraints and modes of development by citing a telling anecdote about Herbert Spencer's fingerprints:

Much has been written, but the last word has not been said, on the rationale of these curious papillary ridges; why in one man and in one finger they form whorls and in another loops. I may mention a characteristic anecdote of Herbert Spencer in connection with this. He asked me to show him my Laboratory and to take his prints, which I did. Then I spoke of the failure to discover the origin of these patterns, and how the fingers of unborn children had been dissected to ascertain their earliest stages, and so forth. Spencer remarked that this was beginning in the wrong way; that I ought to consider the purpose the ridges had to fulfil, and to work backwards. Here, he said, it was obvious that the delicate mouths of the sudorific glands required the protection given to them by the ridges on either side of them, and therefrom he elaborated a consistent and ingenious hypothesis at great length. I replied that his arguments were beautiful and deserved to be true, but it happened that the mouths of the ducts did not run in the valleys between the crests, but along the crests of the ridges themselves.

We feel that the potential rewards of abandoning exclusive focus on the adaptationist program are very great indeed. We do not offer a counsel of despair, as adaptationists have charged; for nonadaptive does not mean non-intelligible. We welcome the richness that a pluralistic approach, so akin to

Darwin's spirit, can provide. Under the adaptationist program, the great historic themes of developmental morphology and *Bauplan* were largely abandoned; for if selection can break any correlation and optimize parts separately, then an organism's integration counts for little. Too often, the adaptationist program gave us an evolutionary biology of parts and genes, but not of organisms. It assumed that all transitions could occur step by step and underrated the importance of integrated developmental blocks and pervasive constraints of history and architecture. A pluralistic view could put organisms, with all their recalcitrant yet intelligible complexity, back into evolutionary theory.

REFERENCES

Baer, K. E. von. 1828. *Entwicklungsgeschichte der Tiere*, Königsberg: Bornträger.

Barash, D. P. 1976. Male response to apparent female adultery in the mountain-bluebird: an evolutionary interpretation, *Am. Nat.*, 110:1097–1101.

Coon, C. S., Garn, S. M., and Birdsell, J. B. 1950. *Races*, Springfield Ohio, C. Thomas.

Costa, R., and Bisol, P. M. 1978. Genetic variability in deep-sea organisms, *Biol. Bull.*, 155:125–133.

Darwin, C. 1872. *The origin of species*, London, John Murray.

———. 1880. Sir Wyville Thomson and natural selection, *Nature*, London, 23:32.

Davitashvili, L. S. 1961. *Teoriya polovogo otbora* [Theory of sexual selection], Moscow, Akademii Nauk.

Falconer, D. S. 1973. Replicated selection for body weight in mice, *Genet. Res.*, 22:291–321.

Galton, F. 1909. *Memories of my life*, London, Methuen.

Gould, S. J. 1966. Allometry and size in ontogeny and phylogeny, *Biol. Rev.*, 41:587–640.

———. 1971. D'Arcy Thompson and the science of form, *New Literary Hist.*, 2, no. 2, 229–258.

———. 1974. Allometry in primates, with emphasis on scaling and the evolution of the brain. In *Approaches to primate paleobiology*, *Contrib. Primatol.*, 5:244–292.

———. 1977. *Ontogeny and phylogeny*, Cambridge, Ma., Belknap Press.

———. 1978. Sociobiology: the art of storytelling, *New Scient.*, 80:530–533.

Grassé, P. P. 1977. *Evolution of living organisms*, New York, Academic Press.

Gregory, W. K. 1936. Habitus factors in the skeleton fossil and recent mammals, *Proc. Am. phil. Soc.*, 76:429–444.

Harner, M. 1977. The ecological basis for Aztec sacrifice. *Am. Ethnologist*, 4:117–135.

Jerison, H. J. 1973. *Evolution of the brain and intelligence*, New York, Academic Press.

Lande, R. 1976. Natural selection and random genetic drift in phenotypic evolution, *Evolution*, 30:314–334.

———. 1978. Evolutionary mechanisms of limb loss in tetrapods, *Evolution*, 32:73–92.

Lewontin, R. C. 1978. Adaptation, *Scient. Am.*, 239 (3):156–169.

———. 1979. Sociobiology as an adaptationist program, *Behav. Sci.*, 24:5–14.

Morton, E. S., Geitgey, M. S., and McGrath, S. 1978. On bluebird "responses to apparent female adultery." *Am. Nat.*, 112:968–971.

Ortiz de Montellano, B. R. 1978. Aztec cannibalism: an ecological necessity? *Science*, 200:611–617.

Remane, A. 1971. *Die Grundlagen des natürlichen Systems der vergleichenden Anatomie und der Phylogenetik.* Königstein-Taunus: Koeltz.

Rensch, B. 1959. *Evolution above the species level,* New York, Columbia University Press.

Riedel, R. 1975. *Die Ordnung des Lebendigen,* Hamburg, Paul Parey, tr. R. P. S. Jefferies, *Order in living systems: A systems analysis of evolution,* New York, Wiley, 1978.

———. 1977. A systems-analytical approach to macro-evolutionary phenomena, *Q. Rev. Biol.,* 52:351–370.

Romanes, G. J. 1900. The Darwinism of Darwin and of the post-Darwinian schools. In *Darwin, and after Darwin,* vol. 2, new ed., London, Longmans, Green and Co.

Rudwick, M. J. S. 1964. The function of zig-zag deflections in the commissures of fossil brachiopods, *Palaeontology,* 7:135–171.

Sahlins, M. 1978. Culture as protein and profit, *New York Review of Books,* 23:Nov., pp. 45–53.

Schindewolf, O. H. 1950. *Grundfragen der Paläontologie,* Stuttgart, Schweizerbart.

Seilacher, A. 1970. Arbeitskonzept zur Konstruktionsmorphologie, *Lethaia,* 3:393–396.

———. 1972. Divaricate patterns in pelecypod shells, *Lethaia,* 5:325–343.

Shea, B. T. 1977. Eskimo craniofacial morphology, cold stress and the maxillary sinus, *Am. J. Phys. Anthrop.,* 47:289–300.

Stanley, S. M. 1970. Relation of shell form to life habits in the Bivalvia (Mollusca). *Mem. Geol. Soc. Am.,* no. 125, 296 pp.

Sweeney, B. W., and Vannote, R. L. 1978. Size variation and the distribution of hemimetabolous aquatic insects: two thermal equilibrium hypotheses. *Science,* 200:444–446.

Thompson, D. W. 1942. *Growth and form,* New York, Macmillan.

Waddington, C. H., and Cowe, J. R. 1969. Computer simulation of a molluscan pigmentation pattern, *J. Theor. Biol.,* 25:219–225.

Wallace, A. R. 1899. *Darwinism,* London, Macmillan.

Wilson, E. O. 1978. *On human nature,* Cambridge, Ma., Harvard University Press.

5 Optimization Theory in Evolution

John Maynard Smith

INTRODUCTION

In recent years there has been a growing attempt to use mathematical methods borrowed from engineering and economics in interpreting the diversity of life. It is assumed that evolution has occurred by natural selection, hence that complex structures and behaviors are to be interpreted in terms of the contribution they make to the survival and reproduction of their possessors—that is, to Darwinian fitness. There is nothing particularly new in this logic, which is also the basis of functional anatomy, and indeed of much physiology and molecular biology. It was followed by Darwin himself in his studies of climbing and insectivorous plants, of fertilization mechanisms and devices to ensure cross-pollination.

What is new is the use of such mathematical techniques as control theory, dynamic programming, and the theory of games to generate a priori hypotheses, and the application of the method to behaviors and life history strategies. This change in method has led to the criticism (e.g., Lewontin, 54, 55) that the basic hypothesis of adaptation is untestable and therefore unscientific, and that the whole program of functional explanation through optimization has become a test of ingenuity rather than an inquiry into truth. Related to this is the criticism that there is no theoretical justification for any maximization principles in biology, and therefore that optimization is no substitute for an adequate genetic model.

My aim in this review is not to summarize the most important conclusions reached by optimization methods, but to discuss the methodology of the program and the criticisms that have been made of it. In doing so, I have taken as my starting point two articles by Lewontin (54, 55). I disagree with some of the views he expresses, but I believe that the development of evolution theory could benefit if workers in optimization paid serious attention to his criticisms.

I first outline the basic structure of optimization arguments, illustrating this with three examples, namely the sex ratio, the locomotion of mammals,

From *Annual Review of Ecology and Systematics*, 1978, 9:31–56.

and foraging behavior. I then discuss the possibility that some variation may be selectively neutral, and some structures maladaptive. I summarize and comment on criticisms made by Lewontin. The most damaging, undoubtedly, is the difficulty of testing the hypotheses that are generated. The next section therefore discusses the methodology of testing; in this section I have relied heavily on the arguments of Curio (23). Finally I discuss mathematical methods. The intention here is not to give the details of the mathematics, but to identify the kinds of problems that have been attacked and the assumptions that have been made in doing so.

THE STRUCTURE OF OPTIMIZATION MODELS

In this section I illustrate the argument with three examples: (*a*) the sex ratio, based on Fisher's (28) treatment and later developments by Hamilton (34), Rosado and Robertson (85), Trivers and Willard (96), and Trivers and Hare (95); (*b*) the gaits of mammals—given a preliminary treatment by Maynard Smith and Savage (66), and further analyzed in several papers in Pedley (78); (*c*) foraging strategies. Theoretical work on them originated with the papers of Emlen (27) and MacArthur and Pianka (57). I have relied heavily on a recent review by Pyke et al. (81). These authors suggest that models have in the main been concerned with four problems: choice by the animal of which types of food to eat (optimal diet); choice of which patch type to feed in; allocation of time to different patches; pattern and speed of movement. In what follows I shall refer only to two of those—optimal diet and allocation of time to different patches.

All optimization models contain, implicitly or explicitly, an assumption about the "constraints" that are operating, an optimization criterion, and an assumption about heredity. I consider these in turn.

The Constraints: Phenotype Set and State Equations

The constraints are essentially of two kinds. In engineering applications, they concern the "strategy set," which specifies the range of control actions available, and the "state equations," which specify how the state of the system being controlled changes in time. In biological applications, the strategy set is replaced by an assumption about the set of possible phenotypes on which selection can operate.

It is clearly impossible to say what is the "best" phenotype unless one knows the range of possibilities. If there were no constraints on what is possible, the best phenotypes would live forever, would be impregnable to predators, would lay eggs at an infinite rate, and so on. It is therefore necessary to specify the set of possible phenotypes, or in some other way describe the limits on what can evolve. The "phenotype set" is an assumption about what can evolve and to what extent; the "state equations" describe features of the situation that are assumed not to change. This distinction will become clearer when particular examples are discussed. Let us consider the three problems in turn.

Sex Ratio   For the sex ratio, the simplest assumption is that a parent can produce a fixed number N of offspring, and that the probability S that each birth will be a male can vary from parent to parent over the complete range from 0 to 1; the phenotype set is then the set of values of S over this range. Fisher (28) extended this by supposing that males and females "cost" different amounts; i.e. he supposed that a parent could produce α males and β females, where α and β are constrained to lie on or below the line $\alpha + \beta k = N$, and k is the cost of a female relative to that of a male. He then concluded that the parent should equalize expenditure on males and females. MacArthur (56) further broadened the phenotype set by insisting only that α and β lie on or below a line of arbitrary shape, and concluded that a parent should maximize $\alpha\beta$. A similar assumption was used by Charnov et al. (11) to analyze the evolution of hermaphroditism as opposed to dioecy. Finally, it is possible to ask (97) what is the optimal strategy if a parent can choose not merely a value of S, hence of the expected sex ratio, but also the variance of the sex ratio.

The important point in the present context is that the optimal solution depends on the assumption made. For example, Crow and Kimura (21) conclude that the sex ratio should be unity, but they do so for a model that assumes that $N = \alpha + \beta$ is a constant.

Gaits   In the analysis of gaits, it is assumed that the shapes of bones can vary but the mechanical properties of bone, muscle, and tendon cannot. It is also assumed that changes must be gradual; thus the gaits of ostrich, antelope, and kangaroo are seen as different solutions to the same problem, not as solutions to different problems—i.e., they are different "adaptive peaks" (101).

Foraging Strategy   In models of foraging behavior, a common assumption is that the way in which an animal allocates its time among various activities (e.g., consuming one prey item rather than another, searching in one kind of patch rather than another, moving between patches rather than continuing to search in the same one) can vary, but the efficiency with which it performs each act cannot. Thus, for example, the length of time it takes to "handle" (capture and consume) a given item, the time and energy spent in moving from place to place, and the time taken to find a given prey item at a given prey density are taken as invariant. Thus the models of foraging so far developed treat the phenotype set as the set of possible behavioral strategies, and treat structure and locomotory or perceptual skills as constants contributing to the state equations (which determine how rapidly an animal adopting some strategy acquires food). In principle there is no reason why optimization models should not be applied to the evolution of structure or skill also; it is simply a question of how the phenotype set is defined.

The Optimization Criterion

Some assumption must then be made concerning what quantity is being maximized. The most satisfactory is the inclusive fitness (see the section Games between Relatives, below); in many contexts the individual fitness (expected number of offspring) is equally good. Often, as in the second and third of my examples, neither criterion is possible, and some other assumption is needed. Two points must be made. First, the assumption about what is maximized is an assumption about what selective forces have been responsible for the trait; second, this assumption is part of the hypothesis being tested.

In most theories of sex ratio the basic assumption is that the ratio is determined by a gene acting in a parent, and what is maximized is the number of copies of that gene in future generations. The maximization has therefore a sound basis. Other maximization criteria have been used. For example, Kalmus and Smith (41) propose that the sex ratio maximizes the probability that two individuals meeting will be of different sexes; it is hard to understand such an eccentric choice when the natural one is available.

An equally natural choice—the maximization of the expected number of offspring produced in a lifetime—is available in theories of the evolution of life history strategies. But often no such easy choice is available.

In the analysis of gaits, Maynard Smith and Savage (66) assumed that the energy expenditure at a given speed would be minimized (or, equivalently, that the speed for a given energy expenditure was maximized). This led to the prediction that the proportion of time spent with all four legs off the ground should increase with speed and decrease with size.

In foraging theory, the common assumption is that the animal is maximizing its energy intake per unit time spent foraging. Schoener (87) points out that this is an appropriate choice, whether the animal has a fixed energy requirement and aims to minimize the time spent feeding so as to leave more time for other activities ("time minimizers"), or has a fixed time in which to feed during which it aims to maximize its energy gain ("energy maximizers"). There will, however, be situations in which this is not an appropriate choice. For example, there may be a higher risk of predation for some types of foraging than for others. For some animals the problem may be not to maximize energy intake per unit time, but to take in a required amount of energy, protein, etc. without taking an excess of any one of a number of toxins (S. A. Altmann, personal communication).

Pyke et al. (81) point out that the optimal strategy depends on the time scale over which optimization is carried out, for two reasons. First, an animal that has sole access to some resource e.g., a territory-holder) can afford to manage that resource so as to maximize its yield over a whole season. Second, and more general, optimal behavior depends on a knowledge of the environment, which can be acquired only by experience; this means that in order to acquire information of value in the long run, an animal may have to behave in a way that is inefficient in the short run.

Having considered the phenotype set and the optimization criterion, a word must be said about their relationship to Levins's (51) concept of a fitness set. Levins was explicitly concerned with defining fitness "in such a way that interpopulation selection would be expected to change a species towards the optimum (maximum fitness) structure." This essentially group-selectionist approach led him to conclusions (e.g., for the conditions for a stable polymorphism) different from those reached from the classic analysis of gene frequencies (93). Nevertheless, Levins's attempt to unite ecological and genetic approaches did lead him to recognize the need for the concept of a fitness set—i.e., the set of all possible phenotypes, each phenotype being characterized by its (individual) fitness in each of the environments in which it might find itself.

Levins's fitness set is thus a combination of what I have called the phenotype set and of a measure of the fitness of each phenotype in every possible environment. It did not allow for the fact that fitnesses may be frequency-dependent (see the section on Games, below). The valuable insight in Levins's approach is that it is possible to discuss what course phenotypic evolution may take only if one makes explicit assumptions about the constraints on what phenotypes are possible. It may be better to use the term "phenotype set" to define these constraints, both because a description of possible phenotypes is a process prior to and separable from an estimation of their fitnesses, and because of the group-selectionist associations of the term "fitness set."

An Assumption about Heredity

Because natural selection cannot produce adaptation unless there is heredity, some assumption, explicit or otherwise, is always present. The nature of this assumption can be important. Fisher (28) assumed that the sex ratio was determined by autosomal genes expressed in the parent, and that mating was random. Hamilton (34) showed that the predicted optima are greatly changed if these assumptions are altered. In particular, he considered the effects of inbreeding, and of genes for meiotic drive. Rosado and Robertson (85), Trivers and Willard (96), and Trivers and Hare (95) have analyzed the effects of genes acting in the children and (in Hymenoptera) in the sterile castes.

It is unusual for the way in which a trait is inherited to have such a crucial effect. Thus in models of mammalian gaits no explicit assumption is made; the implicit assumption is merely that like begets like. The same is true of models of foraging, although in this case "heredity" can be cultural as well as genetic —e.g. (72), for the feeding behavior of oyster-catchers.

The question of how optimization models can be tested is the main topic of the next three sections. A few preliminary remarks are needed. Clearly, the first requirement of a model is that the conclusions should follow from the assumptions. This seems not to be the case, for example, for Zahavi's (102) theory of sexual selection (61). A more usual difficulty is that the conclusions depend on unstated assumptions. For example, Fisher does not state that his sex ratio argument assumes random mating, and this was not noticed until

Hamilton's 1967 paper (34). Maynard Smith and Price (65) do not state that the idea of an ESS (evolutionary stable strategy) assumes asexual inheritance. It is probably true that no model ever states all its assumptions explicitly. One reason for writing this review is to encourage authors to become more aware of their assumptions.

A particular model can be tested either by a direct test of its assumptions or by comparing its predictions with observation. The essential point is that in testing a model we are testing *not* the general proposition that nature optimizes, but the specific hypotheses about constraints, optimization criteria, and heredity. Usually we test whether we have correctly identified the selective forces responsible for the trait in question. But we should not forget hypotheses about constraints or heredity. For example, the weakest feature of theories concerning the sex ratio is that there is little evidence for the existence of genetic variance of the kind assumed by Fisher—for references, see (63). It may be for this reason that the greatest successes of sex ratio theory (34, 95) have concerned Hymenoptera, in which it is easy to see how genes in the female parent can affect the sex of her children.

NEUTRALITY AND MALADAPTATION

I have said that when testing optimization models, one is not testing the hypothesis that nature optimizes. But if it is not the case that the structure and behavior of organisms are nicely adapted to ensure their survival and reproduction, optimization models cannot be useful. What justification have we for assuming this?

The idea of adaptation is older than Darwinism. In the form of the argument from design, it was a buttress of religious belief. For Darwin the problem was not to prove that organisms were adapted but to explain how adaptation could arise without a creator. He was quite willing to accept that some characteristics are "selectively neutral." For example, he says (26) of the sterile dark red flower at the center of the umbel of the wild carrot: "That the modified central flower is of no functional importance to the plant is almost certain." Indeed, Darwin has been chided by Cain (8) for too readily accepting Owen's argument that the homology between bones of limbs of different vertebrates is nonadaptive. For Darwin the argument was welcome, because the resemblance could then be taken as evidence for genetic relationship (or, presumably, for a paucity of imagination on the part of the creator). But Cain points out that the homology would not have been preserved if it were not adaptive.

Biologists differ greatly in the extent to which they expect to find a detailed fit between structure and function. It may be symptomatic of the times that when, in conversation, I raised Darwin's example of the carrot, two different functional explanations were at once suggested. I suspect that these explanations were fanciful. But however much one may be in doubt about the function of the antlers of the irish elk or the tail of the peacock, one can hardly suppose them to be selectively neutral. In general, the structural and behavioral traits chosen for functional analysis are of a kind that rules out neutrality as a plausible explanation. Curio (23) makes the valid point that the ampullae of Lorenzini in elasmobranchs were studied for many years before their role in enabling a fish to locate prey buried in the mud was demonstrated (40), yet the one hypothesis that was never entertained was that the organ was functionless. The same could be said of Curio's own work (24) on the function of mobbing in birds; behavior so widespread, so constant, and so apparently dangerous calls for a functional explanation.

There are, however, exceptions to the rule that functional investigations are carried out with the aim of identifying particular selective forces, and not of demonstrating that traits are adaptive. The work initiated by Cain and Sheppard (9) on shell color and banding in *Cepaea* was in part aimed at refuting the claim that the variation was selectively neutral and explicable by genetic drift. To that extend the work was aimed at demonstrating adaptation as such; it is significant, however, that the work has been most successful when it has been possible to identify a particular selection pressure (e.g., predation by thrushes).

At present, of course, the major argument between neutral and selective theories concerns enzyme polymorphism. I cannot summarize the argument here, but a few points on methodology are relevant. The argument arose because of the formulation by Kimura (43) and King and Jukes (44) of the "neutral" hypothesis; one reason for proposing it was the difficulty of accounting for the extensive variation by selection. Hence the stimulus was quite different from that prompting most functional investigations; it was the existence of widespread variation in a trait of no obvious selective significance.

The neutral hypothesis is a good "Popperian" one; if it is false, it should be possible to show it. In contrast, the hypothesis of adaptation is virtually irrefutable. In practice, however, the statistical predictions of the neutral theory depend on so many unknowns (mutation rates, the past history of population number and structure, hitch-hiking from other loci) that it has proved hard to test (53). The difficulties have led some geneticists (e.g., 14) to propose that the only way in which the matter can be settled is by the classical methods of ecological genetics—i.e., by identifying the specific selection pressures associated with particular enzyme loci. The approach has had some success but is always open to the objection that the loci for which the neutral hypothesis has been falsified are a small and biased sample.

In general, then, the problems raised by the neutral mutation theory and by optimization theory are wholly different. The latter is concerned with traits that differ between species and that can hardly be selectively neutral but whose selective significance is not fully understood.

A more serious difficulty for optimization theory is the occurrence of maladaptive traits. Optimization is based on the assumption that the population is adapted to the contemporary environment, whereas evolution is a process of continuous change. Species lag behind a changing environment. This is particularly serious when studying species in an environment that has recently been drastically changed by man. For example, Lack (48) argued that the number of eggs laid by a bird maximizes the number of surviving young. Although there is much supporting evidence, there are some apparent exceptions. For example, the gannet *Sula bassana* lays a single egg. Studying gannets on the Bass Rock, Nelson (71) found that if a second egg is added, the pair can successfully raise two young. The explanation can hardly be a lack of genetic variability, because species nesting in the Humboldt current off Peru lay two or even three eggs and successfully raise the young.

Lack (48) suggests that the environment for gannets may recently have improved, as evidenced by the recent increase in the population on the Bass Rock. Support for this interpretation comes from the work of Jarvis (39) on the closely related *S. capensis* in South Africa. This species typically lays one egg, but 1 percent of nests contain two. Using methods similar to Nelson's, Jarvis found that a pair can raise two chicks to fledgings, but that the average weight of twins was lower than singles, and in each nest one twin was always considerably lighter than its fellow. There is good evidence that birds fledging below the average weight are more likely to die soon after. Difficulties of a similar kind arise for the glaucous gull (see 45).

The undoubted existence of maladaptive traits, arising because evolutionary change is not instantaneous, is the most serious obstacle to the testing of optimization theories. The difficulty must arise; if species were perfectly adapted, evolution would cease. There is no easy way out. Clearly a wholesale reliance on evolutionary lag to save hypotheses that would otherwise be falsified would be fatal to the whole research program. The best we can do is to invoke evolutionary lag sparingly, and only when there are independent grounds for believing that the environment has changed recently in a relevant way.

What then is the status of the concept of adaptation? In the strong form —that all organs are perfectly adapted—it is clearly false; the vermiform appendix is sufficient to refute it. For Darwin, adaptation was an obvious fact that required an explanation; this still seems a sensible point of view. Adaptation can also be seen as a necessary consequence of natural selection. The latter I regard as a refutable scientific theory (60); but it must be refuted, if at all, by genetic experiment and not by the observation of complex behavior.

CRITIQUES OF OPTIMIZATION THEORY

Lewontin (55) raises a number of criticisms, which I discuss in turn.

Do Organs Solve Problems?

Most organs have many functions. Therefore, if a hypothesis concerning function fails correctly to predict behavior, it can always be saved by proposing an additional function. Thus hypotheses become irrefutable and metaphysical, and the whole program merely a test of ingenuity in conceiving possible functions. Three examples follow: the first is one used by Lewontin.

Orians and Pearson (73) calculated the optimal food item size for a bird, on the assumption that food intake is to be maximized. They found that the items diverged from random in the expected direction, but did not fit the prediction quantitatively. They explained the discrepancy by saying that a bird must visit its nest frequently to discourage predators. Lewontin (54) comments:

This is a paradigm for adaptive reconstruction. The problem is originally posed as efficiency for food-gathering. A deviation of behavior from random, in the direction predicted, is regarded as strong support for the adaptive explanation of the behavior and the discrepancy from the predicted optimum is accounted for by an ad hoc secondary problem which acts as a constraint on the solution to the first.... By allowing the theorist to postulate various combinations of "problems" to which manifest traits are optimal "solutions", the adaptationist programme makes of adaptation a metaphysical postulate, not only incapable of refutation, but necessarily confirmed by every observation. This is the caricature that was immanent in Darwin's insight that evolution is the product of natural selection.

It would be unfair to subject Orians alone to such criticism, so I offer two further examples from my own work.

First, as explained earlier, Maynard Smith and Savage (66) predicted qualitative features of mammalian gaits. However, their model failed to give a correct quantitative prediction. I suspect that if the model were modified to allow for wind resistance and the visco-elastic properties of muscle, the quantitative fit would be improved; at present, however, this is pure speculation. In fact, it looks as if a model that gives quantitiatively precise predictions will be hard to devise (1).

Second, Maynard Smith and Parker (64) predicted that populations will vary in persistence or aggressiveness in contest situations, but that individuals will not indicate their future behavior by varying levels of intensity of display. Rohwer (84) describes the expected variability in aggressivity in the Harris sparrow in winter flocks, but also finds a close correlation between aggressivity and a signal (amount of black in the plumage). I could point to the first observation as a confirmation of our theory, and explain how, by altering the model (by changing the phenotype set to permit the detection of cheating), one can explain the second.

What these examples, and many others, have in common is that a model gives predictions that are in part confirmed by observation but that are contradicted in some important respect. I agree with Lewontin that such discrepancies are inevitable if a simple model is used, particularly a model that assumes each organ or behavior to serve only one function. I also agree that if the investigator adds assumptions to his model to meet each discrepancy, there is no way in which the hypothesis of adaptation can be refuted. But the hypothesis of adaptation is not under test.

What is under test is the specific set of hypotheses in the particular model. Each of the three example models above has been falsified, at least as a complete explanation of these particular data. But since all have had qualitative success, it seems quite appropriate to modify them (e.g., by allowing for predation, for wind resistance, for detection of cheating). What is not justified is to modify the model and at the same time to claim that the model is confirmed by observation. For example, Orians would have to show that his original model fits more closely in species less exposed to predation. I would have to show that Rohwer's data fit the "mixed ESS" model in other ways—in particular, that the fitness of the different morphs is approximately equal. If, as may well be the case, the latter prediction of the ESS model does not hold, it is hard to see how it could be saved.

If the ESS model proves irrelevant to the Harris sparrow, it does not follow, however, that it is never relevant. By analogy, the assertion is logically correct that there will be a stable polymorphism if the heterozygote at a locus with two alleles is fitter than either homozygote. The fact that there are polymorphisms not maintained by heterosis does not invalidate the logic. The (difficult) empirical question is whether polymorphisms are often maintained by heterosis. I claim a similar logical status for the prediction of a mixed ESS.

In population biology we need simple models that make predictions that hold qualitatively in a number of cases, even if they are contradicted in detail in all of them. One can say with some confidence, for example, that no model in May's *Stability and Complexity in Model Ecosystems* describes exactly any actual case, because no model could ever include all relevant features. Yet the models do make qualitative predictions that help to explain real ecosystems. In the analysis of complex systems, the best we can hope for are models that capture some essential feature.

To summarize my comments on this point, Lewontin is undoubtedly right to complain if an optimizer first explains the discrepancy between theory and observation by introducing a new hypothesis, and then claims that his modified theory has been confirmed. I think he is mistaken in supposing that the aim of optimization theories is to confirm a general concept of adaptation.

Is There Genetic Variance?

Natural selection can optimize only if there is appropriate genetic variance. What justification is there for assuming the existence of such variance? The main justification is that, with rare exceptions, artificial selection has always proved effective, whatever the organism or the selected character (53).

A particular difficulty arises because genes have pleiotropic effects, so that selection for trait A may alter trait B; in such cases, any attempt to explain the changes in B in functional terms is doomed to failure. There are good empirical grounds for doubting whether the difficulty is as serious as might be expected from the widespread nature of pleiotropy. The point can best be illustrated by a particular example. Lewontin (54) noted that in primates there is a constant allometric relationship between tooth size and body size. It would be a waste of time, therefore, to seek a functional explanation of the difference between the tooth size of the gorilla and of the rhesus monkey, since the difference is probably a simple consequence of the difference in body size.

It is quite true that for most teeth there is a constant allometric relationship between tooth and body size, but there is more to it than that (36). The canine teeth (and the teeth occluding with them) of male primates are often larger than those of females, even when allowance has been made for the difference in body size. This sex difference is greater in species in which males compete for females than in monogamous species, and greater in ground-living species (which are more exposed to predation) than in arboreal ones. Hence, there is sex-limited genetic variance for canine tooth size, independent of body size, and the behavioral and ecological correlations suggest that this variance has been the basis of adaptation. It would be odd if there were tooth-specific, sex-limited variance, but no variance for the relative size of the teeth as a whole. However, there is some evidence for the latter. The size of the cheek teeth in females (relative to the size predicted from their body size) is significantly greater in those species with a higher proportion of leaves (as opposed to fruit, flowers, or animal matter) in their diets.

Thus, although at first sight the data on primate teeth suggest that there may be nothing to explain in functional terms, a more detailed analysis presents quite a different picture. More generally, changes in allometric relationships can and do occur during evolution (30).

I have quoted Lewontin as a critic of adaptive explanation, but it would misinterpret him to imply that he rejects all such explanations. He remarks (54) that "the serious methodological difficulties in the use of adaptive arguments should not blind us to the fact that many features of organisms are adaptations to obvious environmental 'problems,'" He goes on to argue that if natural selection is to produce adaptation, the mapping of character states into fitnesses must have two characteristics: "continuity" and "quasi-independence." By continuity is meant that small changes in a character result in small changes in the ecological relations of the organism; if this were not so, it would be hard to improve a character for one role without ruining it for another. By quasi-independence is meant that the developmental paths are such that a variety of mutations may occur, all with the same effect on the primary character, but with different effects on other characters. It is hard to think of better evidence for quasi-independence than the evolution of primate canines.

To sum up this point, I accept the logic of Lewontin's argument. If I differ from him (and on this point he is his own strongest critic), it is in thinking that genetic variance of an appropriate kind will usually exist. But it may not always do so.

It has been an implicit assumption of optimization models that the optimal phenotype can breed true. There are two kinds of reasons why this might not be true. The first is that the optimal phenotype may be produced by a heterozygote. This would be a serious difficulty if one attempted to use optimization methods to analyze the genetic structure of populations, but I think that would be an inappropriate use of the method. Optimization models are useful for analyzing phenotypic evolution, but not the genetic structuring of populations. A second reason why the optimal phenotype may not breed true is more serious: the evolutionarily stable population may be phenotypically variable. (This point is discussed further in the section on Games, below).

The assumption concerning the phenotype set is based on the range of variation observable within species, the phenotypes of related species, and on plausible guesses at what phenotypes might arise under selection. It is rare to have any information on the genetic basis of the phenotypic variability. Hence, although it is possible to introduce specific genetic assumptions into optimization models (e.g., 2, 89), this greatly complicates the analysis. In general, the assumption of "breeding true" is reasonable in particular applications; models in which genes appear explicitly need to be analyzed to decide in what situations the assumption may mislead us.

The Effects of History

If, as Wright (101) suggested, there are different "adaptive peaks" in the genetic landscape, then depending on initial conditions, different populations faced with identical "problems" may finish up in different stable states. Such divergence may be exaggerated if evolution takes the form of a "game" in which the optimal phenotype for one individual depends on what others are doing (see the section on Games, below). An example is Fisher's (28) theory of sexual selection, which can lead to an "autocatalytic" exaggeration of initially small differences. Jacob (38) has recently emphasized the importance of such historical accidents in evolution.

As an example of the difficulties that historical factors can raise for functional explanations, consider the evolution of parental care. A simple game-theory model (62) predicts that for a range of ecological parameters either of two patterns would be stable: male parental care only, or female care only. Many fish and amphibia show one or the other of these patterns. At first sight, the explanation of why some species show one pattern and others the other seems historical; the reasons seem lost in an unknown past. However, things may not be quite so bad. At a a recent discussion of fish behavior at See-Wiesen the suggestion emerged that if uniparental care evolved from no parental care, it would be male care, whereas if it evolved from biparental care it would be female care. This prediction is plausible in the light of the original game-theory model, although not a necessary consequence of it. It is, however, testable by use of the comparative data; if it is true, male care should occur in families that also include species showing no care, and female care in families that include species showing biparental care. This may not prove to be the case; the example is given to show that even if there are alternative adaptive peaks, and in the absence of a relevant fossil record, it may still be possible to formulate testable hypotheses.

What Optimization Criterion Should One Use?

Suppose that, despite all difficulties, one has correctly identified the "problem." Suppose, for example, that in foraging it is indeed true that an animal should maximize E, its rate of energy intake. We must still decide in what circumstances to maximize E. If the animal is alone in a uniform environment, no difficulty arises. But if we allow for competition and for a changing environment, several choices of optimization procedure are possible. For example, three possibilities arise if we allow just for competition:

1. The "maximum" solution: Each animal maximizes E on the assumption that other individuals behave in the least favorable way for it.

2. The "Pareto" point: The members of the population behave so that no individual can improve its intake without harming others.

3. The ESS: The members of the population adopt feeding strategy I such that no mutant individual adopting a strategy other than I could do better than typical members.

These alternatives are discussed further in the section on Games, below. For the moment, it is sufficient to say that the choice among them is not arbitrary, but follows from assumptions about the mode of inheritance and the population structure. For individual selection and parthenogenetic inheritance, the ESS is the appropriate choice.

Lewontin's criticism would be valid if optimizers were in the habit of assuming the truth of what Haldane once called "Pangloss's theorem," which asserts that animals do those things that maximize the chance of survival of their species. If optimization rested on Pangloss's theorem it would be right to reject it. My reason for thinking that Lewontin regards optimization and Pangloss's theorem as equivalent is that he devotes the last section of his paper to showing that in *Drosophila* a characteristic may be established by individual selection and yet may reduce the competitive ability of the population relative to others. The point is correct and important, but in my view does not invalidate most recent applications of optimization.

THE METHODOLOGY OF TESTING

The crucial hypothesis under test is usually that the model correctly incorporates the selective forces responsible for the evolution of a trait. Optimization models sometimes make fairly precise quantitiative predictions that can be tested. However, I shall discuss the question how functional explanations can be tested more generally, including cases in which the predictions are only qualitative. It is convenient to distinguish comparative, quantitative, and individual-variation methods.

Comparative Tests

Given a functional hypothesis, there are usually testable predictions about the development of the trait in different species. For example, two main hypotheses have been proposed to account for the greater size of males in many mammalian species: It is a consequence of competition among males for females; or it arises because the two sexes use different resources. If the former hypothesis is true, dimorphism should be greater in harem-holding and group-living species, whereas if the latter is true it should be greater in monogamous ones, and in those with a relatively equal adult sex ratio.

Clutton-Brock et al. (16) have tested these hypotheses by analyzing 42 species of primates (out of some 200 extant species) for which adequate breeding data are available. The data are consistent with the sexual selection hypothesis, and show no sign of the trend predicted by the resource differentiation hypothesis. The latter can therefore be rejected, at least as a major cause of sexual dimorphism in primates. It does not follow that intermale competition is the only relevant selective factor (82). Nor do their observations say anything about the causes of sexual dimorphism in other groups. It is interesting (though not strictly relevant at this point) that the analysis also showed a strong correlation between female body size and degree of dimorphism. This trend, as was first noted by Rensch (83), occurs in a number of taxa, but has never received an entirely satisfactory explanation.

The comparative method requires some criterion for inclusion of species. This may be purely taxonomic (e.g., all primates, all passerine birds), or jointly taxonomic and geographic (e.g., all African ungulates, all passerines in a particular forest). Usually, some species must be omitted because data are not available. Studies on primates can include a substantial proportion of extant species (16, 68); in contrast, Schoener (86), in one of the earliest studies of this type, included all birds for which data were available and which also met certain criteria of territoriality, but he had to be content with a small fraction of extant species. It is therefore important to ask whether the sample of species is biased in ways likely to affect the hypothesis under test. Most important is that there be some criterion of inclusion, since otherwise species may be included simply because they confirm (or contradict) the hypothesis under test.

Most often, limitations of data will make it necessary to impose both taxonomic and geographic criteria. This need not prevent such data from being valuable, either in generating or in testing hypotheses; examples are analyses of flocking in birds (7, 31) and of breeding systems in forest plants (3, 4).

A second kind of difficulty concerns the design of significance test. Different species cannot always be treated as statistically independent. For example, all gibbons are monogamous, and all are arboreal and frugivorous, but since all may be descended from a single ancestor with these properties, they should be treated as a single case in any test of association (not that any is suspected). To take an actual example of this difficulty, Lack (49) criticized Verner and Willson's (98) conclusion that polygamy in passerines is associated with marsh and prairie habitats on the grounds that many of the species concerned belong to a single family, the Icteridae.

Statistical independence and other methodological problems in analyzing comparative data are discussed by Clutton-Brock and Harvey (17). In analyzing the primate data, they group together as a single observation all congeneric species belonging to the same ecological category. This is a conservative procedure, in that it is unlikely to find spurious cases of statistical significance. Their justification for treating genera, but not families, as units is that for their data there are significant differences between genera within families for seven of the eight ecological and behavioral variables, but significant additional variation between families for only two of them. It may be, however, that a more useful application of statistical methods is their use (17) of partial regression, which enables them to examine the effects of a particular variable when the effects of other variables have been removed, and to ask how much of the total variation in some trait is accounted for by particular vaiables.

Quantitative Tests

Quantitative tests can be illustrated by reference to some of the predictions of foraging theory. Consider first the problem of optimal diet. The following model situation has been widely assumed. There are a number of different kinds of food items. An animal can search simultaneously for all of them. Each item has a characteristic food value and "handling time" (the time taken to capture and consume it). For any given set of densities and hence frequencies of encounter, the animal must only decide which items it should consume and which ignore.

Pyke et al. (81) remark that no fewer than eight authors have independently derived the following basic result. The animal should rank the items in order of $V =$ food value/handling time. Items should be added to the diet in rank order, provided that for each new item the value of V is greater than the rate of food intake for the diet without the addition. This basic result leads to three predictions:

1. Greater food abundance should lead to greater specialization. This qualitative prediction was first demonstrated by Ivlev (37) for various fish species in the laboratory, and data supporting it have been reviewed by Schoener (87). Curio (25) quotes a number of cases that do not fit.

2. For fixed densities, a food type should either be always taken, or never taken.

3. Whether a food item should be taken is independent of its density, and depends on the densities of food items of higher rank.

Werner and Hall (100) allowed blue-gill sunfish to feed on *Daphnia* of three different size classes; the diets observed agreed well with the predictions of the model. Krebs et al. (47) studies great tits foraging for parts of mealworms on a moving conveyor belt. They confirmed prediction 3 but not 2; that is, they found that whether small pieces were taken was independent of the density of small pieces, but, as food abundance rose, small pieces were dropped only gradually from the diet. Goss-Custard (29) has provided field evidence confirming the model from a study of redshank feeding on marine worms of different sizes, and Pulliam (80) has confirmed it for chipping sparrows feeding on seeds.

Turning to the problem of how long an animal should stay in a patch before moving to another, there is again a simple prediction, which Charnov (10) has called the "Marginal Value Theorem" (the same theorem was derived independently by Parker and Stuart [77] in a different context). It asserts that an animal should leave a patch when its rate of intake in the patch (its "marginal" rate) drops to the average rate of intake for the habitat as a whole. It is a corollary that the marginal rate should be the same for all patches in the habitat. Two laboratory experiments on tits (20, 46) agree well with the prediction.

A more general problem raised by these experiments is discussed by Pyke et al. (81). How does an animal estimate the parameters it needs to know before it can perform the required optimization? How much time should it spend acquiring information? Sometimes these questions may receive a simple answer. Thus the results of Krebs et al. (46) suggest that a bird leaves a patch if it has not found an item of food for some fixed period r (which varied with the overall abundance of food). The bird seems to be using r, or rather $1/r$, as an estimate of its marginal capture rate. But not all cases are so simple.

Individual Variation

The most direct way of testing a hypothesis about adaptation is to compare individuals with different phenotypes, to see whether their fitnesses vary in the way predicted by the hypothesis. This was the basis of Kettlewell's (42) classic demonstration of selection on industrial melanism in moths. In principle, the individual differences may be produced by experimental interference (Curio's [23] "method of altering a character") or they may be genetic or of unknown origin (Curio's "method of variants"). Genetic differences are open to the objection that genes have pleiotropic effects, and occasionally are components of supergenes in which several closely linked loci affecting the same function are held in linkage disequilibrium, so that the phenotypic difference responsible for the change in fitness may not be the one on which attention is concentrated. This difficulty, however, is trivial compared to that which arises when two species are compared.

The real difficulty in applying this method to behavioral differences is that suitable individual differences are often absent and experimental interference is impractical. Although it is hard to alter behavior experimentally, it may be possible to alter its consequences. Tinbergen et al. (94) tested the idea that gulls remove egg shells from the nest because the shells attract predators to their eggs and young; they placed egg shells close to eggs and recorded a higher predation rate.

However, the most obvious field of application of this method arises when a population is naturally variable. Natural variation in a phenotype may be maintained by frequency-dependent selection; in game-theoretical terms, the stable state may be a mixed strategy. If a particular case of phenotypic variability (genetic or not) is thought to be maintained in this way, it is important to measure the fitnesses of individuals with different phenotypes. At a mixed

ESS (which assumes parthenogenetic inheritance) these fitnesses are equal; with sexual reproduction, exact equality is not guaranteed, but approximate equality is a reasonable expectation (91). If the differences are not genetic, we still expect a genotype to evolve that adopts the different strategies with frequencies that equalize their payoffs.

The only test of this kind known to me is Parker's (76) measurement of the mating success of male dungflies adopting different strategies. His results are consistent with a "mixed ESS" interpretation; it is not known whether the differences are genetic. The importance of tests of this kind lies in the fact that phenotypic variability can have other explanations; for example, it may arise from random environmental effects, or from genes with heterotic effects. In such cases, equality of fitness between phenotypes is not expected.

MATHEMATICAL APPROACHES TO OPTIMIZATION

During the past twenty years there has been a rapid development of mathematical techniques aimed at solving problems of optimization and control arising in economics and engineering. These stem from the concepts of "dynamic programming" (5) and of the "maximum principle" (79). The former is essentially a computer procedure to seek the best control policy in particular cases without the hopelessly time-consuming task of looking at every possibility. The latter is an extension of the classic methods of the calculus of variations that permits one to allow for "inequality" constraints on the state and control variables (e.g., in the resource allocation model discussed below, the proportion u of the available resources allocated to seeds must obey the constraint $u < 1$).

This is not the place to describe these methods, even if I were competent to do so. Instead, I shall describe the kinds of problems that can be attacked. If a biologist has a problem of one of these kinds, he would do best to consult a mathematician. For anyone wishing to learn more of the mathematical background, Clark (12) provides an excellent introduction.

I discuss in turn "optimization," in which the problem is to choose an optimal policy in an environment without competitors; "games," in which the environment includes other "players" who are also attempting to optimize something; and "games of inclusive fitness," in which the "players" have genes in common. I shall use as an illustration the allocation of resources between growth and reproduction.

Optimization

Choice of a Single Value   The simplest type of problem, which requires for its solution only the technique of differentiation, is the choice of a value for a single parameter. For example, in discussing the evolution of gaits, Maynard Smith and Savage (66) found an expression for P, the power output, as a function of the speed V, of size S, and of J, the fraction of time for which all four legs are off the ground. By solving the equation $dP/dJ = 0$, an equation

$J = f(V, S)$ was obtained, describing the optimum gait as a function of speed and size.

Few problems are as simple as this, but some more complex cases can be reduced to problems of this kind, as will appear below.

A Simple Problem in Sequential Control Most optimization theory is concerned with how a series of sequential decisions should be taken. For example, consider the growth of an annual plant (19, 69). The rate at which the plant can accumulate resources depends on its size. The resources can be allocated either for further growth, or to seeds, or divided between them. For a fixed starting size and length of season, how should the plant allocate its resources so as to maximize the total number of seeds produced?

In this problem the "state" of the system at any time is given simply by the plant's size, x; the "control variable" $u(t)$ is the fraction of the incoming resource allocated to seeds at time t; the "constraints" are the initial size, the length of the season, the fact that $u(t)$ must lie between 0 and 1, and the "state equation,"

$$dx/dt = F[x(t), u(t)], \tag{1}$$

which describes how the system changes as a function of its state and of the control variable.

If equation 1 is linear in u, it can be shown that the optimal control is "bang-bang"—that is, $u(t) = 0$ up to some critical time t^*, and subsequently $u(t) = 1$. The problem is thus reduced to finding the single value, t^*. But if equation 1 is nonlinear, or has stochastic elements, the optimal control may be graded.

More Complex Control Problems Consider first the "state" of the system. This may require description by a vector rather than by a single variable. Thus suppose the plant could also allocate resources to the production of toxins that increased its chance of survival. Then its state would require measures of both size and toxicity. The state description must be sufficient for the production of a state equation analogous to equation 1. The state must also include any information used in determining the control function $u(t)$. This is particularly important when analyzing the behavior of an animal that can learn. Thus suppose that an animal is foraging and that its decisions on whether to stay in a given patch or to move depend on information it has acquired about the distribution of food in patches; then this information is part of the state of the animal. For a discussion, see (20).

Just as the state description may be multidimensional, so may the control function; for example, for the toxic plant the control function must specify the allocation both to seeds and to toxins.

The state equation may be stochastic. Thus the growth of a plant depends on whether it rains. A plant may be supposed to "know" the probability of rain (i.e., its genotype may be adapted to the frequency of rain in previous generations) but not whether it will actually rain. In this case, a stochastic state equation may require a graded control. This connection between stochasticity and a "compromise" response as opposed to an all-or-none one is a common feature of optimal control. A second example is the analysis by Oster and Wilson (75) of the optimal division into castes in social insects: A predictable environment is likely to call for a single of worker, while an uncertain one probably calls for a division into several castes.

Reverse Optimality   McFarland (67) has suggested an alternative approach. The typical one is to ask how an organism should behave in order to maximize its fitness. Mathematically, this requires that one define an "objective function" that must be maximized" ("objective" here means "aim" or "goal"); in the plant example, the objective function is the number of seeds produced, expressed as a function of x and $u(t)$. But a biologist may be faced with a different problem. Suppose that he knew, by experiment, how the plant actually allocates its resources. He could then ask what the plant is actually maximizing. If the plant is perfectly adapted, the objective function so obtained should correspond to what Sibly and McFarland (88) call the "cost function"—that is, the function that should be maximized if the organism is maximizing its fitness. A discrepancy would indicate maladaptation.

There are difficulties in seeing how this process of reverse optimality can be used. Given that the organism's behavior is "consistent" (i.e., if it prefers A to B and B to C, it prefers A to C), it is certain that its behavior maximizes *some* objective function; in general there will be a set of functions maximized. Perfect adaptation then requires only that the cost function correspond to one member of this set. A more serious difficulty is that it is not clear what question is being asked. If a discrepancy is found, it would be hard to say whether this was because costs had been wrongly measured or because the organism was maladapted. This is a particular example of my general point that it is not sensible to test the hypothesis that animals optimize. But it may be that the reverse optimality approach will help to analyze how animals in fact make decisions.

Games

Optimization of the kind just discussed treats the environment as fixed, or as having fixed stochastic properties. It corresponds to that part of population genetics that assumes fitnesses to be independent of genotype frequencies. A number of selective processes have been proposed as frequency-dependent, including predation (13, 70) and disease (15, 32). The maintenance of polymorphism in a varied environment (50) is also best seen as a case of frequency-dependence (59). The concept can be applied directly to phenotypes.

The problem is best formulated in terms of the theory of games, first developed (99) to analyze human conflicts. The essence of a game is that the best strategy to adopt depends on what one's opponent will do; in the context of evolution, this means that the fitness of a phenotype depends on what others are present; i.e., fitnesses are frequency dependent.

The essential concepts are those of a "strategy" and a "payoff matrix." A strategy is a specification of what a "player" will do in every situation in which it may find itself; in the plant example, a typical strategy would be to allocate all resources to growth for twenty days, and then divide resources equally between growth and seeds. A strategy may be "pure" (i.e., without chance elements) or "mixed" (i.e., of the form "do A with probability p and B with probability $1 - p$," where A and B are pure strategies).

The "payoff" to an individual adopting strategy A in competition to one adopting B is written $E(A, B)$, which expresses the expected *change* in the fitness of the player adopting A if his opponent adopts B. The evolutionary model is then of a population of individuals adopting different strategies. They pair off at random, and their fitnesses change according to the payoff matrix. Each individual then produces offspring identical to itself, in numbers proportional to the payoff it has accumulated. Inheritance is thus parthenogenetic, and selection acts on the population is infinite, so that the chance of meeting an opponent adopting a particular strategy is independent of one's own strategy.

The population will evolve to an evolutionarily stable strategy, or ESS, if one exists (64). An ESS is a strategy that, if almost all individuals adopt it, no rare mutant can invade. Thus let I be an ESS, and J a rare mutant strategy of frequency $p \ll 1$. Writing the fitnesses of I and J as $W(I)$ and $W(J)$,

$$W(I) = C + (1 - p)E(I, I) + pE(I, J);$$

$$W(J) = C + (1 - p)E(J, I) + pE(J, J).$$

In these equations C is the fitness of an individual before engaging in a contest. Since I is an ESS, $W(I) > W(J)$ for all $J \neq I$; that is, remembering that p is small, either

$$E(I, I) > E(J, I), \text{ or}$$

$$E(I, I) = E(J, I) \text{ and } E(I, J) > E(J, J). \tag{2}$$

These conditions (expressions 2) are the definition of an ESS.

Consider the matrix in table 5.1. For readers who prefer a biological interpretation, A is "Hawk" and B is "Dove"; thus A is a bad strategy to adopt against A, because of the risk of serious injury, but a good strategy to adopt against B, and so on.

Table 5.1. Payoff matrix for a game

	Player 2	
	A	B
Player 1		
A	1	5
B	2	4

Note: The values in the matrix give the payoff to Player 1.

The game has no pure ESS, because $E(A, A) < E(B, A)$ and $E(B, B) < E(A, B)$. It is easy to show that the mixed strategy—playing A and B with equal probability—is an ESS. It is useful to compare this with other "solutions," each of which has a possible biological interpretation:

The Maximin Solution This is the pessimist's solution, playing the strategy that minimizes your losses if your opponent does what is worst for you. For our matrix, the maximin strategy is always to play B. Lewontin (52) suggested that this strategy is appropriate if the "player" is a species and its opponent nature: The species should minimize its chance of extinction when nature does its worst. This is the "existential game" of Slobodkin and Rapoport (92). It is hard to see how a species could evolve this strategy, except by group selection. (Note that individual selection will not necessarily minimize the chance of death: A mutant that doubled the chance that an individual would die before maturity, but that quadrupled its fecundity if it did survive, would increase in frequency.)

The Nash Equilibrium This is a pair of strategies, one for each player, such that neither would be tempted to change his strategy as long as the other continues with his. If in our matrix, player 1 plays A and 2 plays B, we have a Nash equilibrium; this is also the case if 1 plays B and 2 plays A. A population can evolve to the Nash point if it is divided into two classes, and if members of one class compete only with members of the other. Hence it is the appropriate equilibrium in the "parental investment" game (62), in which all contests are between a male and a female. The ESS is subject to the added constraint that both players must adopt the same strategy.

The Group Selection Equilibrium If the two players have the same genotype, genes in either will be favored that maximize the sum of their payoffs. For our matrix both must play strategy B. The problem of the stable strategy when the players are related but not identical is discussed in the section on Games between Relatives, below.

It is possible to combine the game-theoretical and optimization approaches. Mirmirani and Oster (69) make this extension in their model of resource allocation in plants. They ask two questions. What is the ESS for a plant growing in competition with members of its own species? What is the ESS when two species compete with one another?

Thus consider two competing plants whose sizes at time t are P_1 and P_2. The effects of competition are allowed for by writing

$$dP_1/dt = (r_1 - e_1 P_2)(1 - u_1)P_1,$$
$$dP_2/dt = (r_2 - e_2 P_1)(1 - u_2)P_2, \tag{3}$$

where u_1 and u_2 are the fractions of the available resources allocated to seeds. Let $J_1[u_1(t), u_2(t)]$ be the total seed production of plant 1 if it adopts the allocation strategy $u_1(t)$ and its competitor adopts $u_2(t)$. Mirmirani and Oster seek a stable pair of strategies $u_1{}^*(t)$, $u_2{}^*(t)$, such that

$$J_1[u_1(t), u_2{}^*(t)] \leqslant J_1[u_1{}^*(t), u_2{}^*(t)], \text{ and}$$

$$J_2[u_1{}^*(t), u_2(t)] \leqslant J_2[u_1{}^*(t), u_2{}^*(t)]. \tag{4}$$

That is, they seek a Nash equilibrium, such that neither competitor could benefit by unilaterally altering its strategy. They find that the optimal strategies are again "bang-bang," but with earlier switching times than in the absence of competition. Strictly, the conditions indicated by expressions 4 are correct only when there is competition between species, and when individuals of one species compete only with individuals of the other; formally this would be so if the plants grew alternately in a linear array. The conditions indicated by expressions 4 are not appropriate for intraspecific competition, since they permit $u_1{}^*(t)$ and $u_2{}^*(t)$ to be different, which could not be the case unless individuals of one genotype competed only with individuals of the other. For intraspecific competition ($r_1 = r_2, e_1 = e_2$), the ESS is given by

$$J_1[u_1(t), u_1{}^*(t)] \leqslant J_1[u_1{}^*(t), u_1{}^*(t)]. \tag{5}$$

As it happens, for the plant growth example equations 4 and 5 give the same control function, but in general this need not be so.

The ESS model assumes parthenogenetic inheritance, whereas most interesting populations are sexual. If the ESS is a pure stategy, no difficulty arises; a genetically homogeneous sexual population adopting the strategy will also be stable. If the ESS is a mixed strategy that can be achieved by a single individual with a variable behavior, there is again no difficulty. If the ESS is a mixed one that can be achieved only by a population of pure strategists in the appropriate frequencies, two difficulties arise:

1. Even with the parthenogenetic model, the conditions expressed in expressions 2 do not guarantee stability. (This was first pointed out to me by Dr. C. Strobeck.) In such cases, therefore, it is best to check the stability of the equilibrium, if necessary by simulation; so far, experience suggests that stability, although not guaranteed, will usually be found.

2. The frequency distribution may be one that is incompatible with the genetic mechanism. This difficulty, first pointed out by Lewontin (52), has recently been investigated by Slatkin (89–91) and by Auslander et al. (2). It is hard to say at present how serious it will prove to be; my hope is that a sexual population will usually evolve a frequency distribution as close to the ESS as its genetic mechanism will allow.

Games between Relatives

The central concept is that of "inclusive fitness" (33). In classical population genetics we ascribe to a genotype I a "fitness" W, corresponding to the expected number of offspring produced by I. If, averaged over environments and genetic backgrounds, the effect of substituting allele A for a is to increase W, allele A will increase in frequency. Following Oster et al. (74) but ignoring unequal sex ratios, Hamilton's proposal is that we should replace W_i by the inclusive fitness, Z_i, where

$$Z_i = \sum_{j=1}^{R} r_{ij} W_j, \tag{6}$$

where the summation is over all R relatives of I; r_{ij} is the fraction of J's genome that is identical by descent to alleles in I; and W_j is the expected number of offspring of the jth relative of I. (If $J = I$, then equation 6 refers to the component of inclusive fitness from an individual's own offspring.)

An allele A will increase in frequency if it increases Z rather than just W. Three warnings are needed:

1. It is usual to calculate r_{ij} from the pedigree connecting I and J (as carried out, for example, by Malécot (58)). However, if selection is occurring, r_{ij} so estimated is only approximate, as are predictions based on equation 6 (35).

2. Some difficulties arose in calculating appropriate values of r_{ij} for haplodiploids; these were resolved by Crozier (22).

3. If the sex ratio is not unity, additional difficulties arise (74).

Mirmirani and Oster (69) have extended their plant-growth model along these lines to cover the case when the two competitors are genetically related. They show that as r increases, the switching time becomes earlier and the total yield higher.

CONCLUSION

The role of optimization theories in biology is not to demonstrate that organisms optimize. Rather, they are an attempt to understand the diversity of life.

Three sets of assumptions underlie an optimization model. First, there is an assumption about the kinds of phenotypes or strategies possible (i.e., a "phenotype set"). Second, there is an assumption about what is being maximized; ideally this should be the inclusive fitness of the individual, but often one must be satisfied with some component of fitness (e.g., rate of energy intake while foraging). Finally, there is an assumption, often tacit, about the mode of inheritance and the population structure; this will determine the type of equilibrium to which the population will move.

In testing an optimization model, one is testing the adequacy of these hypotheses to account for the evolution of the particular structures or patterns of behavior under study. In most cases the hypothesis that variation in the relevant phenotypes is selectively neutral is not a plausible alternative, because of the nature of the phenotypes chosen for study. However, it is often a plausible alternative that the phenotypes are not well adapted to current circumstances because the population is lagging behind a changing environment; this is a serious difficulty in testing optimization theories.

The most damaging criticism of optimization theories is that they are untestable. There is a real danger that the search for functional explanations in biology will degenerate into a test of ingenuity. An important task, therefore, is the development to an adequate methodology of testing. In many cases the comparative method is the most powerful; it is essential, however, to have clear criteria for inclusion or exclusion of species in comparative tests, and to use statistical methods with the same care as in the analysis of experimental results.

Tests of the quantitative predictions of optimization models in particular populations are beginning to be made. It is commonly found that a model correctly predicts qualitative features of the observations, but is contradicted in detail. In such cases the Popperian view would be that the original model has been falsified. This is correct, but it does not follow that the model should be abandoned. In the analysis of complex systems it is most unlikely that any simple model, taking into account only a few factors, can give quantitatively exact predictions. Given that a simple model has been falsified by observations, the choice lies between abandoning it and modifying it, usually by adding hypotheses. There can be no simple rule by which to make this choice; it will depend on how persuasive the qualitative predictions are, and on the availability of alternative models.

Mathematical methods of optimization have been developed with engineering and economic applications in mind. Two theoretical questions arise in applying these methods in biology. First, in those cases in which the fitnesses of phenotypes are frequency-dependent, the problem must be formulated in game-theoretical terms; some difficulties then arise in deciding to what type of equilibrium a population will tend. A second and related set of questions arise when specific genetic assumptions are incorporated into the model, because it may be that a population with the optimal phenotype cannot breed true. These questions need further study, but at present there is no reason to doubt the adequacy of the concepts of optimization and of evolutionary stability for studying phenotypic evolution.

ACKNOWLEDGMENTS

My thanks are due Dr. R. C. Lewontin for sending me two manuscripts that formed the starting point of this review, and Drs. G. Oster and R. Pulliam for their comments on and earlier draft. I was also greatly helped by preliminary discussions with Dr. E. Curio.

REFERENCES

1. Alexander, R. M. 1977. Mechanics and scaling of terrestrial locomotion. In *Scale Effects in Animal Locomotion*, ed. T. J. Pedley, London, Academic Press, pp. 93–110.

2. Auslander, D., J. Guckenheimer, and G. Oster. 1978. Random evolutionarily stable strategies. *Theor. Pop. Biol.*, 13(2):276–293.

3. Baker, H. G. 1959. Reproductive methods as factors in speciation in flowering plants. *Cold Spring Harbor Symp. Quant. Biol.*, 24:177–191.

4. Bawa, K. S., and P. A. Opler. 1975. Dioecism in tropical forest trees. *Evolution*, 29:167–179.

5. Bellman, R. 1957. *Dynamic Programming*, Princeton, N.J., Princeton University Press.

6. Bishop, D. T., and C. Cannings. 1978. A generalized war of attrition. *J. Theor. Biol.*, 70:85–124.

7. Buskirk, W. H. 1976. Social systems in tropical forest avifauna. *Am. Nat.*, 110:293–310.

8. Cain, A. J. 1964. The perfection of animals. In *Viewpoints in Biology*, ed. J. D. Carthy and C. L. Duddington, 3:36–63.

9. ———, and P. H. Sheppard. 1954. Natural selection in *Cepaea*. *Genetics*, 39:89–116.

10. Charnov, E. L. 1976. Optimal foraging, the marginal value theorem. *Theor. Pop. Biol.* 9:129–136.

11. ———, J. Maynard Smith, and J. J. Bull. 1976. Why be an hermaphrodite? *Nature*, 263:125–126.

12. Clark, C. W. 1976. *Mathematical Bioeconomics*, N.Y., Wiley.

13. Clarke, B. 1962. Balanced polymorphism and the diversity of sympatric species. In *Taxonomy and Geography*, ed. d. Nichols, London, Syst. Assoc. Publ., 4:47–70.

14. ———. 1975. The contribution of ecological genetics to evolutionary theory: detecting the direct effects of natural selection on particular polymorphic loci. *Genetics*, 79:101–113.

15. ———. 1976. The ecological genetics of host-parasite relationships. In *Genetic Aspects of Host-Parasite Relationships*, ed. A. E. R. Taylor and R. Muller, Oxford, Blackwell, pp. 87–103.

16. Clutton-Brock, T. H., and P. H. Harvey. 1977. Primate ecology and social organisation. *J. Zool.*, London, 183:1–39.

17. ———. 1977. Species differences in feeding and ranging behaviour in primates. In *Primate Ecology*, ed. T. H. Clutton-Brock, London, Academic, pp. 557–584.

18. ———, and B. Rudder. 1977. Sexual dimorphism, socioeconomic sex ratio and body weight in primates. *Nature*, 269:797–800.

19. Cohen, D. 1971. Maximising final yield when growth is limited by time or by limiting resources. *J. Theor. Biol.*, 33:299–307.

20. Cowie, R. J. 1977. Optimal foraging in great tits (*Parus major*). *Nature*, 268:137–139.

21. Crow, J. F., and M. Kimura. 1970. *An Introduction to Population Genetics Theory*, N.Y., Harper & Row.

22. Crozier, R. H. 1970. Coefficients of relationship and the identity of genes by descent in the Hymenoptera. *Am. Nat.*, 104:216–217.

23. Curio, E. 1973. Towards a methodology of teleonomy. *Experientia*, 29:1045–1058.

24. ———. 1975. The functional organization of anti-predator behaviour in the pied flycatcher: a study of avian visual perception. *Anim. Behav.*, 23:1–115.

25. ———. 1976. *The Ethology of Predation*, Berlin, Springer-Verlag.

26. Darwin, C. 1877. *The Different Forms of Flowers on Plants of the Same Species*, London, John Murray.

27. Emlen, J. M. 1966. The role of time and energy in food preference. *Am. Nat.*, 100:611–617.

28. Fisher, R. A. 1930. *The Genetical Theory of Natural Selection*, London: Oxford Univ. Press, 291 pp.

29. Goss-Custard, J. D. 1977. Optimal foraging and the size selection of worms by redshank, *Tringa totanus*, in the field. *Anim. Behav.* 25:10–29.

30. Gould, S. J. 1971. Geometric scaling in allometric growth: a contribution to the problem of scaling in the evolution of size. *Am. Nat.*, 105:113–116.

31. Grieg-Smith, P. W. 1978. The formation, structure and feeding of insectivorous bird flocks in West African savanna woodland. *Ibis*, 121(3):284–297.

32. Haldane, J. G. S. 1949. Disease and evolution. *Ric. Sci.*, Suppl., 19:68–76.

33. Hamilton, W. D. 1964. The genetical theory of social behavior. I and II. *J. Theor. Biol.*, 7:1–16; 17–32.

34. ———. 1967. Extraordinary sex ratios. *Science*, 156:477–488.

35. ———. 1972. Altruism and related phenomena, mainly in social insects. *Ann. Rev. Ecol. Syst.*, 3:193–232.

36. Harvey, P. H., M. Kavanagh, and T. H. Clutton-Brock. 1978. Sexual dimorphism in primate teeth. *J. Zool.*, 186:475–485.

37. Ivlev, V. S. 1961. *Experimental Ecology of the Feeding of Fishes*, New Haven, Yale Univ. Press.

38. Jacob, F. 1977. Evolution and tinkering, *Science*, 196:1161–1166.

39. Jarvis, M. J. F. 1974. The ecological significance of clutch size in the South African gannet [*Sula capensis.* (Lichtenstein)]. *J. Anim. Ecol.*, 43:1–17.

40. Kalmijn, A. J. 1971. The electric sense of sharks and rays. *J. Exp. Biol.*, 55:371–383.

41. Kalmus, H., and C. A. B. Smith. 1960. Evolutionary origin of sexual differentiation and the sex-ratio. *Nature*, 186:1004–1006.

42. Kettlewell, H. B. D. 1956. Further selection experiments on industrial melanism in the Lepidoptera. *Heredity*, 10:287–301.

43. Kimura, M. 1968. Evolutionary rate at the molecular level. *Nature*, 217:624–626.

44. King, J. L., and T. H. Jukes. 1969. Non-Darwinian evolution: random fixation of selectively neutral mutations. *Science*, 164:788–798.

45. Krebs, C. J. 1972. *Ecology*, N.Y., Harper & Row.

46. Krebs, J. R., J. C. Ryan, and E. L. Charnov. 1974. Hunting by expectation or optimal foraging? A study of patch use by chickadees. *Anim. Behav.*, 22:953–964.

47. ———, J. T. Ericksen, M. I. Webber, and E. L. Charnov. 1977. Optimal prey selection in the great tit (*Parus major*). *Anim. Behav.*, 25:30–38.

48. Lack, D. 1966. *Population Studies of Birds*. Oxford, Clarendon Press.

49. ———. 1968. *Ecological Adaptiations for Breeding in Birds*, London, Methuen.

50. Levene, H. 1953. Genetic equilibrium when more than one ecological niche is available. *Am. Nat.*, 87:131–133.

51. Levins, R. 1962. Theory of fitness in a heterogeneous environment. I. The fitness set and adaptive function. *Am. Nat.*, 96:361–373.

52. Lewontin, R. C. 1961. Evolution and the theory of games. *J. Theor, Biol.*, 1:382–403.

53. ———. 1974. *The Genetic Basis of Evolutionary Change*, N.Y., Columbia Univ. Press.

54. ———. 1977. Adaptation. In *The Encyclopedia Einaudi*, Torino, Giulio Einaudi Edition.

55. ———. 1978. Fitness, survival and optimality. In *Analysis of Ecological Systems*, ed. D. H. Horn, R. Mitchell, G. R. Stairs, columbus, Oh., Ohio State Univ. Press.

56. MacArthur, R. H. 1965. Ecological consequences of natural selection. In *Theoretical and Mathematical Biology*, ed. T. Waterman, H. Morowritz, N.Y., Blaisdell.

57. ———, and E. R. Pianka. 1966. On optimal use of a patch environment. *Am. Nat.*, 100:603–609.

58. Malécot, G. 1969. *The Mathematics of Heredity*, transl. D. M. Yermanos. San Francisco, W. H. Freeman, 88 pp.

59. Maynard Smith, J. 1962. Disruptive selection, polymorphism and sympatric speciation. *Nature*, 195:60–62.

60. ———. 1969. The status of neo-Darwinism. In *Towards a Theoretical Biology. 2: Sketches*, ed. C. H. Waddington, Edinburgh, Edinburgh Univ. Press, pp. 82–89.

61. ———. 1976. Sexual selection and the handicap principle. *J. Theor. Biol.*, 57:239–242.

62. ———. 1977. Parental investment—a prospective analysis. *Anim. Behav.*, 25:1–9.

63. ———. 1978. *The Evolution of Sex*, London, Cambridge Univ. Press.

64. ———, and G. A. Parker. 1976. The logic of asymmetric contests. *Anim. Behav.*, 24:159–175.

65. ———, and G. R. Price. 1973. The logic of animal conflict. *Nature*, 246:15–18.

66. ———, and R. J. G. Savage. 1956. Some locomotory adaptations in mammals. *Zool, J. Linn. Soc.*, 42:603–622.

67. McFarland, D. J. 1977. Decision making in animals. *Nature*, 269:15–21.

68. Milton, K., and M. L. May. 1976. Body-weight, diet and home range area in primates. *Nature*, 259:459–462.

69. Mirmirani, M., and G. Oster. 1978. Competition, kin selection and evolutionarily stable strategies. *Theor. Pop. Biol.*, 13(3):304–339.

70. Moment, G. 1962. Reflexive selection: a possible answer to an old puzzle. *Science*, 136:262–263.

71. Nelson, J. B. 1964. Factors influencing clutch size and chick growth in the North Atlantic Gannet, *Sula bassana*. *Ibis*, 106:63–77.

72. Norton-Griffiths, M. 1969. The organization, control and development of parental feeding in the oystercatcher (*Haematopus ostralegus*). *Behavior*, 34:55–114.

73. Orians, G. H., and N. E. Pearson. 1978. On the theory of central place foraging. In *Analysis of Ecological Systems*, ed. D. H. Horn, R. Mitchell, G. R. Stairs, Columbus, Ohio State Univ. Press.

74. Oster, G., I. Eshel, and D. Cohen. 1977. Worker-queen conflicts and the evolution of social insects. *Theor. Pop. Biol.*, 12:49–85.

75. ———, and E. O. Wilson. 1978. *Caste and Ecology in the Social Insects*, Princeton, N.J., Princeton Univ. Press.

76. Parker, G. A. 1974. The reproductive behaviour and the nature of sexual selection in *Scatophaga stercoraria* L. IX. Spatial distribution of fertilization rates and evolution of male search strategy within the reproductive area. *Evolution*, 28:93–108.

77. ———, and R. A. Stuart. 1976. Animal behaviour as a strategy optimizer: evolution of resource assessment strategies and optimal emigration thresholds. *Am. Nat.*, 110:1055–1076.

78. Pedley, T. J. 1977. *Scale Effects in Animal Locomotion*, London, Academic Press.

79. Pontryagin, L. S., V. S. Boltyanskii, R. V. Gamkrelidze, and E. F. Mishchenko. 1962. *The Mathematical Theory of Optimal Processes*, N.Y., Wiley.

80. Pulliam, H. R. 1978. Do chipping sparrows forage optimally? A test of optimal foraging theory in nature. *Am. Nat.* In press.

81. Pyke, G. H., H. R. Pullian, and E. L. Charnov. 1977. Optimal foraging: a selective review of theory and tests. *Q. Rev. Biol.*, 52:137–154.

82. Ralls, K. 1976. Mammals in which females are larger than males. *Q. Rev. Biol.*, 51:245–276.

83. Rensch, B. 1959. *Evolution above the Species Level*, New York, Columbia Univ. Press.

84. Rohwer, S. 1977. Status signaling in Harris sparrows: some experiments in deception. *Behaviour*, 61:107–129.

85. Rosado, J. M. C., and A. Robertson. 1966. The genetic control of sex ratio. *J. Theor. Biol.*, 13:324–329.

86. Schoener, T. W. 1968. Sizes of feeding territories among birds. *Ecology*, 49:123–141.

87. ———. 1971. Theory of feeding strategies. *Ann. Rev. Ecol. Syst.* 2:369–404.

88. Sibly, R. and D. McFarland. 1976. On the fitness of behaviour sequences. *Am. Nat.*, 110:601–617.

89. Slatkin, M. 1978. On the equilibration of fitnesses by natural selection. *Am. Nat.*, 112:845–859.

90. ———. 1979. The evolutionary response to frequency- and density-dependent interactions, *Am. Nat.*, 114:384–398.

91. ———. 1979. Frequency- and density-dependent selection on a quantitative character, *Genetics*, 93:755–771.

92. Slobodkin, L. B., and A. Rapoport. 1974. An optimal strategy of evolution. *Q. Rev. Biol.*, 49:181–200.

93. Strobeck, C. 1975. Selection in a fine-grained environment. *Am. Nat.*, 109:419–425.

94. Tinbergen, N., G. J. Broekhuysen, F. Feekes, J. C. W. Houghton, H. Kruuk, and E. Szule. 1963. Egg shell removal by the black-headed gull, *Larus ribidundus* L.: a behaviour component of camouflage. *Behaviour*, 19:74–117.

95. Trivers, R. L., and H. Hare. 1976. Haplodiploidy and the evolution of social insects. *Science*, 191:249–263.

96. ———, and D. E. Willard. 1973. Natural selection of parental ability to vary the sex ratio of offspring. *Science*, 179:90–92.

97. Verner, J. 1965. Selection for sex ratio. *Am. Nat.*, 19:419–421.

98. ———, and M. F. Willson. 1966. The influence of habitats on mating systems of North American passerine birds. *Ecology*, 47:143–147.

99. Von Neumann, J., and O. Morgenstern. 1953. *Theory of Games and Economic Behavior*, Princeton, N.J., Princeton Univ. Press.

100. Werner, E. E., and D. J. Hall. 1974. Optimal foraging and size selection of prey by the bluegill sunfish (*Lepomis mochrochirus*). *Ecology*, 55:1042–1052.

101. Wright, S. 1932. The roles of mutation, inbreeding, crossbreeding and selection in evolution. *Proc. Sixth Int. Congr. Genet.*, 1:356–366.

102. Zahavi, A. 1975. Mate selection—a selection for a handicap. *J. Theor. Biol.*, 53:205–214.

IV Units of Selection

6 Excerpts from *Adaptation and Natural Selection*

George C. Williams

I hope that this book will help to purge biology of what I regard as unnecessary distractions that impede the progress of evolutionary theory and the development of a disciplined science for analyzing adaptation. It opposes certain of the recently advocated qualifications and additions to the theory of natural selection, such as genetic assimilation, group selection and cumulative progress in adaptive evolution. It advocates a ground rule that should reduce future distractions and at the same time facilitate the recognition of really justified modifications of the theory. The ground rule—or perhaps *doctrine* would be a better term—is that adaptation is a special and onerous concept that should be used only where it is really necessary. When it must be recognized, it should be attributed to no higher a level of organization than is demanded by the evidence. In explaining adaptation, one should assume the adequacy of the simplest form of natural selection, that of alternative alleles in Mendelian populations, unless the evidence clearly shows that this theory does not suffice....

Benefits to groups can arise as statistical summations of the effects of individual adaptations. When a deer successfully escapes from a bear by running away, we can attribute its success to a long ancestral period of selection for fleetness. Its fleetness is responsible for its having a *low probability* of death from bear attack. The same factor repeated again and again in the herd means not only that it is a herd of fleet deer, but also that it is a fleet herd. The group therefore has a *low rate* of mortality from bear attack. When every individual in the herd flees from a bear, the result is effective protection of the herd.

As a very general rule, with some important exceptions, the fitness of a group will be high as a result of this sort of summation of the adaptations of its members. On the other hand, such simple summations obviously cannot produce collective fitness as high as could be achieved by an adaptive organization of the group itself. We might imagine that mortality rates from predation by bears on a herd of deer would be still lower if each individual, instead

From *Adaptation and Natural Selection*. Princeton University Press, 1966, 4–5, 16–19, 22–25, 92–101, 108–124, 208–212.

of merely running for its life when it saw a bear, would play a special role in an organized program of bear avoidance. There might be individuals with especially well-developed senses that could serve as sentinels. Especially fleet individuals could lure bears away from the rest, and so on. Such individual specialization in a collective function would justify recognizing the herd as an adaptively organized entity. Unlike individual fleetness, such group-related adaptation would require something more than the natural selection of alternative alleles as an explanation.

It may also happen that the incidental effects of individual activities, of no functional significance in themselves, can have important statistical consequences, sometimes harmful, sometimes beneficial. The depletion of browse is a harmful effect of the feeding activities of each member of a dense population of deer. If browse depletion were beneficial, I suspect that someone, sooner or later, would have spoken of the feeding behavior of deer as a mechanism for depleting browse. A statement of this sort should not be based merely on the evidence that the statistical effect of eating is beneficial; it should be based on an examination of the causal mechanisms to determine whether they cannot be adequately explained as individual adaptations for individual nourishment.

The feeding activities of earthworms would be a better example, because here the incidental statistical effects *are* beneficial, from the standpoint of the population and even of the ecological community as a whole. As the earthworm feeds, it improves the physical and chemical properties of the soil through which it moves. The contribution from each individual is negligible, but the collective contribution, cumulative over decades and centuries, gradually improves the soil as a medium for worm burrows and for the plant growth on which the earthworm's feeding ultimately depends. Should we therefore call the causal activities of the earthworm a soil-improvement mechanism? Apparently Allee (1940) believed that some such designation is warranted by the fact that soil improvement is indeed a result of the earthworm's activities. However, if we were to examine the digestive system and feeding behavior of an earthworm, I assume that we would find it adequately explained on the assumption of design for individual nutrition. The additional assumption of design for soil improvement would explain nothing that is not also explainable as a nutritional adaptation. It would be a violation of parsimony to assume both explanations when one suffices. Only if one denied that some benefits can arise by chance instead of by design, would there be a reason for postulating an adaptation behind every benefit.

On the other hand, suppose we did find some features of the feeding activities of earthworms that were inexplicable as trophic adaptations but were exactly what we should expect of a system designed for soil improvement. We would then be forced to recognize the system as a soil-modification mechanism, a conclusion that implies a quite different level of adaptive organization from that implied by the nutritional function. As a digestive system, the gut of a worm plays a role in the adaptive organization of that worm and nothing else, but as a soil-modification system it would play a role in the adaptive organization of the whole community. This, as I will argue at length in later chapters, is a reason for rejecting soil-improvement as a purpose of the worm's activities if it is possible to do so. Various levels of adaptive organization, from the subcellular to the biospheric, might conceivably be recognized, but the principle of parsimony demands that we recognize adaptation at the level necessitated by the facts and no higher.

It is my position that adaptation need almost never be recognized at any level above that of a pair of parents and associated offspring. As I hope to show in the later chapters, this conclusion seldom has to rest on appeals to parsimony alone, but is usually supported by specific evidence.

The most important function of this book is to echo a plea made many years ago by E. S. Russell (1945) that biologists must develop an effective set of principles for dealing with the general phenomenon of biological adaptation. This matter is considered mainly in the final chapter.

The essence of the genetical theory of natural selection is a statistical bias in the relative rates of survival of alternatives (genes, individuals, etc.). The effectiveness of such bias in producing adaptation is contingent on the maintenance of certain quantitative relationships among the operative factors. One necessary condition is that the selected entity must have a high degree of permanence and a low rate of endogenous change, relative to the degree of bias (differences in selection coefficients). Permanence implies reproduction with a potential geometric increase.

Acceptance of this theory necessitates the immediate rejection of the importance of certain kinds of selection. The natural selection of phenotypes cannot in itself produce cumulative change, because phenotypes are extremely temporary manifestations. They are the result of an interaction between genotype and environment that produces what we recognize as an individual. Such an individual consists of genotypic information and information recorded since conception. Socrates consisted of the genes his parents gave him, the experiences they and this environment later provided, and a growth and development mediated by numerous meals. For all I know, he may have been very successful in the evolutionary sense of leaving numerous offspring. His phenotype, nevertheless, was utterly destroyed by the hemlock and has never since been duplicated. If the hemlock had not killed him, something else soon would have. So however natural selection may have been acting on Greek phenotypes in the fourth century B.C. it did not of itself produce any cumulative effect.

The same argument also holds for genotypes. With Socrates' death, not only did his phenotype disappear, but also his genotype. Only in species that can maintain unlimited clonal reproduction is it theoretically possible for the selection of genotypes to be an important evolutionary factor. This possibility is not likely to be realized very often, because only rarely would individual clones persist for the immensities of time that are important in evolution. The loss of Socrates' genotype is not assuaged by any consideration of how prolifically he may have reproduced. Socrates' genes may be with us yet, but not his genotype, because meiosis and recombination destroy genotypes as surely as death.

It is only the meiotically dissociated fragments of the genotype that are transmitted in sexual reproduction, and these fragments are further fragmented by meiosis in the next generation. If there is an ultimate indivisible fragment it is, by definition, "the gene" that is treated in the abstract discussions of population genetics. Various kinds of suppression of recombination may cause a major chromosomal segment or even a whole chromosome to be transmitted entire for many generations in certain lines of descent. In such cases the segment or chromosome behaves in a way that approximates the population genetics of a single gene. In this book I use the term *gene* to mean "that which segregates and recombines with appreciable frequency." Such genes are potentially immortal, in the sense of there being no physiological limit to their survival, because of their potentially reproducing fast enough to compensate for their destruction by external agents. They also have a high degree of qualitative stability. Estimates of mutation rates range from about 10^{-4} to 10^{-10} per generation. The rates of selection of alternative alleles can be much higher. Selection among the progeny of individuals heterozygous for recessive lethals would eliminate half the lethal genes in one generation. Aside from lethal and markedly deleterious genes in experimental populations, there is abundant evidence (e.g., Fisher and Ford 1947; Ford 1956; Clarke, Dickson, and Sheppard 1963) for selection coefficients in nature that exceed mutation rates by one to many multiples of ten. There can be no doubt that the selective accumulation of genes can be effective. In evolutionary theory, a gene could be defined as any hereditary information for which there is a favorable or unfavorable selection bias equal to several or many times its rate of endogenous change. The prevalence of such stable entities in the heredity of populations is a measure of the importance of natural selection.

Natural selection would produce or maintain adaptation as a matter of definition. Whatever gene is favorably selected is better adapted than its unfavored alternatives. This is the reliable outcome of such selection, the prevalence of well-adapted genes. The selection of such genes of course is mediated by the phenotype, and to be favorably selected, a gene must augment phenotypic reproductive success as the arithmetic mean effect of its activity in the population in which it is selected. . . .

This [work] is a rejoinder to those who have questioned the adequacy of the traditional model of natural selection to explain evolutionary adaptation. The topics considered in the preceding chapters relate mainly to the adequacy of this model in the realms of physiological, ecological, and developmental mechanisms, matters of primary concern to individual organisms. At the individual level the adequacy of the selection of alternative alleles has been challenged to only a limited degree. Many more doubts on the importance of such selection have been voiced in relation to the phenomenon of interactions among individuals. Many biologists have implied, and a moderate number have explicitly maintained, that groups of interacting individuals may be adaptively organized in such a way that individual interests are compromised by a functional subordination to group interests.

It is universally conceded by those who have seriously concerned themselves with this problem (e.g., Allee *et al.* 1949; Haldane 1932; Lewontin 1958, 1962; Slobodkin 1954; Wynne-Edwards 1962; Wright, 1945) that such group-related adaptations must be attributed to the natural selection of alternative *groups* of individuals and that the natural selection of alternative alleles within populations will be opposed to this development. I am in entire agreement with the reasoning behind this conclusion. Only by a theory of between-group selection could we achieve a scientific explanation of group-related adaptations. However, I would question one of the premises on which the reasoning is based. Chapters 5 to 8 [of *Adaptation and Natural Selection*] will be primarily a defense of the thesis that group-related adaptations do not, in fact, exist. A *group* in this discussion should be understood to mean something other than a family and to be composed of individuals that need not be closely related.

The present chapter examines the logical structure of the theory of selection between groups, but first I wish to consider an apparent exception to the rule that the natural selection of individuals cannot produce group-related adaptations. This exception may be found in animals that live in stable social groups and have the intelligence and other mental qualities necessary to form a system of personal friendships and animosities that transcend the limits of family relationship. Human society would be impossible without the ability of each of us to know, individually, a variety of neighbors. We learn that Mr. X is a noble gentleman and that Mr. Y is a scoundrel. A moment of reflection should convince anyone that these relationships may have much to do with evolutionary success. Primitive man lived in a world in which stable interactions of personalities were very much a part of his ecological environment. He had to adjust to this set of ecological factors as well as to any other. If he was socially acceptable, some of his neighbors might bring food to himself and his family when he was temporarily incapacitated by disease or injury. In time of dearth, a stronger neighbor might rob our primitive man of food, but the neighbor would be more likely to rob a detestable primitive Mr. Y and his troublesome family. Conversely, when a poor Mr. X is sick our primitive man will, if he can, provide for him. Mr. X's warm heart will know the emotion of gratitude and, since he recognizes his benefactor and remembers the help provided, will probably reciprocate some day. A number of people, including Darwin (1896, Chap. 5), have recognized the importance of this factor in human evolution. Darwin speaks of it as the "lowly motive" of helping others in the hope of future repayment. I see no reason why a conscious motive need be involved. It is necessary that help provided to others be occasionally reciprocated if it is to be favored by natural selection. It is not necessary that either the giver or the receiver be aware of this.

Simply stated, an individual who maximizes his friendships and minimizes his antagonisms will have an evolutionary advantage, and selection should favor those characters that promote the optimization of personal relationships. I imagine that this evolutionary factor has increased man's capacity for altruism and compassion and has tempered his ethically less acceptable heritage of sexual and predatory aggressiveness. There is theoretically no limit to the extent and complexity of group-related behavior that this factor could produce, and the immediate goal of such behavior would always be the well-being of some other individual, often genetically unrelated. Ultimately, however, this would not be an adaptation for group benefit. It would be developed by the differential survival of individuals and would be designed for the perpetuation of the genes of the individual providing the benefit to another. It would involve only such immediate self-sacrifice for which the probability of later repayment would be sufficient justification. The natural selection of alternative alleles can foster the production of individuals willing to sacrifice their lives for their offspring, but never for mere friends.

The prerequisites for the operation of this evolutionary factor are such as to confine it to a minor fraction of the Earth's biota. Many animals form dominance hierarchies, but these are not sufficient to produce an evolutionary advantage in mutual aid. A consistent interaction pattern between hens in a barnyard is adequately explained without postulating emotional bonds between individuals. One hen reacts to another on the basis of the social releasers that are displayed, and if individual recognition is operative, it merely adjusts the behavior towards another individual according to the immediate results of past interactions. There is no reason to believe that a hen can harbor grudges against or feel friendship toward another hen. Certainly the repayment of favors would be out of the question.

A competition for social goodwill cannot fail to have been a factor in human evolution, and I would expect that it would operate in many of the other primates. Altman (1962) described the formation of semipermanent coalitions between individuals within bands of wild rhesus monkeys and cited similar examples from other primates. Members of such coalitions helped each other in conflicts and indulged in other kinds of mutual aid. Surely an individual that had a better than average ability to form such coalitions would have an evolutionary advantage over its competitors. Perhaps this evolutionary factor might operate in the evolution of porpoises. This seems to be the most likely explanation for the very solicitous behavior that they sometimes show toward each other (Slijper 1962, pp. 193–197). I would be reluctant, however, to recognize this factor in any group but the mammalia, and I would imagine it to be confined to a minority of this group. For the overwhelming mass of the Earth's biota, friendship and hate are not parts of the ecological environment, and the only way for socially beneficial self-sacrifice to evolve is through the biased survival and extinction of populations, not by selective gene substitution within populations.

To minimize recurrent semantic difficulties, I will formally distinguish two kinds of natural selection. The natural selection of alternative alleles in a Mendelian population will henceforth be called *genic selection*. The natural selection of more inclusive entities will be called *group selection*, a term introduced by Wynne-Edwards (1962). *Intrademic* and *interdemic*, and other terms with the same prefixed, have been used to make the same distinction. It has been my experience, however, that the repeated use in the same discussion of "inter" and "intra" for specifically contrasted concepts is a certain cause of confusion, unless a reader exerts an inconvenient amount of attention to spelling, or a speaker indulges in highly theatrical pronunciation.

The definitions of other useful terms, and the conceptual relations between the various creative evolutionary factors and the production of adaptation are indicated in figure 6.1. Genic selection should be assumed to imply the current conception of natural selection often termed *neo-Darwinian*. An *organic adaptation* would be a mechanism designed to promote the success of an individual organism, as measured by the extent to which it contributes genes to later generations of the population of which it is a member. It has the individual's *inclusive fitness* (Hamilton 1964) as its goal. Biotic evolution is any change in a biota. It can be brought about by an evolutionary change in one or more of the constituent populations, or merely by a change in their relative numbers. A *biotic adaptation* is a mechanism designed to promote the success of a biota, as measured by the lapse of time to extinction. The biota considered would have to be restricted in scope so as to allow comparison with other biotas. It could be a single biome, or community, or taxonomic group, or, most often, a single population. A change in the fish-fauna of a lake would be considered biotic evolution. It could come about through some change in the characters of one or more of the constituent populations or through a change in the relative numbers of the populations. Either would result in a changed fish-fauna, and such a change would be biotic evolution. A biotic adaptation could be a mechanism for the survival of such a group as the fish-fauna of a lake, or of any included population, or of a whole species that lives in that lake and elsewhere.

I believe that it is useful to make a formal distinction between biotic and organic evolution, and that certain fallacies can be avoided by keeping the

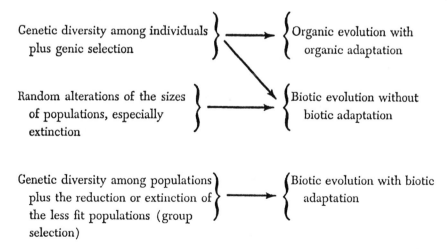

Figure 6.1 Summary comparison of organic and biotic evolution and of organic and biotic adaptation.

Excerpts from *Adaptation and Natural Selection*

distinction in mind. It should be clear that, in general, the fossil record can be a direct source of information on organic evolution only when changes in single populations can be followed through a continuous sequence of strata. Ordinarily the record tells us only that the biota at time t' was different from that at time t and that it must have changed from one state to the other during the interval. An unfortunate tendency is to forget this and to assume that the biotic change must be ascribed to appropriate organic change. The horse-fauna of the Eocene, for instance, was composed of smaller animals than that of the Pliocene. From this observation, it is tempting to conclude that, at least most of the time and on the average, a larger than mean size was an advantage to an individual horse in its reproductive competition with the rest of its population. So the component populations of the Tertiary horse-fauna are presumed to have been evolving larger size most of the time and on the average. It is conceivable, however, that precisely the opposite is true. It may be that at any given moment during the Tertiary, most of the horse populations were evolving a smaller size. To account for the trend towards larger size it is merely necessary to make the additional assumption that group selection favored such a tendency. Thus, while only a minority of the populations may have been evolving a larger size, it could have been this minority that gave rise to most of the populations of a million years later. Figure 6.2 shows how the same observations on the fossil record can be rationalized on two entirely different bases. The unwarranted assumption of organic evolution as an explanation for biotic evolution dates at least from Darwin. In *The*

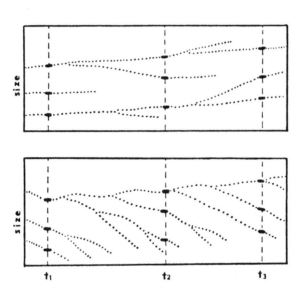

Figure 6.2 Alternative ways of interpreting the same observation of the fossil record. Average sizes in hypothetical horse species at three different times are indicated by boldface marks on the vertical time-scale at t_1, t_2, and t_3. Upper and lower diagrams show the same observations. In the upper, hypothetical phylogenies explain the observations as the result of the organic evolution of increased size and of occasional chance extinction. In the lower, the hypothetical phylogenies indicate the organic evolution mainly of decreased size but with effective counteraction by group selection so that the biota evolves a larger average size.

George C. Williams

Origin of Species he dealt with a problem that he termed "advance in organization." He interpreted the fossil record as indicating that the biota has evolved progressively "higher" forms from the Cambrian to Recent, clearly a change in the biota. His explanation, however, is put largely in terms of the advantage that an individual might have over his neighbors by virtue of a larger brain, greater histological complexity, etc. Darwin's reasoning here is analogous to that of someone who would expect that if the organic evolution of horses proceeded toward larger size during the Tertiary, most equine mutations during this interval must have caused larger size in the affected individuals. I suspect that most biologists would tend toward the opposite view, and expect that random changes in the germ plasm would be more likely to curtail growth than to augment it. Organic evolution would normally run counter to the direction of mutation pressure. There is a formally similar relation between organic evolution and group selection. Organic evolution provides genetically different populations, the raw material on which group selection acts. There is no necessity for supposing that the two forces would normally be in precisely the same direction. It is conceivable that at any given moment since the Cambrian, the majority of organisms were evolving along lines that Darwin would consider retrogression, degeneration, or narrow specialization, and that only a minority were progressing. If the continued survival of populations were sufficiently biased in favor of this minority, however, the biota as a whole might show "progress" from one geologic period to the next. I expect that the fossil record is actually of little use in evaluating the relative potency of genic and group selection.

In another respect the analogy between mutation and organic evolution as sources of diversity may be misleading. Mutations occur at random and are usually destructive of any adaptation, whereas organic evolution is largely concerned with the production or at least the maintenance of organic adaptation. Any biota will show a system of adaptations. If there is no group selection, i.e., if extinction is purely by chance, the adaptations shown will be a random sample of those produced by genic selection. If group selection does operate, even weakly, the adaptations shown will be a biased sample of those produced by genic selection. Even with such bias in the kinds of adaptations actually represented, we would still recognize genic selection as the process that actually produced them. We could say that the adaptations were produced by group selection only if it was so strong that it constantly curtailed organic evolution in all but certain favored directions and was thus able, by its own influence, to accumulate the functional details of complex adaptations. This distinction between the production of a biota with a certain set of organic adaptations and the production of the adaptations of a biota will be emphasized again in a number of contexts.

... It is essential, before proceeding further with the discussion, that the reader firmly grasp the general meaning of biotic adaptation. He must be able to make a conceptual distinction between a population of adapted insects and an adapted population of insects. The fact that an insect population survives through a succession of generations is not evidence for the existence of biotic adaptation. The survival of the population may be merely an incidental consequence of the organic adaptations by which each insect attempts to survive and reproduce itself. The survival of the population depends on these individual efforts. To determine whether this survival is the proper function or merely an incidental by-product of the individual effort must be decided by a critical examination of the reproductive processes. We must decide: Do these processes show an effective design for maximizing the number of descendants of the individual, or do they show an effective design for maximizing the number, rate of growth, or numerical stability of the population or larger system? Any feature of the system that promotes group survival and cannot be explained as an organic adaptation can be called a biotic adaptation. If the population has such adaptations it can be called an adapted population. If it does not, if its continued survival is merely incidental to the operation of organic adaptations, it is merely a population of adapted insects.

Like the theory of genic selection, the theory of group selection is logically a tautology and there can be no sane doubt about the reality of the process. Rational criticism must center on the importance of the process and on its adequacy in explaining the phenomena attributed to it. An important tenet of evolutionary theory is that natural selection can produce significant cumulative change only if selection coefficients are high relative to the rates of change of the selected entity. Since genic selection coefficients are high relative to mutation rates, it is logically possible for the natural selection of alternative alleles to have important cumulative effects. It was pointed out [above] that there can be no effective selection of somata. They have limited life spans and (often) zero biotic potential. The same considerations apply to populations of somata. I also pointed out that genotypes have limited lives and fail to reproduce themselves (they are destroyed by meiosis and recombination), except where clonal reproduction is possible. This is equally true of populations of genotypes. All of the genotypes of fruit-fly populations now living will have ceased to exist in a few weeks. Within a population, only the gene is stable enough to be effectively selected. Likewise in selection among populations, only populations of genes (gene pools) seem to qualify with respect to the necessary stability. Even gene pools will not always qualify. If populations are evolving rapidly and have a low rate of extinction and replacement, the rate of endogenous change might be too great for group selection to have any cumulative effect. This argument precisely parallels that which indicates that mutation rates must be low relative to selection coefficients for genic selection to be effective.

If a group of adequately stable populations is available, group selection can theoretically produce biotic adaptations, for the same reason that genic selection can produce organic adaptations. Consider again the evolution of size among Tertiary horses. Suppose that at one time there was a genus of two species, one that averaged 100 kilograms when full grown and another that averaged 150 kilograms. Assume that genic selection in both species favored a smaller size so that a million years later the larger of the two averaged only

130 kilograms and the smaller had become extinct, but had lost 20 kilograms before it did so. In this case we could say that the genus evolved an increased size, even though both of the included species evolved a decreased size. If the extinction of the smaller species is not just a chance event but is attributable to its smaller size, we might refer to large size as a biotic adaptation of a simple sort. However, it is the origin of complex adaptations, for which the concept of functional design would be applicable, that is the important consideration.

If alternative gene pools are not themselves stable, it is still conceivable that group selection could operate among more or less constant rates of change. A system of relatively stable rates of change in the gene frequencies of a population might be called an evolutionary trajectory. It could be described as a vector in n-dimensional space, with n being the number of relevant gene frequencies. In a given sequence of a few generations a gene pool may be undergoing certain kinds of change at a certain rate. This is only one of an infinite number of other evolutionary trajectories that might conceivably be followed. Some trajectories may be more likely to lead to extinction than others, and group selection will then operate by allowing different kinds of evolutionary change to continue for different average lengths of time. There is paleontological evidence that certain kinds of evolutionary change may continue for appreciable lengths of time on a geological scale. Some of the supposed examples disappear as the evidence accumulates and shows that actual courses of evolution are more complex than they may have seemed at first. Other examples are apparently real and are attributed by Simpson (1944, 1953) to continuous genic selection in certain directions, a process he terms "ortho-selection."

Wright (1945) proposed that group selection would be especially effective in a species that was divided up into many small populations that were almost but not quite isolate from each other. Most of the evolutionary change in such a species would be in accordance with genic selection coefficients, but the populations are supposed to be small enough so that genes would occasionally be fixed by drift in spite of adverse selection within a population. Some of the genes so fixed might benefit the population as a whole even though they were of competitive disadvantage within the population. A group so favored would increase in size (regarded as a benefit in Wright's discussion) and send out an augmented number of emigrants to neighboring populations. These migrants would partly or wholly counteract the adverse selection of the gene in neighboring populations and give them repeated opportunity for the chance fixation of the gene. The oft-repeated operation of this process eventually would produce complex adaptations of group benefit, but of competitive disadvantage to an individual. According to this theory, selection not only can act on preexisting variation, but also can help to produce the variation on which it acts, by repeatedly introducing the favored gene into different populations.

Wright formally derived this model in a review of a book by G. G. Simpson. Later, Simpson (1953, pp. 123, 164–165) briefly criticized Wright's theory by pointing out that it leaves too much to a rather improbable concatenation of the population parameters of size, number, degree of isolation, and the balance of genic and group selection coefficients. The populations have to be small enough for genetic drift to be important, but not so small that they are in danger of extinction, and they have to be big enough for certain gene substitutions to be more important than chance factors in determining size and rate of emigration. The unaugmented rates of immigration must be too small to reestablish the biotically undesirable gene after it is lost by drift. The populations must be numerous enough for the postulated process to work at a variety of loci, and each of the populations must be within the necessary size range. Lastly, the balance of these various factors must persist long enough for an appreciable amount of evolutionary change to take place. At the moment, I can see no hope of achieving any reliable estimate of how frequently the necessary conditions have been realized, but surely the frequency of such combinations of circumstances must be relatively low and the combinations quite temporary when they do occur. Simpson also expressed doubts on the reality of the biotic adaptations that Wright's theory was proposed to explain.

A number of writers have since postulated a role for the selection of alternative populations within a species in the production of various supposed "altruistic" adaptations. Most of these references, however, have completely ignored the problem that Wright took such pains to resolve. They have ignored the problem of how whole populations can acquire the necessary genes in high frequency in the first place. Unless some do and some do not, there is no set of alternatives for group selection to act upon. Wright was certainly aware, as some later workers apparently were not, that even a minute selective disadvantage to a gene in a population of moderate size can cause an almost deterministic reduction of the gene to a negligible frequency. This is why he explicitly limited the application of his model to those species that are subdivided into many small local populations with only occasional migrants between them. Others have postulated such group selection as an evolutionary factor in species that manifestly do not have the requisite population structures. Wynne-Edwards (1962), for example, postulated the origin of biotic adaptations of individual disadvantage, by selection among populations of smelts, in which even a single spawning aggregation may consist of tens of thousands of individuals. He envisioned the same process for marine invertebrates that may exist as breeding adults by the million per square mile and have larval stages that may be dispersed many miles from their points of origin.

A possible escape from the necessity of relying on drift in small populations to fix the genes that might contribute to biotic adaptation, is to assume that such genes are not uniformly disadvantageous in competitive individual relationships. If such a gene were, for some reason, individually advantageous in one out of ten populations, group selection could work by making the descendants of that population the sole representatives of the species a million years later. However, this process also loses plausibility on close examination. Low rates of endogenous change relative to selection coefficients are a necessary precondition for any effective selection. The necessary stability is the general rule for genes. While gene pools or evolutionary trajectories can persist little altered through a long period of extinction and replacement of populations, there is no indication that this is the general rule. Hence the effectiveness of group selection is open to question at the axiomatic level for almost any group of organisms. The possibility of effective group selection can be dismissed for any species that consists, as many do, of a single population. Similarly the group selection of alternative species cannot direct the evolution of a monotypic genus, and so on.

Even in groups in which all of the necessary conditions for group selection might be demonstrated, there is no assurance that these conditions will continue to prevail. Just as the evolution of even the simplest organic adaptation requires the operation of selection at many loci for many generations, so also would the production of biotic adaptation require the selective substitution of many groups. This is a major theoretical difficulty. Consider how rapid is the turnover of generations in even the slowest breeding organisms, compared to the rate at which populations replace each other. The genesis of biotic adaptation must for this reason be orders of magnitude slower than that of organic adaptation. Genic selection may take the form of the replacement of one allele by another at the rate of 0.01 per generation, to choose an unusually high figure. Would the same force of group selection mean that a certain population would be 0.01 larger, or be growing 0.01 faster, or be 0.01 less likely to become extinct in a certain number of generations, or have a 0.01 greater emigration rate than another population? No matter which meaning we assign, it is clear that what would be a powerful selective force at the genic level would be trivial at the group level. For group selection to be as strong as genic selection, its selection coefficients would have to be much greater to compensate for the low rate of extinction and replacement of populations.

The rapid turnover of generations is one of the crucial factors that makes genic selection such a powerful force. Another is the large absolute number of individuals in even relatively small populations, and this brings us to another major difficulty in group selection, especially at the species level. A species of a hundred different populations, sufficiently isolated to develop appreciable genetic differences, would be exceptional in more groups of organisms. Such a complexly subdivided group, however, might be in the same position with respect to a bias of 0.01 in the extinction and replacement of groups, as a population of fifty diploid individuals with genic selection coefficients that differ by 0.01. In the population of fifty we would recognize genetic drift, a chance factor, as much more important than selection as an evolutionary force. Numbers of populations in a species, or of taxa in higher categories, are usually so small that chance would be much more important in determining group survival than would even relatively marked genetic differences among the groups. By analogy with the conclusions of population genetics, group selection would be an important creative force only where there were at least some hundreds of populations in the group under consideration.

Excerpts from *Adaptation and Natural Selection*

Obviously the comments above are not intended to be a logically adequate evaluation of group selection. Analogies with the conclusions on genic selection are only analogies, not rigorously reasoned connections. I would suggest, however, that they provide a reasonable basis for skepticism about the effectiveness of this evolutionary force. The opposite tendency is frequently evident. A biologist may note that, logically and empirically, the evolutionary process is capable of producing adaptations of great complexity. He then assumes that these adaptations must include not only the organic but also the biotic, usually discussed in such terms as "for the good of the species." A good example is provided by Montagu (1952), who summarized the modern theory of natural selection and in so doing presented an essentially accurate picture of selective gene substitution by the differential reproductive survival of individuals. Then in the same work he states, "We begin to understand then, that evolution itself is a process which favors cooperating rather than disoperating groups and that 'fitness' is a function of the group as a whole rather than separate individuals." This kind of evolution and fitness is attributed to the previously described natural selection of individuals. Such an extrapolation from conclusions based on analyses of the possibilities of selective gene substitutions *in* populations to the production of biotic adaptations *of* populations is entirely unjustified. Lewontin (1961) has pointed out that population genetics as it is known today relates to genetic processes in populations, not of populations.

Lewontin (1962; Lewontin and Dunn 1960) has produced what seems to me to be the only convincing evidence for the operation of group selection. There is a series of alleles symbolized by *t* in house-mouse populations that produces a marked distortion of the segregation ratio of sperm. As much as 95 percent of the sperm of a heterozygous male may bear such a gene, and only 5 percent bear the wild-type allele. This marked selective advantage is opposed by other adverse effects in the homozygotes, either an embryonic lethality or male sterility. Such characters as lethality, sterility, and measurable segregation ratios furnish an excellent opportunity for calculating the effect of selection as a function of gene frequency in hypothetical populations. Such calculations, based on a deterministic model of selection, indicate that these alleles should have certain equilibrium frequencies in the populations in which they occur. Studies of wild populations, however, consistently give frequencies below the calculated values. Lewontin concludes that the deficiency must be ascribed to some force in opposition to genic selection, and that group selection is the likely force. He showed that by substituting a stochastic model of natural selection, so as to allow for a certain rate of fixation of one or another allele in family groups and small local populations, he could account for the observed low frequencies of the *t*-alleles.

It should be emphasized that this example relates to genes characterized by lethality or sterility and extremely marked segregation distortions. Selection of such genes is of the maximum possible intensity. Important changes in frequency can occur in a very few generations as a result of genic selection, and no long-term isolation is necessary. Populations so altered would then be subject to unusually intense group selection. A population in which a segregation distorter reaches a high frequency will rapidly become extinct. A small population that has such a gene in low frequency can lose it by drift and thereafter replace those that have died out. Only one locus is involved. One cannot argue form this example that group selection would be effective in producing a complex adaptation involving closely adjusted gene frequencies at a large number of loci. Group selection in this example cannot maintain very low frequencies of the biotically deleterious gene in a population because even a single heterozygous male immigrant can rapidly "poison" the gene pool. The most important question about the selection of these genes is why they should produce such extreme effects. The segregation distortion makes the genes extremely difficult to keep at low frequency by either genic or group selection. Why has there not been an effective selection of modifiers that would reduce this distortion? Why also has there not been effective selection for modifiers that would abolish the lethality and sterility. The *t*-alleles certainly must constitute an important part of the genetic environment of every other gene in the population. One would certainly expect the other genes to become adapted to their presence.

Segregation distortion is something of a novelty in natural populations. I would be inclined to attribute the low frequency of such effects to the adjustment of each gene to its genetic environment. When distorter genes appear they would be expected to replace their alleles unless they produced, like the *t*-alleles, drastic reductions in fitness at some stage of development. When such deleterious effects are mild, the population would probably survive and would gradually incorporate modifiers that would reduce the deleterious effects. In other words, the other genes would adjust to their new genetic environment. It is entirely possible, however, that populations and perhaps entire species could be rendered extinct by the introduction of such genes as the *t*-alleles of mice. Such an event would illustrate the production, by genic selection, of characters that are highly unfavorable to the survival of the species. The gene in question would produce a high phenotypic fitness in the gamete stage. It might have a low effect on some other stage. The selection coefficient would be determined by the mean of these two effects relative to those of alternative alleles, regardless of the effect on population survival. I wonder if anyone has thought of controlling the mouse population of an area by flooding it with *t*-carriers.

I am entirely willing to concede that the kinds of adaptations evolved by a population, for instance segregation distortion, might influence its chance for continued survival. I question only the effectiveness of this extinction-bias in the production and maintenance of any adaptive mechanisms worthy of the name. This is not the same as denying that extinction can be an important factor in biotic evolution. The conclusion is inescapable that extinction has been extremely important in producing the Earth's biota as we know it today. Probably only on the order of a dozen Devonian vertebrates have left any

Recent descendants. If it had happened that some of these dozen had not survived, I am sure that the composition of today's biota would be profoundly different.

Another example of the importance of extinction can be taken from human evolution. The modern races and various extinct hominids derive from a lineage that diverged from the other Anthropoidea a million or perhaps several million years ago. There must have been a stage in which man's ancestors were congeneric with, but specifically distinct from, the ancestors of the modern anthropoid apes. At this time there were probably several and perhaps many other species in this genus. All but about four, however, became extinct. One that happened to survive produced the gibbons, another the orang, another the gorilla and chimpanzee, and another produced the hominids. These were only four (or perhaps three or five) of an unknown number of contemporary Pliocene alternatives. Suppose that the number had been one less, with man's ancestor being assigned to the group that became extinct! We have no idea how many narrow escapes from extinction man's lineage may have experienced. There would have been nothing extraordinary about his extinction; on the contrary, this is the statistically most likely development. The extinction of this lineage would, however, have provided the world today with a strikingly different biota. This one ape, which must have had a somewhat greater than average tendency toward bipedal locomotion and, according to recent views, a tendency towards predatory pack behavior, was transferred by evolution from an ordinary animal, with an ordinary existence, to a cultural chain reaction. The production and maintenance of such tributary adaptations as an enlarged brain, manual dexterity, the arched foot, etc. was brought about by the gradual shifting of gene frequencies at each genetic locus in response to changes in the genetic, somatic, and ecological environments. It was this process that fashioned a man from a beast. The fashioning was not accomplished by the survival of one animal type and the extinction of others.

I would concede that such matters of extinction and survival are extremely important in biotic evolution. Of the systems of adaptations produced by organic evolution during any given million years, only a small proportion will still be present several million years later. The surviving lines will be a somewhat biased sample of those actually produced by genic selection, biased in favor of one type of adaptive organization over another, but survival will always be largely a matter of historical accident. It may be that some people would not even recognize such chance extinction as important in biotic evolution. Ecologic determinists might attribute more of a role to the niche factor; man occupies an ecologic niche, and if one ancestral ape had failed to fill it, another would have. This sort of thinking probably has some validity, but surely historical contingency must also be an important factor in evolution. The Earth itself is a unique historical phenomenon, and many unique geological and biological events must have had a profound effect on the nature of the world's biota.

There is another example that should be considered, because it has been used to illustrate a contrary point of view. The extinction of the dinosaurs may have been a necessary precondition to the production of such mammalian types as elephants and bears. This extinction, however, was not the creative force that designed the locomotor and trophic specializations of these mammals. That force can be recognized in genic selection in the mammalian populations. There are analogies in human affairs. In World War II there was a rubber shortage due to the curtailment of imports of natural rubber. Scientists and engineers were thereby stimulated to develop suitable substitutes, and today we have a host of their inventions, some of which are superior to natural rubber for many uses. Necessity may have been the mother of invention, but she was not the inventor. I would liken the curtailment of imports, surely not a creative process, to the extinction of the dinosaurs, and the efforts of the scientists and engineers, which certainly were creative, to the selection of alternative alleles within the mammalian populations. In this attitude I ally myself with Simpson (1944) and against Wright (1945), who argued that the extinction of the dinosaurs, since it may have aided the adaptive radiation of the mammals, should be regarded as a creative process.

Group selection is the only conceivable force that could produce biotic adaptation. It was necessary, therefore, in this discussion of biotic adaptation to examine the nature of group selection and to attempt some preliminary evaluation of its power. The issue, however, cannot be resolved on the basis of hypothetical examples and appeals to intuitive judgments as to what seems likely or unlikely. A direct assessment of the importance of group selection would have to be based on an accurate knowledge of rates of genetic change, due to different causes, within populations; rates of proliferation and extinction of populations and larger groups; relative and absolute rates of migration and interbreeding; relative and absolute values of the coefficients of genic and group selection; etc. We would need such information for a large and unbiased sample of present and past taxa. Obviously this ideal will not be met, and some indirect method of evaluation will be necessary. The only method that I can conceive of as being reliable is an examination of the adaptations of animals and plants to determine the nature of the goals for which they are designed. The details of the strategy being employed will furnish indications of the purpose of its employment. I can conceive of only two ultimate purposes as being indicated, genic survival and group survival. All other kinds of survival, such as that of individual somata, will be of the nature of tactics employed in the grand strategy, and such tactics will be employed only when they do, in fact, contribute to the realization of a more general goal.

The basic issue then is whether organisms, by and large, are using strategies for genic survival alone, or for both genic and group survival. If both, then which seems to be the predominant consideration? If there are many adaptations of obvious group benefit which cannot be explained on the basis of genic selection, it must be conceded that group selection has been operative and important. If there are no such adaptations, we must conclude that group selection has not been important, and that only genic selection—natural se-lection in its most austere form—need be recognized as the creative force in evolution. We must always bear in mind that group selection and biotic adaptation are more onerous principles than genic selection and organic adap-tation. They should only be invoked when the simpler explanation is clearly inadequate. Our search must be specifically directed at finding adaptations that promote group survival but are clearly neutral or detrimental to individ-ual reproductive survival in within-group competition. The criteria for the recognition of these biotic adaptations are essentially the same as those for organic adaptations. The system in question should produce group benefit in an economical and efficient way and involve enough potentially independent elements that mere chance will not suffice as an explanation for the beneficial effect.

The examples considered above all related to interactions between indi-viduals, and the important consideration was to find a parsimonious explana-tion of why one individual would expend its own resources or endanger itself in an attempt to aid another. There remain a number of examples of indi-viduals' acting, at their own expense, in a manner that benefits their con-specific neighbors in general, not specific individuals. Such activity can take place only when the animals occur in unrelated groups larger than two. The important initial problem is why animals should exist in groups of several to many individuals.

It is my belief that two basic misconceptions have seriously hampered progress in the study of animals in groups. The first misconception is the assumption that when one demonstrates that a certain biological process produces a certain benefit, one has demonstrated *the* function, or at least *a* function of the process. This is a serious error. The demonstration of a benefit is neither necessary nor sufficient in the demonstration of function, although it may sometimes provide insight not otherwise obtainable. It is both neces-sary and sufficient to show that the process is designed to serve the function. A relevant example is provided by Allee (1931). He observed that a certain marine flatworm, normally found in aggregated groups, can be killed by placement in a hypotonic solution. The harmfulness of such a solution is reduced when large numbers of worms, not just one or a few, are exposed to it. The effect is caused by the liberation of an unknown substance from the worms, especially dead ones, into the water. The substance is not osmotically important in itself, but somehow protects the worms against hypotonicity. Allee saw great significance in this observation, and assumed that he had demonstrated that a beneficial chemical conditioning of the environment is a function of aggregation in these worms. The fallacy of such a conclusion should be especially clear when it relates to very artificial situations like placing large numbers of worms in a small volume of brackish water. The kind of evidence that would be acceptable would be the demonstration that social cohesion increased as the water became hypotonic or underwent some other chemically harmful change; that specific integumentary secretory machinery was activated by the deleterious change; that the substance secreted not only provided protection against hypotonicity, but was an extraordinarily effective substance for this protection. One or two more links in such a chain of circumstances would provide the necessary evidence of functional design and leave no doubt that protection from hypotonicity was a function of aggregation, and not merely an effect.

The second misconception is the assumption that to explain the functional aspects of groups, one must look for group functions. An analogy with human behavior will illustrate the nature of this fallacy. Suppose a visitor from Mars, unseen, observed the social behavior of a mob of panic-stricken people rushing from a burning theatre. If he was burdened with the misconception in question he would assume that the mob must show some sort of an adaptive organization for the benefit of the group as a whole. If he was sufficiently blinded by this assumption he might even miss the obvious conclusion that the observed behavior could result in total survival below what would have resulted from a wide variety of other conceivable types of behavior. He would be impressed by the fact that the group showed a rapid "response" to the stimulus of fire. It went rapidly from a widely dispersed distribution to the formation of dense aggregations that very effectively sealed off the exits.

Someone more conversant with human nature, however, would find the explanation not in a functioning of the group, but in the functioning of individuals. An individual finds himself in a theatre in which a dangerous fire has suddenly broken out. If he is sitting near an exit he may run for it immediately. If he is a bit farther away he sees others running for the exits and, knowing human nature, realizes that if he is to get out at all he must get out quickly; so he likewise runs for the door, and in so doing, intensifies the stimulus that will cause others to behave in the same way. This behavior is clearly adaptive from the standpoint of individual genetic survival, and the behavior of the mob is easily understood as the statistical summation of individual adaptation.

This is an extreme example of damage caused by the social consequences of adaptive behavior, but undoubtedly such effects do occur, and they may be fairly common in some species. There are numerous reports, at least at the anecdotal level, of the mass destruction of large ungulates when individuals in the van of a herd are pushed off cliffs by the press from the rear. Less spectacular examples of harm deriving from social grouping are probably of greater significance. I would imagine the most important damage from social behavior to be the spread of communicable disease.

The statistical summation of adaptive individual reactions, which I believe to underlie all group action, need not be harmful. On the contrary, it may often be beneficial, perhaps more often than not. An example of such a benefit would be the retention of warmth by close groups of mammals or birds in cold weather, but there is no more reason to assume that a herd is designed for the retention of warmth than to assume that it is designed for transmitting diseases. The huddling behavior of a mouse in cold weather is designed to minimize its own heat loss, not that of the group. In seeking warmth from its neighbors it contributes heat to the group and thereby makes the collective warmth a stronger stimulus in evoking the same response from other individuals. The panic-stricken man in the theatre contributed to the panic stimulus in a similar fashion. Both man and mouse probably aid in the spread of disease. Thus the demonstration of effects, good or bad, proves nothing. To prove adaptation one must demonstrate a functional design.

REFERENCES

Allee, W. C. 1931. *Animal Aggregations: A Study in General Sociology*. Chicago: University of Chicago Press.

―――. 1940. "Concerning the origin of sociality in animals." *Scientia* 1940:154–160.

Allee, W. C., Alfred E. Emerson, Orlando Park, Thomas Park, and Karl P. Schmidt. 1949. *Principles of Animal Ecology*. Philadelphia: W. B. Saunders.

Altman, Stuart A. 1962. "A field study of the sociobiology of rhesus monkeys, *Macaca mulatta*." *Ann. N. Y. Acad. Sci.* 102:338–435.

Clarke, C. A., C. G. G. Dickson, P. M. Sheppard. 1963. "Larval color pattern in *Papilio demodocus*." *Evolution* 17:130–137.

Darwin, Charles R. 1896. *The Descent of Man and Selection in Relation to Sex*. New York: D. Appleton.

Fisher, R. A., E. B. Ford. 1947. "The spread of a gene in natural conditions in a colony of the moth, *Panaxia dominula* (L)." *Heredity* 1:143–174.

Ford, E. B. 1956. "Rapid evolution and the conditions which make it possible." *Cold Spring Harbor Symp. Quant. Biol.* 20:230–238.

Haldane, J. B. S. 1932. *The Causes of Evolution*. London: Longmans.

Hamilton, W. D. 1964. "The genetical evolution of social behaviour, I." *J. Theoret. Biol.* 7:1–16.

Lewontin, R. C. 1958. "The adaptations of populations to varying environments." *Cold Spring Harbor Symp. Quant. Biol.* 22:395–408.

―――. 1961. "Evolution and the theory of games." *J. Theoret. Biol.* 1:382–403.

―――. 1962. "Interdeme selection controlling a polymorphism in the house mouse." *Am. Naturalist* 96:65–78.

Lewontin, R. C., and L. C. Dunn. 1960. "The evolutionary dynamics of a polymorphism in the house mouse." *Genetics* 45:705–722.

Montagu, M. F. Ashley. 1952. *Darwin, Competition and Cooperation*. New York: Henry Schuman.

Russell, E. S. 1945. *The Directiveness of Organic Activities*. Cambridge: Cambridge University Press.

Simpson, George Gaylord. 1944. *Tempo and Mode in Evolution*. New York: Columbia University Press.

―――. 1953. *The Major Features of Evolution*. New York: Columbia University Press.

Slijper, E. J. 1962. *Whales*. Trans. A. J. Pomerans, New York: Basic Books.

Slobodkin, L. Basil. 1954. "Population dynamics of *Daphnia obtusa* Kurz." *Ecol. Monog.* 24:69–88.

Williams, George C. 1966. *Adaptation and Natural Selection*. Princeton: Princeton University Press.

Wright, Sewall. 1945. *"Tempo and mode in evolution*: A critical review." *Ecology* 26:415–419.

Wynne-Edwards, V. C. 1962. *Animal Dispersion in Relation to Social Behaviour*. Edinburgh and London: Oliver and Boyd.

7 Levels of Selection: An Alternative to Individualism in Biology and the Human Sciences

David Sloan Wilson

Biology and many branches of the human sciences are dominated by an individualistic tradition that treat groups and communities as collections of organisms without themselves having the properties implicit in the word "organism". In biology, the individualistic tradition achieves generality only by defining self-interest as "anything that evolves by natural selection." A more meaningful definition of self-interest shows that natural selection operates on a hierarchy of units from genetic elements to multispecies communities, and that a unit becomes organismic to the degree that natural selection operates at the level of that unit. I review levels-of-selection theory in biology and sketch a parallel argument for the human sciences.

INTRODUCTION

The related concepts of adaptation, function, intention and purpose are central to both biology and the human sciences. Natural selection endows species with the functional design required to survive and reproduce in their environments. Humans organize their behavior to achieve various proximate goals in their everyday lives.

Biology and the human sciences also share a controversy over the units that can be said to have the properties of adaptation, function, intention, and purpose. Almost everyone would grant these properties to individuals, but some biologists also speak of social groups and multi-species communities as if they were single purposeful organisms. Similarly, some psychologists, anthropologists, and sociologists speak of culture and society as superorganisms in which individuals are mere cells.

In recent decades the hierarchical view of functional organization has fallen on hard times. Larger entities are regarded as mere collections of organisms, without themselves having the properties of organisms. In biology the reductionistic trend has proceeded so far that even individuals are sometimes treated as upper units of the hierarchy, mere collections of "selfish" genes (Dawkins 1976, 1982). The human sciences are more heterogeneous, but many of its branches appear to be dominated by the individualistic view.

Despite its widespread acceptance, the case for individualism as a general prediction that emerges from evolutionary theory, or as a general principle to explain human behavior, actually is very frail. In this chapter I will describe

From *Social Networks*, 1989, 11:257–72.

why functional organization in nature is necessarily hierarchical and then will attempt to sketch a parallel argument for the human sciences.

THE EVOLUTION OF ALTRUISM

In biology, the debate over units of adaptation has centered on the evolution of seemingly altruistic behaviors that benefit others at the expense of the self. Consider a population of N individuals. Two types exist, A and S, in proportions p and $(1 - p)$, respectively. Each A-type expresses a behavior toward a single recipient, chosen at random from the population. As a result, the recipient has an additional number b of offspring while the altruist has c fewer offspring. The average number of offspring, W, can then be calculated for each type.

$$W_A = X - c + b(Np - 1)/(N - 1), \qquad W_S = X + bNp/(N - 1) \qquad (1)$$

X is the number of offspring in the absence of altruistic behaviors, and is the same for both types. In addition to the cost of being an altruist, each A-type can serve as a recipient to the $(Np - 1)$ other altruists who are distributing their benefits among $(N - 1)$ individuals in the group. Selfish S-types have no cost of altruism and can serve as recipients to all Np altruists in the group. S-types have more offspring than A-types whenever $W_S > W_A$, which reduces to the inequality.

$$b/(N - 1) > -c. \qquad (2)$$

This inequality always holds, because b, c, and N are positive numbers and N is greater than 1. Thus, selfish types always have more offspring than altruistic types. To the degree that the behaviors are heritable, selfish types will be found at a greater frequency in the next generation.

A numerical example is shown in table 7.1, in which $N = 100$, $p = 0.5$, $X = 10$, $b = 5$, and $c = 1$. Thus, the altruist bestows an additional 5 offspring on the recipient at a cost of 1 offspring to itself. The average altruist has 11.47 offspring, while the average selfish type has 12.53 offspring. Assume that the types reproduce asexually, such that the offspring exactly resemble the parents. The proportion of altruists among the progeny is then $p' = 0.478$, a decline from the parental value of $p = 0.5$. Since populations cannot grow to infinity, we also assume that mortality occurs equally among the A- and

Table 7.1. Evolution in a single population

$N = 100$, $p = 0.5$, $X = 10$, $b = 5$, $c = 1$
$W_A = X - c + b(Np - 1)/(N - 1) = 10 - 1 + 49(5)/99 = 11.47$
$W_S = X + bNp/(N - 1) = 10 + 50(5)/99 = 12.53$
$N' = N(pW_A + (1 - p)W_S) = 100(0.5(11.47) + 0.5(12.53)) = 1200$
$p' = NpW_A/N' = 100(0.5)(11.47)/1200 = 0.478$

Note: The altruistic type declines from a frequency of $p = 0.5$ before selection to a frequency of $p' = 0.478$ after selection.

S-types, returning the population to a density of $N = 100$. At this point we expect approximately 52 selfish and 48 altruistic types. If this procedure is iterated many times, representing natural selection acting over many generations, the A-types continue to decline in frequency and ultimately become extinct.

This is the paradox that makes altruism such a fascinating subject for evolutionary biologists. As humans we would like to think that altruism can evolve, as biologists we see animal behaviors that appear altruistic in nature, yet almost by definition it appears that natural selection will act against them. This is the sense in which evolution appears to be an inherently selfish theory.

The paradox, however, can be resolved by a simple alteration of the model. Table 7.2 differs from table 7.1 in only two respects: (1) we now have two groups instead of one; and (2) the groups have different proportions of altruistic and selfish types. Looking at each group separately, we reach the same conclusion as for table 7.1; selfish types have more offspring than altruistic types. Adding the individuals from both groups together, however, we get the opposite answer: altruistic types have more offspring than selfish types.[1]

What has happened to produce this interesting (and for many people counterintuitive) result? First, there must be more than one group; there must be a *population of groups*. Second, the groups cannot all have the same proportion of altruistic types, for then the results would not differ from a single group. The groups must *vary* in the proportion of altruistic types. Third, there must be a direct relationship between the proportion of altruists and the total number of offspring produced by the group; groups of altruists must be *more fit* than groups without altruists. These are the necessary conditions for the evolution of altruism in the elaborated model. To be sufficient, the differential

Table 7.2. Evolution in two groups that differ in the proportion of the altruistic type

Group 1	Group 2
$N_1 = 100$, $p_1 = 0.2$	$N_2 = 100$, $p_2 = 0.8$
$W_A = 10 - 1 + 19(5)/99 = 9.96$	$W_A = 10 - 1 + 79(5)/99 = 12.99$
$W_S = 10 + 20(5)/99 = 11.01$	$W_S = 10 + 80(5)/99 = 14.04$
$N'_1 = 1080$	$n'_2 = 1320$
$p'_1 = 0.184$	$p'_2 = 0.787$

Global population
$N = 200$, $P = 0.5$
$N' = N'_1 + N'_2 = 2400$
$P' = (N'_1 p'_1 + N'_2 p'_2)/(N'_1 + N'_2) = 0.516$

Note: Values for X, b, c and the functions for W_A and W_S are provided in Table 1. The altruistic type declines in frequency within each group (compare p'_1 with p_1 and p'_2 with p_2) but increases in frequency when both groups are considered together (compare P' with P). This is because group 2, with the most altruists, is more productive than group 1 (compare N'_2 with N'_1).

fitness of groups—the force favoring the altruists—must be great enough to counter the differential fitness of individuals within groups—the force favoring the selfish types.

Readers familiar with evolutionary theory immediately will recognize a similarity between the above conditions and Darwin's original theory of natural selection, which requires a *population of individuals*, that *vary* in their genetic composition, with some variants *more fit* than others. Thus, natural selection can operate simultaneously at more than one level. Individual selection promotes the fitness of individuals relative to others in the same group. Group selection promotes the fitness of groups, relative to other groups in the global population. These levels of selection are not always in conflict. A single behavior can benefit both the individual performing it and others in the group. Altruistic behaviors by definition are costly to self and beneficial to others, however, and so are favored by group selection and disfavored by individual selection.

This simple numerical example shows that the process of natural selection does not inevitably evolve selfish behaviors. A notion of *group-interest* must be added to the notion of *self-interest*, to the extent that group selection is important in nature.

VALID INDIVIDUALISM AND CHEAP INDIVIDUALISM

Let us now consider the individualistic claim that "virtually all adaptations evolve by individual selection." If by individual selection we mean within-group selection, we are saying that A-types virtually never evolve in nature, that we should observe only S-types. This is a meaningful statement because it identifies a set of traits that conceivably could evolve, but does not, because between-group selection is invariably weak compared to within-group selection. Let us call this *valid individualism*.

There is, however, another way to calculate fitness in the two-group model that leads to another definition of individual selection. Instead of separately considering evolution within groups and the differential fitness of groups, we can directly average the fitness of A- and S-types across all groups. Thus, the 2 A-types in groups one have 9.96 offspring and the 8 A-types in group two have 12.99 offspring, for an average fitness of $0.2(9.96) + 0.8(12.99) = 12.38$. The 8 S-types in group one have 11.01 offspring and the 2 S-types in group two have 14.04 offspring, for an average fitness of $0.8(11.01) + 0.2(14.04) = 11.62$. The average A-type individual is more fit then the average S-type individual, which is merely another way of saying that it evolves.

Let us now return to the individualistic claim that "virtually all adaptations evolve by individual selection." If by individual selection we mean the fitness of individuals averaged across all groups, we have said nothing at all. Since this definition includes both within- and between-group selection, it makes "individual selection" synonymous with "whatever evolves," including either S-types or A-types. It does not identify any set of traits that conceivably could evolve but does not. Let us therefore call it *cheap individualism*.

Cheap individualism is so meaningless that no one would explicitly endorse it. Even the most ardent individualists, such as G. C. Williams (1966, 1985), R. Dawkins (1976, 1982), and J. Maynard Smith (1987), believe that there is something outside individual selection called group selection that in principle can evolve altruistic traits. Nevertheless, the history of individual selection from 1960 to the present has been a slow slide from valid individualism to cheap individualism. Before documenting this claim it is necessary to review three reasons why the slide could occur unnoticed.

First, group-structured population models such as the one described above can be applied to an enormous range of biological phenomena. The single groups can be isolated demes that persist for many generations, groups of parasites interacting within single hosts, clusters of caterpillars interacting on a single leaf, or coalitions of baboons that behaviorally segregate within a larger troop. The groups can be communities whose members are separate species, social units whose members are conspecifics, or even single organisms whose "members" are genes of cell lineages (Crow 1979; Cosmides and Tooby 1981; Buss 1987). Historically, however, the first group selection models focused on a particular conception of isolated demes that persist for many generations. Thus, it has been possible for biologists studying other kinds of groups to assume that they are not invoking group selection, when in fact their models are miniature versions of traditional group selection models.

Second, many biologists today regard group selection as a heretical concept that was discarded twenty years ago and consider their own work to be entirely within the grand tradition of "individual selection." Gould (1982:xv) remembers "the hooting dismissal of Wynne-Edwards and group selection in any form during the late 1960's and most of the 1970's," and even today graduate students tell me how difficult it is for them to think about group selection in a positive light after being taught in their courses that it "just doesn't happen." The vast majority of authors who claim that such-and-such evolves by individual selection do not even include an explicit model of group selection to serve as a possible alternative. Individual selection truly has become the modern synonym for "everything that evolves in my model," and group selection is mentioned only as a bogey man in the introduction or the conclusion of the paper.

Third, averaging the fitness of individual types across groups is a useful, intuitively reasonable procedure that correctly predicts the outcome of natural selection. Biologists commonly average the fitness of types across a range of physical environments, and it seems reasonable to average across social environments in the same way. I emphasize that there is nothing wrong with this procedure—it merely cannot be used to define individual selection because it leaves nothing outside of it.

Now I must document my claim that individualism in biology achieves generality only by averaging the fitness of individuals across groups.

THREE EXAMPLES OF CHEAP INDIVIDUALISM IN BIOLOGY

The Evolution of Avirulence in Parasites and Diseases

Disease organisms provide an excellent real-world example of a group-structured population similar to the model outlined above. Each infected host comprises an isolated group of disease organisms, which compete with other groups to infect new hosts. Natural selection within single hosts is expected to favor strains with high growth rates. Excessively high growth rates tend to kill the host, however, driving the entire group of disease organisms extinct (assuming that transmission requires the host to be alive). Avirulent strains therefore can be envisioned as "altruists" that increase the survival of entire groups, but which nevertheless decline in frequency within every group containing more virulent strains. Lewontin (1970) was the first to recognize that avirulence evolves by between-group selection, and the process has been well documented in a *myxoma* virus that was introduced into Australia to control the European rabbit (Fenner and Ratcliffe 1965). Nevertheless, consider the following account in the first edition of Futuyma's (1979:455) textbook *Evolutionary Biology*:

In many interactions the exploiter cannot evolve to be avirulent; it profits a fox nothing to spare the hare. But if the fitness of an individual parasite or its offspring is lowered by the death of its host, avirulence is advantageous. The *myxoma* virus, introduced into Australia to control European rabbits, at first caused immense mortality. But within a few years mortality levels were lower, both because the rabbits had evolved resistance and because the virus had evolved to be less lethal.... Because the virus is transmitted by mosquitoes that feed only on living rabbits, *virulent virus genotypes are less likely to spread than benign genotypes* [italics mine]. Avirulence evolves not to assure a stable future supply of hosts, but to benefit individual parasites.

Thus, by the simple procedure of comparing the fitness of virulent and avirulent types across all hosts (see italicized portion of text), rather than within single hosts, the evolution of avirulence can be made to appear an individualistic process. Futuyma, incidentally, is sympathetic to the concept of group selection and properly attributes avirulence to between-group selection in the second edition of his textbook (1986:496–497). This example of cheap individualism therefore is inadvertent, and shows how easily selection at multiple levels can be represented as occurring entirely at the lowest level.

Inclusive Fitness Theory

Within the individualistic tradition in biology, natural selection is widely thought to maximize a property called inclusive fitness, which is the sum of an individual's effects on the fitness of others multiplied by the probability that the others will share the genes causing the behavior. As Hamilton (1963: 354–355) originally put it:

Despite the principle of "survival of the fittest" the ultimate criterion which determines whether G [an altruistic allele] will spread is not whether the behavior is to the benefit of the behavior but whether it is to the benefit of the gene G; and this will be the case if the average net result of the behavior is to add to the gene-pool a handful of genes containing G in higher concentration than does the gene-pool itself. With altruism this will happen only if the affected individual is a relative of the altruist, therefore having an increased chance of carrying the gene, and if the advantage conferred is large enough compared to the personal disadvantage to offset the regression, or "dilution," of the altruist's genotype in the relative in question.

In this formulation, individuals evolve to maximize the fitness of "their genes" relative to other genes in the population, regardless of whether "their genes" are located in children, siblings, cousins, parents, and so on. Aid-giving toward relatives therefore ceases to appear altruistic, and becomes part of an individual's "selfish" strategy to maximize its inclusive fitness. Even sterility and death can be inclusive fitness maximizing if the positive effects on relatives are sufficiently great.

Let us pursue this idea by considering an Aa female who mates with an aa male and produces a clutch of ten offspring, five of whom are Aa and the other five aa. The dominant allele A codes for an altruistic behavior that is expressed only toward siblings. The sibling group therefore is equally divided between altruists and nonaltruists, and the fitness of the two genotypes from equation (1) is

$$W_{Aa} = X - c + b(4/9), \qquad W_{aa} = X + b(5/9).$$

The selfish aa genotype is inevitably most fit, which merely reiterates the general conclusion obtained [previously] for evolution in all single groups. The fact that the group in this case consists of full siblings is irrelevant to the conclusion. To see how altruism expressed toward siblings evolves, we must consider a large number of family groups, initiated by all combinations of parental genotypes — AA × AA, AA × Aa, Aa × Aa, AA × aa, Aa × aa, aa × aa. Within-group selection favors the selfish a-allele in all groups containing both altruistic and selfish genotypes. The fitness of entire sibling groups, however, is directly proportional to the frequency of altruistic A-alleles in the group. Thus, Hamilton's conclusions cannot be reached without combining within-group selection and between-group selection into a single measure of "inclusive fitness."

The idea that aid-giving toward relatives is a form of "true" altruism that requires between-group selection has been reached by many authors (reviewed in Wilson 1983). Nevertheless, evolutionists within the individualistic tradition continue to use inclusive fitness theory as their guiding light to explain the evolution of "apparently" altruistic behaviors, "without invoking group selection." This is cheap individualism.

Diploid Population Genetics and Evolutionary Game Theory

My final example involves a comparison between two seemingly different bodies of theory in evolutionary biology. Diploid population genetics models begin with a population of gametic types (A, a) which combine into pairs to form diploid genotypes (AA, Aa, aa). Selection usually is assumed to occur in the diploid stage, after which the genotypes dissociate back into gametes and the process is reiterated. The most common way for selection to occur in these models is for some genotypes to survive and reproduce better than others, the standard process of between-individual selection. In addition, however, it is possible for some alleles to survive and reproduce better than others *within single individuals*. For example, the rules of meiosis usually cause the two chromosome sets to be equally represented in the gametes. Some alleles manage to break the rules of meiosis, however, biasing their own transmission into the sperm and eggs of heterozygotes. The differential fitness of alleles within heterozygotes is termed meiotic drive, and can cause the evolution of genes that have neutral or even deleterious effects on the fitness of individuals (Crow 1979; Cosmides and Tooby 1981). In short, diploid population genetics models are explicitly hierarchical by recognizing the existence of both between- and within-individual selection.

Evolutionary game theory (also called ESS theory for "evolutionarily stable strategy") begins with a population of individual types (A, a) that combine into groups of size N for purposes of interaction. Selection occurs during the grouped stage, after which the groups dissociate back into individuals and the process is reiterated. Usually $N = 2$, which yields three types of groups (AA, Aa, aa). ESS theory was borrowed directly from economic game theory (Maynard Smith and Price 1973; Maynard Smith 1982) but the two are not identical. In particular, economic game theory assumes that the players are rational actors trying to maximize their (absolute) payoff, while ESS theory assumes that natural selection will favor the strategy that delivers the highest payoff relative to other competing strategies in the population.

It should be obvious that the population structure of genes combining into individuals in a diploid model is identical to the population structure of individuals combining into groups of $N = 2$ in an ESS model. Similarly, natural selection in an ESS model can happen in two ways: groups can outperform other groups or individuals can outperform other individuals within groups. In the familiar hawk-dove model, for example, dove-dove groups (in which resources are equitably shared) are more fit than hawk-hawk groups (in which resources are contested), while hawks are more fit than doves within hawk-dove groups. To be consistent with population genetics models we should say that hawks are favored by within-group selection and doves by between-group selection. ESS theorists, however, average the fitness of individual types across groups and call everything that evolves the product of "individual selection." The term "between-group selection" is never used, and Maynard Smith actually borrowed game theory from economics as an alternative to group selection (Maynard Smith and Price 1973; Maynard Smith 1982). As Dawkins (1980:360) puts it: "There is a common misconception that cooperation within a group at a given level of organization must come about through selection between groups.... ESS theory provides a more parsimonious alternative." This one passage provides all the elements of cheap individualism: the fitness of individuals is averaged across groups, everything that evolves is called the product of individual selection, and something else is called group selection, outside the model and completely unspecified, except to say that it need not be invoked.

These three examples show that, despite its widespread acceptance, individualism in biology is on very thin ice. Self-interest defined as "whatever evolves" is meaningless, and yet when self-interest is defined more meaningfully as "within-group selection" it cannot claim to explain everything that evolves in nature. We must therefore accept a hierarchical view of evolution in which the properties of functional organization implicit in the word "organism" need not be restricted to individuals. The differential fitness of genetic elements within individuals ushers us into a bizarre world in which the genetic elements are the purposeful organisms and individuals are mere collections of quarreling genes, the way we usually think of groups. The differential fitness of individuals within groups ushers us into a familiar world in which groups are mere collections of purposeful individuals. The differential fitness of groups ushers us into another bizarre world (for individualists) in which the groups are the organisms whose properties are caused by individuals acting in a coordinated fashion, the way we usually think of genes and the organs they code for. See Wilson and Sober (1989) for a more detailed review of levels-of-selection theory in biology.

A PARALLEL ARGUMENT FOR THE HUMAN SCIENCES

If human behavior is measured against the dual standard of effects on self and effects on others, it appears to show the full range of potential. Individuals have sacrificed their lives for the benefit of others, and they have sacrificed the lives of others for their own trivial gain. Viewed at the society level, some human groups are so well coordinated that they invite comparison to single organisms, while others show all the disorganization of a bar-room brawl.

Humans also are frequently embedded in a complex network of interactions in which single expressions of a behavior affect the actor and a relatively small number of associates. Put another way, human populations are subdivided into clusters of associates similar to the local populations of the evolutionary models outlined above. It seems possible that a theory of human behavior in social networks could be developed that parallels levels-of-selection theory in biology, leading to a similar hierarchical view of functional organization in human affairs.

As with any theory of human behavior, the first step is to specify the rules that cause people to choose among alternative behaviors, which serve as the analog of natural selection in an evolutionary model. Following Axelrod and others (Axelrod and Hamilton 1981; Brown et al. 1982; Pollock 1988), assume that humans adopt behaviors that maximize a given utility, relative to competing behaviors in the population. The utility might be pleasure (to a psychologist), annual income (to an economist), or genetic fitness (to a sociobiologist). The details of the utility are relatively unimportant because the hallmark of a hierarchical model is not the nature of the utility but the way it is partitioned into within- and between-group components. Consider, for example, a behavior that decreases the utility of self and increases the utility of others. If others include the entire population, then the utility of those expressing the behavior will be lower than those that do not, and the behavior will be rejected precisely as it is selected against in the one-group evolutionary model. Now assume that the human population is subdivided into a mosaic of associates in which the expression of behavior is nonrandom; some groups of associates behave primarily one way, other groups the other way. The utility of the behaviors now depends on the frame of comparison. The behavior fares poorly in all groups in which the alternative behavior is expressed, but may still deliver the highest utility when averaged across all groups, exactly as in the multigroup evolutionary model. Adoption of the behavior therefore depends on two factors, the effect on self and others and the interaction structure within which the behavior is embedded.

Theories of behavior in the human sciences frequently consider both factors but combine them into an overarching definition of self-interest as "utility-maximizing behavior"—i.e., all behaviors adopted by rational humans! This is cheap individualism, that achieves generality only by definitional fiat. Levels-of-selection theory keeps the factors separate, defining behaviors as self-interested when they increase relative utility within single groups, and group-interested when they increase the average utility of groups, relative to other groups. This provides a framework in which rational (utility maximizing) humans need not be self-interested by definition.

As for the situation in biology, many human behaviors that are catagorized as selfish by cheap individualism emerge as "groupish" in a levels-of-selection model.[2] The concept of morality, for example, involves rules of conduct that promote the common good. This implies a category of immoral behaviors— frequently termed "selfish" in everyday language—that benefit individuals at the expense of the common good. Since moral behaviors are vulnerable to exploitation, they succeed only if they can be segregated from the expression of immoral behaviors. This is nicely illustrated by the following passage from a seventeenth-century Hutterite document (English translation in Ehrenpreis 1978:67):

The bond of love is kept pure and intact by the correction of the Holy Spirit. People who are burdened with vices that spread and corrupt can have no part in it. This harmonious fellowship excludes any who are not part of the unanimous spirit.... If a man hardens himself in rebellion, the extreme step of separation is unavoidable. Otherwise the whole community would be dragged into his sin and become party to it.... The Apostle Paul therefore says "Drive out the wicked person from among you."

The maintenance of behaviorally pure groups allowed the Hutterites to practice such extreme altruism that their communities are best regarded as the human equivalent of a bee colony (a metaphor that they themselves used to describe themselves). More generally, human societies everywhere possess mechanisms for segregating behaviors, allowing less extreme forms of morally acceptable behavior to be successful. The distinction between moral and immoral behavior, and the mechanisms whereby both can be advantageous, correspond nicely to "groupish" and "selfish" behaviors in a levels-of-selection model. In contrast, cheap individualism is placed in the awkward situation of defining both moral and immoral behavior as brands of self-interest.

Many authors have expressed the idea that higher entities such as biological communities and human societies can be organisms in their own right. Unfortunately, the idea usually is stated as a poetic metaphor or as an axiom that is not subject to disproof. Levels-of-selection theory shows that single-species groups and multispecies communities can become functionally organized by the exact same process of between-unit selection that causes the groups of genes known as individuals to become functionally organized. For the first time, the hierarchical view in biology now enjoys a solid mechanistic foundation. Perhaps this foundation also will be useful within the human sciences to show how people sometimes coalesce into society-level organisms.

NOTES

This research was funded from a J. S. Guggenheim fellowship. I thank G. Pollock, R. Boyd, P. Richerson, and virtually dozens of other people for helpful conversations.

1. Adding the contents of both groups is justified biologically only if the occupants of the groups physically mix during a dispersal stage or compete for the colonization of new groups. See Wilson (1977, 1980, 1983) for a more detailed discussion of the nature of groups in levels-of-selection models.

2. Both cheap individualism and levels-of-selection models define their terms on the basis of utilities, which do not translate easily into psychological definitions of altruism and selfishness based on internal motivation. In outlining his economic theory of human behavior, Becker (1976:7) states that it does not matter how people actually feel or think about what they do as long as the end result of their behavior is utility maximizing. In the same way, behaviors categorized as group interested in a levels-of-selection model do not imply that the actor is internally motivated to help others. This does not mean that psychological definitions of altruism are irrelevant, but only that their relationship with definitions based on utility are complex. I hope to explore the complexities in a future paper.

REFERENCES

Axelrod, R., and W. D. Hamilton. 1981. "The evolution of cooperation." *Science* 211:1390–1396.

Becker, G. S. 1976. *The Economic Approach to Human Behavior*. Chicago: Chicago University Press.

Brown, J. S., M. J. Sanderson, and R. E. Michod. 1982. "Evolution of social behavior by reciprocation." *Journal of Theoretical Biology* 99:319–339.

Buss L. W. 1987. *The Evolution of Individuality*. Princeton: Princeton University Press.

Comides, L. M, and J. Tooby. 1981, "Cytoplasmic inheritance and intragenomic conflict." *Journal of Theoretical Biology* 89:83–129.

Crow. J. F, 1979. "Genes that violate Mendel's rules." *Scientific American* 240:104–113.

Dawkins. R. 1976. *The Selfish Gene*. Oxford: Oxford University Press.

―――. 1980. "Good strategy or evolutionary stable strategy?" In G. W. Barlow and J. Silverberg (eds.), *Sociobiology: Beyond Nature/Nurture?* pp. 331–367. Boulder. CO: Westview Press.

―――. 1982. *The Extended Phenotype.* Oxford: Oxford University Press.

Ehrenpreis, A. 1978. *Brotherly Community: The Highest Command of Love.* Rifton, NY: Plough Publishing Co.

Fenner, F., and F. N. Ratcliffe. 1965. *Myxomatosis.* London: Cambridge University Press.

Futuyma, D. J. 1979. *Evolutionary Biology* (first edn.). Sunderland, MA: Sinauer Press.

―――. 1986. *Evolutionary Biology* (second edition.). Sunderland, MA: Sinauer Press.

Gould, S. J. 1982. *The Uses of Heresy: An Introduction to Richard Goldschmidt's The Material Basis of Evolution,* pp. xiii–xlii. New haven, CT: Yale University Press.

Hamilton, W. D. 1963. "The evolution of altruistic behavior." *American Naturalist* 97 : 354–356.

Lewontin, R. C. 1970. "The units of selection." *Annual Review of Ecology and Systematics* 1 : 1–18.

Maynard Smith, J. 1982. *Evolution and the Theory of Games.* Cambridge: Cambridge, University Press.

―――. 1987. "How to model evolution." In J. Dupre (ed.), *The Latest on the Best: Essays on Evolution and Optimality,* pp. 117–131. Cambridge, MA: MIT Press.

―――, and G. R. Price. 1973. "The logic of animal conflict." *Nature* 246 : 15–18.

Pollock, G. B. 1988. "Population structure, spite, and the iterated prisoner's dilemma." *American Journal of Physical Anthropology* 77 : 459–469.

Williams, G. C. 1966. *Adaptation and Natural Selection.* Princeton: Princeton University Press.

―――. 1985. "A defense of reductionism in evolutionary biology." In R. Dawkins and M. Ridley (eds.), *Oxford Surveys in Evolutionary Biology.* Vol. 2, pp. 1–27. Oxford: Oxford University Press.

Wilson, D. S. 1977. "Structured demes and the evolution of group-advantageous traits." *American Naturalist* 111 : 157–185.

―――. 1980. *The Natural Selection of Populations and Communities.* Menlo Park, CA: Benjamin-Cummings.

―――. 1983. "The groups selection controversy: history and current status." *Annual Review of Ecology and Systematics* 14 : 159–187.

―――, and E. Sober. 1989. "Reviving the superorganism."*Journal of Theoretical Biology* 136 : 337–356.

V   Essentialism and Population Thinking

8 Typological versus Population Thinking

Ernst Mayr

Rather imperceptibly, a new way of thinking began to spread through biology soon after the beginning of the nineteenth century. It is now most often referred to as population thinking. What its roots were is not at all clear, but the emphasis of animal and plant breeders on the distinct properties of individuals was clearly influential. The other major influence seems to have come from systematics. Naturalists and collectors realized increasingly often that there are individual differences in collected series of animals, corresponding to the kind of differences one would find in a group of human beings. Population thinking, despite its immense importance, spread rather slowly, except in those branches of biology that deal with natural populations.

In systematics it became a way of life in the second half of the nineteenth century, particularly in the systematics of the better-known groups of animals, such as birds, mammals, fishes, butterflies, carabid beetles, and land snails. Collectors were urged to gather large samples at many localities, and the variation within populations was studied as assiduously as differences between localities. From systematics, population thinking spread, through the Russian school, to population genetics and to evolutionary biology. By and large it was an empirical approach with little explicit recognition of the rather revolutionary change in conceptualization on which it rested. So far as I know, the following essay, excerpted from a paper originally published in 1959, was the first presentation of the contrast between essentialist and population thinking, the first full articulation of this revolutionary change in the philosophy of biology.

The year of publication of Darwin's *Origin of Species*, 1859, is rightly considered the year in which the modern science of evolution was born. It must not be forgotten, however, that preceding this zero year of history there was a long prehistory. Yet, despite the existence in 1859 of a widespread belief in evolution, much published evidence on its course, and numerous speculations on its causation, the impact of Darwin's publication was so immense that it ushered in a completely new era.

It seems to me that the significance of the scientific contribution made by Darwin is threefold:

1. He presented an overwhelming mass of evidence demonstrating the occurence of evolution.

2. He proposed a logical and biologically well-substantiated mechanism that might account for evolutionary change, namely, natural selection. Muller (1949:459) has characterized this contribution as follows:

From *Evolution and the Diversity of Life*, Harvard University Press, 1975, 26–29.

Darwin's theory of evolution through natural selection was undoubtedly the most revolutionary theory of all time. It surpassed even the astronomical revolution ushered in by Copernicus in the significance of its implications for our understanding of the nature of the universe and of our place and role in it.... Darwin's masterly marshaling of the evidence for this [the ordering effect of natural selection], and his keen-sighted development of many of its myriad facets, remains to this day an intellectual monument that is unsurpassed in the history of human thought.

3. He replaced typological thinking by population thinking.

The first two contributions of Darwin are generally known and sufficiently stressed in the scientific literature. Equally important but almost consistently overlooked is the fact that Darwin introduced into the scientific literature a new way of thinking, "population thinking." What is this population thinking, and how does it differ from typological thinking, the then-prevailing mode of thinking? Typological thinking no doubt had its roots in the earliest efforts of primitive man to classify the bewildering diversity of nature into categories. The *eidos* of Plato is the formal philosophical codification of this form of thinking. According to it, there are a limited number of fixed, unchangeable "ideas" underlying the observed variability, with the *eidos* (idea) being the only thing that is fixed and real, while the observed variability has no more reality than the shadows of an object on a cave wall, as it is stated in Plato's allegory. The discontinuities between these natural "ideas" (types), it was believed, account for the frequency of gaps in nature. Most of the great philosophers of the seventeenth, eighteenth, and nineteenth centuries were influenced by the idealistic philosophy of Plato, and the thinking of this school dominated the thinking of the period. Since there is no gradation between types, gradual evolution is basically a logical impossibility for the typologist. Evolution, if it occurs at all, has to proceed in steps or jumps.

The assumptions of population thinking are diametrically opposed to those of the typologist. The populationist stresses the uniqueness of everything in the organic world. What is true for the human species—that no two individuals are alike—is equally true for all other species of animals and plants. Indeed, even the same individual changes continuously throughout its lifetime and when placed into different environments. All organisms and organic phenomena are composed of unique features and can be described collectively only in statistical terms. Individuals, or any kind of organic entities, form populations of which we can determine only the arithemetic mean and the statistics of variation. Averages are merely statistical abstractions; only the individuals of which the populations are composed have reality. The ultimate conclusions of the population thinker and of the typologist are precisely the opposite. For the typologist, the type (*eidos*) is real and the variation an illusion, while for the populationist the type (average) is an abstraction and only the variation is real. No two ways of looking at nature could be more different.

The importance of clearly differentiating these two basic philosophies and concepts of nature cannot be overemphasized. Virtually every controversy in the field of evolutionary theory, and there are few fields of science with as many controversies, was a controversy between a typologist and a populationist. Let me take two topics, race and natural selection, to illustrate the great difference in interpretation that results when the two philosophies are applied to the same data.

RACE

The typologist stresses that every representative of a race has the typical characteristics of that race and differs from all representatives of all other races by the characteristics "typical" for the given race. All racist theories are built on this foundation. Essentially, it asserts that every representative of a race conforms to the type and is separated from the representatives of any other race by a distinct gap. The populationist also recognizes races but in totally different terms. Race for him is based on the simple fact that no two individuals are the same in sexually reproducing organisms and that consequently no two aggregates of individuals can be the same. If the average difference between two groups of individuals is sufficiently great to be recognizable on sight, we refer to such groups of individuals as different races. Race, thus described, is a universal phenomenon of nature occurring not only in man but in two thirds of all species of animals and plants.

Two points are especially important as far as the views of the populationist on race are concerned. First, he regards races as potentially overlapping population curves. For instance, the smallest individual of a large-sized race is usually smaller than the largest individual of a small-sized race. In a comparison of races the same overlap will be found for nearly all examined characters. Second, nearly every character varies to a greater or lesser extent independently of the others. Every individual will score in some traits above, in others below the average for the population. An individual that will show in all of its characters the precise mean value for the population as a whole does not exist. In other words, the ideal type does not exist.

NATURAL SELECTION

A full comprehension of the difference between population and typological thinking is even more necessary as a basis for a meaningful discussion of the most important and most controversial evolutionary theory—namely, Darwin's theory of evolution through natural selection. For the typologist everything in nature is either "good" or "bad," "useful" or "detrimental." Natural selection is an all-or-none phenomenon. It either selects or rejects, with rejection being by far more obvious and conspicuous. Evolution to him consists of the testing of newly arisen "types." Every new type is put through a screening test and is either kept or, more probably, rejected. Evolution is defined as the preservation of superior types and the rejection of inferior ones, "survival of the fittest" as Spencer put it. Since it can be shown rather easily in any thorough analysis that natural selection does not operate in this described fashion, the typologist comes by necessity to the conclusions: (1) that natural selection does not work, and (2) that some other forces must be in operation to account for evolutionary progress.

The populationist, on the other hand, does not interpret natural selection as an all-or-none phenomenon. Every individual has thousands or tens of thousands of traits in which it may be under a given set of conditions selectively superior or inferior in comparison with the mean of the population. The greater the number of superior traits an individual has, the greater the probability that it will not only survive but also reproduce. But this is merely a probability, because under certain environmental conditions and temporary circumstances, even a "superior" individual may fail to survive or reproduce. This statistical view of natural selection permits an operational definition of "selective superiority" in terms of the contribution to the gene pool of the next generation.

REFERENCE

Muller, H. J. 1949. The Darwinian and modern conceptions of natural selection. *Proc. Amer. Phil. Soc.* 93:459–470.

9 Evolution, Population Thinking, and Essentialism

Elliott Sober

Philosophers have tended to discuss essentialism as if it were a *global* doctrine—a philosophy which, for some uniform reason, is to be adopted by all the sciences, or by none of them. Popper (1972) has taken a negative global view because he sees essentialism as a major obstacle to scientific rationality. And Quine (1953b, 1960), for a combination of semantical and epistemological reasons, likewise wishes to banish essentialism from the whole of scientific discourse. More recently, however, Putnam (1975) and Kripke (1972) have advocated essentialist doctrines and have claimed that it is the task of each science to investigate the essential properties of its constitutive natural kinds.

In contrast to these global viewpoints is a tradition which sees the theory of evolution as having some special relevance to essentialist doctrines within biology. Hull (1965) and Mayr (1959) are perhaps the two best known exponents of this attitude; they are *local* anti-essentialists. For Mayr, Darwin's hypothesis of evolution by natural selection was not simply a new theory, but a new *kind of theory*—one which discredited essentialist modes of thought within biology and replaced them with what Mayr has called "population thinking." Mayr describes essentialism as holding that

[t]here are a limited number of fixed, unchangeable "ideas" underlying the observed variability [in nature], with the *eidos* (idea) being the only thing that is fixed and real, while the observed variability has no more reality than the shadows of an object on a cave well.... [In contrast], the populationist stresses the uniqueness of everything in the organic world.... All organisms and organic phenomena are composed of unique features and can be described collectively only in statistical terms. Individuals, or any kind of organic entities, from populations of which we can determine the arithmetic mean and the statistics of variation. Averages are merely statistical abstractions only the individuals of which the population are composed have reality. The ultimate conclusions of the population thinker and of the typologist are precisely the opposite. For the typologist the type (*eidso*) is real and the variation an illusion, while for the populationist, the type (average) is an abstraction and only the variation is real. No two ways of looking at nature could be more different. (Mayr 1959, 28–9)

From *Philosophy of Science*, 1980, 47:350–383.

A contemporary biologist reading this might well conclude that essentialists had no scientifically respectable way of understanding the existence of variation in nature. In the absence of this, typologists managed to ignore the fact of variability by inventing some altogether mysterious and unverifiable subject matter for themselves. The notion of *types* and the kind of anti-empiricism that seems to accompany it, appear to bear only the most distant connection with modern conceptions of evidence and argument. But this reaction raises a question about the precise relation of evolution to essentialism. How could the *specifics* of a particular scientific theory have mattered much here, since the main obstacle presented by essentialist thinking was just to get people to be scientific about nature by paying attention to the evidence? The problem was to bring people down to earth by rubbing their noses in the diversity of nature. Viewed in this way, Mayr's position does not look much like a form of *local* anti-essentialism.

Other perplexities arise when a contemporary biologist tries to understand Mayr's idea of population thinking as applying to his or her own activity. If "only the individuals of which the population are composed have reality," it would appear that much of population biology has its head in the clouds. The Lotke-Volterra equations, for example, describe the interactions of predator and prey populations. Presumably, population thinking, properly so called, must allow that there is something real over and above individual organisms. Population thinking countenances organisms and populations; typological thinking grants that both organisms and types exist. Neither embodies a resolute and ontologically austere focus on individual organisms alone. That way lies nominalism, which Mayr (1969) himself rejects.

Another issue that arises from Mayr's conception of typological and population thinking is that of how we are to understand his distinction between "reality" and "abstraction." One natural way of taking this distinction is simply to understand reality as meaning existence. But presumably no population thinker will deny that there are such things as averages. If there are groups of individuals, then there are numerous properties that those groups possess. The *average* fecundity within a population is no more a property which we invent by "mere abstraction" than is the fecundity of individual organisms. Individual and group properties are equally "out there" to be discovered. And similarly, it is unclear how one could suggest that typologists held that variability is unreal; surely the historical record shows that typologists realized that differences between individuals *exist*. How, then, are we to understand the difference between essentialism and population thinking in terms of what each holds to be "real" about biological reality?

Answering these questions about the difference between essentialist and population modes of thought will be the main purpose of this chapter. How did essentialists propose to account for variability in nature? How did evolutionary theory undermine the explanatory strategy that they pursued? In what way does post-Darwinian biology embody a novel conception of variability? How has population thinking transformed our conception of what is *real*? The form of local anti-essentialism which I will propound in what follows will be congenial to many of Mayr's views. In one sense, then, our task will be to explicate and explain Mayr's insight that the shift from essentialist to populationist modes of thinking constituted a shift in the concept of biological reality. However, I will try to show why essentialism was a manifestly *scientific* working hypothesis. Typologists did not close their eyes to variation but rather tried to explain it in a particular way. And the failure of their explanatory strategy depends on details of evolutionary theory in ways which have not been much recognized.[1]

The approach to these questions will be somewhat historical. Essentialism about species is today a dead issue, not because there is no conceivable way to defend it, but because the way in which it was defended by biologists was thoroughly discredited. At first glance, rejecting a metaphysics or a scientific research program because one of its formulations is mistaken may appear to be fallacious. But more careful attention vindicates this pattern of evaluation. It is pie-in-the-sky metaphysics and science to hold on to some guiding principle simply because *it is possible* that there might be some substantive formulation and development of it. Thus, Newtonianism, guided by the maxim that physical phenomena can be accounted for in terms of matter in motion, would have been rejected were it not for the success of particular Newtonian explanations. One evaluates regulative principles by the way in which they regulate the actual theories of scientists. At the same time, I will try in what follows to identify precisely what it is in essentialism and in evolutionary theory that makes the former a victim of the latter. It is an open question to what degree the source of this incompatibility struck working biologists as central. As I will argue at the end of this section, one diagnosis of the situation which seems to have been historically important is much less decisive than has been supposed.

The essentialist's method of explaining variability, I will argue, was coherently formulated in Aristotle, and was applied by Aristotle in both his biology and in his physics. Seventeenth- and eighteenth-century biologists, whether they argued for evolution or against it, made use of Aristotle's natural state model. And to this day, the model has not been refuted in mechanics. Within contemporary biology, however, the model met with less success. Twentieth-century population genetics shows that the model cannot be applied in the way that the essentialist requires. But the natural state model is not wholly without a home in contemporary biology; in fact, the way in which it finds an application there highlights some salient facts about what population thinking amounts to.

An essentialist view of a given species is committed to there being some property which all and only the members of that species possess. Since there are almost certainly only finitely many individuals in any given species, we are quite safe in assuming there is some finitely statable condition which all and only the members of the species satisfy.[2] This could trivially be a list of the spatiotemporal locations of the organisms involved. But the fact that such a condition exists is hardly enough to vindicate essentialism. The essentialist thinks that there is a diagnostic property which any *possible* organism must have if it is to be a member of the species. It cannot be the case that the property in question is possessed by all organisms belonging to *Homo sapiens*, even though there might exist a member of *Homo sapiens* who lacked the trait. It must be necessarily true, and not just accidental, that all and only the organisms in *Homo sapiens* have the characteristic.

However, even this requirement of essentialism is trivially satisfiable. Is it not necessarily true that to be a member of *Homo sapiens* an organism must be a member of *Homo sapiens*? This is guaranteed if logical truths are necessary. But essentialism about biology is hardly vindicated by the existence of logical truths. In a similar vein, if it is impossible for perpetual motion machines to exist, then it is necessarily true that something belongs to *Homo sapiens* if and only if it belongs to *Homo sapiens* or is a perpetual motion machine. This necessary truth is not a truth of logic; it is a result of the theory of thermodynamics. But it too fails to vindicate biological essentialism. What more, then, is required?

The key idea, I think, is that the membership condition must be *explanatory*. The essentialist hypothesizes that there exists some characteristic unique to and shared by all members of *Homo sapiens* which explains why they are the way they are. A species essence will be a causal mechanism which works on each member of the species, making it the kind of thing that it is.

The characterization of essentialism just presented is fairly vague. For one thing, a great deal will depend on how one understands the crucial idea of *explanation*. But since explanation is clearly to be a scientific notion, I hope that, on my sketch, essentialism has the appearance of a scientific thesis, although perhaps one that is not terribly precise Although historically prey to obscurantism, essentialism has nothing essentially to do with mystery mongering, or with the irrational injunction that one should ignore empirical data. It is a perfectly respectable claim about the existence of hidden structures which unite diverse individuals into natural kinds.

Besides its stress on the giving of explanations, there is another feature of our characterization of essentialism which will be important in what follows. The essentialist requires that *a species* be defined in terms of the characteristics of the *organisms* which belong to it. We might call this kind of definition a *constituent definition*; wholes are to be defined in terms of their parts, sets are to be defined in terms of their members, and so on. Pre-Darwinian critics of the species concept, like Buffon and Bonnet, argued that species are unreal, because no such characteristics of organisms can be singled out (see Lovejoy 1936), and pre-Darwinian defenders of the species concept likewise agreed that the concept is legitimate only if constituent definitions could be provided. Constituent definitions are *reductionistic*, in that concepts at higher levels of organization (e.g., species) are legitimate only if they are definable in terms of concepts applying at lower levels of organization (e.g., organisms). It is quite clear that if there are finitely many levels of organization, one cannot demand constituent definitions for concepts at every level of organization (Kripke 1978). As we will see in what follows, evolutionary theory emancipated the species concept from the requirement that it be provided with a constituent definition. The scientific coherence of discourse at the population level of organization was to be assured in another way, one to which the label "population thinking" is especially appropriate.

Chemistry is prima facie a clear case in which essentialist thinking has been vindicated. The periodic table of elements is a taxonomy of chemical kinds. The essence of each kind is its atomic number. Not only is it the case that all actual samples of nitrogen happen to have atomic number 14; it is necessarily the case that a thing is made of nitrogen if and only if it is made of stuff having atomic number 14. Moreover, this characteristic atomic number plays a central role in explaining other chemical properties of nitrogen. Although things made of this substance differ from each other in numerous respects, underlying this diversity there is a common feature. It was hardly irrational for chemists to search for this feature, and the working assumption that such essences were out there to be found, far from stifling inquiry, was a principal contributor to that inquiry's bearing fruit.

Can an equally strong case be made for an essentialist view of biological species? One often hears it said that evolution undermined essentialism because the essentialist held that species are static, but from 1859 on we had conclusive evidence that species evolve. This comment makes a straw man of essentialism and is in any case historically untrue to the thinking of many essentialists. For one thing, notice that the discovery of the transmutation of elements has not in the slightest degree undermined the periodic table. The fact that nitrogen can be changed into oxygen does not in any way show that nitrogen and oxygen lack essences. To be nitrogen is to have one atomic number; to be oxygen is to have another. To change from nitrogen into oxygen, a thing must therefore shift from one atomic number to another. The mere fact of evolution does not show that species lack essences.

As a historical matter, some essentialists, like Agassiz (1859), did assert a connection between essentialism and stasis. But others considered the possibility that new species should have arisen on earth since the beginning (if they thought that there was a beginning). Thus, Linnaeus originally hypothesized that all species were created once and for all at the beginning, but later in his career he changed his mind because he thought that he had discovered a species, *Peloria*, which arose through cross-species hybridization (Rabel 1939, Ramsbottom 1938). And in *Generation of Animals* (II 746a30), Aristotle himself speculates about the possibility of new species arising as fertile hybrids. Countenancing such species need have no effect on binomial nomenclature or on deciding which characteristics of organisms to view as diagnostic. The question of when there started to be various kinds of things in the universe seems to be quite independent of what makes for differences between kinds.

Another, more plausible, suggestion, concerning how evolution undermined essentialism, is this: The fact that species evolve *gradually* entails that the boundaries of species are vague. The essentialist holds that there are characteristics which all and only the members of a given species possess. But this is no longer a tenable view; it is just as implausible as demanding that there should be a precise number of dollars which marks the boundary between rich and poor. This is the Sorites problem. Since ancient Greece, we have known that being a heap of stones, being bald, and being rich are concepts beset by line-drawing problems. But, the suggestion goes, it was only since 1859 that we have come to see that *Homo sapiens* is in the same boat. Thus, Hull (1965) has argued that essentialism was refuted because of its Aristotelian theory of *definition*; the requirement that species have nontrivial necessary and sufficient conditions runs afoul of the kind of continuity found in nature.

Unfortunately, this limpid solution to our problem becomes clouded a bit when we consider the historical fact that many essentialists conceded the existence of line-drawing problems. Thus, Aristotle in his *History of Animals*, (5888b4 ff.), remarks:

nature proceeds little by little from inanimate things to living creatures, in such a way that we are unable, in the continuous sequence to determine the boundary line between them or to say which side an intermediate kind falls. Next, after inanimate things come the plants: and among the plants there are differences between one kind and another in the extent to which they seem to share in life, and the whole genus of plants appears to be alive when compared with other objects, but seems lifeless when compared with animals. The transition from them to the animals is a continuous one, as remarked before. For with some kinds of things found in the sea one would be at a loss to tell whether they are animals or plants.

It is unclear exactly how one should interpret this remark. Does it indicate that there are in fact no boundaries in nature, or does it mean that the boundaries are difficult to discern? From the time of Aristotle up to the time of Darwin, the principle of continuity seems to have coexisted peacefully with strong essentialist convictions in the minds of many thinkers (Lovejoy 1936). Bonnet, Akenside, and Robinet are eighteenth-century biologists who exemplify this curious combination of doctrines. Does this coexistence imply that the two doctrines are in fact compatible, or rather, does it show that their conceptual dissonance was a long time in being appreciated? To answer this question, let us return to our analogy with the transmutation of elements.

In what sense are the boundaries between chemical kinds any more definite than those which we encounter in biology? At first glance, there appears to be all the difference in the world: in the periodic table, we have discrete jumps—between atomic number 36 and atomic number 37 there are no intermediate atomic numbers to blur distinctions. But let us reflect for a moment on the mechanism of transmutation. Consider, as an example, the experiment which settled the question of how nitrogen can be transmuted into oxygen (Ihde 1964, 509):

$$\frac{4}{2}He + \frac{14}{7}N \rightarrow \frac{17}{8}O + \frac{1}{1}H.$$

In this reaction, the α-particle is absorbed and a proton is expelled. Let us ask of this process a typical Sorites question: At what point does the bombarded nucleus cease to be a nitrogen nucleus and when does it start being a nucleus of oxygen?

There *may* be a precise and principled answer to this question which is given by the relevant physical theory. But then again there may not.[3] I would suggest that which of these outcomes prevails really does not matter to the question of whether essentialism is a correct doctrine concerning the chemical kinds. It may well be that having a particular atomic number is a vague concept. But this is quite consistent with that (vague) property's being the essence of a chemical kind. This really does not matter, as long as the vagueness of "nitrogen" and that of "atomic number 14" coincide. Essentialism is in principle consistent with *vague essences*.[4] In spite of this, one wonders what the history of chemistry, and its attendant metaphysics, would have looked like, if the transmutation of elements had been a frequent and familiar phenomenon during the second half of the nineteenth century. Just as the fact of evolution at times tempted Darwin to adopt a nominalist attitude toward species, so in chemistry the impressive taxonomy which we now have in the form of the periodic table might never have been arrived at, line-drawing problems having convinced chemists that chemical kinds are unreal.[5]

As a historical matter, Hull (1965) was right in arguing that essentialism was standardly associated with a theory of definition in which vagueness is proscribed. Given this association, nonsaltative evolution was a profound embarassment to the essentialist. But, if I am right, this theory of definition is inessential to essentialism. Our argument that the gradualness of evolution is not the decisive issue in undermining essentialism is further supported, I think, by the fact that contemporary evolutionary theory contains proposals in which evolutionary gradualism is rejected. Eldredge and Gould (1972) have argued that the standard view of speciation (as given, for example, in Ayala 1978 and Mayr 1963) is one in which phylogeny is to be seen as a series of "punctuated equilibria." Discontinuities in the fossil record are not to be chalked up to incompleteness, but rather to the fact that, in geological time, jumps are the norm. I would suggest that this theory of discontinuous speciation is cold comfort to the essentialist. Whether lines are easy or hard to draw is not the main issue, or so I shall argue.[6]

Another local anti-essentialist argument has been developed by Ghiselin (1966, 1969, 1974) and Hull (1976, 1978). They have argued that evolutionary theory makes it more plausible to view species as spatiotemporally extended individuals than as natural kinds. A genuine natural kind like gold may "go extinct" and then reappear; it is quite possible for there to be gold things at one time, for there to be no gold at some later time, and then, finally, for gold to exist at some still later time. But the conception of species given by evolutionary theory does not allow this sort of flip-flopping in and out of existence: once a biological taxon goes extinct, it must remain so. Hull (1978) argues that the difference between chemical natural kinds and biological species is that the latter, but not the former, are historical entities. Like organisms, biological species are individuated in part by historical criteria of spatiotemporal continuity. I am inclined to agree with this interpretation; its impact on pre-Darwinian conceptions of species could hardly be more profound. But what of its impact on essentialism? If essentialism is simply the

Evolution, Population Thinking, and Essentialism

view that species have essential properties (where a property need not be purely qualitative), then the doctrine remains untouched (as Hull himself realizes). Kripke (1972) has suggested that each individual human being has the essential property of being born of precisely the sperm and the egg of which he or she was born. If such individuals as organisms have essential properties, then it will presumably also be possible for individuals like *Drosophila melanogaster* to have essential properties as well. Of course, these essences will be a far cry from the "purely qualitative" characteristics which traditional essentialism thought it was in the business of discovering.

My analysis of the impact of evolutionary theory on essentialism is parallel, though additional. Whether species are natural kinds or spatiotemporally extended individuals, essentialist theories about them are untenable. Two kinds of arguments will be developed for this conclusion. First, I will describe the way in which essentialism seeks to explain the existence of variability, and will argue that this conception is rendered implausible by evolutionary theory. Second, I will show how evolutionary theory has removed *the need* for providing species with constituent definitions; population thinking provides another way of making species scientifically intelligible. This consideration, coupled with the principle of parsimony, provides an additional reason for thinking that species do not have essences.

ARISTOTLE'S NATURAL STATE MODEL

One of the fundamental ideas in Aristotle's scientific thinking is what I will call his natural state model. This model provides a technique for explaining the great diversity found in natural objects. Within the domain of physics, there are heavy and light objects, ones that move violently and ones that do not move at all. How is one to find some order that unites and underlies all this variety? Aristotle's hypothesis was that there is a distinction between the *natural state* of a kind of object and those states which are not natural. These latter are produced by subjecting the object to an *interfering force*. In the sublunar sphere, for a heavy object to be in its natural state is for it to be located where the center of the Earth is now (*On the Heavens*, ii, clr 296b and 310b, 2–5). But, of course, many heavy objects fail to be there. The cause for this divergence from what is natural is that these objects are acted on by interfering forces which prevent them from achieving their natural state by frustrating their natural tendency. Variability within nature is thus to be accounted for as a deviation from what is natural; were there no interfering forces, all heavy objects would be located in the same place (Lloyd 1968).

Newton made use of Aristotle's distinction, but disagreed with him about what the natural state of physical objects is. The first law of motion says that if a body is not acted upon by a force, then it will remain at rest or in uniform motion. And even in general relativity, the geometry of space-time specifies a set of geodesics along which an object will move as long as it is not subjected to a force. Although the terms "natural" and "unnatural" no longer survive in Newtonian and post-Newtonian physics, Aristotle's distinction can clearly be made within those theories. If there are no forces at all acting on an object, then, *a fortiori*, there are no interfering forces acting on it either. A natural state, within these theories, is a zero-force state.

The explanatory value of Aristotle's distinction is fairly familiar. If an object is not in its natural state, we know that the object must have been acted on by a force, and we set about finding it. We do this by consulting our catalog of known forces. If none of these is present, we might augment our catalog, or perhaps revise our conception of what the natural state of the system is. This pattern of analysis is used in population genetics under the rubric of the Hardy-Weinberg law. This law specifies an equilibrium state for the frequencies of genotypes in a panmictic population; this natural state is achieved when the evolutionary forces of mutation, migration, selection, and drift are not at work.

In the biological world, Aristotle sets forth the same sort of explanatory model. Diversity was to be accounted for as the joint product of natural regularities and interfering forces. Aristotle invokes this model when he specifies the regularities governing how organisms reproduce themselves: "[for] any living thing that has reached its normal development and which is unmutilated, and whose mode of generation is not spontaneous, the most natural act is the production of another like itself, an animal producing an animal, a plant a plant" (*De Anima*, 415a26). Like producing like, excepting the case of spontaneous generation, is the natural state, subject to a multitude of interferences, as we shall see.

In the case of spontaneous generation, the natural state of an organism is different. Although in the *Metaphysics* and the *Physics* "spontaneous" is used to mean unusual or random, in the later biological writings, *History of Animals* and *Generation of Animals*, Aristotle uses the term in a different way (Balme 1962, Hull 1967). Spontaneous generation obeys its own laws. For a whole range of organisms classified between the intermediate animals and the plants, like *never* naturally produces like. Rather, a bit of earth will spontaneously generate an earthworm, and the earthworm will then produce an eel. Similarly, the progression from slime to ascarid to gnat and that from cabbage leaf to grub to caterpillar to chrysallis to butterfly likewise counts as the natural reproductive pattern for this part of the living world (*History of Animals*, 570a5, 551b26, 551a13).

So much for the natural states. What counts as an interference for Aristotle? According to Aristotle's theory of sexual reproduction, the male semen provides a set of instructions which dictates how the female matter is to be shaped into an organism.[7] Interference may arise when the form fails to completely master the matter. This may happen, for example, when one or both parents are abnormal, or when the parents are from different species, or when there is trauma during fetal development Such interferences are anything but rare, according to Aristotle. Mules—sterile hybrids—count as deviations from the natural state (*Generation of Animals*, ii, 8). In fact, the females of a species do too, even though they are necessary for the species to reproduce itself (*Generation of Animals*, ii, 732a; ii, 3, 737a27; iv, 3, 767b8; iv, 6,

775a15). In fact, reproduction that is completely free of interference would result in an offspring which exactly resembles the father.[8] So failure to exactly resemble the male parent counts as a departure from the natural state. Deviations from type, whether mild or extreme, Aristotle labels "*terata*"—monsters. They are the result of interfering forces (*biaion*) deflecting reproduction from its natural pattern.

Besides trying to account for variation within species by using the natural state model, Aristotle at time seems to suggest that there are entire species which count as monsters (Preuss 1975, 215–16; Hull 1968). Seals are deformed as a group because they resemble lower classes of animals, owing to their lack of ears. Snails, since they move like animals with their feet cut off, and lobsters, because they use their claws for locomotion, are likewise to be counted as monsters (*Generation of Animals*, 19, 714b, 18–19; *Parts of Animals*, iv, 8, 684a35). These so-called dualizing species arise because they are the best possible organisms that can result from the matter out of which they are made. The scale of nature, it is suggested, arises in all its graduated diversity because the quality of the matter out of which organisms are made also varies—and nature persists in doing the best possible, given the ingredients at hand.

One cannot fault Aristotle for viewing so much of the biological domain as monstrous. Natural state models habitually have this characteristic; Newton's first law of motion is not impugned by the fact that no physical object is wholly unaffected by an outside force. Even so, Aristotle's partition of natural state and non-natural state in biology sounds to the modern ear like a reasonable distinction run wild. "Real terrata are one thing," one might say, "but to call entire species, and all females, and all males who don't exactly resemble their fathers monsters, seems absurd." Notice that our "modern" conceptions of health and disease and our notion of normality as something other than a statistical average enshrine Aristotle's model. We therefore are tempted to make only a conservative criticism of Aristotle's biology: we preserve the form of model he propounded, but criticize the applications he made of it. Whether this minimal critique of Aristotle is possible in the light of evolutionary theory remains to be seen.

The natural state model constitutes a powerful tool for accounting for variation. Even when two species seem to blend into each other continuously, it may still be the case that all the members of one species have one natural tendency while the members of the other species have a quite different natural tendency. Interfering forces may, in varying degrees, deflect the individuals in both species from their natural states, thus yielding the surface impression that there are no boundaries between the species. This essentialist response to the fact of diversity has the virtue that it avoids the ad hoc maneuver of contracting the boundaries of species so as to preserve their internal homogeneity.[9] This latter strategy was not unknown to the essentialist, but its methodological defects are too well known to be worth recounting here. Instead of insisting that species be defined in terms of some surface morphological feature, and thereby having each species shrink to a point, the essentialist can countenance unlimited variety in, and continuity between, species, as long as underlying this plenum one can expect to find discrete natural tendencies. The failure to discover such underlying mechanisms is no strong reason to think that none exists; but the development of a theory which implies that natural tendencies are not part of the natural order is another matter entirely.

Aristotle's model was a fixed point in the diverse conjectures to be found in pre-Darwinian biology. Preformationists and epigeneticists, advocates of evolution and proponents of stasis, all assumed that there is a real difference between natural states and states caused by interfering forces. The study of monstrosity—teratology—which in this period made the transition from unbridled speculation to encyclopedic catalogues of experimental oddities (Meyer 1939), is an especially revealing example of the power exerted by the natural state model. Consider, for example, the eighteenth-century disagreement between Maupertuis and Bonnet over the proper explanation of polydactyly. Both had at their fingertips a genealogy; it was clear to both that somehow or other the trait regularly reappeared through the generations. Maupertuis conjectured that defective hereditary material was passed along, having originally made its appearance in the family because of *an error in nature* (Glass 1959b, 62–67). Maupertuis, a convinced Newtonian, thought that traits, both normal and anomalous, resulted from the lawful combination of hereditary particles (Roger 1963). When such particles have normal quantities of attraction for each other, normal characteristics result. However, when particles depart from this natural state, either too many or too few of them combine, thus resulting in *monstres par exces* or *monstres par defaut*. Bonnet, a convinced ovist, offered a different hypothesis. For him, polydactyly is never encoded in the germ, but rather results from abnormal interuterine conditions or from male sperm interfering with normal development (Glass 1959a, 169). Thus whether polydactyly is "naturalized" by Maupertuis's appeal to heredity or by Bonnet's appeal to environment, the trait is never regarded as being completely natural. Variability in nature—in this case variability as to the number of digits—is a deviation from type.

In pre-Darwinian disputes over evolution, natural states loom equally large. Evolutionary claims during this period mainly assumed that living things were programmed to develop in a certain sequence, and that the emergence of biological novelty was therefore in conformity with some natural plan. Lovejoy (1936) discusses how the Great Chain of Being was "temporalized" during the eighteenth century; by this, he has in mind the tendency to think that the natural ordering of living things from those of higher type down to those of lower type also represented a historical progression. Such programmed, directed evolution—in which some types naturally give rise to others—is very much in the spirit of the natural state model. Whether species are subject to historical unfolding, or rather exist unchanged for all time, the concept of species was inevitably associated with that of type; on either view, variation is deviation caused by interfering forces.

It was generally presupposed that somewhere within the possible variations that a species is capable of, there is a privileged state—a state which has a special causal and explanatory role. The laws governing a species will specify this state, just as the laws which make sense of the diversity of kinematic states found in physics tell us what is the natural state of a physical object. The diversity of individual organisms is a veil which must be penetrated in the search for invariance. The transformation in thinking which we will trace in the next two sections consisted in the realization that this diversity itself constituted an invariance, obeying its own laws.

THE LAW OF ERRORS AND THE EMERGENCE OF POPULATION THINKING

So far, I have sketched several of the applications that have been made of Aristotle's model within biology. This strategy for explaining variation, I will argue in the next section, has been discredited by modern evolutionary theory. Our current theories of biological variation provide no more role for the idea of natural state than our current physical theories do for the notion of absolute simultaneity. Theories in population genetics enshrine a different model of variation, one which only became possible during the second half of the nineteenth century. Some brief account of the evolution within the field of statistics of our understanding of the law of errors will lay the groundwork for discussing the modern understanding of biological variation.

From its theoretical formulation and articulation in the eighteenth century, up until the middle of the nineteenth century, the law of errors was understood as a law about *errors*. Daniel Bernouilli, Lagrange, and Laplace each tried to develop mathematical techniques for determining how a set of discordant observations was to be interpreted (Todhunter 1865). The model for this problem was, of course, that there is a single true value for some observational variable, and a multiplicity of inconsistent readings that have been obtained. Here we have a straightforward instance of Aristotle's model: interfering forces cause variation in opinion; in nature there is but one true value. The problem for the theory of errors was to penetrate the veil of variability and to discover behind it the single value which was the constant cause of the multiplicity of different readings. Each observation was thus viewed as the causal upshot of two kinds of factors: part of what determines an observational outcome is the real value of the variable, but interfering forces which distort the communication of this information from nature to mind, also play a role. If these interfering forces are random—if they are as likely to take one value as any other—then the mean value of the readings is likely to represent the truth, when the number of observations is large. In this case, one reaches the truth by ascending to the summit of the bell curve. It is important to notice that this application of the natural state model is epistemological, not ontological. One seeks to account for variation in our observations of nature, not variation in nature itself. The decisive transition, from this epistemological to an ontological application, was made in the 1830s by the influential Belgian statistician Adolphe Quetelet.

Quetelet's insight was that the law of errors could be given an ontological interpretation by invoking a distinction which Laplace had earlier exploited in his work in Newtonian mechanics.[10] Laplace decomposed the forces at work in the solar system into two kinds. First, there are the *constant causes* by which the planets are affected by the sun's gravitation; second, there are the particular *disturbing causes* which arise from the mutual influences of the planets, their satellites, and the comets. Laplace's strategy was a familiar analytic one. He tried to decompose the factors at work in a phenomenon into components, and to analyze their separate contributions to the outcome. The character of this decomposition, however, is of special interest: one central, causal agent is at work on the components of a system, but the effects of this force are complicated by the presence of numerous interferences which act in different directions.

In his book of 1835, *Sur l'homme et le développement de ses facultés, ou essai de physique social*, Quetelet put forward his conception of the *average man* which for him constituted the true subject of the discipline of social physics. By studying the average man, Quetelet hoped to filter out the mutifarious and idiosyncratic characteristics which make for diversity in a population, and to focus on the central facts which constitute the social body itself. Like Weber's later idea of an ideal type, Quetelet's conception of the average man was introduced as a "fiction" whose utility was to facilitate a clear view of social facts by allowing one to abstract from the vagaries of individual differences. But unlike Weber, Quetelet quickly came to view his construct as real—a subject matter in its own right. Quetelet was struck by the analogy between a society's average man and a physical system's center of gravity. Since the latter could play a causal role, so too could the former; neither was a mere abstraction. For Quetelet, variability within a populations is *caused* by deviation from type. When the astronomer John Herschel reviewed Quetelet's *Lettres sur les probabilités* in 1850, he nicely captured Quetelet's idea that the average man is no mere artifact of reflection:

An average may exist of the most different objects, as the heights of houses in a town, or the sizes of books in a library. It may be convenient to convey a general notion of the things averaged; but it involves no conception of a natural and recognizable central magnitude, all differences from which ought to be regarded as deviations from a standard. The notion of a mean, on the other hand, does imply such a conception, standing distinguished from an average by this very feature, *viz*. The regular marching of the groups, increasing to a maximum and thence again diminishing. An average gives us no assurance that the future will be like the past. A mean may be reckoned on with the most implicit confidence. (Hilts 1973, 217)

Quetelet found little theoretical significance in the fact of individual differences. Concepts of correlation and amount of variation were unknown to him. For Quetelet, the law of errors is still a law about errors, only for him the mistakes are made by nature, not by observers. Our belief that there is variation in a population is no mistake on our part. Rather, it is the result of interferences confounding the expression of a prototype. Were interfering forces not to occur, there would be no variation.

It may strike the modern reader as incredible that anyone could view a trait like girth on this mode. However, Quetelet, who was perhaps the most influential statistician of his time, did understand biological differences in this way. He was impressed, not to say awe struck, by the fact that the results of accurately measuring the waists of a thousand Scottish soldiers would assume the same bell-shaped distribution as the results of inaccurately measuring the girth of a single, average, soldier a thousand times. For Quetelet, the point of attending to variation was to *see through* it—to render it transparent. Averages were the very antitheses of artifacts; they alone were the true objects of inquiry.[11]

Frances Galton, Darwin's cousin, was responsible for fundamental innovations in the analysis of individual differences.[12] He discovered the standard deviation and the correlation coefficient. His work on heredity was later claimed by both Mendelians and biometricians as seminal, and thus can be viewed as a crucial step toward the synthetic theory of evolution (Provine 1971). But his interest to our story is more restricted. Galton, despite his frequently sympathetic comments about the concept of type, helped to displace the average man and the idea of deviation from type.[13] He did this, not by attacking these typological constructs directly, but by developing an alternative model for accounting for variability. This model is a nascent form of the kind of population thinking which evolutionary biologists today engage in.

One of Galton's main intellectual goals was to show that heredity is a central cause of individual differences. Although the arguments which Galton put forward for his hereditarian thesis were weak, the conception of variability he exploited in his book *Hereditary Genius* (1869) is of great significance. For Galton, variability is not to be explained away as the result of interference with a single prototype. Rather, variability within one generation is explained by appeal to variability in the previous generation and to facts about the transmission of variability. Galton used the law of errors, but no longer viewed it as a law *about* errors. As Hilts (1973, 223–24) remarks: "Because Galton was able to associate the error distribution with individual differences caused by heredity, the distinction between constant and accidental causes lost much of its meaning." At the end of his life, Galton judged that one of his most important ideas was that the science of heredity should be concerned with deviations measured in statistical units. Quetelet had earlier denied that such units exist. Galton's discovery of the standard deviation gave him the mathematical machinery to begin treating variability as obeying its own laws, as something other than an idiosyncratic artifact.

Eight years after the publication of *Hereditary Genius*, Galton was able to sketch a solution for the problem he had noted in that work: What fraction of the parental deviations from the norm is passed on to offspring? Galton described a model in which hereditary causes and nonhereditary causes are partitioned. Were only the former of these at work, he conjectured, each child would have traits that are intermediate between those of its parents. In this case, the amount of variation would decrease in each generation. But Galton suspected that the amount of variation is constant across generations. To account for this, he posited a second, counteracting force which causes variability within each family. Were this second force the only one at work, the amount of variation would increase. But in reality, the centrifugal and centripetal forces combine to yield a constant quantity of variability across the generations. An error distribution is thus accounted for by way of a hypothesis which characterizes it as the sum of two other error distributions.

In his *Natural Inheritance* of 1889, Galton went on to complete his investigations of the correlation coefficient, and introduced the name "normal law" as a more appropriate label for what had traditionally been called the law of errors.[14] Bell curves are normal; they are found everywhere, Galton thought. This change in nomenclature crystallized a significant transformation in thinking. Bell curves need not represent mistakes made by fallible observers or by sportive nature. Regardless of the underlying etiology, they *are real*; they enter into explanations because the variability they represent is lawful and causally efficacious.

The transition made possible by statistical thinking from typological to population thinking was not completed by Galton.[15] Although his innovations loosened the grip of essentialism, he himself was deeply committed to the idea of racial types and believed that evolutionary theory presupposes the reality of types. Both Galton and Darwin (1859, ch. 5; 1868, ch. 13) spoke sympathetically about the ideas of unity of type and of reversion to type, and sought to provide historical justifications of these ideas in terms of common descent. Unity of type was just similarity owing to common ancestry; reversion to type was the reappearance of latent ancestral traits. But the presence of these ideas in their writings should not obscure the way in which their theorizing began to undermine typological thinking.

Darwin and Galton focused on the population as a unit of organization. The population is an entity, subject to its own forces, and obeying its own laws. The details concerning the individuals who are parts of this whole are pretty much irrelevant. Describing a single individual is as theoretically peripheral to a populationist as describing the motion of a single molecule is to the kinetic theory of gases. In this important sense, population thinking involves *ignoring individuals*: it is holistic, not atomistic. This conclusion contradicts Mayr's (1959, 28) assertion that for the populationist, "the individual alone is real."

Typologists and populationists agree that averages exist; and both grant the existence of variation. They disagree about the explanator character of these. For Quetelet, and for typologists generally, variability does not explain anything. Rather it is something to be explained or explained away. Quetelet posited a process in which uniformity gives rise to diversity; a single prototype—the average man—is mapped onto a variable resulting population. Galton, on the other hand, explained diversity in terms of an earlier diversity and constructed the mathematical tools to make this kind of analysis possible.

Both typologists and populationists seek to transcend the blooming, buzzing confusion of individual variation. Like all scientists, they do this by trying to identify properties of systems which remain constant in spite of the system's changes. For the typologist, the search for invariances takes the form of a search for natural tendencies. The typologist formulates a causal hypothesis about the forces at work on each individual within a population. The invariance underlying this diversity is the possession of a particular natural tendency *by each individual organism*. The populationist, on the other hand, tries to identify invariances by ascending to a different level of organization. For Galton, the invariant property across generations within a lineage is the amount of variability, and this is a property *of populations*. Again we see a way in which the essentialist is more concerned with individual organisms than the populationist is. Far from ignoring individuals, the typologist, *via* his use of the natural state model, resolutely focuses on individual organisms as the entities which possess invariant properties. The populationist, on the other hand, sees that it is not just individual organisms which can be the bearers of unchanging characteristics. Rather than looking for a reality that *underlies* diversity, the populationist can postulate a reality *sustained* by diversity.

I have just argued that there is an important sense in which typologists are more concerned with individual organisms than populationists are. However, looked at in another way, Mayr's point that populationists assign a more central role to organisms than typologists do can be established. In models of natural selection in which organisms enjoy different rates of reproductive success because of differences in fitness, natural selection is a force that acts on individual (organismic) differences. This standard way of viewing evolution assigns a causal role to individual idiosyncrasies. Individual differences are not *the effects* of interfering forces confounding the expression of a prototype; rather they are *the causes* of events that are absolutely central to the history of evolution. It is in this sense that Mayr is right in saying that evolutionary theory treats individuals as real in a way that typological thought does not (see also Lewontin 1974, 5–6). Putting my point and Mayr's point, thus interpreted, together, we might say that population thinking endows individual organisms with more reality *and* with less reality than typological thinking attributes to them.

To be real is to have causal efficacy; to be unreal is to be a mere artifact of some causal process. This characterization of what it is to be real, also used by Hacking (1975), is markedly different from the one used in traditional metaphysical disputes concerning realism, verificationism, and idealism (Sober 1980b). There, the problem is not how things are causally related, but rather it concerns what in fact *exists*, and whether what exists exists "independently" of us. The causal view of what it is to be real offers an explanation of a peculiar fact that is part of the more traditional metaphysical problem. Although two predicates may name real physical properties, natural kinds, theoretical magnitudes, or physical objects, simple operations on that pair of predicates may yield predicates which fail to name anything real. Thus, for example, "mass" and "charge" may name real physical magnitudes, even though "mass^2/

charge[3]" fails to name anything real. This is hard to explain, if reality is simply equated with existence (or with existence-that-is-independent-of-us). After all, if an object has a mass and if it has a charge, then there must be such a thing as what the square of its mass over the cube of its charge is. While this is quite true, it is *not* similarly correct to infer that because an object's mass causes some things and its charge causes other things, then there must be something which is caused by appeal to the square of its mass divided by the cube of its charge. Realism, in this case at least, is a thesis about what is cause and what is effect.

If we look forward in time, from the time of Galton and Darwin to the Modern Synthesis and beyond, we can see how population models have come to play a profoundly important role in evolutionary theorizing. In such models, properties of populations are identified and laws are formulated about their interrelations. Hypotheses in theoretical ecology and in island biogeography, for example, *generalize over populations* (see, for example, Wilson and Bossert 1971, chs. 3 and 4). The use of population concepts is not legitimized in those disciplines by defining them in terms of concepts applying at some lower level of organization. Rather, the use of one population concept is vindicated by showing how it stands in lawlike relations with other concepts *at the same level of organization*. It is in this way that we can see that there is an alternative to constituent definition. Here, then, is one way in which evolutionary theorizing undermined essentialism: Essentialism requires that species concepts be legitimized by constituent definition, but evolutionary theory, in its articulation of population models, makes such demands unnecessary. Explanations can proceed without this reductionistic requirement's being met.

If this argument is correct, there is a standard assumption made in traditional metaphysical problems having to do with identity which needs to be reevaluated. There could hardly be a more central category in our metaphysics, both scientific and everyday, than that of an enduring physical object. The way philosophers have tried to understand this category is as follows: Imagine a collection of instantaneous objects—i.e., objects at a moment in time. How are these various instantaneous objects united into the temporally enduring objects of our ontology? What criteria do we use when we lump together some time slices, but not others? This approach to the problem is basically that of looking for a constituent definition: enduring objects are to be defined out of their constituent time-slices. But, if populations can be scientifically legitimized in ways other than by using constituent definitions, perhaps the same thing is true of the category of physical object itself. I take it that Quine's (1953a) slogan "no entity without identity" is basically a demand for constituent definitions; this demand, which has been so fruitful in mathematics, should not be generalized into a universal maxim (nor can it be, if there are finitely many levels of organization. See Kripke 1978).

If constituent definitions for population concepts are theoretically unnecessary, then we have one argument, via the principle of parsimony (Sober 1980a), for the view that species do not have essences. However, there are equally pressing problems which essentialism faces when the natural state model is evaluated in the light of our current understanding of the origins of variability. It is to these problems that we now turn.

THE DISAPPEARANCE OF A DISTINCTION

The fate of Aristotle's model at the hands of population biology bears a striking resemblance to what happened to the notion of absolute simultaneity with the advent of relativity theory. Within classical physics, there was a single, well-defined answer to the question, "What is the temporal separation of two events x and y?" However, relativity theory revealed that answering this question at all depends on one's choice of a rest frame; given different rest frames, one gets different answers. We might represent the way the temporal separation of a pair of events may depend on a choice of frame as in the graph in figure 9.1. As is well known, the classical notions of temporal separation and spatial separation gave way in relativity theory to a magnitude that is not relative at all: this is the spatiotemporal separation of the two events. How large this quantity is does not depend on any choice of rest frame; it is frame invariant. Minkowski (1908) took this fact about relativity theory to indicate that space and time are not real physical properties at all, since they depend for their values on choices that are wholly arbitrary. For Minkowski, to be real is to be invariant, and space and time become mere shadows.

Special relativity fails to discriminate between the various temporal intervals represented in figure 9.1; they are all on a par. No one specification of the temporal separation is any more correct than any other. It would be utterly implausible to interpret this fact as indicating that there is a physically real distinction which special relativity fails to make. The fact that our best theory fails to draw this distinction gives us a very good reason for suspecting that the distinction is unreal, and this is the standard view of the matter which was crystallized in the work of Minkowski.

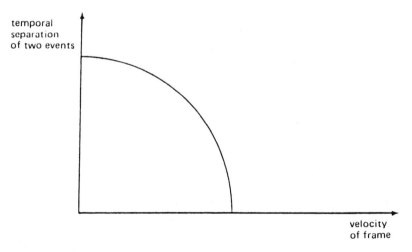

Figure 9.1 The temporal separation of a pair of events, relative to choices of rest frame.

Elliott Sober

According to the natural state model, there is one path of fetal development which counts as the realization of the organism's natural state, while other developmental results are consequences of unnatural interferences. Put slightly differently, for a given genotype, there is a single phenotype which it can have that is the natural one. Or, more modestly, the requirement might be that there is some restricted range of phenotypes which count as natural. But when one looks to genetic theory for a conception of the relation between genotype and phenotype, one finds no such distinction between natural state and states which are the results of interference. One finds, instead, the *norm of reaction*, which graphs the different phenotypic results that a genotype can have in different environments.[16] Thus the height of a single corn plant genotype might vary according to environmental differences in temperature, as is shown in figure 9.2. How would one answer the question: "Which of these phenotypes is the natural one for the corn plant to have?" One way to take this obscure question is indicated by the following answer: Each of the heights indicated in the norm of reaction is as "natural" as any other, since each happens in nature. Choose an environment, and relative to that choice we know what the phenotypic upshot in that environment is. But, of course, if the question we are considering is understood in terms of the natural state model, this sort of answer will not do. The natural state model presupposes that there is some phenotype which is the natural one *which is independent of a choice of environment*. The natural state model presupposes that there is some environment which is the natural environment for the genotype to be in, which determines, in conjunction with the norm of reaction, what the natural phenotype for the genotype is. But these presuppositions find no expression

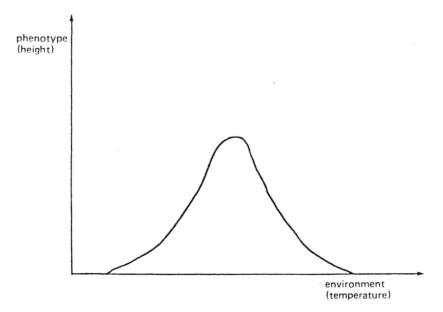

Figure 9.2 The norm of reaction of a given corn plant genotype, showing height as a function of temperature.

Evolution, Population Thinking, and Essentialism

in the norm of reaction: all environments are on a par, and all phenotypes are on a par. The required distinctions simply are not made.

When one turns from the various phenotypes that a single genotype might produce, to the various genotypes that a population might contain the same result obtains. Again, according to the natural state model, there is a single genotype or restricted class of genotypes, which count as the natural states of the population or species, all other genotypes being the result of interfering forces. But again, statistical profiles of genotypic variance within a population enshrine no such difference. Genotypes differ from each other in frequency; but unusual genotypes are not in any literal sense to be understood as deviations from type.

When a corn plant of a particular genotype withers and dies, owing to the absence of trace elements in the soil, the natural state model will view this as an outcome that is not natural. When it thrives and is reproductively successful, one wants to say that *this* environment might be the natural one. Given these ideas, one might try to vindicate the natural state model from a selectionist point of view by identifying the natural environment of a genotype with the environment in which it is fittest.[17]

This suggestion fails to coincide with important intuitions expressed in the natural state model. First of all, let us ask the question: What is the range of environments relative to which the fittest environment is to be understood? Shall we think of the natural state as that which obtains when the environment is the fittest *of all possible environments*? If so, the stud bull, injected with medications, its reproductive capacities boosted to phenomenal rates by an efficient artificial insemination program, has achieved its natural state. And in similar fashion, the kind of environment that biologists use to characterize the intrinsic rate of increase (r) of a population—one in which there is no disease, no predation, no limitations of space or food supplies—will likewise count as the natural environment. But these optimal environments are *not natural*, the natural state model tells us. They involve "artificially boosting" the fitness of resulting phenotypes by placing the genotypes in environments that are more advantageous than the natural environment.

Let us consider another, perhaps more plausible, way to understand the range of environments with respect to which the fittest environment is to be calculated. Instead of taking the best of all possible environments, why not, more modestly, consider the best of all environments that have been historically represented? This suggestion evades the second, but not the first, counterexample mentioned above. However, other problems present themselves. The natural state of a genotype is often understood to be one which has yet to occur. Perhaps every environment that a species has historically experienced is such that a given genotype in that environment results in a *diseased* phenotypes, or one which is developmentally impaired in some way. The natural state of a genotype is often taken to be some sort of ideal state which may or may not be closely approximated in the history of the species.

I have just argued that the idea of a fittest environment does not allow one to impose on the norm of reaction the kind of distinction that the natural state

Elliott Sober

model requires. Precisely the same reasons count against construing the idea of a genotype's being the natural state of a species in terms of maximal fitness. It is part of the natural state model that the natural genotypes for a species can be less fit (in some range of environments) than the best of all possible genotypes. And the natural genotype can likewise fail to be historically represented.

Aristotle is typical of exponents of the natural state model in holding that variation is introduced into a population by virtue of interferences with normal sexual reproduction. Our current understanding of the mechanisms of reproduction shows that precisely the opposite is the case. Even if one dismisses mutations as "unnatural interferences," the fact of genetic recombination in meiosis looms large. Generally, the number of total genotypes that a gene pool can produce by recombination is the product of the number of diploid genotypes that can be constructed at each locus. For species like *Homo sapiens* and *Drosophila melanogaster*, the number of loci has been estimated to be about 10,000 or more. What this means is that the number of genotypes that can be generated by recombination is greater than the number of atoms in the visible universe (Wilson and Bossert 1971, 39). For species with this number of loci, even a single male and a single female can themselves reproduce a significant fraction of the variation found in a population from which they are drawn. All sorts of deleterious phenotypes may emerge from the recombination process initiated by a founder population.

A doctrinaire advocate of the natural state model may take these facts to show that recombination has the status of an interference with what is natural. But this desperate strategy conflicts with the received evolutionary view of the function of sexuality. The deploying of prodigious quantities of variability is not a dysfunction which sexual organisms are vulnerable to. Rather it is the principal advantage of sexuality; it is standardly construed to be *what sexuality is for* (but see Williams 1975 for a dissenting opinion). If the notion of a natural state is to make any sense at all, then variability must be viewed as the upshot of natural forces.

The natural state model is a *causal*, and thereby a *historical*, *hypothesis*. The essentialist attempts to understand variation within a species as arising through a process of deviation from type. By tracing back the origins of this variability we discover the natural state of a species. To do this is to uncover that natural tendency possessed by each member of the species. But the science which describes the laws governing the historical origins of variation within species—population genetics—makes no appeal to such "natural tendencies." Rather, this frame invariant "natural tendency"—this property that an organism is supposed to have regardless of what environment it might be in—has been replaced by a frame relative property—namely, the phenotype that a genotype will produce *in a given environment*. The historical concept of a natural state is discredited in much the same way that the kinematic concept of absolute simultaneity was.

Our current concepts of function and dysfunction, of disease and health, seem to be based on the kinds of distinctions recommended by the natural state model. And both of these distinctions resist characterization in terms of maximum fitness. For virtually any trait you please, there can be environments in which that trait is selected for, or selected against. Diseases can be rendered advantageous, and health can be made to represent a reproductive cost. And even if we restrict our attention to historically actual environments, we still encounter difficulties. A perfectly healthy phenotype may be historically nonexistent; the optimum actually attained might still be some diseased state.

The functional notions just mentioned make distinctions which are sanctioned by the natural state model. Given the inadequacy of this model, does this show that the difference between disease and health and the difference between function and dysfunction are mere illusions? I do not think that this follows. What we should conclude is that these functional notions of normality are not to be characterized in terms of a historical notion of fitness. Perhaps they can be understood in some other way; that remains to be seen.

In addition to the influence that the natural state model continues to exert in scientific thinking, perhaps even more pervasive is the way that notions of naturalness have had, and continue to have, an influence in politics and in popular culture. Political theorists of both the left and the right have appealed to something called "human nature" (Lewontin 1977, Hull 1978).[18] Political optimists see human nature as essentially good; the evil that human beings have done is to be chalked up to interferences on the part of civilization, or of the state, or of particular economic institutions. Pessimists, on the other hand, see in human beings a natural tendency toward evil, which the restraints made possible by civilization can perhaps correct. The common presupposition here is that each human being has a particular dispositional property—a natural tendency—whose manifestation is contingent on whether environmental forces facilitate the expression of what is natural, or, on the other hand, go against nature by imposing unnatural interferences.

A more recent manifestation of the same habit of mind is to be found in debates about "environmental policy." Current environmental controversy, both on the part of those who want further industrialization to take its course and on the part of those who want to check or alter the way in which industry impinges on wildlife, tends to picture nature as something apart from us. The question before us, both sides imply, is how we should behave toward this separate sphere. We are not part of what is natural, and what we do has the character of an intervention from the outside into this natural domain. Our pollution of lakes, disruption of ecosystems, and extinction of species is just not natural. Natural, it would seem, is a good thing to be nowadays. Civilization is more often than not an interfering force, deflecting us from what is natural.

The Victorians, too, had their unnatural acts, thus hoping to find their ethics at least consistent with, and possibly vindicated by, the natural order. But they, at least, maintained some distance from the automatic equation of natural and good. Although some unnatural acts were wrong, others were decidedly right: here natural tendencies had to be checked if morally desirable qualities were to emerge. Perhaps it is a sign of our crumbling moral confidence that we no longer find it possible to separate questions of what is natural from what is good. By equating the two, we hope to read off our ethics directly from what happens in nature, and this gives us the illusion of needing to make no moral decisions for ourselves. This moral buck-passing is incoherent. What happens in nature is simply everything that happens. There is no other sense of "natural." Human society is not external to nature but a special part of it. It is no more a part of human nature to be healthy than to be diseased. Both kinds of phenotypes are to be found, and the norm of reaction makes no distinction between them. If we prefer one and wish to create environments in which it is encouraged, let us say so. But our reasons cannot be given in terms of allowing what is natural to occur unimpeded—by letting nature take its course, as if it has only one. Our activity, and inactivity, requires a more substantive justification than this.

CONCLUSION

Essentialism is as much entitled to appeal to the principle of tenacity as any other scientific hypothesis or guiding principle. It was hardly irrational for nineteenth-century research on the chemical elements to persist in its assumption that chemical kinds exist and have essential properties. The same holds true for those who hold that species are natural kinds and have essential properties; repeated failure to turn up the postulated items may be interpreted as simply showing that inquiry has not proceeded far enough. Matters change, however, when theoretical reasons start to emerge which cast doubt on the existence claim. For example, if the existence claim is shown to be theoretically superfluous, that counts as one reason for thinking that no such thing exists, or so the principle of parsimony would suggest (Sober 1980a). In another vein, if the causal mechanism associated with the postulated entity is cast in doubt, that too poses problems for the rationality of the existence claim. Our discussion of how population thinking emancipated biology from the need for constituent definitions of species is an argument of the first kind. Our examination of the theory of variation presupposed by essentialism is an argument of the second kind.

No phenotypic characteristic can be postulated as a species essence; the norm of reaction for each genotype shows that it is arbitrary to single out as privileged one phenotype as opposed to any other. Similar considerations show that no genotypic characteristic can be postulated as a species essence; the genetic variability found in sexual populations is prodigious and, again, there is no biologically plausible way to single out some genetic characteristics as natural while viewing others as the upshot of interfering forces. Even if a species were found in which some characteristic is shared by all and only the organisms that are in the species, this could not be counted as a species essence. Imagine, for example, that some novel form of life is created in the laboratory and subjected to some extreme form of stabilizing selection. If the number of organisms is kept small, it may turn out that the internal homogeneity of the species, as well as its distinctness from all other species, has been assured. However, the explanation of this phenomenon would be given in terms of the selection pressures acting on the population. If the universal property were a species essence, however, explaining why it is universal would be like explaining why all acids are proton donors, or why all bachelors are unmarried, or why all nitrogen has atomic number 14. These latter necessary truths, if they are explainable at all, are not explained by saying that some contingent causal force acted on acids, bachelors, or samples of nitrogen, thereby endowing them with the property in question. Characteristics possessed by all and only the extant members of a species, if such were to exist, would not be species essences. It is for this reason that hypotheses of discontinuous evolution like that proposed by Eldredge and Gould (1972) in no way confirm the claims of essentialism.

The essentialist hoped to penetrate the veil of variability found within species by discovering some natural tendency which each individual in the species possesses. This natural tendency was to be a dispositional property which would be manifest, were interfering forces not at work. Heterogeneity is thus the result of a departure from the natural state. But, with the development of evolutionary theory, it turned out that no such property was available to the essentialist, and in fact our current model of variability radically differs from the essentialist's causal hypothesis about the origins of variability.

At the same time that evolutionary theory undermined the essentialist's model of variability, it also removed the need for discovering species essences. Characteristics of populations do not have to be defined in terms of characteristics of organisms for population concepts to be coherent and fruitful. Population biology attempts to formulate generalizations about kinds of populations. In spite of the fact that species cannot be precisely individuated in terms of their constituent organisms, species undergo evolutionary processes, and the character of such processes is what population biology attempts to describe. Laws generalizing over population will, of course, include the standard *ceteris paribus* rider: they will describe how various properties and magnitudes are related, as long as no other forces affect the system. At least one such law describes what happens when *no* evolutionary force is at work in a panmictic Mendelian population. This is the Hardy-Weinberg equilibrium law. This law describes an essential property—a property which is necessary for a population to be Mendelian. But, of course, such laws do not pick out *species'* essences. Perhaps essentialism can reemerge as a thesis, not about species, but about *kinds* of species. The natural state model arguably finds an application at that level of organization in that the Hardy-Weinberg zero-force state is distinguished from other possible population configurations.

The transposition of Aristotle's distinction is significant. The essentialist searched for a property *of individual organisms* which is invariant across the organisms in a species. The Hardy-Weinberg law and other more interesting population laws, on the other hand, identify properties of *populations* which are invariant across all populations of a certain kind. In this sense, essentialism pursued an individualistic (organismic) methodology, which population thinking supplants by specifying laws governing objects at a higher level of organization.[19] From the individualistic (organismic) perspective assumed by essentialism, species are real only if they can be delimited in terms of membership conditions applying to individual organisms. But the populationist point of view made possible by evolutionary theory made such reductionistic demands unnecessary. Since populations and their properties are subject to their own invariances and have their own causal efficacy, it is no more reasonable to demand a species definition in terms of the properties of constituent organisms than it is to require organismic biology to postpone its inquiries until a criterion for sameness of organism is formulated in terms of relations between constituent cells. Essentialism lost its grip when populations came to be thought of as real.[20] And the mark of this latter transformation in thought was the transposition of the search for invariances to a higher level of organization.[21]

NOTES

Suggestions made by William Coleman, James Crow, Joan Kung, David Hull, Geoffrey Joseph, Steven Kimbrough, Richard Lewontin, Ernst Mayr, Terrence Penner, William Provine, Robert Stauffer, Dennis Stampe, and Victor Hilts helped me considerably in writing this paper.

1. Mayr (1963) has argued additionally that essentialist errors continue to be made in population biology in the form of the distortions of "bean-bag genetics." The assumption that the fitness of single genes is independent of their genetic context is and has been known to be mistaken; but how this simplifying assumption is essentialist in character is obscure to me.

2. If species are *individuals*—spatiotemporally extended lineages—as Ghiselin (1966, 1969, 1974), and Hull (1976, 1978) have argued, then we have our assurance of finitude. If, on the other hand, species are kinds of things, which may in principle be found anywhere in the universe at any time, then a slightly different argument is needed for the claim that the same species is overwhelmingly unlikely to have evolved twice. Such an argument is provided by considering the way in which speciation depends on the coincidence of a huge number of initial conditions. See Ayala (1978) for a summary of the received view of this matter.

3. I would suggest that quantum mechanical considerations show that the concept of being a nucleus with a particular atomic number is a vague one. Presumably, a collection of protons constitutes a nucleus when the strong force which causes them to attract each other overcomes their mutual electromagnetic repulsion. Whether this happens or not is a function of the distances between the protons. But *this* concept—that of "the" distance between particles—is indeterminate. Hence, the question of whether something is or is not a nucleus with a particular atomic number can only be answered probabilistically.

4. It is probably a mistake to talk about concepts being vague *simpliciter*. Rather, one should formulate matters in terms of concepts being vague relative to particular application. The issue of whether a concept is vague seems to reduce to the issue of whether there are cases in which it is indeterminate whether the concept applies. I would guess that practically every concept applying to physical objects is vague in this sense. Thus, even such concepts as "being two in number" are such that circumstances can be described in which it is indeterminate whether they apply to the objects in question. Degrees of vagueness can be partially defined as follows: If the set of circumstances in which concept P is indeterminate in its application is properly included in the set of circumstances in which concept Q is indeterminate, then Q is more vague than P.

5. Thus in his (1859), p. 52, Darwin says: "From these remarks it will be seen that I look at the term species, as one arbitrarily given for the sake of convenience to a set of individuals closely resembling each other, and that it does not essentially differ from the term variety, which is

Evolution, Population Thinking, and Essentialism

given to less distinct and more fluctuating forms. The term variety, again, in comparison with mere individual differences, is also applied arbitrarily, and for mere convenience sake." Elsewhere in (1859, e.g., pp. 432–33), Darwin espouses his perhaps more dominant populationist view that, in spite of line-drawing problems, species are real.

6. I am not arguing that Hull (1965) and others have misidentified the essence of essentialism and that their criticisms thereby fail to get to the heart of the matter. Essentialism, like most isms which evolve historically, probably does not even have an essence. Rather, I am trying to construe essentialism as a fairly flexible doctrine which, in at least some circumstances, can be seen to be quite consistent with the existence of insoluble line-drawing problems.

7. This characterization of Aristotle's view in terms of some information bearing entity is not completely anachronistic, as Delbrück (1971) points out when he (in jest) suggests that Aristotle should receive a Nobel Prize for having discovered DNA.

8. In this discussion of Aristotle's view of *terrata*, I have been much helped by Furth's (1975, section 11).

9. If one views Aristotle as excluding monstrous forms from membership on any species category, then one will have an extreme instance of this ad hoc strategy; *no* organism will belong to any species. Hull (1973, 39–40) sees Aristotle and scholastic science as hopelessly committed to this futile strategy. However, on the view I would attribute to Aristotle, most, if not all, monstrous forms are members of the species from which they arose. They, like Newtonian particles which fail to be at rest or in uniform motion, fail to achieve their natural states because of identifiable causal forces.

10. Hilts (1973, 209–210). My discussion of Quetelet and Galton in what follows leans heavily on Hilts (1973). It has a number of points in common with Hacking's (1975).

11. Boring (1929, 477) brings out the Aristotelian teleology contained in Quetelet's ideas quite well when he characterizes Quetelet as holding that "we might regard such human variation as if it occurred when nature aimed at an ideal and missed by varying amounts."

12. Although Galton found *The Origin of Species* an encouragement to pursue his own ideas, he indicates that his interest in variation and inheritance were of long standing. See Hilts (1973, 220).

13. In his *Hereditary Genius*, Galton compared the development of species with a many-faceted spheroid tumbling over from one facet or stable equilibrium to another. See Provine (1971, 14–15). This saltative process ensured unity of type. In spite of Galton's adherence to the idea of discontinuous evolution and certain other essentialist predilections (Lewontin 1974, 4), his innovations in population thinking were anti-essentialist in their consequences, or so I will argue.

14. Hilts (1973, 228). Walker (1929, 185) claims that the origin of the name "normal curve" is obscure. It occurs in Lexis and, she says, "It is not improbable that the term goes back to Quetelet." As natural and inevitable as Quetelet found his interpretation of the bell curve in terms of the Natural State Model, by the time Galton's *Natural Inheritance* appeared in 1889, there was growing sentiment that this interpretation was acceptable, if at all, only as a special case. Thus we find Galton, in that work (p. 58), saying that "the term Probable Error is absurd when applied to the subjects now in hand, such as Stature, Eye-colour, Artistic Faculty, or Disease." A year earlier, Venn, in his *The Logic of Chance* (p. 42), made a similar comment: "When we perform an operation ourselves with a clear consciousness of what we are aiming at, we may quite correctly speak of every deviation from this as being an error; but when Nature presents us with a group of objects of every kind, it is using a rather bold metaphor to speak in this case also of a law of error, as if she had been aiming at something all the time, and had like the rest of us missed her mark more or less in every instance." Quotations are drawn from Walker (1929, 53).

15. It would be important to trace the development of statistical ideas from Galton through Pearson and his circle to R. A. Fisher, and to see whether Pearson's positivistic convictions had the effect of further proscribing the idea of types on the grounds that it is "unscientific." Cohen (1972) sees Galton as already adopting some positivistic attitudes in his idea that heredity was to be understood in terms of correlations, and not in terms of causal forces. Also, see Hacking (1975) for a bold attempt to link Galton's innovations to other developments in nineteenth-century thought. I should point out that a fuller treatment of the emergence of population thinking would have to ascribe a central role to Mendel. He, much more than Galton, provided the central elements of our present conception of the relation of heredity and variation. I have stressed Galton, however, because of his interpretation of statistics and because of his view of the population as a unit of explanation.

16. The discussion of the norm of reaction in what follows depends heavily on some points made in Lewontin (1977).

17. This selectionist suggestion needs to be made more precise by specifying the notion of fitness used. I will not lay out these different conceptions here. Rather, I invite the reader to choose the one that he or she finds most plausible. the upshot of my argument does not seem to depend on which biologically plausible characterization is chosen.

18. Lewontin (1977, 11) has argued that the idea of a "natural phenotype" has been used in some hereditarian thinking in the IQ controversy. He quotes Herrnstein (1971, 54) as talking about "artificially boosting" an individual's IQ score. The presupposition seems to be that each human genotype has associated with it an IQ score (or range of such scores) which counts as its natural phenotype. As in Aristotle, the individual can be deflected from what is natural by environmental interference.

19. It is significant that biologists to this day tend to use "individual" and "organism" interchangeably. For arguments that populations, and even species, are to be construed as individuals, see Ghiselin (1966, 1969, 1974), and Hull (1976, 1978).

20. I borrow this way of putting matters from Hacking (1975) in which he describes the series of transformations in thought which resulted in "chance becoming real."

21. The group selection controversy provides an interesting example of the question of whether, and in what respects, it is appropriate to view populations as objects. In some ways, this debate recapitulates elements of the dispute between methodological holism and methodological individualism in the social sciences. See Sober (1980c) for details.

REFERENCES

Agassiz, L. (1859). *Essay on Classification*. Cambridge, Mass: Harvard University Press.

Ayala, F. (1978). "The Mechanisms of Evolution." *Scientific American* 239, 3:56–69.

Balme, D. (1962). "Development of Biology in Aristotle and Theophrastus: Theory of Spontaneous Generation." *Phronesis* 2, 1:91–104.

Boring, E. (1929). *A History of Experimental Psychology*. New York: Appleton-Century-Crofts.

Buffon, L. (1749). *Histoire Naturelle*. Paris.

Cohen, R. (1972). "Francis Galton's Contribution to Genetics." *Journal of the History of Biology* 5, 2:389–412.

Darwin, C. (1859). *On the Origin of Species*. Cambridge, Mass.: Harvard University Press.

———. (1868). *The Variation of Animals and Plants under Domestication*. London: Murray.

Delbrück, M. (1971). "Aristotle-totle-totle." In Monod, J., and Borek, J. (eds), *Microbes and Life* 50–55. New York: Columbia University Press.

Eldredge, N., and Gould, S. (1972). "Punctuated Equilibria: An alternative to Phyletic/ Gradualism." In T. Schopf (ed.), *Models in Paleobiology*, 82–115. San Francisco: Freeman Cooper.

Furth, M. (1975). *Essence and Individual: Reconstruction of an Aristotelian Metaphysics*, chapter 11, duplicated for the meeting of the Society for Ancient Greek Philosophy, unpublished.

Ghiselin, M. (1966). "On Psychologism in the Logic of Taxonomic Controversies." *Systematic Zoology* 15:207–15.

———. (1969). *The Triumph of the Darwinian Method*. Berkeley: University of California Press.

———. (1974). "A Radical Solution to the Species Problem." *Systematic Zoology* 23:536–44.

Glass, B. (1959a). "Heredity and Variation in the eighteenth Century Concept of the Species." In Glass, B., et al. (eds.), *Forerunners of Darwin*, 144–72. Baltimore: The Johns Hopkins Press.

———. (1959b). "Maupertuis, Pioneer of Genetics and Evolution." in Glass, B., et al. (eds.), *Forerunners of Darwin*, 51–83. Baltimore: The Johns Hopkins Press.

Hacking, I. (1975). "The Autonomy of Statistical Law." Talk delivered to American Philosophical Association, Pacific Division, unpublished.

Herrnstein, R. (1971). "IQ." *Atlantic Monthly* 228, 3:43–64.

Hilts, V. (1973). "Statistics and Social Science." In Giere, R. and Westfall, R. (eds.), *Foundations of Scientific Method in the Nineteenth Century*, 206–33. Bloomington: Indiana University Press.

Hull, D. (1965). "The Effect of Essentialism on Taxonomy: 2000 Years of Stasis." *British Journal for the Philosophy of Science* 15:314–16; 16:1–18.

———. (1967). "The Metaphysics of Evolution." *British Journal for the History of Science* 3, 12:309–37.

———. (1968). "The Conflict between Spontaneous Generation and Aristotle's Metaphysics." *Proceedings of the Seventh Inter-American Congress of Philosophy* 2 (1968): 245–50. Quebec City: Les Presses de l'Université Laval.

———. (1973). *Darwin and His Critics*. Cambridge, Mass.: Harvard University Press.

———. (1976). "Are Species Really Individuals?" *Systematic Zoology* 25:174–91.

———. (1978). "A Matter of Individuality." *Philosophy of Science* 45:335–60.

Ihde, A. (1964). *The Development of Modern Chemistry*. New York: Harper & Row.

Kripke, S. (1972). "Naming and Necessity." In Davidson, D., and Harman, G. (eds.), *Semantics of Natural Languages*, 253–355; 763–9. Dordrecht: Reidel.

———. (1978). "Time and Identity." Lectures given at Cornell University, unpublished.

Lewontin, R. (1974). *The Genetic Basis of Evolutionary Change*. New York: Columbia University Press.

———. (1977). "Biological Determinism as a Social Weapon." In Ann Arbor Science for the People Editorial Collective, *Biology as a Social Weapon*, 6–20. Minneapolis: Burgess.

Lloyd, G. (1968). *Aristotle: The Growth and Structure of His Thought*. Cambridge: Cambridge University Press.

Lovejoy, A. (1936). *The Great Chain of Being*. Cambridge, Mass.: Harvard University Press.

Mayr, E. (1959). "Typological versus Population Thinking." In *Evolution and Anthropology: A Centennial Appraisal*, 409–12. Washington: Anthropological Society of Washington.

———. (1963). *Animal Species and Evolution*. Cambridge, Mass.: Belknap Press of Harvard University Press.

————. (1969). "The Biological Meaning of Species." *Biology Journal of the Linnean Society* 1:311–20.

————. (1976). *Evolution and the Diversity of Life*. Cambridge, Mass.: Harvard University Press.

Meyer, A. (1939). *The Rise of Embryology*. Stanford, Calif.: Stanford University Press.

Minkowski, H. (1908). "Space and Time." In Lorentz, H., Einstein, A., et al., *The Principle of Relativity* 73–91. New York: Dover.

Popper, K. (1972). *Objective Knowledge*. Oxford: Oxford University Press.

Preuss, A. (1975). *Science and Philosophy in Aristotle's Biological Works*. New York: Georg Olms.

Provine, W. (1971). *The Origins of Theoretical Population Genetics*. Chicago: University of Chicago Press.

Putnam, H. (1975), "The Meaning of 'Meaning,'" In *Mind, Language and Reality*, 215–71. Cambridge: Cambridge University Press.

Quetelet, A. (1842). *A Treatise on Man and the Development of His Faculties*. Edinburgh.

Quine (1953a). "Identity, Ostension, Hypostasis." In *From a Logical Point of View*, 65–79. New York: Harper Torchbooks.

————. (1953b). "Reference and Modality." In *From a Logical Point of View*, 139–59. New York: Harper Torchbooks

————. (1960). *Word and Object*. Cambridge, Mass.: MIT Press.

Rabel, G. (1939). "Long before Darwin: Linne's Views on the Origin of Species," *Discovery* n.s., 2:121–75.

Ramsbottom, J. (1938). "Linnaeus and the Species Concept." *Proceedings of the Linnean Society of London*: 192–219.

Roger, J. (1963). *Les Sciences de la vie dans la pensée française du XVIII siècle*. Paris: Armand Colin.

Sober, E. (1980a), "The Principle of Parsimony." *British Journal for Philosophy of Science*.

————. (1980b). "Realism and Independence." *Noûs*. 41:369–386.

————. (1980c). "Holism, Individualism, and the Units of Selection." *Proceedings of the Biennial Meeting of the Philosophy of Science Association*: 93–101. E. Lansing, Michigan. Philosophy of Science Association.

Todhunter, I. (1865). *History of the Theory of Probability to the Time of Laplace*. New York: Chelsea Publishing.

Walker, H. (1929), *Studies in the History of Statistical Method*. Baltimore: Williams & Wilkins.

Williams, G. C. (1975). *Sex and Evolution*. Princeton, N.J.: Princeton University Press.

Wilson, E., and Bossert, W. (1971). *A Primer of Population Biology*. Sunderland, Mass.: Sinauer.

VI Species

10   A Matter of Individuality

David L. Hull

Biological species have been treated traditionally as spatiotemporally unrestricted classes. If they are to perform the function which they do in the evolutionary process, they must be spatiotemporally localized individuals, historical entities. Reinterpreting biological species as historical entities solves several important anomalies in biology, in philosophy of biology, and within philosophy itself. It also has important implications for any attempt to present an "evolutionary" analysis of science and for sciences such as anthropology which are devoted to the study of single species.

INTRODUCTION

The terms "gene," "organism," and "species" have been used in a wide variety of ways in a wide variety of contexts. Anyone who attempts merely to map this diversity is presented with a massive and probably pointless task. In this chapter I consciously ignore "the ordinary uses" of these terms, whatever they might be, and concentrate on their biological uses. Even within biology the variation and conflicts in meaning are sufficiently extensive to immobilize all but the most ambitious ordinary language philosopher. Thus I have narrowed my focus even further to concentrate on the role which these terms play in evolutionary biology. In doing so, I do not mean to imply that this usage is primary or that all other biological uses which conflict with it are mistaken. Possibly evolutionary theory is *the* fundamental theory in biology, and all other biological theories must be brought into accord with it. Possibly all biological theories, including evolutionary theory, eventually will be reduced to physics and chemistry. But regardless of the answers to these global questions, at the very least various versions of evolutionary theory are sufficiently important in biology to warrant an investigation of the implications which they have for the biological entities which they concern.

Genes are the entities which are passed on in reproduction and which control the ontogenetic development of the organism. Organisms are the complex systems which anatomists, physiologists, embryologists, histologists, and others analyze into their component parts. Species have been treated traditionally as the basic units of classification, the natural kinds of the living world,

From *Philosophy of Science*, 1978, 45:335–360.

comparable to the physical elements. But these entities also function in the evolutionary process. Evolution consists in two processes (mutation and selection) which eventuate in a third (evolution). Genes provide the heritable variation required by the evolutionary process. Traditionally organisms have been viewed as the primary focus of selection, although considerable disagreement currently exists over the levels at which selection takes place. Some biologists maintain that selection occurs exclusively at the level of genes; others that supragenic, even supraorganismic units can also be selected. As one might gather from the title of Darwin's book, species are the things which are supposed to evolve. Whether the relatively large units recognized by taxonomists as species evolve or whether much less extensive units such as populations are the effective units of evolution is an open question. In This chapter when I use the term "species," I intend to refer to those supraorganismic entities which evolve regardless of how extensive they might turn out to be.

The purpose of this chapter is to explore the implications which evolutionary theory has for the ontological status of genes, organisms, and species. The only category distinction I discuss is between individuals and classes. By "individuals" I mean spatiotemporally localized cohesive and continuous entities (historical entities.) By "classes" I intend spatiotemporal unrestricted classes, the sorts of things which can function in traditionally defined laws of nature. The contrast is between Mars and planets, the Weald and geological strata, Gargantua and organisms. The terms used to mark this distinction are not important; the distinction is. For example, one might distinguish two sorts of sets: those that are defined in terms of a spatiotemporal relation to a spatiotemporally localized focus, and those that are not. On this view, historical entities such as Gargantua become sets. But they are sets of a very special kind—sets defined in terms of a spatiotemporal relation to a spatiotemporally localized focus. Gargantua, for instance, would be the set of all cells descended from the zygote which gave rise to Gargantua.

The reason for distinguishing between historical entities and genuine classes is the differing roles which each plays in science according to traditional analyses of scientific laws. Scientific laws are supposed to be spatiotemporally unrestricted generalizations. No uneliminable reference can be made in a genuine law of nature to a spatiotemporally individuated entity. To be sure, the distinction between accidentally true generalizations (such as all terrestrial organisms using the same genetic code) and genuine laws of nature (such as those enshrined in contemporary versions of celestial mechanics) is not easy to make. Nor are matters helped much by the tremendous emphasis placed on laws in traditional philosophies of science, as if they were the be-all and end-all of science. Nevertheless, I find the distinction between those generalizations that are spatiotemporally unrestricted and those that are not fundamental to our current understanding of science. Whether one calls the former "laws" and the latter something else, or whether one terms both sorts of statements "laws" is of little consequence. The point I wish to argue is that genes, organisms, *and* species, as they function in the evolutionary process, are necessarily spatiotemporally localized individuals. They could not perform the functions which they perform if they were not.

The argument presented in this chapter is metaphysical, not epistemological. Epistemologically red light may be fundamentally different from infrared light and mammals from amoebae. Most human beings can see with red light and not infrared light. Most people can see mammals; few if any can see amoebae with the naked eye. Metaphysically they are no different. Scientists know as much about one as the other. Given our relative size, period of duration, and perceptual acuity, organisms appear to be historical entities, species appear to be classes of some sort, and genes cannot be seen at all. However, after acquainting oneself with the various entities which biologists count as organisms and the roles which organisms and species play in the evolutionary process, one realizes exactly how problematic our commonsense notions actually are. The distinction between an organism and a colony is not sharp. If an organism is the "total product of the development of the impregnated embryo," then as far back as 1899 T. H. Huxley was forced to conclude that the medusae set free from a hydrozoan "are as much organs of the latter as the multitudinous pinnules of a *Comatula*, with their genital glands, are organs of the Echinoderm. Morphologically, therefore, the equivalent of the individual *Comatula* is the Hydrozoic stock and all the Medusae which proceed from it" (24). More recently, Daniel Janzen (25) has remarked that the "study of dandelion ecology and evolution suffers from confusion of the layman's 'individual' with the 'individual' of evolutionary biology. The latter individual has 'reproductive fitness' and is the unit of selection in most evolutionary conceptualizations" (see also 2). According to evolutionists, units of selection, whether they be single genes, chromosomes, organisms, colonies, or kinship groups, are individuals. In this chapter I intend to extend this analysis to units of evolution.

If the ontological status of space-time in relativity theory is philosophically interesting in and of itself (and God knows enough philosophers have written on that topic), then the ontological status of species in evolutionary theory should also be sufficiently interesting philosophically to discuss without any additional justification. However, additional justification does exist. From Socrates and Plato to Kripke and Putnam, organisms have been paradigm examples of primary substances, particulars, and/or individuals, while species have served as paradigm examples of secondary substances, universals, and/or classes. I do not think that this chapter has any necessary implications for various solutions to the problem of universals, identity, and the like. However, if my main contention is correct, if species are as much spatiotemporally localized individuals as organisms, then some of the confusion among philosophers over these issues is understandable. One of the commonest examples used in the philosophical literature is inappropriate. Regardless of whether one thinks that "Moses" is a proper name, a cluster concept, or a rigid designator, "*Homo sapiens*" must be treated in the same way.

THE EVOLUTIONARY JUSTIFICATION

Beginning with the highly original work of Michael Ghiselin (12, 13, 14), biologists in increasing numbers are beginning to argue that species as units of evolution are historical entities (15, 20, 21, 22, 23, 34, 38). The justification for such claims would be easier if there were one set of propositions (presented preferably in axiomatic form) which could be termed *the* theory of evolution. Unfortunately, there is not. Instead there are several, incomplete, partially incompatible versions of evolutionary theory currently extant. I do not take this state of affairs to be unusual, especially in periods of rapid theoretical change. In general the myth that some one set of propositions exists which can be designated unequivocally as Newtonian theory, relativity theory, etc., is an artifact introduced by lack of attention to historical development and unconcern with the primary literature of science. The only place one can find *the* version of a theory is in a textbook written long after the theory has ceased being of any theoretical interest to scientists.

In this section I set out what it is about the evolutionary process which results in species being historical entities, not spatiotemporally unrestricted classes. In doing so I have not attempted to paper over the disagreements which currently divide biologists working on evolutionary theory. For example, some disagreement exists over how abruptly evolution can occur. Some biologists have argued that evolution takes place saltatively, in relatively large steps. Extreme saltationists once claimed that in the space of a single generation new species can arise which are so different from all other species that they have to be placed in new genera, families, classes, etc. No contemporary biologist to my knowledge currently holds this view. Extreme gradualists, on the other side, argue that speciation *always* occurs very slowly, over periods of hundreds of generations, either by means of a single species changing into a new species (phyletic evolution) or else by splitting into two large subgroups which gradually diverge (speciation). No contemporary biologist holds this view either. Even the most enthusiastic gradualists admit that new species can arise in a single generation, e.g., by means of polyploidy. In addition, Eldredge and Gould (11), building on Mayr's founder principle (36, 37), have recently argued that speciation typically involves small, peripheral isolates which develop quite rapidly into new species. Speciation is a process of "punctuated equilibria."

However, the major dispute among contemporary evolutionary theorists is the level (or levels) at which selection operates. Does selection occur *only* and *literally* at the level of genes? Does selection take place *exclusively* at the level of organisms, the selection of genes being only a consequence of the selection of organisms? Can selection also take place at levels of organization more inclusive than the individual organism, e.g., at the level of kinship groups, populations, and possibly even entire species? Biologists can be found opting for every single permutation of the answers to the preceding questions. I do not propose to go through all the arguments which are presented to support these various conclusions. For my purposes it is sufficient to show that the points of dispute are precisely those which one might expect if species are being interpreted as historical entities, rather than as spatiotemporally unrestricted classes. Richard Dawkins puts the crucial issue as follows:

Natural selection in its most general form means the differential survival of entities. Some entities live and others die but, in order for this selective death to have any impact on the world, an additional condition must be met. Each entity must exist in the form of lots of copies, and at least some of the entities must be *potentially* capable of surviving—in the form of copies—for a significant period of evolutionary time.

The results of evolution by natural selection are *copies* of the entities being selected, not *sets*. Elements in a set must be characterized by one or more common characteristics. Even fuzzy sets must be characterized by at least a "cluster" of traits. Copies need not be.[1] A particular gene is a spatiotemporally localized individual which either may or may not replicate itself. In replication the DNA molecule splits down the middle producing two new molecules composed *physically* of half of the parent molecule while *largely* retaining its structure. In this way genes form lineages, ancestor-descendant copies of some original molecule. The relevant genetic units in evolution are not *sets* of genes defined in terms of structural similarity but lineages formed by the imperfect copying process of replications.[2] Genes can belong to the same lineage even though they are structurally different from other genes in that lineage. What is more, continued changes in structure can take place indefinitely. If evolution is to occur, not only *can* such indefinite structural variation take place within gene lineages, but it *must*. Single genes are historical entities, existing for short periods of time. The more important notion is that of a *gene lineage*. Gene lineages are also historical entities persisting while changing indefinitely through time. As Dawkins puts this point:

Genes, like diamonds, are forever, but not quite in the same way as diamonds. It is an individual diamond crystal which lasts, as an unaltered pattern of atoms. DNA molecules don't have that kind of permanence. The life of any one physical DNA molecule is quite short—perhaps a matter of months, certainly not more than one lifetime. But a DNA molecule could theoretically live on in the form of *copies* of itself for a hundred million years. (8, p. 36)

Exactly the same observations can be made with respect to organisms. A particular organism is a spatiotemporally localized individual which either may or may not reproduce itself. In asexual reproduction, part of the parent organism buds off to produce new individuals. The division can be reasonably equitable, as in binary fission, or extremely inequitable, as in various forms of parthenogenesis. In sexual reproduction gametes are produced which unite to form new individuals. Like genes, organisms form lineages. The relevant organismal units in evolution are not sets of organisms defined in terms of structural similarity but lineages formed by the imperfect copying processes of reproduction. Organisms can belong to the same lineage even though they are structurally different from other organisms in that lineage. What is more, continued changes in structure can take place indefinitely. If evolution is to occur, not only *can* such indefinite structural variation take place within organism lineages, but it *must*. Single organisms are historical entities, existing for short periods of time. Organism lineages are also historical entities persisting while changing indefinitely through time.

Both replication and reproduction are spatiotemporally localized processes. There is no replication or reproduction at a distance. Spatiotemporal continuity through time is required. Which entities at which levels of organization are sufficiently cohesive to function as units of selection is more problematic. Dawkins presents one view:

In sexually reproducing species, the individual [the organism] is too large and too temporary a genetic unit to qualify as a significant unit of natural selection. The group of individuals is an even larger unit. Genetically speaking, individuals and groups are like clouds in the sky or dust-storms in the desert. They are temporary aggregates of federations. They are not stable through evolutionary time. Populations may last a long while, but they are constantly blending with other populations and so losing their identity. They are subject to evolutionary change from within. A population is not a discrete enough entity to be a unit of natural selection, not stable and unitary enough to be "selected" in preference to another population. (8, p. 37).

From a commonsense perspective, organisms are paradigms of tightly organized, hierarchically stratified systems. Kinship groups such as hives also seem to be internally cohesive entities. Populations and species are not. Dawkins argues that neither organisms (in sexually reproducing species) nor populations in any species are sufficiently permanent and cohesive to function as units in selection. In asexual species, organisms do not differ all that much from genes. They subdivide in much the same way that genes do, resulting in progeny which are identical (or nearly identical) with them. In sexual species, however, organisms must pool their genes to reproduce. The resulting progeny contain a combined sample of parental genes, Populations lack even this much cohesion.

Other biologists are willing to countenance selection at levels more inclusive than the individual gene, possibly parts of chromosomes, whole chromosomes, entire organisms, or even kinship groups (32). The issues, both empirical and conceptual, are not simple. For example, G. C. Williams in his classic work (61) argues that selection occurs only at the level of individuals. By "individual," biologists usually mean "organism." However, when Williams is forced to admit that kinship groups can also function as units of selection, he promptly dubs them "individuals." One of the commonest objections to E. O. Wilson's (62) equally classic discussion of evolution is that he treats kin selection as a special case of group selection. According to the group selectionists, entities more inclusive than kinship groups can also function as units of selection (63).[3] Matters are not improved much by vagueness over what is meant by "units of selection." Gene frequencies are certainly altered from generation to generation, but so are genotype frequencies. Genes cannot be selected in isolation. They depend on the success of the organism which contains them for survival. Most biologists admit that similar observations hold for certain kinship groups. Few are willing to extend this line of reasoning to include populations and entire species.

Although the dispute over the level(s) at which selection takes place is inconclusive, the points at issue are instructive. In arguing that neither organisms nor populations function as units of selection in the same sense that genes do, Dawkins does not complain that the cells in an organism or the organisms in a population are phenotypically quite diverse, though they frequently are. Rather he denigrates their cohesiveness and continuity through time, criteria which are relevant to individuating historical entities, not spatiotemporally unrestricted classes. Difficulties about the level(s) at which selection can operate to one side, the issue with which we are concerned is the ontological status of species. Even if entire species are not sufficiently well integrated to function as units of selection, they are the entities which evolve as a result of selection at lower levels. The requirements of selection at these lower levels place constraints on the manner in which species can be conceptualized. Species as the results of selection are necessarily lineages, not sets of similar lineages, not sets of similar organisms. In order for differences in gene frequencies to build up in populations, continuity through time must be maintained. To some extent genes in sexual species are reassorted each generation, but the organisms which make up populations cannot be. To put the point in the opposite way, if such shuffling of organisms were to take place, selection would be impossible.

The preceding characteristic of species as evolutionary lineages by itself is sufficient to preclude species being conceptualized as spatiotemporally unrestricted sets or classes. However, if Eldredge and Gould are right, the case for interpreting species as historical entities is even stronger. They ask why species are so coherent, why groups of relatively independent local populations continue to display fairly consistent, recognizable phenotypes, and why reproductive isolation does not arise in every local population if gene flow is the only means of preventing differentiation:

The answer probably lies in a view of species and individuals [organisms] as homeostatic systems—as amazingly well-buffered to resist change and maintain stability in the face of disturbing influences.... In this view, the importance of peripheral isolates lies in their small size and the alien environment beyond the species border that they inhabit—for only here are selective pressures strong enough and the inertia of large numbers sufficiently reduced to produce the "genetic revolution" (Mayr, 1963, p. 533) that overcomes homeostasis. The coherence of a species, therefore, is not maintained by interaction among its members (gene flow). It emerges, rather, as an historical consequence of the species' origin as a peripherally isolated population that acquired its own powerful homeostatic system. (11, p. 114)

Eldredge and Gould argue that, from a theoretical point of view, species appear so amorphous because of a combination of the gradualistic interpretation of speciation and the belief that gene exchange is the chief (or only) mechanism by which cohesion is maintained in natural populations. However, in the field, species of both sexual and asexual organisms seem amazingly coherent and unitary. If gene flow were the only mechanism for the maintenance of evolutionary unity, asexual species should be as diffuse as duststorms in the desert. According to Eldredge and Gould, new species arise through the budding off of peripheral isolates which succeed in establishing new equilibria in novel environments. Thereafter they remain largely unchanged during the course of their existence and survive only as long as they maintain this equilibrium.

Another possibility is that evolutionary unity is maintained by both internal and external means. Gene flow and homeostasis within a species are internal mechanisms of evolutionary unity. Perhaps the external environment in the form of unitary selection pressures also contributes to the integrity of the entities which are evolving (10). For example, Jews have remained relatively distinct from the rest of humankind for centuries, in part by internal means (selective mating, social customs, etc.) but also in part by external means (discrimination, prejudice, laws, etc.). An ecological niche is a relation between a particular species and key environmental variables. A different species in conjunction with the same environmental variables could define quite a different niche. In the past biologists have tended to play down the integrating effect of the environment, attributing whatever unity and coherence which exists in nature to the integrating effect of gene complexes. At the very least, if the coherence of asexual species is not illusory, mechanisms other than gene flow must be capable of bringing about evolutionary unity.

INDIVIDUATING ORGANISMS AND SPECIES

By and large, the criteria which biologists use to individuate organisms are the same as those suggested by philosophers—spatiotemporal continuity, unity, and location. Differences between these two analyses have three sources: first, philosophers have been most interested in individuating persons, the hardest case of all, while biologists have been content to individuate organisms; second, when philosophers have discussed the individuation of organisms, they have usually limited themselves to adult mammals, while biologists have attempted to develop a notion of organism adequate to handle the wide variety of organisms which exist in nature; and finally, philosophers have felt free to resort to hypothetical, science fiction examples to test their conceptions, while biologists rely on actual cases. In each instance, I prefer the biologists' strategy. A clear notion of an individual organism seems an absolute prerequisite for any adequate notion of a person, and this notion should be applicable to all organisms, not just a minuscule fraction. But most important, real examples tend to be much more detailed and bizarre than those made up by philosophers. Too often the example is constructed for the sole purpose of supporting the preconceived intuitions of the philosophers and has no life of its own. It cannot force the philosopher to improve his analysis the way that real examples can. Biologists are in the fortunate position of being able to test their analyses against a large stock of extremely difficult, extensively documented actual cases.

Phenotypic similarity is irrelevant in the individuation of organisms. Identical twins do not become one organism simply because they are phenotypically indistinguishable. Conversely, an organism can undergo massive phenotypic change while remaining the same organism. The stages in the life cycles of various species of organisms frequently are so different that biologists have placed them in different species, genera, families, and even classes—until the continuity of the organism was discovered. If a caterpillar develops into a butterfly, these apparently different organisms are stages in the life cycle of a single organism regardless of how dissimilar they might happen to be (figure 10.1a). In ontogenetic development, a single lineage is never divided successively in time into separate organisms; some sort of splitting is required. In certain cases, such as transverse fission in paramecia, a single organism splits equally into two new organisms (figure 10.1b). In such cases, the parent organism no longer exists, and the daughter organisms are two new individuals. Sometimes a single individual will bud off other individuals which are roughly its own size but somewhat different in appearance, for example, strobilization in certain forms of Scyphozoa (figure 10.1c). At the other extreme, sometimes a small portion of the parent organism buds off to form a new individual, as in budding in Hydrozoa (figure 10.1d). In the latter two cases, the parent organism continues to exist while budding off new individuals. The relevant consideration is how much of the parent organism is lost and its internal organization disrupted.

Fusion also takes place at the level of individual organisms. For example, when presented with a prey too large for a single individual to digest, two amoebae will fuse cytoplasmically in order to engulf and digest it. However, the nuclei remain distinct and the two organisms later separate, genetically unchanged. The commonest example of true fusion occurs when germ cells unite to form a zygote. In such cases, the germ cells as individuals cease to exist and are replaced by a new individual (figure 10.2a). Sometimes one organism will invade another and become part of it. Initially, these organisms, even when they become obligate parasites, are conceived of as separate organisms, but sometimes they can become genuine parts of the host organism. For example, one theory of the origin of certain cell organelles is that they began as parasites. Blood transfusions are an unproblematic case of part of one organism's becoming part of another; conjugation is another (figure 10.2b).

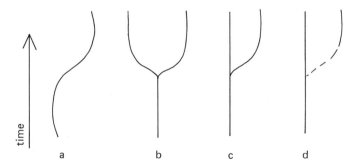

Figure 10.1 Diagrams that can be interpreted alternately as organisms undergoing ontogenetic change and the production of new organisms and as species undergoing phylogenetic change and speculation.

Sometimes parts of two different organisms can merge to form a third. Again, sexual reproduction is the commonest example of such an occurrence (figure 10.2c). In each of these cases, organisms are individuated on the basis of the amount of material involved and the effect of the change on the internal organization of the organisms. For example, after conjugation two paramecia are still two organisms and the same two organisms even though they have exchanged some of their genetic material.

If species are historical entities, then the same sorts of considerations which apply in the individuation of organisms should also apply to them, and they do (35). The only apparent discrepancy results from the fact that not all biologists have been totally successful in throwing off the old preevolutionary view of species as classes of similar organisms and replacing it with a truly evolutionary view. However, even these discrepancies are extremely instructive. For example, G. G. Simpson (50) maintains that a single lineage which changes extensively through time without speciating (splitting) should be divided into separate species (see figure 10.1a). Willi Hennig (17) disagrees: new species should be recognized only upon splitting. This particular debate has been involved, touching upon both conceptual and empirical issues. For example, how can a gradually evolving lineage be divided into discrete species in an objective, nonarbitrary way? Are later organisms considered to belong to different species from their ancestors because they are sufficiently dissimilar or because they can no longer interbreed with them even if they coexisted? Can such extensive change take place in the absence of speciation?

I cannot attempt to answer fully all of these questions here. Instead, I must limit myself to the remark that, on Simpson's view, species and organisms are quite different sorts of things. An organism undergoes limited change, constrained by its largely unchanging genotype. A single species is capable of indefinite, open-ended development. Although the course of a species' development is constrained from generation to generation by its gene pool, this gene pool is indefinitely modifiable. However, if Eldredge and Gould are right, species are more like organisms than anyone has previously supposed. Both

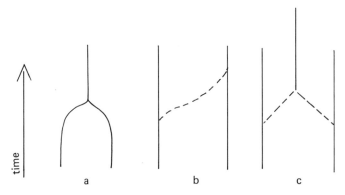

Figure 10.2 Diagrams that can be interpreted alternately as organisms merging totally or partially to give rise to new organisms and as species merging totally or partially to give rise to new species.

David Hull

are finite and can undergo only limited change before ceasing to exist. Significant evolutionary change can take place only through a series of successive species, not within the confines of a single species. Species lineages, not species, are the things which evolve. On this view, Hennig's refusal to divide a single lineage into two or more species is preferable to Simpson's alternative.

No disagreement exists between Simpson and Hennig over the situation depicted in figure 10.1b, a single species splitting equally into two. Both agree that the ancestor species is extinct, having given rise to two new daughter species. However, this figure is drawn as if divergence always takes place upon speciation. When this diagram was interpreted as depicting the splitting of one organism into two, divergence was not presupposed. Two euglenae resulting from binary fission are two organisms and not one even though they may be phenotypically and genotypically identical. The same is true of species. Sometimes speciation takes place with no (or at least extremely minimal) divergence; e.g., sibling species are no less two species simply because they look alike. The assumption is, however, that in reproductively isolated species some divergence, at least in the mechanisms of reproduction, must have taken place, even if we cannot detect it. The role of similarity becomes controversial once again when speciation takes place and one species remains unchanged, while the other diverges from the parental type (see figure 10.1c). According to Hennig (17), when speciation occurs, the ancestor species must be considered extinct regardless of how similar it might be to one of its daughter species. Simpson (50) disagrees.

The factor which is causing the confusion in the preceding discussion is the role of similarity in the individuation of species. If species are classes defined by sets (or clusters) of traits, then similarity should be relevant. At one extreme, the pheneticists (54) argue that all that matters is phenetic similarity and dissimilarity, regardless of descent, reproduction, evolutionary cohesiveness, etc. Highly polytypic species such as dogs must be considered numerous different "species" because of the existence of so many reasonably discrete clusters. Sibling species must be considered a single "species" because they form a single cluster. At the other extreme, the Hennigians (commonly termed "cladists") concentrate solely on the splitting of phylogenetic lineages regardless of phenetic similarity. Polytypic species are single species because they form a single clade; sibling species are separate species because they form more than one clade. The evolutionists, represented by Simpson and Mayr, argue that somehow the two considerations must be balanced against each other.

However, on the historical entity interpretation, similarity is a red herring; it is not the issue at all. What really matters is how many organisms are involved and how much the internal organization of the species involved is disrupted. If speciation takes place when a small, peripheral isolate succeeds in bringing about a genetic revolution (see figure 10.1d), then the parent species can still be said to persist unchanged. It has not lost significant numbers of organisms, nor has its internal organization been affected much. One

Hennigian, at least, has come to this conclusion for precisely these reasons (60). If, however, the species is split into two or more relatively large subgroups, then it is difficult to see how the ancestral species can still be said to exist, unless one of these subgroups succeeds in retaining the same organization and internal cohesion of the ancestral species. Incidentally, it would also be phenetically similar to the ancestral species, but that would be irrelevant.

Fusion can also take place at the level of species. The breaking down of reproductive isolation sufficient to permit two entire species to merge into one is extremely unlikely (see figure 10.2a). If it did occur, the consideration would be the same as those raised in connection with figure 10.1b. However, introgression and speciation by polyploidy are common (see figures 10.2b and 10.2c). In Such cases, a few organisms belonging to separate species mate and produce fertile offspring. Contrary to popular opinion, the production of an occasional fertile hybrid is not enough for biologists to consider two species one. What matters is how extensive the introgression becomes—exactly the right consideration if species are historical entities. As Dobzhansky remarks, "What matters is not whether hybrids can be obtained but whether the Mendelian populations do or do not exchange genes, and if they do whether at a rate which destroys the adaptive equilibrium of the populations concerned" (9, p. 586).

One final parallel between organisms and species warrants mentioning. Organisms are unique. When an organism ceases to exist, numerically that same organism cannot come into existence again. For example, if a baby were born today who was identical in every respect to Adolf Hitler, including genetic makeup, he still would not be a Adolf Hitler. He would be as distinct and separate a human being as ever existed because of his unique "insertion into history," to use Vendler's propitious phrase (58; see also 57). But the same observation can be made with respect to species. If a species evolved which was identical to a species of extinct pterodactyl save origin, it would still be a new, distinct species. Darwin himself notes, "When a species has once disappeared from the face of the earth, we have reason to believe that the same identical form never reappears" (7, p. 313). Darwin presents this point as if it were a contingent state of affairs, when actually it is conceptual. Species are segments of the phylogenetic tree. Once a segment is terminated, it cannot reappear somewhere else in the phylogenetic tree. As Griffiths observes, the "reference of an individual to a species is determined by its parentage, not by any morphological attribute" (15, p. 102).

If species were actually spatiotemporally unrestricted classes, this state of affairs would be strange. If all atoms with atomic number 79 ceased to exist, gold would cease to exist, although a slot would remain open in the periodic table. Later when atoms with the appropriate atomic number were generated, they would be atoms of gold regardless of their origins. But in the typical case, to *be* a horse one must be *born* of horse. Obviously, whether one is a gradualist or saltationist, there must have been instances in which nonhorses (or borderline horses) gave rise to horses. The operative term is still "gave rise to." But what of the science fiction examples so beloved to philosophers?

What if a scientist made a creature from scratch identical in every respect to a human being including consciousness, emotionality, a feeling of personhood, etc.? Wouldn't it be included in *Homo sapiens*? It all depends. If all the scientist did was to make such a creature and then destroy it, it was never part of our species. However, if it proceeded to mate with human beings born in the usual way and to produce offspring, introducing its genes into the human gene pool, then it would become part of our species. The criterion is precisely the same one used in cases of introgression. In the evolutionary world view, unlike the Aristotelian world view, an organism can change its species while remaining numerically the same individual (see 19).

One might complain that being born of human beings and/or mating with human beings are biological criteria, possibly good enough for individuating *Homo sapiens*, but inadequate for the humanistic notion of a human being. We are a social species. An entity which played the role of a human being in a society would *be* a "human being," even if it was not born of human beings or failed to mate with human beings. I'm not sure how one makes such decisions, but the conclusion is not totally incompatible with the position being presented in this chapter. Species as they are commonly thought of are not the only things which evolve. Higher levels of organization also exist. Entities can belong to the same cultural system or ecosystem without belonging to the same biological species. As Eugene Odum has put it, "A human being, for example, is not only a hierarchical system composed of organs, cells, enzyme systems, and genes as subsystems, but is also a component of supraindividual hierarchical systems such as populations, cultural systems, and ecosystems" (44, p. 1289). If pets or computers function as human beings, then from certain perspectives they might well count as human beings even though they are not included in the biological species *Homo sapiens*.

BIOLOGICAL AND PHILOSOPHICAL CONSEQUENCES

Empirical evidence is usually too malleable to be very decisive in conceptual revolutions. The observation of stellar parallax, the evolution of new species right before our eyes, the red shift, etc. are the sorts of things which are pointed to as empirical reasons for accepting new scientific theories. However, all reasonable people had accepted the relevant theories in the absence of such observations. Initial acceptance of fundamentally new ideas leans more heavily on the increased coherence which the view brings to our general world picture. If the conceptual shift from species being classes to species being historical entities is to be successful, it must eliminate longstanding anomalies both within and about biology. In this section, I set out some of the implications of viewing species as historical entities, beginning with those that are most strictly biological, and gradually working my way toward those that are more philosophical.

The role of type specimens in biological systematics puzzles philosophers and biologists alike. As R. A. Crowson remarks, "The current convention that a single specimen, the Holotype, is the only satisfactory basic criterion for a species would be difficult to justify logically on any theory but Special Creation" (5, p. 29). According to all three codes of biological nomenclature, a particular organism, part of an organism, or trace of an organism is selected as the type specimen for each species. In addition, each genus must have its type species, and so on. Whatever else one does with this type and for whatever resson, the name goes with the type.[4] The puzzling aspect of the type method on the class interpretation is that the type need not be typical. In fact, it can be a monster. The following discussion by J. M. Schopf is representative:

It has been emphasized repeatedly, for the benefit of plant taxonomists, at least, that the nomenclatural type (holotype) of a species is not to be confused or implicated in anyone's concept of what is "typical" for a taxon. A nomenclatural type is simply *the specimen*, or other element, with which a name is permanently associated. This element need not be "typical" in any sense; for organisms with a complicated life cycle, it is obvious that no single specimen could physically represent all the important characteristics, much less could it be taken to show many features near the mean of their range of variation. (see also 6, 39, 50, 51; 49, p. 1043)

Species are polymorphic. Should the type specimen for *Homo sapiens*, for instance, be male or female? Species are also polytypic. What skin color, blood type, etc. should the type specimen for *Homo sapiens* have? Given the sort of variability characteristic of biological species, no one specimen could possibly be "typical" in even a statistical sense (37, p. 369). On the class interpretation, one would expect at the very least for a type specimen to have many or most of the more important traits characteristic of its species (16, p. 465–466), but on the historical entity interpretation, no such similarity is required. Just as a heart, kidneys, and lungs are included in the same organism because they are part of the same ontogenetic whole, parents and their progeny are included in the same species because they are part of the same genealogical nexus, no matter how much they might differ phenotypically. The part/whole relation does not require similarity.

A taxonomist in the field sees a specimen of what he takes to be a new species. It may be the only specimen available or else perhaps one of a small sample which he gathers. The taxonomist could not possibly select a typical specimen, even if the notion made sense, because he has not begun to study the full range of the species' variation. He selects a specimen, any specimen, and names it. Thereafter, if he turns out to have been the first to name the species of which this specimen is part, that name will remain firmly attached to that species. A taxon has the name it has *in virtue of* the naming ceremony, not *in virtue of* any trait or traits it might have. If the way in which taxa are named sounds familiar, it should. It is the same way in which people are baptized.[5] They are named in the same way because they are the same sort of thing—historical entities (see Ghiselin, 13, 14).

But what, then, is the role of all those traits which taxonomists include in their monographs? For example, Article 13 of the Zoological Code of Nomenclature states that any name introduced after 1930 must be accompanied by a statement that "purports to give characteristics differentiating the taxon."

Taxonomists distinguish between descriptions and diagnoses. A description is a lengthy characterization of the taxon, including reference to characteristics which are easily recognizable and comparable, to known variability within a population and from population to population, to various morphs, and to traits which can help in distinguishing sibling species. A diagnosis is a much shorter and selective list of traits chosen primarily to help differentiate a taxon from its nearest neighbors of the same rank. As important as the traits listed in diagnoses and descriptions may be for a variety of purposes, they are not definitions. Organisms could possess these traits and not be included in the taxon; conversely, organisms could lack one or more of these traits and be clear-cut instances of the taxon. They are, as the name implies, *descriptions*. As descriptions, they change through time as the entities which they describe change. Right now all specimens of *Cygnus olor* are white. No doubt the type specimen of this species of swan is also white. However, if a black variety were to arise, *Cygnus olor* would not on that account become a new species. Even if this variety were to become predominant, this species would remain the same species and the white type specimen would remain the type specimen. The species description would change but that is all. Organisms are not included in the same species *because* they are similar to the type specimen or to each other but *because* they are part of the same chunk of the genealogical nexus (Ghiselin, 13, 14).

On the class interpretation, the role of particular organisms as type specimens is anomalous. The role of lower taxa as types for higher taxa is even more anomalous. On the class interpretation, organisms are members of their taxa, while lower taxa are included in higher taxa (3). How could entities of two such decidedly different logical types play the same role? But on the historical entity interpretation, both organisms and taxa are of the same logical type. Just as organisms are part of their species, lower taxa are part of higher taxa. Once again, parts do not have to be similar, let alone typical, to be part of the same whole.

A second consequence of treating species as historical entities concerns the nature of biological laws. If species are actually spatiotemporally unrestricted classes, then they are the sorts of things which can function in laws. "All swans are white," if true, might be a law of nature, and generations of philosophers have treated it as such. If statements of the form "species X has the property Y" were actually laws of nature, one might rightly expect biologists to be disturbed when they are proven false. To the contrary, biologists expect exceptions to exist. At any one time, a particular percentage of a species of crows will be non-black. No one expects this percentage to be universal or to remain fixed. Species may be classes, but they are not very important classes because their names function in no scientific laws. Given the traditional analyses of scientific laws, statements which refer to particular species do not count as scientific laws, as they should not if species are spatiotemporally localized individuals (20, 21).

Hence, if biologists expect to find any evolutionary laws, they must look at levels of organization higher than particular taxa. Formulations of evolutionary theory will no more make explicit reference to *Bos bos* than celestial mechanics will refer to Mars. Predictions about these entities should be derivable from the appropriate theories but no uneliminable reference can be made to them. In point of fact, no purported evolutionary laws refer to particular species. One example of such a law is the claim that in diploid sexually reproducing organisms, homozygotes are more specialized in their adaptive properties than heterozygotes (31, p. 397). Evolutionary theory deals with the rise of individual homeostasis as an evolutionary mode, the waxings and wanings of sexuality, the constancy or variability of extinction rates, and so on. People are dismayed to discover that evolutionists can make no specific predictions about the future of humankind *qua* humankind. Since that's all they are interested in, they conclude that evolutionary theory is not good for much. But dismissing evolutionary theory because it cannot be used to predict the percentage of people who will have blue eyes in the year 2000 is as misbegotten as dismissing celestial mechanics because it cannot be used to predict the physical make-up of Mars. Neither theory is designed to make such predictions.

The commonest objection raised by philosophers against evolutionary theory is that its subject matter—living creatures—is spatiotemporally localized (52, 53; see also 42). They exist here on earth and nowhere else. Even if the earth were the only place where life had arisen (and that is unlikely), this fact would not count in the least against the spatiotemporally unrestricted character of evolutionary theory. "Hitler" refers to a particular organism, a spatiotemporally localized individual. As such, Hitler is unique. But organisms are not. Things which biologists would recognize as organisms could develop (and probably have developed) elsewhere in the universe. "*Homo sapiens*" refers to a particular species, a spatiotemporally localized individual. As such it is unique. but species are not. Things which biologists would recognize as species could develop (and probably have developed) elsewhere in the universe. Evolutionary theory refers explicitly to organisms and species, not to Hitler and *Homo sapiens* (see 43, 48).

One advantage to biologists of the historical entity interpretation of species is that it frees them of any necessity of looking for any lawlike regularities at the level of particular species. Both "Richard Nixon has hair" and "most swans are white" may be true, but they are hardly laws of nature. If forces them to look for evolutionary laws at higher levels of analysis, at the level of *kinds* of species. It also can explain certain prevalent anomalies in philosophy. From the beginning, a completely satisfactory explication of the notion of a natural kind has eluded philosophers. One explanation for this failure is that the traditional examples of natural kinds were a mixed lot. The three commonest examples of natural kinds in the philosophical literature have been geometric figures, biological species, and the physical elements. By now it should be clear that all three are very different sorts of things. No wonder a general anaylsis, applicable equally to all of them, has eluded us.

Some of the implications of treating species as historical entities are more philosophical in nature. For example, one of Ludwig Wittgenstein's most famous (or infamous) contributions to philosophy is that of family resemblances, a notion which itself has a family resemblance to cluster concepts and multivariate analysis (64). Such notions have found their most fertile ground in ethics, aesthetics, and the social sciences. Hence, critics have been able to claim that defining a word in terms of statistical covariation of traits merely results from ignorance and informality of context. If and when these areas become more rigorous, cluster concepts will give way to concepts defined in the traditional way. The names of biological species have been the chief counter-example to these objections. Not only are the methods of contemporary taxonomists rigorous, explicit, objective, etc., but also good reasons can be given for the claim that the names of species can never be defined in classical terms. They are inherently cluster concepts (18). On the analysis presented in this chapter, advocates of cluster analysis lose their best example of a class term which is, nevertheless, a cluster concept. If "*Homo sapiens*" is or is not a cluster concept, it will be for the same reason that "Moses" is or (more likely) is not.

A second philosophical consequence of treating species as historical entities concerns the nature of scientific theories. Most contemporary philosophers view scientific theories as atemporal conceptual objects. A theory is a timeless set of axioms and that is that. Anyone who formulates a theory consisting of a particular set of axioms has formulated that theory period. Theories in this sense cannot change through time. Any change results in a new theory. Even if one decides to get reasonable and allow for some variation in axioms, one still must judge two versions of a theory to be versions of the "same" theory because of similarity of axioms. Actual causal connections are irrelevant. However, several philosophers have suggested that science might profitably be studied as an "evolutionary" phenomenon (4, 21, 27, 28, 29, 45, 46, 56). If one takes these claims seriously and accepts the analysis of biological species presented in this chapter, then it follows that whatever conceptual entities are supposed to be analogous to species must also be historical entities. Theories seem to be the most likely analog to species. Because biological species cannot be characterized intelligibly in terms of timeless essences, it follows that theories can have no essences either. Like species, theories must be individuated in terms of some sort of descent and cohesiveness, not similarity.

The relative roles of similarity and descent in individuating scientific theories go a long way in explaining the continuing battle between historians and philosophers of science. Philosophers individuate theories in terms of a set (or at least a cluster) of axioms. Historians tend to pay more attention to actual influence. For example, we all talk about contemporary Mendelian genetics. If theories are to be individuated in terms of a single set (or even cluster) of axioms, it is difficult to see the justification of such an appellation. Mendel's paper contained three statements which he took to be basic. Two of these statements were rapidly abandoned at the turn of the century when Mendel's so-called laws were rediscovered. The third has been modified since. If overlap in substantive claims is what makes two formulations versions of the "same" theory, then it is difficult to see the justification for interpreting all the various things which have gone under the title of "Mendelian genetics" versions of the same theory. Similar observations are appropriate for other theories as well, including Darwin's theory of evolution. The theory that was widely accepted in Darwin's day differed markedly from the one he originally set out. Modern theories of evolution differ from his just as markedly. Yet some are "Darwinian" and others not.

When presented with comparable problems, biologists resort to the type specimen. One organism is selected as the type. Any organism related to it in the appropriate ways belongs to its species, regardless of how aberrant the type specimen might turn out to be or how dissimilar other organisms may be. Males and females belong to the same species even though they might not look anything like each other. A soldier termite belongs in the same species with its fertile congeners even though it cannot mate with them. One possible interpretation of Kuhn's notion of an exemplar (27) is that it is designed to function as a type specimen. Even though scientific change is extremely complicated and at times diffuse, one still might be able to designate particular theories by reference to "concrete problem-solutions," as long as one realizes that these exemplars have a temporal index and need not be in any sense typical.[6] Viewing theories as sets (or clusters) of axioms does considerable damage to our intuitions about scientific theories. On this interpretation, most examples of scientific theories degenerate into unrelated formulations. Viewing scientific theories as historical entities also results in significant departures from our usual modes of conception. Perhaps scientific theories really cannot be interpreted as historical entities. If so, then this is just one more way in which conceptual evolution differs from biological evolution. The more these disanalogies accumulate, the more doubtful the entire analogy becomes.

Finally, and most controversially, treating species as historical entities has certain implications for those sciences which are limited to the study of single species. For instance, if enough scientists were interested, one might devote an entire science to the study of *Orycteropus afer*, the African aardvark. Students of aardvarkology might discover all sorts of truths about aardvarks; that it is nocturnal, eats ants and termites, gives birth to its young alive, etc. Because aardvarks are highly monotypic, aardvarkologists might be able to discover sets of traits possessed by all and only extant aardvarks. But could they discover the essence of aardvarks, the traits which aardvarks must have necessarily to be aardvarks? Could there be scientific laws which govern aardvarks necessarily and exclusively? When these questions are asked of aardvarks or any other nonhuman species, they sound frivolous, but they are exactly the questions that students of human nature treat with utmost seriousness. What is human nature and its laws?

Early in the history of learning theory, Edward L. Thorndike (55) claimed that learning performance in fishes, chickens, cats, dogs, and monkeys differed only quantitatively, not qualitatively. Recent work tends to contradict his claim (1). Regardless of who is right, why does it make a difference? Learning, like any other trait, has evolved. It may be universally distributed among all species of animals or limited to a few. It may be present in all organisms included in the same species or distributed less than universally. In either case, it may have evolved once or several times. If "learning" is defined in terms of its unique origin, if all instances of learning must be evolutionarily homologous, then "learning" is limited by definition to one segment of the phylogenetic tree. Any regularities which one discovers are necessarily descriptive. If, on the other hand, "learning" is defined so that it can apply to any organism (or machine) which behaves in appropriate ways, then it *may* be limited to one segment of the phylogenetic tree. It *need* not be. Any regularities which one discovers are at least candidates for laws of learning. What matters is whether the principles are generalizable. Learning may be species specific, but if learning theory is to be a genuine scientific theory, it cannot be limited *necessarily* to a single species the way that Freud's and Piaget's theories seem to be. As important as descriptions are in science, they are not theories.

If species are interpreted as historical entities, then particular organisms belong in a particular species because they are part of that genealogical nexus, not because they possess any essential traits. No species has an essence in this sense. Hence there is no such thing as human nature. There may be characteristics which all and only extant human beings possess, but this state of affairs is contingent, depending on the current evolutionary state of *Homo sapiens*. Just as not all crows are black (even potentially), it may well be the case that not all people are rational (even potentially). On the historical entity interpretation, retarded people are just as much instances of *Homo sapiens* as are their brighter congeners. The same can be said for women, blacks, homosexuals, and human fetuses. Some people may be incapable of speaking or understanding a genuine language; perhaps bees can. It makes no difference. Bees and people remain biologically distinct species. On other, nonbiological interpretations of the human species, problems arise (and have arisen) with all of the groups mentioned. Possibly women and blacks are human beings but do not "participate fully" in human nature. Homosexuals, retardates, and fetuses are somehow less than human. And if bees use language, then it seems we run the danger of considering them human. The biological interpretation has much to say in its favor, even from the humanistic point of view.

NOTES

The research for this chapter was supported by NSF grant Soc 75 03535. I am indebted to the following people for reading and criticizing early versions of it: Michael Ghiselin, Stephen Gould, G. C. D. Griffiths, John Koethe, Ernst Mayr, Bella Selan, W. J. van der Steen, Gareth Nelson, Michael Perloff, Mark Ridley, Michael Ruse, Thomas Schopf, Paul Teller, Leigh Van Valen, Linda Wessels, Mary Williams, and William Wimsatt. Their advice and criticisms are much appreciated.

1. Once again I am excluding from the notion of class those "classes" defined by means of a spatiotemporal relation to a spatiotemporally localized individual. Needless to say, I am also excluding such constructions as "similar in origin" from the classes of similarities. I wish the need to state the obvious did not exist, but from past experience it does.

2. In population genetics the distinction between structurally similar genes forming a single lineage and those which do not is marked by the terms "identical" and "independent"; see (41), pp. 56–57.

3. Until recently even the most ardent group selectionists admitted that the circumstances under which selection can occur at the level of populations and/or entire species are so rare that group selection is unlikely to be a major force in the evolutionary process (30, 32, 33). Michael Wade (59), however, has presented a convincing argument to the effect that the apparent rarity of group selection may be the result of the assumptions commonly made in constructing mathematical models for group selection and not an accurate reflection of the actual state of nature. In his own research, the differential survival of entire populations has produced significant divergence.

4. The three major codes of biological nomenclature are (1) the International Code of Botanical Nomenclature, 1966, International Bureau for Plant Taxonomy and Nomenclature, Utrecht; (2) the International Code of Nomenclature of Bacteria, 1966, *International Journal of Systematic Bacteriology*, 16:459–490; and (3) the *International Code of Zoological Nomenclature*, 1964, International Trust for Zoological Nomenclature, London. In special circumstances the priority rule is waived, usually because the earlier name is discovered only long after a later name has become firmly and widely established.

5. Although the position on the names of taxa argued for in this chapter might sound as if it supported S. Kripke's (26) analysis of general terms, it does not. Taxa names are very much like "rigid designators," as they should be if taxa are historical entities. However, Kripke's analysis is controversial because it applies to *general* terms. It is instructive to note that during the extensive discussion of the applicability of Kripke's notion of a rigid designator to such terms as "tiger," no one saw fit to see how those scientists most intimately concerned actually designated tigers. According to Putnam's principle of the linguistic division of labor (47), they should have. If they had, they would have found rules explicitly formulated in the various codes of nomenclature which were in perfect accord with Kripke's analysis—but for the wrong reason. That no one bothered tells us something about the foundations of conceptual analysis.

6. Kuhn himself (28) discusses taxa names such as "*Cygnus olor*" and the biological type specimen. Unfortunately, he thinks swans are swans because of the distribution of such traits as the color of feathers.

REFERENCES

1. Bitterman, M. E. 1975. The comparative analysis of learning. *Science* 188:699–709.

2. Boyden, A. 1954. The significance of asexual reproduction. *Systematic Zoology* 3:26–37.

3. Buck, R. C., and D. L. Hull. 1966. The logical structure of the Linnaean hierarchy. *Systematic Zoology* 15:97–111.

4. Burian, R. M. 1977. More than a marriage of convenience: On the inextricability of history and philosophy of science. *Philosophy of Science* 44:1–42.

5. Crowson, R. A. 1970. *Classification and Biology*. New York, Atherton Press.

6. Davis, P. H., and V. H. Heywood. 1963. *Principles of Angiosperm Taxonomy*. Princeton, Van Nostrand.

7. Darwin, C. 1966. *On the Origin of Species*. Cambridge, Mass., Harvard University Press.

8. Dawkins, R. 1976. *The Selfish Gene*, New York and Oxford, Oxford University Press.

9. Dobzhansky, T. 1951. Mendelian populations and their evolution. In L. C. Dunn (ed.), *Genetics in the 20th Century*, New York, Macmillan, pp. 573–589.

10. Ehrlich, P. R., and P. H. Raven. 1969. Differentiation of populations, *Science* 165:1228–1231.

11. Eldredge, N., and S. J. Gould. 1972. Punctuated equilibria: An alternative to phyletic gradualism. In T. J. M. Schopf (ed.), *Models in Paleobiology*. San Francisco, Freeman, Cooper and Company, pp. 82–115.

12. Ghiselin, M. T. 1966. On psychologism in the logic of taxonomic controversies. *Systematic Zoology* 15:207–215.

13. ———. 1969. *The Triumph of the Darwinian Method*. Berkeley and London, University of California Press.

14. ———. 1974. A radical solution to the species problem. *Systematic Zoology* 23:536–544.

15. Griffiths, G. C. D. 1974. On the foundations of biological systematics. *Acta Biotheoretica* 23:85–131.

16. Heise, H., and M. P. Starr. 1968. Nomenifers: Are they christened or classified? *Systematic Zoology* 17:458–467.

17. Hennig, W. 1966. *Phylogenetic Systematics*. Urbana, Illinois, University of Illinois Press.

18. Hull, D. L. 1965, 1966. The effect of essentialism on taxonomy. *British Journal for the Philosophy of Science* 15:314–326; 16:1–18.

19. ———. 1968. The conflict between spontaneous generation and Aristotle's metaphysics. *Proceedings of the Seventh Inter-American Congress of Philosophy*, Québec City, Les Presses de l'Université Laval, 2:245–250.

20. ———. 1974. *Philosophy of Biological Science*. Englewood Cliffs, Prentice-Hall.

21. ———. 1975. Central subjects and historical narratives. *History and Theory* 14:253–274.

22. ———. 1976. Are species really individuals? *Systematic Zoology* 25:174–191.

23. ———. 1976. The ontological status of biological species. In R. Butts and J. Hintikka (eds.), *Boston Studies in the Philosophy of Science*, vol. 32, Dordrecht, D. Reidel, pp. 347–358.

24. Huxley, T. H. 1889. Biology. *Encyclopedia Britannica*.

25. Janzen, Daniel. 1977. What are dandelions and aphids? *American Naturalist* 111:586–589.

26. Kripke, S. S. 1972. Naming and necessity. In D. Davidson and H. Harman (eds.), *Semantics and Natural Language*. Dordrecht, Holland, D. Reidel, pp. 253–355.

27. Kuhn, T. S. 1969. *The Structure of Scientific Revolutions*, Chicago, University of Chicago Press, 2nd ed.

28. ———. 1974. Second Thoughts on Paradigms. In F. Suppe (ed.), *The Structure of Scientific Theory*. Urbana, Illinois, University of Illinois Press.

29. Laudan, L. 1977. *Progress and Its Problems*. Berkeley and London, University of California Press.

30. Levins, R. 1968. *Evolution in Changing Environments*. Princeton, Princeton University Press.

31. Lewontin, R. C. 1961. Evolution and the theory of games. *Journal of Theoretical Biology* 1:382–403.

32. ———. 1970. The units of selection. *Annual Review of Ecology and Systematics* 1:1–18.

33. ———. 1974. *The Genetic Basis of Evolutionary Change*. New York, Columbia University Press.

34. Löther, R. 1972. *Die Beherrschung der Mannigfaltigkeit*. Jena, Gustav Fisher.

35. Mayr, E. 1957 (ed.). *The Species Problem*. Washington, D.C., American Association for the Advancement of Science, Publication Number 50.

36. ———. 1959. Isolation as an evolutionary factor. *Proceedings of the American Philosophical Society* 103:221–230.

37. ———. 1963. *Animal Species and Evolution*. Cambridge, Mass., Belknap Press of Harvard University Press.

38. ———. 1976. Is the species a class or an individual? *Systematic Zoology* 25:192.

39. Mayr, E., E. G. Linsley, and R. L. Usinger. 1953. *Methods and Principles of Systematic Zoology*. New York, McGraw-Hill Book Company.

40. Meglitsch, P. A. 1954. On the nature of species. *Systematic Zoology* 3:49–65.

41. Mettler, L. E., and T. G. Gregg. 1969. *Population Genetics and Evolution*. Englewood Cliffs, Prentice-Hall.

42. Monod, J. L. 1975. On the molecular theory of evolution. In R. Harré (ed.), *Problems of Scientific Revolution*. Oxford, Clarendon Press, pp. 11–24.

43. Munson, R. 1975. Is biology a provincial science? *Philosophy of Science* 42:428–447.

44. Odum, E. P. 1977. The emergence of ecology as a new integrative discipline. *Science* 195:1289–1293.

45. Popper, K. R. 1972. *Objective Knowledge*, Oxford, Clarendon Press.

46. ———. 1975. The rationality of scientific revolutions. In R. Harré (ed.), *Problems of Scientific Revolution*. Oxford, Clarendon Press, pp. 72–101.

47. Putnam, H. 1974. The meaning of meaning. In K. Gunderson (ed.), *Minnesota Studies in the Philosophy of Science*, vii. Minneapolis, University of Minnesota Press, pp. 131–193.

48. Ruse, M. J. 1973. *The Philosophy of Biology*. London, Hutchinson University Library.

49. Schopf, J. M. 1960. Emphasis on holotype. *Science* 131:1043.

50. Simpson, G. G. 1945. The principles of classification and a classification of mammals. *Bulletin of the American Museum of Natural History* 85:1–350.

51. ———. 1961. *Principles of Animal Taxonomy*. New York, Columbia University Press.

52. Smart, J. J. C. 1963. *Philosophy and Scientific Realism*. London, Routledge and Kegan Paul.

53. ———. 1968. *Between Science and Philosophy*. New York, Random House.

54. Sneath, P. H. A., and R. R. Sokal. 1973. *Numerical Taxonomy*. San Francisco, W. H. Freeman and Company.

55. Thorndike, E. L. 1911. *Animal Intelligence*. New York, Macmillan.

56. Toulmin, S. 1972. *Human Understanding*. Princeton, Princeton University Press.

57. Van Fraassen, Bas. 1972. Probabilities and the problem of individuation. In S. A. Luckenbach (ed.), *Probabilities, Problems and Paradoxes*. Encino, Calif., Dickinson Publishing Co., pp. 121–138.

58. Vendler, Z. 1976. On the possibility of possible worlds. *Canadian Journal of Philosophy* 5:57–72.

59. Wade, M. J. 1978. A critical review of the models of group selection. *Quarterly Review of Biology* 53:101–114.

60. Wiley, E. O. 1978. The evolutionary species concept reconsidered. *Systematic Zoology* 27:17–26.

61. Williams, G. C. 1966. *Adaptation and Natural Selection*. Princeton, Princeton University Press.

62. Wilson, E. O. 1975. *Sociobiology: The New Synthesis*. Cambridge, Mass., Belknap Press of Harvard University Press.

63. Wynne-Edwards, V. C. 1962. *Animal Dispersion in Relation to Social Behaviour*. Edinburgh and London, Oliver and Boyd.

64. Wittgenstein, L. 1953. *Philosophical Investigations*. New York, Macmillan.

11 Species Concepts: A Case for Pluralism

Brent D. Mishler and Michael J. Donoghue

We must resist at all costs the tendency to superimpose a false simplicity on the exterior of science to hide incompletely formulated theoretical foundations.
—Hull 1970, 37

It has often been argued that it is empirically true and/or theoretically necessary that "species," as units in nature, are fundamentally and universally different from taxa at all other levels. Species are supposed to be unique because they are individuals (in the philosophical sense, as opposed to classes)— integrated, cohesive units, with a real existence in space and time (Ghiselin 1974, Hull 1978) Interbreeding among the members (parts) of a species and reproductive isolation between species are generally believed to account for their individuality. These reproductive criteria are supposed to provide the greater objectivity of the species category and have been suggested as *the* criteria by which species taxa are to be delimited in nature.

Wake (1980) has pointed out that this conception of species forms the basis upon which Eldredge and Cracraft (1980) have built their formulation of evolutionary process and phylogenetic analysis. In fact, this notion of species seems to underlie much of the recent and growing body of theory which, for convenience, could be called macroevolutionary theory (Eldredge and Gould 1972, Stanley 1975, Gould 1982). Moreover, most recent texts in systematics and ecology are predicated on the idea that species taxa are unique and fundamental (e.g., White 1978, Ricklefs 1979, Wiley 1981). It is therefore important to assess carefully any claim that species do or should possess the properties of individuals, and whether breeding criteria are adequate indicators of individuality.

The "species problem" has yielded an enormous quantity of literature, and it is not the purpose of this chapter to provide a review (for which see Mayr 1957, Wiley 1978, and papers cited therein). Instead, we will (1) briefly characterize prevailing species concepts, (2) summarize some empirical observations that bear on the species problem, (3) consider the respects in which species taxa as currently delimited by systematists do and do not have the properties

From *Systematic Zoology*, 1982, 31:491–503.

of individuals, and (4) discuss several choices with which we are faced if all the criteria of individuality are not always met.

We will argue that current species concepts are theoretically oversimplified. Empirical studies show that patterns of discontinuity in ecological, morphological, and genetical variation are generally more complex than are represented by these concepts. Criteria for what constitutes "important" discontinuity appear to vary in response to the vast differences in biology between groups of organisms. In our view, no single and universal level of fundamental evolutionary units exists; in most cases species taxa have no special reality in nature. We urge explicit recognition and acceptance of a more pluralistic conception of species, one that recognizes the evident variety and complexity of "species situations." We will conclude by exploring important consequences of this view for ecology, paleontology, and systematics.

PREVAILING SPECIES CONCEPTS

A consensus appears to have been reached that species are integrated unique entities. The so-called biological species concept emphasizes that species are reproductive communities within which genes are (or can be) freely exchanged, but between which gene flow does not occur or at least is very rare (e.g., Mayr 1970). According to this view a species is a group of organisms with a common gene pool that is reproductively isolated from other such groups.

The evolutionary species concept (Simpson 1961, Grant 1971, Wiley 1978, 1981) is an important extension of the concept of biological species, an attempt to broaden the definition to include all sorts of organisms (not just sexually reproductive ones) and to portray the existence of species through time. According to this view species are separate ancestor-descendant lineages with their own evolutionary roles, tendencies, and fates. The ecological species concept of Van Valen (1976) is similar (but see Wiley 1981); however, it emphasizes the "adaptive zone" occupied by a lineage.

Ghiselin (1974) and Hull (1976, 1978) have examined the status of species from a philosophical standpoint. They contend that if species are to play the role required of them in current systematic and evolutionary theory, they must be "individuals" (i.e., integrated and cohesive entities with a restricted spatiotemporal location) rather than "classes" (i.e., spatiotemporally unrestricted sets with defining characteristics). Hull (1980), Wiley (1980, 1981), and Ghiselin (1981) argue that species are fundamentally different from genera, families, and other higher taxa, because they are the most inclusive entities that are "actively evolving."

In general, then, species are considered to be the most objectively defined taxonomic and evolutionary units. As Mayr (1970, 374) put it, they are "the real units of evolution, as the temporary incarnation of harmonious, well-integrated gene complexes." They differ from taxa at all other levels, which are considered to be arbitrarily defined and more subjective categories (e.g., Mayr 1969, 91).

For many workers, these views are not only theoretically satisfying but also seem sufficiently unproblematical in application. Many biologists (especially zoologists) seem to be satisfied that, with the exception of some sibling species complexes and rassenkreisse, the application of biological/evolutionary species concepts will yield the same sets of organisms that would be recognized as "species" by a competent taxonomist in a museum, or by a person on the street.[1]

It must be pointed out, however, that the prevailing species concepts are based on relatively few well-studied groups such as birds and *Drosophila*, groups in which discontinuities in the ability to interbreed are relatively complete, and discontinuities in morphological and ecological variation coincide well with the inability to breed in nature. It also must be pointed out that even though relatively few groups have been studied in detail, a correspondence between morphological, ecological, and breeding discontinuities is often simply assumed.

The acceptance of biological/evolutionary species concepts has not been universal. In particular, the botanical community has not wholeheartedly taken them up, and alternatives have been proliferated.[2] It seems clear that the group of organisms on which one specializes strongly influences the view of "species" that one develops. It also seems clear that in order to fully appreciate biological diversity (for purposes of developing general concepts), it is essential to study a variety of different kinds of organisms, or at least take seriously those who have.[3]

Numerous attacks have been leveled at the biological/evolutionary species concepts. Many of these have been concerned primarily with whether they are operational (e.g., Sokal and Crovello 1970). However, as Hull (1968, 1970) has pointed out, a concept cannot be completely operational and still be useful for the growth of science. The critical question is whether a concept is operational *enough* to be useful as a conceptual framework. Considerations of operationality, while certainly of interest, are not central to the argument developed below, which primarily concerns the theoretical adequacy of prevailing species concepts.

EMPIRICAL CONSIDERATIONS

In our view, a theoretically satisfactory species concept must bear some specifiable relationship to observed patterns of variation among organisms. It is *not* acceptable to adopt a definition of species simply because it conveniently fits into some more inclusive theory, e.g., a theory of evolutionary process. A species concept is, in effect, a low-level hypothesis about the nature of that variation, itself subject to empirical tests. Therefore, in this section, we summarize some relevant empirical findings, many of which have not been generally recognized.

The Noncorrespondence of Discontinuities

The reason for discontent among botanists and other workers is not that they have been unable to perceive discontinuities in nature. Instead, it has become apparent that there are many kinds of discontinuities, all of which may be of interest (Davis and Heywood 1963, 91). The question is, how well do various discontinuities correspond; i.e., are the same sets of organisms delimited by discontinuities when we look at morphology, as when we look at ecology, or breeding? The answer appears to be that there is no necessary correspondence. Stebbins (1950), Grant (1957, 1971, 1981), Stace (1978), and many others have discussed hybridization, apomixis, poly-ploidy, and anomalous breeding systems in plants and have clearly documented the frequent noncorrespondence of different kinds of discontinuities. In some groups there is complete reproductive isolation between populations that would be recognized as one species on morphological grounds (i.e., "sibling species," as in some groups of *Gilia* [Grant 1964], and *Clarkia* [Small 1971]), and in many other groups of plants the interbreeding unit encompasses two to many morphological units (e.g., *Quercus* [Burger 1975]).

It has also become clear that discontinuities in morphological variation or in the ability to interbreed do not necessarily correspond to differences in ecology ("niche"?). The early work of Turesson (1922a, 1922b) in Europe, and of Clausen, Keck, and Hiesey (1939, 1940) in North America, demonstrated that ecotypes "may or may not possess well-marked morphological differences which enable them to be recognized in the field" (Stebbins 1950, 49). The great extent to which local populations of the same biological or morphological species are physiologically differentiated and adapted to their particular environments is only now being realized (Mooney and Billings, 1961, Antonovics et al. 1971, Antonovics 1972, Bradshaw 1972, Kiang 1982).

If noncorrespondence is prevalent, then strict biological species will not necessarily have anything in common but reproductive isolation. It might be argued that a species concept that unambiguously reflects one aspect of variation may be preferable to one that ambiguously reflects several things. But why should we necessarily pin species names on sets of organisms delimited by reproductive barriers? Why not choose, for example, to name morphological units instead?

One argument for pinning species names on reproductively isolated groups is that breeding discontinuities are thought to be more clear-cut than morphological ones and therefore less arbitrary. However, Ornduff (1969) has summarized the complexity of the reproductive biology of flowering plants and pointed out the difficulty of applying rigid species delimitations based on interfertility. When variation in the ability to interbreed is examined in detail, we find discontinuities of many different degrees and kinds. Groups of organisms range from completely interfertile to completely reproductively isolated. Hierarchies or networks of breeding groups vary in complex ways in space and time. Therefore, even if we were to decide that breeding discontinuities

Brent D. Mishler and Michael J. Donoghue

were theoretically the most important kind of discontinuity, and the ones that species names should reflect, the choice of what constitutes a significant discontinuity remains problematical.

A second argument for the importance of reproductive barriers is that gene flow prevents significant divergence while a lack of gene flow allows it. However, this now appears not to be the case. If a population is subjected to disruptive selection, there can be divergence even in the face of gene flow (Jain and Bradshaw 1966). In these instances it appears that some means of reproductive isolation will usually evolve, but such isolation follows initial divergence. Moreover, allopatric populations can remain morphologically similar for very long periods or they can diverge morphologically (see discussion of this point by Bremer and Wanntorp 1979a). This morphological divergence may or may not be accompanied by reproductive isolation, though it appears likely that eventually a reproductive barrier will result. The point is that morphological divergence and the attainment of means of reproductive isolation can be uncoupled events in time and space. Levin (1978, 288–289) concluded: "If we adhere to the biological species concept—the integrated reproductive communities—described by Mayr, then speciation is capricious. . . . Isolating mechanisms are not the cause of divergent evolution, nor are they essential for it to occur."

A related, larger-scale argument for the importance of reproductive barriers is that groups that are reproductively isolated for long periods of time are at least evolutionarily independent (whether or not they diverge morphologically), making them effectively separate entities. Reproductive barriers indeed may often be important in this way, but other factors such as ecological role and homeostatic "inertia" are important as well. Because of the complex nature of variation in each of these factors, and because different factors may be "most important" in the evolution of different groups, a *universal* criterion for delimiting fundamental, cohesive evolutionary entities does not exist.

Questionable Internal Genetic Cohesion

The notion of integration and internal cohesion is central to biological/evolutionary/individualistic species concepts. In this chapter we will follow the common assumption that "cohesion" means genetic cohesion maintained via gene flow, a notion that has recently been explicitly formulated (Wiley and Brooks 1982). However, Hull (1978) has pointed out that other factors such as internal homeostasis and "external environment in the form of unitary selection pressures" (p. 344) may contribute to or confer cohesion. It seems to us likely that "cohesion," and the factors responsible for it, will differ from one group of organisms to another and from one level in the hierarchy to another.

Ehrlich and Raven (1969) pointed out that the extent of gene flow seems to be very limited in many organisms and may not account for the apparent integrity of the morphological units we recognize in nature. Bradshaw (1972, 42) suggested that "effective population size in plants is to be measured in meters and not in kilometers." Endler (1973) studied clinal variation and concluded that "gene flow may be unimportant in the differentiation of populations along environmental gradients" (p. 249). Levin and Kerster (1974) thoroughly reviewed and analyzed the literature concerning gene flow in seed plants and concluded that "the numbers [of individuals] within panmictic units are to be measured in tens and not hundreds" (p. 203). These same points were reiterated by Sokal (1973), Raven (1976), and Levin (1978, 1981). Levin (1979, 383) stated: "The idea that plant species are Mendelian populations wedded by the bonds of mating is most difficult to justify given our knowledge about gene flow. Indeed a contrary viewpoint is supported. Populations separated by several kilometers may rarely, if ever, exchange genes and as such may evolve independently in the absence of strong or even weak selective differentials."

Lande (1980) has stressed that there has been an overemphasis on the genetic cohesion of widespread species and argued that "of the major forces conserving phenotypic uniformity in time and space stabilizing selection is by far the most powerful" (p. 467). Grant (1980, 167) suggested that "the homogeneity of species is due more to descent from a common ancestor than to gene exchange across significant parts of the species area."

Jackson and Pound (1979) critically reviewed much of this literature and rightly pointed out that there is little rigorous evidence in animals to support or to reject the generality of any statement about gene flow because detailed studies are rare. They concluded, however, that data "seem sufficient to indicate that gene flow in plants can be limited due to local or leptokurtic dispersal of pollen and seeds" (p. 78). It is important to keep in mind that population genetic theory predicts that a small amount of migration between populations may be sufficient to maintain genetic similarity in the absence of differential selection (Lewontin 1974, 212–216). Clearly, determining the relative importance of factors such as gene flow, developmental homeostasis, and selection in nature will require rigorous population genetic theory (e.g., Lande 1980) and careful quantification of empirical data, rather than qualitative, anecdotal arguments.

Evolutionary biologists are just beginning to understand gene flow in plants and animals, but have hardly begun to address the complicated patterns of gene exchange present in the fungi, bacteria, and "protists." A kind of chauvinism has so far restricted discussions of gene flow to comparisons of biparental sexual organisms and asexual ones. Complex patterns of sexuality are present in the fungi (Clémençon 1977); intricate incompatibility systems, as well as incompletely understood parasexuality cycles, make the simplistic application of the biological species concept impossible in most cases. The existence of discrete, integrated genetic lineages is even less likely in the "Monera" (Cowan 1962). There probably are very few absolute barriers to genetic exchange in bacteria, because of the phenomena of DNA-mediated transformation, phage-mediated transduction, and bacterial conjugation (Bodmer 1970).

ARE SPECIES TAXA INDIVIDUALS?

In our view, the empirical considerations discussed above indicate that in many (perhaps most) major groups of organisms, actual patterns of variation are such that the species taxa *currently recognized* by taxonomists cannot be considered discrete, primary, and comparable individuals, integrated and cohesive via the exchange of genes, fundamentally different from taxa at other levels. Variation in morphology, ecology, and breeding is enormous and complex; there are discontinuities of varying degree in each of these factors, and the discontinuities are often not congruent. There may often be roughly continuous reduction in the degree of cohesion due to gene flow as more inclusive groups of organisms are considered. The acquisition of reproductive isolating mechanisms appears in many cases to be fortuitous and such isolation is neither the cause of morphological or ecological divergence nor is it necessary for divergence to occur.

Although many currently recognized species do not meet one important criterion of "individuality," namely, cohesion and integration of parts, another important criterion often is met, namely, restricted spatiotemporal location (i.e., units united by common descent). These units are not strictly "individuals" or "classes," but clearly they can function in evolutionary theory and phylogeny construction. Wiley (1980) called such units "historical entities", but applied this term only to taxa above the species level.

We should mention, as a disclaimer, that although many species taxa (as currently delimited) cannot be considered unique, individualistic units, this does not mean that all species taxa are not. In some groups of organisms, biological species may conform in all respects to the philosophical concept of individual. We simply suggest that this condition is a special case, and that unwarranted extrapolations have been made from a very few groups of organisms to organisms generally.

SOME OPTIONS

As discussed previously, in many plant and some animal groups, evolutionary processes (i.e., replication and interaction in the sense of Hull 1980) occur primarily on a small scale (even when extrapolated over many generations) relative to the traditional species level. In such groups, the units in nature that are more like individuals are actually interbreeding local populations, and therefore, the basal taxonomic unit (the species) is currently more inclusive than the basal evolutionary units (the populations). This means that many currently recognized species taxa are, at best, historical entities. If this is the case, and if we want species taxa that are more fully individuals, can we bring taxonomic practices in line with our theoretical desires, and at what cost? If we cannot, or if the costs are too great, are there any theoretically acceptable alternatives, and what would they entail?

We formulate here three options with which we are faced and reject the first two. In the next section we explore some implications of the third alternative.

1. Alter the usage of "species" to equal "evolutionary unit," i.e., attempt to locate all of the effectively isolated and independently evolving populations and apply species names to them.

2. Alter the usage of "species" to equal the "cenospecies" or "comparium" (see Stebbins 1950, Grant 1971), i.e., recognize as the basic taxonomic units only those taxa that are *completely* intersterile.

3. Apply species names at about the same level as we have in the past, and decouple the basal taxonomic unit from notions of "basic" evolutionary units.

We reject choice 1 for several reasons, some practical and some theoretical. In a practical sense, formally naming whatever the truly genetically integrated units turned out to be would be disastrous. There are certainly very many such units, they are at best very difficult to perceive even with the most sophisticated techniques and in the most studied organisms, and these units are continuously changing in size and membership from one generation to the next. At any one time we can never know which units will diverge forever.

Rosen (1978, 1979) has discussed and adopted a species concept quite similar to choice 1. While we would generally agree with him that populations with apomorphous character states are units of evolutionary significance (1978, 176), we could not agree that species should be "the smallest natural aggregation of individuals with a specifiable geographic integrity that can be defined by any current set of analytical techniques" (1979, 277). Since we could probably distinguish each individual organism, or very small groups of organisms, on the basis of apomorphies (if we looked hard enough), why shouldn't each of these units be given a Linnaean binomial?

There is a more important, theoretical reason for rejecting alternative 1, one that we have alluded to above. A pervasive confusion runs through much discussion of species: the erroneous notion that *a* single basal evolutionary unit is somewhere to be found among all the possible units that could be recognized. There are *many* evolutionary, genealogical units within a given lineage (Hull 1980)—a rough hierarchy or network of units, which may be temporally and spatially overlapping. Thus, in the search to find *the* evolutionary unit, one is on a very "slippery slope" indeed. Units all along this slope may be of interest to evolutionists, depending on the level of focus of the particular investigator. These units do require some sort of designation in order to be studied, but a formal, hierarchical Linnaean name is not necessary.

Option 2, in many instances, would represent the opposite extreme (an attempt to locate the "top" of the slippery slope). Absolute reproductive isolation would be used as the overriding ranking criterion. If two organisms could potentially exchange genes, either directly or through intermediates, they would be placed in the same species taxon. There are several reasons why we reject this alternative.

First, it is unclear that reproductive criteria necessarily provide species taxa that are useful for purposes of phylogeny reconstruction and historical biogeography. As Rosen (1978, 1979) and Bremer and Wanntorp (1979a) have pointed out, "biological species" may be paraphyletic assemblages of popula-

Brent D. Mishler and Michael J. Donoghue

tions united only by a plesiomorphy, i.e., all those organisms that have not acquired a means of reproductive isolation. If reproductive criteria are to be useful for cladistic analysis, it is necessary to determine which modes of isolation arose as evolutionary novelties in a group.

Our second objection to option 2 has to do with the problem of measuring "potentiality." There have been numerous comments on the inadequacy of potential interbreeding as a ranking criterion, and even strong proponents of the biological species concept have rejected potential interbreeding as a part of their species definitions. Under certain conditions, very disparate organisms can be made to cross. If we adopted this option, the family Orchidaceae, with approximately 20,000 species at present (covering a great range of variation), might be lumped into just a few species because horitculturalists have produced so many bi- and pluri-genene hybrids. the universal application of any one criterion will undoubtedly obscure important patterns of variation in other parameters.

SPECIES LIKE GENERA

If we adopted alternative 3, what would happen to the species category? Would species taxa necessarily be theoretically meaningless entities? Are all alternatives to biological/evolutionary/individual species concepts devoid of theoretical interest, as implied by Eldredge and Cracraft (1980, 94)?

We would agree that if species were simply phenetically similar groups of populations they might indeed be unsatisfactory for many purposes. The application of species concepts like those of Cronquist (1978) and of Nelson and Platnick (1981) may yield species taxa that are not useful from the standpoint of reconstructing phylogenies (see discussion by Beatty 1982).[4]

However, we think that one form of option 3 may provide theoretically meaningful units. In groups where the actually interbreeding units are small relative to the morphologically delimited units, species can be considered to be like genera or families or higher taxa at all levels. That is, they are assemblages of populations united by descent just as genera are assemblages of species united by descent, etc. If we required that species be monophyletic assemblages of populations (to the extent that this could be hypothesized), then they could play a role in evolutionary and phylogenetic theory just as monophyletic taxa at all levels can. Theoretical significance does not reside solely in the basal taxonomic units or in units that are "fully individuals."

If we recognize that species are like genera, and insist that they be monophyletic, then we are faced with the problems of assessing monophyly and of ranking, problems that plague systematists working at all levels. Several different concepts of monophyly have been employed by systematists, but none of them explicitly at the species level (see discussion by Holmes 1980). We favor Hennig's (1966) concept of monophyly (except explicitly applied at the species level) but are fully aware of the difficulties in its application at low taxonomic levels (Arnold 1981, Hill and Crane 1982). In particular, the difficulty posed by reticulation (hybridization) (Bremer and Wanntorp 1979b) may be especially acute at lower taxonomic levels. Using synapomorphy as evidence of monophyly requires that the polarity of character states be determined, and again this may be an especially difficult problem near the species level. Polarity assessments will be possible to a greater or lesser extent depending on the certainty with which out-groups are known (Stevens 1980b).

As noted previously, in order to use reproductive isolation as evidence of monophyly, it would be necessary to determine which means of reproductive isolation are apomorphies at a given level, and which are not. An example of the difficulty of applying a Hennigian concept of monophyly is the very real possibility of "paraphyletic speciation." If speciation by peripheral isolation happens frequently, then a population (geographically defined), which has developed some apomorphic feature (such as a morphological novelty or an isolating mechanism) with respect to its "parent" species, may often be cladistically more closely related to some part of the parent species than to the remainder (see discussion and example in Bremer and Wanntorp 1979a). In such a case, we would take the (perhaps controversial) position that if the population is to be recognized as a formal species taxon, and if the phylogenetic relationships of the populations in the parent species can be resolved, then the taxonomist should not formally name the parent "species" (which has now been found to be a paraphyletic group), but instead name monophyletic groups discerned within it. Conversely, however, if cladistic structure within the parent species cannot be resolved, then in our view it would be acceptable to provisionally name it as a species (even if the populations included within shared no apomorphy).

This example illustrates the fact that even when monophyletic groups are delimited, the problem of ranking remains since monophyletic groups can be found at many levels within a clade. Species ranking criteria could include group size, gap size, geological age, ecological or geographical criteria, degree of intersterility, tradition, and possibly others. The general problem of ranking is presently unresolved, and we suspect that an absolute and universally applicable criterion may never be found and that, instead, answers with have to be developed on a group by group basis.

SOME CONSEQUENCES OF PLURALISM

We have outlined a concept of species (i.e., "species like genera") that may be appropriate for groups of organisms in which certain conditions obtain. However, we think that a variety of species concepts are necessary to adequately capture the complexity of variation patterns in nature. To subsume this variation under the rubric of any one concept leads to confusion and tends to obscure important evolutionary questions. As Hull (1970, see epigraph) has argued, we must resist the urge to superimpose false simplicity. If "species situations" are diverse, then a variety of concepts may be necessary and desirable to reflect this complexity.

Many theories in biology appear to lack the universality of theories in other natural sciences. Often the problem is to decide which one of several

Brent D. Mishler and Michael J. Donoghue

theories (not necessarily mutually exclusive) applies to a particular situation (for a specific application of this theoretical pluralism to evolutionary biology, see Gould and Lewontin 1979). A satisfactory general theory is one in which the number of subtheories is kept to a minimum, but not reduced to the point where important patterns and processes are obscured. The evaluation of how well a theoretical system "accounts for" patterns in the world is problematical, and we cannot offer any generally applicable criterion for making such an evaluation. However, in the case of species, we think that the search for a universal species concept, wherein the basal unit in evolutionary biology and in taxonomy is the same, is misguided. In our opinion, it is time for "species" to suffer a fate similar to that of the classical concept of "gene."[5]

We should recognize that species taxa have never been, and very probably cannot be made, readily comparable units. This observation has a number of important theoretical implications. Ecologists must consider the extent to which "species" can be considered equivalent and comparable from one group of organisms to another. Population sizes and structures, gene flow, social organization, the nature of selective factors, and developmental constraints differ in multifarious ways. This means that it is imperative that systematists be explicit about the nature of variation in, and the properties of, the species that they recognize in the groups they study. In turn, the users of species names must at all times be aware that "species are only equivalent by designation, and not by virtue of the nature or extent of their evolutionary differentiation" (Davis and Heywood 1963, 92). As obfuscatory as this may seem, comparative biologists must not make inferences from a species name without consulting the systematic literature to see what patterns of variation the name purports to represent.

These considerations are also important to paleontologists, who make inferences about, and from, "fossil species," and imply correspondences between variation in morphology, ecology, and breeding. It is perplexing that some quite innovative paleontologists, such as Eldredge and Gould, have uncritically retained the biological species concept in their work. As we have shown, there are many reasons why species should not be treated as particles or quanta. Paleontologists should consider exactly what macroevolutionary theories require species to be. For many purposes they may not require species that are completely individuals, but simply monophyletic lineages. If units that are cohesive via gene flow are an absolute requirement, then fossils may not provide appropriate evidence.

Finally, what are the implications for the systematist of a pluralistic outlook on species? Systematists working on relatively little known organisms should not assume that concepts derived from other groups of organisms are necessarily applicable. Instead, in each group the systematists is obligated to study patterns of variation in morphology, ecology, and breeding, and to detail the nature of the correspondences among these patterns. It is essential that the ways in which names are applied to taxa at all levels be stated explicitly.

If we adopt a case-by-case approach and urge specialists to unabashedly develop concepts for their particular groups, are we saying that "anything goes"? Of course, the answer is no. We are only suggesting pluralism within limits. Taxa (including species) recognized by systematists must have a specifiable relationship to theoretically important variation; more specifically, we have argued that species taxa should be phylogenetically meaningful units. There may not be a *universal* criterion to arbitrate between conflicting species classifications of a given genus, but through the complex process that is science, the community of involved workers can and will hammer out criteria for making such decisions.

NOTES

We are indebted to J. Beatty, E. Coombs, S. Fink, W. Fink, C. Hill, D. Hull, E. Mayr, N. Miller, P. Stevens, and five anonymous reviewers for criticizing this manuscript at one stage or another during its ontogeny; however, they are not to blame for its contents.

1. Gould (1979) and others have defended the biological species concept on the grounds that the same taxa recognized by Western taxonomists are recognized by tribespeople in New Guinea, etc. There are several problems with this kind of argument. First, it is not clear that this finding constitutes an independent test because, after all, New Guinea tribespeople are human too, with similar cognitive principles and limitations of language. It should also be born in mind that the observer is by no means neutral. Folk taxonomies have been collected by people with a knowledge of evolution and modern systematic concepts. Second, it is generally not a strong argument to show that a prescientific society has recognized something that modern science currently accepts. Surely a modern astronomer would not consider it very strong evidence that a primitive mythology supported one cosmological theory over another. Finally, the taxa recognized by Western taxonomists (and often by natives at some level of their linguistic hierarchy) in these instances are not known to be biological species—for the most part they are morphological units that are *believed* to be reproductively isolated from other such units.

2. Initially, the biological species concept was embraced and promulgated by plant systematists interested in evolution (Stebbins 1950, Grant 1957). Cronquist (1978) detailed Grant's efforts (from 1956 to 1966) to apply the biological species concept in *Gilia* (Polemoniaceae). It very soon became apparent that the biological species concept was fraught with difficulties, but Grant chose to amend the concept (rather than abandon it altogether), first (1957) with the notion of the syngameon (i.e., the unit of interbreeding higher than the species), later (1971) by adopting an evolutionary species concept. Finally, in the second edition of his classic book on plant speciation, Grant (1981) treats species in a more flexible and pluralistic manner. Some botanists (e.g., Stebbins 1979, 25) continue to feel that the biological species concept, or some modification of it, is the only suitable framework for understanding plant diversity. However, many (perhaps most) botanical systematists remain rather skeptical about the general applicability of the concept in botany (Davis and Heywood 1963, Raven 1976, Cronquist 1978, Levin 1979, Stevens 1980a).

The different attitudes of zoologists and botanists toward the concept of species may be of interest to historians, sociologists, and philosophers of science. For organismic and evolutionary biology the "modern synthesis" of the 1930s and 1940s may have represented a revolution in the sense of Kuhn (1970). For systematists, the principal outcome was the biological species concept. Zoologists (especially vertebrate systematists) appear to have largely accepted the new paradigm and to have entered a period of "normal science," applying the concept in particular cases ("puzzle solving"). While problems like sibling species, semispecies, and subspecies have become apparent, these have generally not prompted a critical evaluation of the paradigm or a proliferation of alternatives. In contrast, in the botanical community the biological species concept was soon found to be inapplicable or of difficult application and likely to lead to confusion. This resulted in a groping for alternatives and a defense of older concepts.

Brent D. Mishler and Michael J. Donoghue

In this regard, the historical development of species concepts in botany seems to fit better Feyerabend's (1970) characterization of scientific change as the simultaneous practice of normal science and the proliferation of alternative theories.

3. The zoologists initially responsible for developing the biological species concept were aware of the difficulties in applying the concept in some groups of animals and many groups of plants. Dobzhansky (1937, 1972) consistently pointed out the diversity of "species situations" observable in nature. Mayr (1942, 122) was careful to point out differences between plants and animals, and difficulties in the practical application of the biological species concept in some cases. Particularly rigid versions of the biological species concept have been promulgated more recently, in attempted generalizations that have shown a startling lack of concern for the biology of the majority of organisms on earth. Mayr (1982) has examined the resistance of botanists to the biological species concept and concluded that "the concept does not describe an exceptional situation" (p. 280). But he grants some justification to the ideas of "certain botanists" who question "whether the wide spectrum of breeding systems that can be found in plants can all be subsumed under the single concept (and term) 'species'" (p. 278).

4. The species concepts of Cronquist and of Nelson and Platnick are as follows:

Cronquist (1978, 15) "the smallest groups that are consistently and persistently distinct, and distinguishable by ordinary means."

Nelson and Platnick (1981, 12): "the smallest detected samples of self-perpetuating organisms that have unique sets of characters."

5. Initially the "gene" was considered to be *the* unit of heredity, but the classical concept of gene has been replaced by several concepts which stand in a complex relation to one another (Hull 1965). The use of a disjunctive definition (Hull 1965) allows a single term to designate a complex of concepts. However, this can become so confusing that it may be desirable to replace (at least in part) an old terminology with a new set of terms with more precise meanings.

REFERENCES

Antonovics, J. 1972. Population dynamics of the grass *Anthoxanthum oderatum* on a zinc mine. *J. Ecol.* 60:351–365.

Antonovics, J., A. D. Bradshaw, and R. G. Turner. 1971. Heavy metal tolerance in plants. *Adv. Ecol. Res.* 7:1–85.

Arnold, E. N. 1981. Estimating phylogenies at low taxonomic levels. *Z. Zool. Syst. Evolutforsch.* 19:1–35.

Beatty, J. 1982. Classes and cladists. *Syst. Zool.* 31:25–34.

Bodmer, W. F. 1970. The evolutionary significance of recombination in prokaryotes. *Soc. Gen. Microb. Symp.* 20:279–294.

Bradshaw, A. D. 1972. Some of the evolutionary consequences of being a plant. *Evol. Biol.* 5:25–47.

Bremer, K., and H.-E. Wanntorp. 1979a. Geographic populations or biological species in phylogeny reconstruction? *Syst. Zool.* 28:220–224.

———. 1979b. Hierarchy and reticulation in systematics. *Syst. Zool.* 28:624–627.

Burger, W. C. 1975. The species concept in *Quercus. Taxon.* 24:45–50.

Clausen, J., D. D. Keck, and W. M. Hiesey. 1939. The concept of species based on experiment. *Amer. J. Bot.* 26:103–106.

———. 1940. Experimental studies on the nature of species. I. The effect of varied environments on Western North American plants. Camegie Inst. Wash., Publ. No. 520.

Clémençon, H. (ed.) 1977. *The species concept in Hymenomycetes*. J. Cramer, Vaduz, Liechtenstein.

Cowan, S. T. 1962. The microbial species—a macromyth? *Soc. Gen. Microb. Symp.* 12:433–455.

Cronquist, A. 1978. Once again, what is a species? Pp. 3–20, in *Biosystematics in agriculture* (J. A. Romberger, ed.). Allanheld & Osmun, Montclair, N.J.

Davis, P. H., and V. H. Heywood. 1963. *Principles of angiosperm taxonomy*. Oliver and Boyd, Edinburgh.

Dobzhansky, T. 1937. *Genetics and the origin of species*. Columbia University Press, New York.

———. 1972. Species of *Drosophila*, *Science* 177:664–669.

Ehrlich, P. R., and P. H. Raven. 1969. Differentiation of populations. *Science* 165:1228–1232.

Eldredge, N., and J. Cracraft. 1980. *Phylogenetic patterns and the evolutionary process*. Columbia University Press, New York.

Eldredge, N., and S. J., Gould. 1972. Punctuated equilibria: an alternative to phyletic gradualism. Pp. 82–115, in *Models in paleobiology* (T. J. M. Schopf, ed.). Freeman, Cooper and Co., San Francisco.

Endler, J. A. 1973. Gene flow and population differentiation. *Science* 179:243–250.

Feyerabend, P. 1970. Consolations for the specialist. Pp. 197–230, in *Criticism and the growth of knowledge* (I. Lakatos and A. Musgrove, eds.). Cambridge University Press, London.

Ghiselin, M. T. 1974. A radical solution to the species problem. *Syst. Zool.* 23:536–544.

———. 1981. The metaphysics of phylogeny. Review of Eldredge, N., and J. Cracraft. 1980. *Phylogenetic patterns and the evolutionary process*. *Paleobiology*, 7:139–143.

Gould, S. J. 1979. A quahog is a quahog. *Nat. Hist.*, 88:18–26.

———. 1982. Darwinism and the expansion of evolutionary theory. *Science* 216:380–387.

Gould, S. J., and R. C. Lewontin. 1979. The spandrels of San Marco and the Panglossian paradigm: A critique of the adaptionist programme. *Proc. Roy. Soc. Lond.* (B) 205:581–598.

Grant, V. 1957. The plant species in theory and practice, Pp. 39–80, in *The species problem* (E. Mayr, ed.). Amer. Assoc. Adv. Sci., Publ. 50, Washington, D. C.

———. 1964. The biological composition of a taxonomic species in *Gilia*. *Adv. Genet.* 12:281–328.

———. 1971. *Plant speciation*. First edition. Columbia University Press, New York.

———. 1980. Gene flow and the homogeneity of species populations. *Biol. Zbl.* 99:157–169.

———. 1981. *Plant speciation*. Second edition Columbia University Press, New York.

Hennig, W. 1966. *Phylogenetic systematics*. University of Illionis Press, Urbana, Ill.

Hill, C. R., and P. R. Crane. 1982. Evolutionary cladistics and the origin of angiosperms. Pp. 269–361, in *Problems of phylogenetic reconstruction* (K. A. Joyse and A. E. Friday, eds.). Systematics Association Special Volume No. 21. Academic Press, London and New York.

Holmes, E. B. 1980. Reconsideration of some systematic concepts and terms. *Evol. Theory*. 5:35–87.

Hull, D. L. 1965. The effect of essentialism on taxonomy—two thousand years of stasis (II). *British J. Phil. Sci.* 16:1–18.

———. 1968. The operational imperative: Sense and nonsense in operationism. *Syst. Zool.* 17:438–457.

Brent D. Mishler and Michael J. Donoghue

————. 1970. Contemporary systematic philosophies. *Ann. Rev. Ecol. Syst.* 1:19–54.

————. 1976. Are species really individuals? *Syst. Zool.* 25:174–191.

————. 1978. A matter of individuality. *Phil. Sci.* 45:335–360.

————. 1980. Individuality and selection. *Ann. Rev. Ecol. Syst.* 11:311–332.

Jackson, J. F., and J. A. Pound 1979. Comments on assessing the dedifferentiating effect of gene flow. *Syst. Zool.* 28:78–85.

Jain, S. K., and A. D. Bradshaw. 1966. Evolutionary divergence among adjacent plant populations. I. The evidence and its theoretical analysis. *Heredity* 21:407–441.

Kiang, Y. T. 1982. Local differentiation of *Anthoxanthum odoratum* L. populations and roadsides. *Amer. Midl. Nat.* 107:340–350.

Kuhn, T. S.. 1970. *The structure of scientific revolutions.* Second enlarged edition. University of Chicago Press, Chicago.

Lande, R. 1980. Genetic variation and phenotypic evolution during allopatric speciation. *Amer. Nat.* 116:463–479.

Levin, D. A. 1978. The origin of isolating mechanisms in flowering plants. *Evol. Biol.* 11:185–317.

————. 1979. The nature of plant species. *Science* 204:381–384.

————. 1981. Dispersal versus gene flow in plants. *Ann. Missouri Bot. Gard.* 68:233–253.

Levin, D. A., and H. W. Kerster. 1974. Gene flow in seed plants. *Evol. Biol.* 7:139–220.

Lewontin, R. C. 1974. *The genetic basis of evolutionary change.* Columbia University Press, New York.

Mayr, E. 1942. *Systematics and the origin of species: From the viewpoint of a zoologist.* Columbia University Press, New York.

————. 1957. Species concepts and definitions. Pp. 1–22, in *The species problem* (E. Mayr. ed.). Amer. Assoc. Adv. Sci., Publ. 50, Washington, D. C.

————. 1969. *Principles of systematic zoology.* McGraw-Hill Book Co., New York.

————. 1970. *Populations, species, and evolution.* Harvard University Press, Cambridge, Mass.

————. 1982. *The growth of biological thought.* Harvard University Press, Cambridge, Mass.

Mooney, H. A., and W. D. Billings. 1961. Comparative physiological ecology of arctic and alpine populations of *Oxyria digyna*. *Ecol. Monogr.* 31:1–29.

Nelson, G., and N. Platnick. 1981. *Systematics and biogeography: Cladistics and vicariance.* Columbia University Press, New York.

Ornduff, R. 1969. Reproductive biology in relation to systematics. *Taxon* 18:121–133.

Raven, P. H. 1976. Systematics and plant population biology. *Syst. Bot.* 1:284–316.

Ricklefs, R. E. 1979. *Ecology.* Second edition. Chiron Press, New York.

Rosen, D. E. 1978. Vicariant patterns and historical expalanations in biogeography. *Syst. Zool.* 27:159–188.

————. 1979. Fishes from the uplands and intermontane basins of Guatemala: Revisionary studies and comparative geography. *Bull. Amer. Mus. Nat. Hist.*, 162:267–376.

Simpson, G. G. 1961. *Principles of animal taxonomy.* Columbia University Press, New York.

Small, E. 1971. The evolution of reproductive isolation in *Clarkia*, section *Myxocarpa*. *Evolution* 25:330–346.

Sokal, R. R. 1973. The species problem reconsidered. *Syst. Zool.* 22:360–374.

Sokal, R. R., and T. J. Crovello. 1970. The biological species concept: A critical evaluation. *Amer. Nat.* 104:127–153.

Stace, C. A. 1978. Breeding systems, variation patterns and species delimitation. Pp. 57–78, in *Essays in plant taxonomy* (H. E. Street, ed.). Academic Press, New York.

Stanley, S. M. 1975. A theory of evolution above the species level. *Proc. Nat. Acad. Sci. U.S.A.* 72:646–650.

Stebbins, G. L. 1950. *Variation and evolution in plants.* Columbia University Press, New York.

———. 1979. Fifty years of plant evolution. Pp. 18–41, in *Topics in plant population biology* (O. T. Solbrig, S. Jain, G. B. Johnson, and P. H. Raven, eds.). Columbia Univerity Press, New York.

Stevens, P. F. 1980a. A revision of the Old World species of *Calophyllum* L. (Guttiferae). *J. Arnold Arb.* 61:117–699.

———. 1980b. Evolutionary polarity of character states. *Ann. Rev. Ecol. Syst.* 11:333–358.

Turesson, G. 1922a. The species and the variety as ecological units. *Hereditas* 3:100–113.

———. 1922b. The genotypical response of the plant species to the habitat. *Hereditas* 3:211–350.

Van Valen, L. 1976. Ecological species, multispecies, and oaks. *Taxon* 25:233–239.

Wake, D. B. 1980. A view of evolution. Review of Eldredge, N., and J. Cracraft. 1980. *Phylogenetic patterns and the evolutionary process. Science* 210:1239–1240.

White, M. J. D. 1978. *Modes of speciation.* W. H. Freeman and Co., San Francisco.

Wiley, E. O. 1978. The evolutionary species concept reconsidered. *Syst. Zool.* 27:17–26.

———. 1980. Is the evolutionary species fiction?—A consideration of classes, individuals, and historical entities. *Syst. Zool.* 29:76–80.

———. 1981. *Phylogenetics: The theory and practice of phylogenetic systematics.* John Wiley, New York.

Wiley, E. O., and D. R. Brooks. 1982. Victims of history—a nonequilibrium approach to evolution. *Syst. Zool.* 31:1–24.

VII Systematic Philosophies

12 The Continuing Search for Order

Robert R. Sokal

ORDER AND DIVERSITY

The subject matter of this essay, taxonomy, was probably the most frequent topic of discussion during the early years of the American Society of Naturalists. Various aspects of taxonomy and discussions of particular taxonomies were regular features of the meetings of the society since its founding in the nineteenth century and well into the present century. But in these days when molecular genetics provides us with new and exciting discoveries on a regular basis—findings that are of profound importance for an understanding of the evolution of organisms—one may well wonder why scientists should still bother with taxonomy. Is this not an outdated science, practiced by a few unreconstructed museum types, whose ideas are as dusty as some of the cases and specimens with which they surround themselves?

Any such attitude reveals a lack of understanding and appreciation of the fundamental role that taxonomy continues to play in modern biology. There are few papers, if any, published in the *American Naturalist* or presented at any meeting of the society that are not founded on comparative biology. That this is more than a propagandistic claim can be shown by a detailed analysis of any series of articles in the journal. Even if the thrust of a given paper is along lines quite remote from taxonomy, this science is implicated whenever a comparative approach is taken in which a biological phenomenon is compared over different groups of organisms. Alternatively, a taxonomic hypothesis is implied when the generality of a phenomenon is assumed. When we study transposable elements in *Drosophila*, it is believed that these are not exclusively restricted to *Drosophila melanogaster*, but that other *Drosophila*, and presumably other flies (indeed, other organisms as well), will exhibit the phenomenon. It is this generality that justifies the effort invested into an elucidation of such complex biological processes whose investigation is so time-consuming and expensive as to make a comprehensive analysis in numerous organisms prohibitive. Of course, whenever evolutionary hypotheses are tested, the importance of taxonomy is demonstrated directly. For without a taxonomic

From *American Naturalist*, 1985, 126:729–749.

framework in which these hypotheses are considered, evolutionary hypotheses such as punctuated-equilibrium evolution, constant-rate evolution, species selection, or vicariance biogeography could not be tested. Thus, classifications are necessary if we are to engage in biology. But how are these classifications to be constructed?

The classificatory system set up by Linnaeus is hierarchical. At the lower-rank levels—species and genus—the Linnaean system was a direct continuation of folk taxonomy. Primitive peoples have been shown to recognize many of the species in their immediate surroundings, and they resemble the European tradition preceding Linnaeus in being aware of the similarity of some species and assembling these conceptually and linguistically into groups of similar species—the genera (see, e.g., Breedlove and Raven 1974). Linnaeus extended the method by establishing higher-ranking taxa and defining them on the basis of group characteristics that distinguished one taxon from another at the same rank. These differentiating characteristics were to all practical purposes Aristotelian, defining essences of the established groups (Cain 1958). It has been pointed out repeatedly that the nature of the classificatory system has changed little since Linnaeus's time, although, of course, the classifications themselves have undergone drastic changes and expansions. It is also remarkable to what extent the classificatory system remained unaffected by the advent of evolutionary theory in the latter half of the nineteenth century. The classificatory system was soundly established and successful well before Darwin's time. As numerous authors have shown, the early taxonomists attempted to establish a natural system, and whereas evolutionary theory furnished a new explanation for the natural system, the thought processes concerning its establishment and the actual taxonomic practices changed little (e.g., Remane 1956).

But what is a natural system? The common answer is that it is a system that reflects the state of nature. Early biologists and indeed the common people, as reflected in their various languages, seem to have recognized a roughly hierarchical arrangement of natural diversity. So, to the degree that nature is truly hierarchical, a system should be the more natural, the more it conforms to the true innate hierarchy of nature (if only we could know it). An alternative view is that a natural system is one that is natural to the human mind. Do we find it easier to pigeonhole objects into mutually exclusive classes that are subsumed under ever more inclusive nonoverlapping classes? Cognitive psychologists tend to think that we do, and that the human mind cannot easily perceive membership in overlapping classes and in continua (Smith and Medin 1981). It has even been suggested that because we are evolving organisms and because the human mind has to deal with an evolved (and hence hierarchical) universe, the structure of our thinking developed so as to favor hierarchical interpretations (Riedl 1976, p. 230). In actual work with arranging organisms in a natural system, it was eventually discovered that group characteristics were difficult to define categorically and that exceptional organisms could always be found. Such organisms would in most respects clearly be members of a given taxon but would lack one or another defining characteristic. Groups with such members were termed polythetic (Beckner 1959; Sneath 1962), and most taxonomists have recognized that natural taxa are largely polythetic.

We may visualize a polythetic taxon as arrayed in a space defined by the characters describing members. In figure 12.1 the horizontal and vertical dimensions define the space spanned by two characters. The parallel lines indicate different states of the two characters, and the two clusters of points are two taxa. Most of their members share a specific combination of character states for the two characters and are shown within one of the squares defined by the lattice. But some members possess an alternative state for one or the other character, and yet belong to the taxon (the cloud of points). Such members are shown as points in neighboring squares. When three characters are considered simultaneously, the character space can be represented as a cubic lattice (figure 12.2); polythetic taxa largely reside inside cubes representing the combinations of states of three characters within this lattice. Here again, some of the cluster members do not share all three "defining" character states.

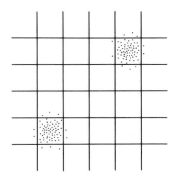

Figure 12.1   Two polythetic taxa defined by two characters. (For explanation, see text.)

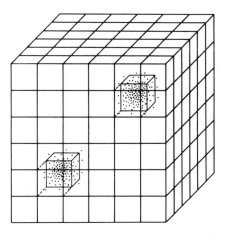

Figure 12.2   Two polythetic taxa defined by three characters. (For explanation, see text.)

In recent years, cognitive psychologists (Smith and Medin 1981) have recognized that the representation of a concept is a summary description of an entire class and cannot be restricted to a set of necessary and sufficient conditions; rather, it provides some measure of a central tendency of the patterns of its members. This probabilistic view of concepts corresponds to polythetic classes in biological taxonomy. If humans generally process information and form concepts on the basis of polythetic classes, it is not surprising that such classes have been developed for biological classification. Thus, it is also possible that taxonomists have been reinforced in the formation of hierarchical and polythetic classifications because they have an inborn tendency toward such arrangements.

Nevertheless, it is not necessarily obvious that a hierarchical system is the most faithful representation of organic diversity. Continua may exist in character space, which would make it difficult and rather arbitrary to decide how to arrange the taxa hierarchically (figure 12.3). This is most apparent at the population level, at which any hierarchical division can be shown to be inadequate. This can easily be demonstrated in human populations. Where should one draw the boundary between Europid and Mongolid populations in Siberia? Or between the Europid Weddids of India and the Southern Mongolids of Malaysia and Indonesia? Or between the Chinese and the Southern Mongolids of the Indochinese peninsula? Clearly all such boundaries are more or less arbitrary and may falsely suggest differences between equally ranked populations, even where such differences do not exist.

Do such continua also exist at higher taxonomic ranks? In numerous taxonomic groups, not only genera, but families as well, grade into one another. Whereas the biological processes resulting in a phenetic continuum of populations within species are easily understood, one might expect the divergent dendritic nature of the phylogenetic process to result in discontinuities that would easily submit to a hierarchical arrangement. Yet parallelism, conver-

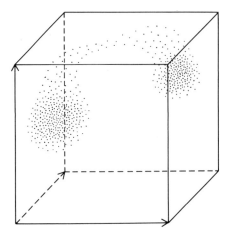

Figure 12.3   Two clusters (taxa) that show a narrow continuum in a three-dimensional character space.

Robert R. Sokal

gence, and hybridization readily produce results that will appear as continua in phenetic space.

There have been proposals from time to time to deal with this problem by abandoning the traditional Linnaean classificatory system in favor or one that is better adapted to reflect the actual taxonomic diversity (see, e.g., DuPraw 1964). All such proposals for non-Linnaean taxonomies, however, have foundered on the shoals of tradition, and it may indeed be true that humans, including taxonomists, have an innate bias toward hierarchical systems.

SCHOOLS OF TAXONOMY

In this century, biologists have attempted to provide a theoretical basis for taxonomy that goes beyond the considerations of the bare outline given above. There are three currently active schools of taxonomy. These are known as phenetic taxonomy, cladistics, and evolutionary systematics. Phenetic taxonomy (Sneath and Sokal 1973) is a system of classification based on the overall similarity of the organisms being classified. The similarity is expressed in terms of phenotypic characters (Cain and Harrison 1960). The goal of phenetic taxonomy is to arrange objects or operational taxonomic units (OTUs) in a stable and convenient classification. It is believed that basing classifications on similarity will result in such stability and convenience. The measurement of similarity is made on the basis of numerous, equally weighted characteristics. The degree of belonging to a class is based on its constituent properties. Following Gilmour's dictum (1937, 1940, 1951, 1961) that a system of classification is the more natural the more propositions there are that can be made regarding its constituent classes, affinity in a polythetic taxon is based on the greatest number of shared character states. No single state is either essential to group membership or sufficient to make an organism a member of the group.

In cladistic classification (Eldredge and Cracraft 1980; Wiley 1981), one establishes classes based on estimated cladograms, or branching trees, of phylogenetic relationships. Although there is currently considerable debate among cladists concerning the goals of cladistic classification (some recent workers—pattern cladists—appear to seek internal consistency of patterns divorced from the true genealogy of the organisms under study; for a discussion of this issue, see Beatty 1982; Brooks and Wiley 1985; Platnick 1985), I concur with Cracraft (1983) in considering the estimation of the phylogenetic relationship or genealogical affinities to be the ultimate purpose of establishing cladistic classifications. The cladograms are estimated by postulating monophyletic sister groups which, in turn, are based on putative synapomorphies, shared derived character states believed to have arisen in a common ancestor. Monophyly is defined by a strict criterion: all members of a taxon have a common ancestor, all of whose descendants are members of the taxon.

The evolutionary systematists (Mayr 1969, 1982, p. 233; Bock 1977; Ashlock 1980) also seek classes that share a common ancestor, but they permit some groups to be established in a classification that are not entirely

monophyletic as defined by the cladists. As has been pointed out by pheneticists, as well as by cladists (Sneath and Sokal 1973; Cracraft 1983), evolutionary systematists have not attempted to establish a series of objective and
quantitative procedures that result in classifications desirable from their point
of view. In fact, Estabrook (1978) has provided a useful definition through his
concepts of character state trees and convex taxa which, however, have not
been adopted by other advocates of evolutionary systematics.

The differing classifications of these schools of taxonomy are illustrated in
figure 12.4. Phenetic and cladistic relations for five OTUs are approximated
by the dendrogram. The horizontal axis represents phenetic dissimilarity; the
presumed phylogenetic branching sequence is shown by the bifurcations; and
the vertical axis indicates time. Because OTUs D and E have diverged greatly
from the other three and are mutually close, they would be placed in their
own taxon by pheneticists. OTU C has not diverged much from the ancestral
stem, and some convergence in B has made B and C quite similar. The phenetic classification, therefore, erects a taxon BC, which at a higher-rank level
adds A to form the taxon ABC, which ultimately joins DE to include all
OTUs. This classification is shown by "roofs" over the OTUs. Cladists establish the taxa AB and DE because these pairs of OTUs share a most recent
common ancestor that is not also shared by other OTUs. In the cladistic
classification (again shown by roofs), OTU C joins taxon DE because these
three share a common ancestor before taxa AB and CDE are united. It is not
possible to specify how a taxonomist employing evolutionary systematics
would classify these taxa, since no explicit classificatory rules are given by

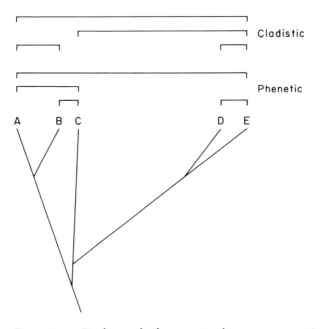

Figure 12.4 Dendogram for five operational taxonomic units (OTUs) illustrating how the
three schools of taxonomy would arrive at different classifications on the basis of the information furnished. (For further explanation, see text.)

that school. Depending on whether the divergence of DE and convergence of B and C are considered more striking or whether more importance is given to the common ancestry of C and DE, the phenetic or cladistic classification may be adopted. For this reason, no roofs are shown for evolutionary systematics.

What are the assumptions that are used by the various schools of taxonomy? The Linnaean system assumes a nonoverlapping hierarchical arrangement of nature. All three schools of classification meet the assumption of the Linnaean system in attempting to create such an arrangement. Phenetic taxonomy assumes that similarity can be measured and that nature is not continuous, so that mutually most similar taxa can be defined and placed into the nonoverlapping hierarchy. Cladistic taxonomy postulates a bifurcating, divergent topology and further assumes that these bifurcations are characterized by synapomorphies and that these synapomorphies can be recognized. Evolutionary systematics has no clearly stated special assumptions. If the character-state tree model of Estabrook (1978) is adopted, then one could state that classifications acceptable by evolutionary systematics should be convex, but, as already stated, this is a technical, not generally accepted definition.

Some of the assumptions are shared by the schools of taxonomy. The assumption that a hierarchical arrangement of taxa is appropriate is common to all three. The assumptions that an overall similarity can be quantified and that there are discontinuities among taxa are essential to phenetic numerical taxonomy; they must be applied in evolutionary systematics as well, because the departures of evolutionary classifications from strictly cladistic classifications are based on the amount of phenetic differentiation of various subgroups of a taxonomic tree. Cladistic numerical taxonomic methods do not specifically require this assumption, but in fact, a cladogram with the character states superimposed on it can result in a quantification of similarity of the terminal OTUs, which may exhibit discontinuities if given bifurcation is supported by numerous synapomorphies. The notion of the length of a tree, commonly employed when estimating cladograms by parsimony methods, also implies a measure of dissimilarity between OTUs and their ancestors. The bifurcation and synapomorphy assumptions in classification are specific to cladistics, including numerical cladistics, and are not an essential assumption of any other method. The convexity of character-state trees as defined by Estabrook is a requirement of evolutionary classifications only. It is not rigorous enough for numerical cladistics, and it is not applicable to phenetic classifications.

All these assumptions have been subjected to serious questioning. We have already seen that the hierarchical nonoverlapping arrangement of organized nature is not necessarily the best representation of the actual diversity. The measurement of similarity by pheneticists has been criticized as being ambiguous, subject to variations in character coding, scaling, and similarity coefficients. Even if these effects can be ignored, differences in the clustering algorithms produce differences in classifications (Sneath and Sokal 1973, p. 427). The operational assumptions of cladists that cladograms should be entirely bifurcating trees can surely not be true in nature; nor will the presence of shared derived character states in the two sister species be inevitable in a given data set. There has additionally been considerable controversy concerning the methods for recognizing the direction or polarity of character-state change, a necessary step in cladistic analysis (Eldredge and Cracraft 1980, p. 54; Brooks and Wiley 1985; Kluge 1985).

The assumptions made by the various taxonomic schools are not appropriate at all systematic levels. Thus, the hierarchical assumption is clearly not applicable for populations within a species (and even less for individuals within a population). Except for a few specialized instances in which populations split as a result of a dendritic branching process, the notion of a hierarchical system of populations is not supportable. One of these rare exceptions is that of the villages of the Yanomama Indians in South America, which exhibit a true branching system. Because each newly branched population occupies a new habitat, the process continues with little gene flow and without the replacement of one terminal population by another (Neel 1978). There should be analogies of this phenomenon in the plant and animal world, perhaps when there are introductions of new populations into unoccupied habitats, but this situation is not typical of populations in an existing, reasonably stable environment.

The assumption that similarity can be quantified holds for all taxonomic levels, but that concerning discontinuities does not. Discontinuities possibly do not occur among populations within species, and perhaps not even among species within a genus, since gene flow will tend to diminish the distinctness of their phenetic boundaries. The cladistic assumptions of bifurcations and of synapomorphies marking these branches are not tenable for populations within a species, but they may apply to all higher levels. Similar relations hold for the evolutionary systematic assumption of the convexity of taxa. Phenetic taxonomy can handle all of the levels, although because of a lack of discontinuities at the intraspecific level, the techniques for populations within a species have to be tailored to the nature of the material and generally involve ordinations, rather than cluster analyses. The other schools have difficulties incorporating populations within a species into their systems.

The emergence of the modern schools of taxonomy was accompanied by the development of numerical methods for obtaining the classifications (Sokal and Sneath 1963; Camin and Sokal 1965; Fitch and Margoliash 1968; Farris et al. 1970; Sneath and Sokal 1973). Characters were precisely defined, expressed or coded numerically, and subjected to appropriate algorithms intended to achieve the goals of a given school. In practice, however, the methods developed until now only approximate these goals at best. These developments in zoology and botany, with parallel trends in anthropology, have been given the general name of numerical taxonomy with subdivisions of numerical phenetics, numerical cladistics, and numerical systematics, corresponding roughly to their applications to the respective taxonomic schools.

Regardless of one's classificatory philosophy, the trend toward precise definitions of characters and their states and toward an explicit presentation of complete data sets, which was brought about by the development of numerical taxonomy, is undoubtedly a considerable improvement over the subjective traditional methods still practiced by many taxonomists. The taxonomic process is fraught with considerable risk of subjective bias. Battles among individual taxonomists regarding the importance and suitability of certain characters are legendary. Paleoanthropology still reflects this trend today, each new find seemingly the cause of a new argument. Different taxonomists using subjective methods can obtain quite different classifications of the same taxa by stressing different characters (Sokal and Rohlf 1980). Yet, it can be shown that attempts by different persons to obtain explicit data matrices of the same group result in largely identical classifications, regardless of subjective divergences in character coding and terminology (Sokal and Rohlf 1970). Different populations, such as Europeans and Chinese, do not agree on the characteristics they employ to distinguish their groups (Sokal 1974): Whereas Europeans are struck by "slanted" eyes, skin color, and prominent cheekbones (in that order), Chinese emphasize wavy and light-colored hair and the prominent noses of Europeans. Similar cultural and personal biases surely affect the work of taxonomists working with other species.

In lectures to my classes I have for years cited the example of the intergradation between the Mongolid and Indian populations of Southeast Asia to point out the inadequacy of a hierarchical arrangement of these groups. Lately I have recognized that this view is also biased, from the perspective of the larger population groups at the ends of this spectrum: the Chinese and the Indians. A Khmer, Thai, or Burmese doubtlessly does not consider himself or herself as an intergrade, but rather as a clearly defined central type with the individuals in these more numerous but peripheral populations being extreme departures from one's own norm. In a similar way, greater familiarity with one of several related groups, or a greater number of species in some taxa than in others, may distort the taxonomic judgment. A related phenomenon is the greater perception of diversity in one's own group or in a familiar group with respect to a less familiar one. Europeans tend to stress the great diversity of Europid types even when these are restricted to those found in Europe alone. There is no reason to doubt that a comparable diversity exists in the population of China; yet this is rarely perceived by European scientists. Although even the intelligent layman is aware of the great diversity of the African population, differences are nevertheless deemphasized by comparison with one's own group. Similar biases exist in the taxonomy of all organisms and can be minimized only by requiring explicit and clearly defined data matrices, such as those used in numerical taxonomy.

Let us next examine the purposes of taxonomies. There is, first of all, the need to provide a "system of nature," a classification for laymen as well as other scientists. This is the classificatory system that exists today. It has been established generally on an ill-defined basis, typically without quantitative or other objective criteria. The importance of a consistent and stable classificatory system for reference purposes cannot be overemphasized. The public at large and scientists who are not systematists need a system that assigns an organism to a taxon, gives it name, associates it with other related taxa (the meaning of related here is, of course, the catchword being defined variously by different schools of taxonomy), and enables statements and inferences to be made about the characteristics of a given group. Biologists of all kinds need such a system of classifcation to test the generality of phenomena they observe; they need to know what other organisms exist that are closely related, so that a phenomenon observed in one member of the taxon might be looked for in other members.

A second purpose is to generate hypotheses about evolutionary relationships. Are there laws about the number of species to be found in a genus? About the density with which species fill niche spaces; about the conformation of such space; about the diversity in shape and width of niches; and so forth? To this end, more than a formal classification is required. A phenetic classification quantifying the amount of variation of OTUs within and among taxa is necessary for studying these phenomena.

A third purpose of a classification is to serve as a model or benchmark against which to test hypotheses about evolutionary phenomena, such as phyletic evolution, gradualism versus punctuated-equilibrium evolution, convergence, reversed evolution, and the like. For this purpose, one requires an estimated cladogram of the group and a mapping of the character-state changes that are believed to have occurred along its edges. If we wish to estimate the true phylogenetic tree of a taxon to test various evolutionary hypotheses, cladistic methods would seem indicated. But, as we shall see, methods taken from other schools of taxonomy may frequently yield better results.

The relations of the various taxonomic methods to these taxonomic purposes can be summarized as follows: all methods aim at representing a "natural" system. Phenetics and evolutionary systematics may serve to test hypotheses about patterns of diversity of analyses of niche hypervolume. Evolutionary systematics can furnish evolutionary phenomena as described above. Cladistic taxonomy can only provide such a benchmark. These purposes are met, of course, only to the degree to which the methods of each school succeed in meeting their goals.

OPTIMALITY CRITERIA

In recent controversies among contending schools and methods, three criteria of the quality of classifications have figured prominently: stability, predictive value, and fit to the true cladistic relationship (Rohlf and Sokal 1981). Taxonomic stability includes the robustness of classifications to the addition of new characters. Methods exhibit greater *character stability* when classifications based on different subsets of a suite of characters, or on different kinds of characters, are more congruent. Robustness of classifications to the addition (or deletion) of OTUs is known as *OTU stability*. A third type of stability is robustness of classfications to methods of coding characters and their states and to the numerical algorithms employed. *Predictive value* is a measure of how similar OTUs are with respect to their character states to the other members of their taxon (at various rank levels). *Fit to the true cladistic relationship* is an optimality criterion only for cladistic classifications. We shall take up this criterion first.

The accuracy of an estimated cladogram can be tested in only those few cases in which the true phylogeny is known. Although a few studies (e.g., Baum and Estabrook 1978; Baum 1983) have been reported as comparisons of true phylogenies with estimated ones, these comparisons are more accurately described as being of better-documented estimates with less well documented ones. In general, phylogenies of real organisms are unknown. For this reason, tests of the accuracy of cladogram estimation have to be carried out with artificial data sets.

One such data set, which has received considerable attention by systematists, is the Caminalcules (Sokal 1983a). This group of organisms has the great advantage of a known phylogeny. Its disadvantage, however, is that the group is artificial, owing its existence to the fantasy and inventiveness of the late Professor Joseph Camin. For this reason, it has been claimed that inferences made on the basis of the Caminalcules are not valid for the systematics of living organisms. Doubtlessly, any group of real organisms can be differentiated in innumerable ways from the Caminalcules. The comparisons to which I shall refer, however, have been made only with respect to a subset of propertics relevant to the analysis of classifications. In a study of such measurable properites of the Caminalcules. I found that for none of the properties examined (homoplasy, symmetry, adequacy of the character states for resolving the cladogram, evolutionary rates, species longevities, and the ratio of speciation to extinction) do the characters of the Caminalcules differ from those of living organisms (Sokal 1983a). In fact, none of the statistics calculated for the Caminalcules is beyond the range of those observed for 19 zoological data sets. In view of these findings, I maintain that, with respect to the properties important for classification, the Caminalcules behave similarly to real organisms, and the burden of proof falls on those critics who wish to ignore the Caminalcules as relevant for biological systematics. Recently, E. W. Holman (pers. comm.) has shown that the Caminalcules deviate from real organisms in not following a hollow curve distribution with respect to the number of Recent species per genus. He concluded that this is due to a variation in evolutionary rates among the lineages of these creatures.

Subjective taxonomists, ranging from distinguished professors to grade school students, have divided the Caminalcules into five major groups, so-called genera, and this classification has been supported by numerical phenetic analysis of the data (Sokal and Rohlf 1980). Figure 12.5 shows one representative species of each genus. The characters of the Caminalcules were described by persons unaware of the true phylogeny of these organisms. There are 29 Recent species of Caminalcules and 48 fossil species, making 77 species in all for this group. Eighty-five characters were described on the basis of the 29 Recent forms, but 106 characters are necessary to describe all of the species including the fossils (Sokal 1983a).

To discover how well various numerical methods estimate the true phylogeny, we can try numerical phenetic classifications, as well as numerical cladistic

Figure 12.5 Representative species from the five genera of the Caminalcules. The five genera A, B, C, DE, and F are represented, respectively, by the following five numbered species: 7, 11, 17, 5, 29.

estimates of the true cladogram. Phenetic classifications are achieved by means of the usual procedures. Characters are standardized, correlations and taxonomic distances are computed between all pairs of OTUs, and the resemblance matrices are clustered by the UPGMA algorithm (for these methods, see Sneath and Sokal 1973). Cladistic estimates are obtained by common numerical cladistic procedures, which either are based on a hypothesis of parsimony, resulting in the shortest obtainable trees, or are based on tree structures that permit the greatest compatibility among character states. I discuss only Wagner parsimony (Farris 1970; Felsenstein 1982), having found this to be the relatively best of the cladistic methods available.

Both the phenograms and the estimated cladograms are compared with the true cladogram. The latter is featured here only in reduced form in figure 12.7, but can be inspected in Sokal (1983a,b, or 1984). To compare the estimated cladograms with the true cladogram, one needs a measure of consensus between the two. This subject is currently an active area of research, and there is a multiplicity of proposed indexes (Rohlf 1982; Shao 1983; Stinebrickner 1984). Although I have applied several of these to the Caminalcules, I present here only the simplest and perhaps also the most conservative index, the strict consensus index CI_c (Colless 1980; Rohlf 1982). When all characters of the Caminalcules are used, one finds that cladistic methods (Wagner trees) estimate the true cladogeny better than phenetic methods (UPGMA trees). With fewer characters, however, phenetic methods give closer estimates to the true cladogram than cladistic methods (Sokal 1983b), and this corresponds to the findings of other authors (Colless 1970; Sokal and Rohlf 1981; Tateno et al.

1982; Fiala and Sokal 1985), who have shown that phenetic classifications in many cases are as good as cladistic classifications for estimating the true cladogeny and in some cases provide superior estimates. We shall return to the relation between character number an goodness of estimate presently.

A second important finding is that none of the numerical methods correctly estimates the entire phylogenetic tree (Sokal 1983b). If such work were done on real organisms, one could not expect even the relatively weak consensus that was found, since I assisted the estimate in various ways that could not be repeated in the absence of knowledge of the true cladogeny. An attempt to obtain an estimate of the true cladogeny by conventional, nonnumerical cladistics resulted in a relatively poor approximation of the true tree (Sokal 1983b).

Numerical cladistic estimates of the true cladogram are seriously affected by the input order of the OTUs to the computer program; many different combinations yield an array of different tree lengths for any one data set. Only rarely does one know that the shortest length obtained is indeed a tree of minimum length. In the Caminalcules one notes in general that shorter trees tend to give better estimates of the true cladogeny, but this is only a trend. Some of the shortest trees deviate appreciably from the best estimates, and conversely, some of the best estimates of trees are longer than other, poorer estimates. This should not surprise us greatly. The true length of the cladogram of the Caminalcules is 321 evolutionary units, but when the computer optimizes the distribution of character states on the true tree so as to obtain the shortest possible length, one obtains a tree with only 217 units, considerably shorter than the true length. A wagner-tree program, constructing hypothetical taxonomic units (HTUs) as ancestors in order to minimize the total length of the tree, will, however, produce trees as short as 211 units. Such trees cannot possibly be topologically correct.

It is possible to analyze in detail the differences in results between the true cladogram and various estimates of it obtained by numerical cladistic and numerical phenetic methods. One can examine these relationships by means of a standard three-taxon cladogram in which B and C join before their common stem joins A (figure 12.6). The results of such an endeavor can be summarized as follows. In the absence of homoplasy and additional divergence, phenetic relationships fully correspond to cladistic ones. Phenograms also correspond to the true cladogram in a three-taxon case whenever there is divergence in the outgroup species, divergence in the stems subtending the ingroup, or parallelism between the two ingroup species. Phenetic-cladistic agreement diminishes in response to divergence in either or both ingroup species, parallelism between outgroup and one or both ingroup species, or reversals in one or both ingroup species.

One finds empirically that estimated cladograms obtained by various numerical cladistic methods are also affected by homoplasy and unusual divergence, although not to an equally great extent. When one compares the actual phenetic classifications and estimated cladograms of the Caminalcules with the true cladogram, one notices good correspondence in all such cases in which

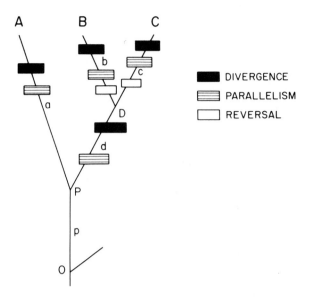

Figure 12.6 A three-taxon cladogram to illustrate the effects of divergence and homoplasy on differences obtained by numerical phenetic and numerical cladistic methods. Capital letters indicate Recent OTUs and ancestors; lowercase letters, the lengths of the corresponding internodes measured in units of evolutionary change. The bars across the internodes show the potential for evolutionary factors that could enhance or diminish the congruence between classifications obtained by phenetic and cladistic methods.

there is considerable divergence in the stem, setting off the taxon from other taxonomic units that might compete for phenetic affiliation. Thus, in figure 12.6, whenever length *d* is substantial by comparison with lengths *b* and *c*, we would expect good fits by either phenograms or estimated cladograms to the true tree. Discrepancies from the true cladogeny arise from one of two kinds of situations: parallelisms in the cladogram affect the phenetic similarities of the OTUs; and divergence of cladistically closer relatives increases the relative phenetic similarity of cladistically more distant relatives. These two phenomena occur in more or less equal frequency in the Caminalcules, and they affect both the numerical phenetic and the numerical cladistic estimates. These two situations can be illustrated quite easily with examples from the Caminalcules (figure 12.7).

In genus B, OTUs 11 and 21 are cladistically closest, but 21 diverges from 11 by 5 evolutionary units, whereas 11 is only 2 units distant from OTU 10, although the latter branches off earlier. That is why 10 and 11 are closer in a phenogram than 11 is to its sister species, 21. This type of argument can be continued for OTUs 6 and 9, which in a phenogram would next join the nuclear cluster 10–11 in that order. All are closer to 11 and to each other in terms of evolutionary units than they are to OTU 21. That is why they join the phenetic cluster before OTU 21 does. In contrast to these results, those for genus C show complete agreement among the true cladogram figure 12.7), a phenogram, and an estimated cladogram (Sokal 1983b). The evolutionary

Robert R. Sokal

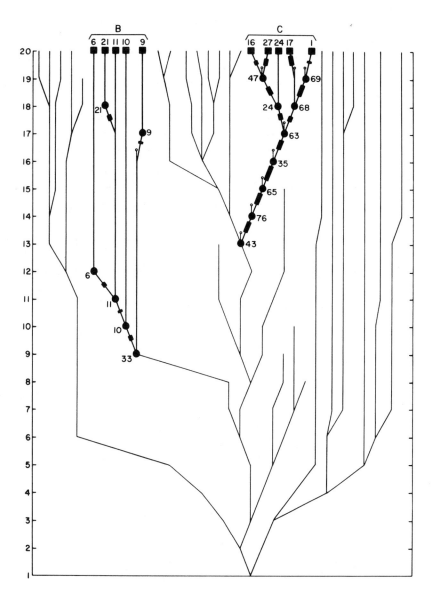

Figure 12.7 Subtrees for genera B and C from the true cladogram of the Caminalcules. Morphological change (slanted lines) in these genera occurred from time period 8(B) or 13(C) onward; vertical lines indicate periods without such change. The amount of change (path length of the internode) based on eighty-five characters is shown by the lengths of the thickened bars along the slanted lines. Species are identified by numbers; squares identify Recent species; circles, fossil species. Small, hollow circles represent extinct species whose lineages continue with evolutionary change.

The Continuing Search for Order

changes in the stems leading to the bifurcations in this genus are so great that they ensure the phenetic similarity of sister species.

An additional recent finding in our laboratory (Fiala and Sokal 1985) supports these conclusions. In a series of simulation studies in which different tree topologies were given different patterns of character-state evolution based on various evolutionary assumptions, we found that, on the average, phenograms gave estimates of the true cladogeny that were as accurate as numerically estimated cladograms. The relative successes of the methods have to do with the topology of the trees. Those trees that have relatively long basal stems vis-à-vis short issuing branches are estimated much better than those with short joint stems and long issuing branches. This "stemminess" has a marked effect in our simulation work. It therefore appears that the results of cladistic estimation methods greatly depend on the unknown topology of the true evolutionary tree. Additional problems with estimating cladograms by parsimony methods have recently been described by Rohlf (1984). Peculiarities in character-state codes result in dramatic instabilities of estimated cladograms by means of minimum-length Wagner trees.

It appears common wisdom among systematists that if only fossils were available, it would be easy to reconstruct phylogenetic trees and to establish classifications. When we include all 48 fossil species with the 29 Recent ones in the Caminalcules, the results are surprisingly no better than the analysis of Recent OTUs alone (Sokal 1983c). Even the best estimated cladogram has a strict consensus index with the true cladogram of only 0.667. The phenetic classification of these OTUs introduces some nonconvex taxa at higher phenetic levels, but unites phenetically homogeneous groups of mixed Recent and fossil composition. There is good correspondence of phenetics with phylogenetic sequences. All but one of the mutually closest pairs in the phenogram are ancestor-descendant pairs. Yet, it would not be possible to piece together a true phylogenetic tree of the Caminalcules by using these short sequences, even if their polarity were known, which is not always the case.

The first of the purposes of taxonomy, enumerated earlier, is the establishment of a natural system, available for reference to the scientific public and to laymen. Stability is clearly desirable for such a natural system. Actual tests of taxonomic stability are not usually carried out by adding either characters or OTUs to a data matrix. Rather, they are accomplished by taking the most complete available data matrix for the group and randomly subsampling characters or OTUs from it. From each subsample one computes a phenogram and also an estimated cladogram, and compares each with other such subsamples or with a standard dendrogram, that is, with the phenogram or the cladogram based on the entire data set.

Tests of stability carried out in this manner in the Caminalcules are of considerable interest. When we analyze the entire number of characters, estimated cladograms are more stable than phenograms when OTU numbers are small; phenograms are more stable than cladograms when OTU numbers become large (Sokal 1983d). These results can be inspected in figure 12.8,

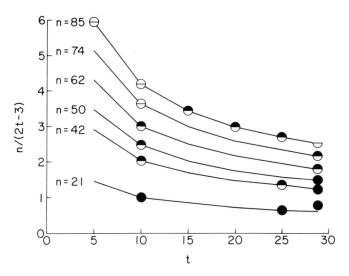

Figure 12.8 Phenograms and estimated cladograms for the Caminalcules compared in two ways: their stability tested against a standard, and their congruence compared with the true cladogeny. (For explanation, see the text.)

where each of the points is computed on the basis of at least 100 random samples. The ordinate indicates the ratio $n'/(2t - 3)$, where n' is the number of characters (in binary equivalents) and t, the number of OTUs, also defines the abscissa. For some selected sample sizes of characters (n, not binary), curves show the ratio as a function of OTU number t. The circles indicate the results of the comparisons. White denotes a superiority of estimated cladograms; black, a superiority of phenorgrams. The upper semicircle furnishes results on stability against a standard; the lower semicircle shows the results of comparisons of classifications with the true cladogram. For the combination 29 OTUs and 85 characters, the upper semicircle is missing because a stability test is meaningless for the full data set. When the number of characters in the sample is small, or the OTU number is large, phenograms are more stable and provide better estimates of the true cladogeny. But as the number of characters increases and/or the number of OTUs decreases, estimated cladograms improve and ultimately overtake phenograms in terms of stability and as estimators of the true cladogeny.

It was noted eventually that the stability is a function of the ratio of the number of characters and OTUs (Sokal et al. 1984). If t is the number of OTUs and n the number of characters, cladograms have $2t - 3$ edges, which must be estimated, whereas phenograms need only $t - 1$ parameters to determine the junction levels. The ratio $n/(2t - 3)$ is a coarse indicator of the adequacy of the data matrix for cladistic estimates. When this value is above 1, stability is higher for classifications based on estimated cladograms; below 1, classifications based on phenograms are more stable. These results parallel our experiences with estimates of the true cladogram. For all 29 Recent OTUs and 50 or fewer characters, one can estimate the true cladogram better by means of

The Continuing Search for Order

a phenogram; for more than 50 characters the relation is reversed. Similar experiences were obtained on a group of real organisms, bugs of the genus *Leptopodomorpha* (Sokal et al. 1984). here, of course, one can measure only stabilities, since the true cladogram is unknown. Similar results are arrived at for various other data sets: a study of proteins in mice (W. M. Fitch and W. R. Atchley, pers. comm.), two simulation studies (Fiala and Sokal 1985; J. Sourdis and C. Krimbas, MS), and an assemblage of 39 random and real data sets (Sokal and Shao 1985).

Taxonomic congruence is a special case of character stability. It is a measure of the similarity of classifications based on different kinds or classes of characters, such as external versus internal characters, morphological versus behavioral characters, and so forth. here, too, the results of critical tests show varying outcomes in favor of phenetic or cladistic approaches (Rohlf et al. 1983). The outcome of congruence tests is undoubtedly also affected by the ratio of the numbers of characters and OTUs and by stemminess.

The measurement of predictive value in classifications has not so far been uniformly successful. It is difficult to decide whether predictive value should be computed separately for each taxon or globally for the entire classification. The hierarchical nature of biological classifications has also complicated the development of measures of predictive value. A good measure must allow for the several taxonomic levels of a given study, since any one character might be highly predictive at one categoric level but of little value at another level. A second important consideration is whether the predictive value should measure only errors of inclusion or both errors of inclusion and of exclusion. Errors of inclusion imply inhomogeneity of character states within a taxon. Errors of exclusion mean that character states found within a given taxon are also found in other taxa and, to this degree, decrease the predictive value of the given character state. Unanimity on this issue has been difficult to achieve, and the debate has been clouded by the controversy between pheneticists and cladists. In a recent study, Archie (1984) has clarified some of the concepts related to predictive value and developed new indexes for measuring this property of classifications. It is probably fair to say that at this time it is not clear which school of taxonomy produces classifications with higher predictive value.

CONCLUSIONS

How can all these results be summarized? Numerically estimated cladograms are not good estimates of the true phylogeny of a group of organisms. The shortest trees are not necessarily closest to the true tree. Differences between true cladograms and phenograms or between phenograms and estimated cladograms can be explained as the results of homoplasy or divergence. Estimated cladograms are affected almost as much by homoplasy as are phenograms. As the number of characters decreases or the number of OTUs increases, phenograms become better estimates of the true cladogeny than estimated cladograms. These same relations exist for taxonomic stability based on either characters or OTUs. Even the inclusion of fossils in the data matrix does not substantially increase the quality of the estimate of the phylogeny. The topology of the true tree is a critical factor in determining the quality of its estimate. Such results are not causes for optimism for those who wish to estimate phylogenies.

One should be cautious about establishing classifications on the basis of few characters, as is often done by conventional (i.e., nonnumerical) cladists. Under less than optimal conditions (i.e., with few characters or many OTUs), it would appear that a phenetic classification would continue to be the most desirable system for establishing general classifications because it permits greater stability and possibly greater predictive value than a cladistic classification. For those whose major purpose is to estimate phylogenies, phenograms would in many cases give estimates of the true cladogeny that are no worse, and possibly better, than those obtained by current cladistic methods.

Finally, it is now entirely obvious that almost all scientists who so desire have access to a computer. Data matrices in systematics can therefore be analyzed in a great variety of ways, with little effort and at relatively low cost. Alternative systems of classification are clearly feasible. Scientists can analyze their data phenetically as well as cladistically, and then synthesize the two to arrive at evolutionary classifications. Nevertheless, a general system is still desirable; and for the reasons stated above, it would seem that such a system should be phenetic.

SUMMARY

In this chapter I review the principles for forming biological classifications and summarize recent findings concerning optimality criteria for classifications. Natural taxa are recognized as polythetic and related to concept formation in cognitive psychology. The three currently advocated schools of taxonomy are reviewed and their assumptions and purposes compared. Three criteria of optimality—predictive value, stability, and fit to the true cladistic relationship—are discussed, and evidence from recent numerical taxonomic studies of these criteria is reviewed. Numerical classifications based on phenetic and cladistic computer programs differ in their taxonomic stability and fit to the true cladogram. There is no universally superior approach, but the relative advantage of phenetic versus cladistic algorithms is a function of the ratio of characters to OTUs in the data. Accuracy of cladogram estimation is also affected by tree topology.

ACKNOWLEDGMENTS

This chapter is an expanded version of the presidential address to the American Society of Naturalists presented in June 1984 at Crested Butte, Colorado. The assistance of K. Fiala and K. T. Shao, G. Hart, and B. Thomson in the research leading up to this work is gratefully acknowledged. F. J. Rohlf read the manuscript and furnished useful suggestions for improvements. R. Chapey prepared the manuscript and J. Schirmer drew the figures. The research upon which this chapter is based was supported by National Science Foundation grant BSR 83-06004. Contribution no. 556 in Ecology and Evolution from the State University of New York at Stony Brook.

REFERENCES

Archie, J. W. 1984. A new look at the predictive value of numerical classifications. *Syst. Zool.* 33:30–51.

Ashlock, P. H. 1980. An evolutionary taxonomist's view of classification. *Syst. Zool.* 28:441–450.

Baum, B. R. 1983. Relationships between transformation series and some numerical cladistic methods at the infraspecific level, when genealogies are known. Pages 340–345 in J. Felsenstein, ed. *Numerical taxonomy*. Springer-Verlag. Berlin.

Baum, B. R., and G. F. Estabrook. 1978. Application of compatibility analysis in numerical cladistics at the infraspecific level. *Can. J. Bot.* 56:1130–1135.

Beatty, J. 1982. Classes and cladists. *Syst. Zool.* 31:25–34.

Beckner, M. 1959. *The biological way of thought*. Columbia University Press. New York.

Bock, W. J. 1977. Foundations and methods of evolutionary classification. Pages 851–895 in M. K. Hecht, P. C. Goody, and B. M. Hecht, eds. *Major patterns of vertebrate evolution*. Plenum. New York.

Breedlove, D. E., and P. H. Raven. 1974. *Principles of Tzeltal plant classification: An introduction to the botanical ethnography of a Mayan speaking people of highland Chiapas*. Academic Press, New York.

Brooks, D. R., and E. O. Wiley, 1985. Theories and methods in different approaches to phylogenetic systematics. *Cladistics* 1:1–11.

Cain, A. J. 1958. Logic and memory in Linnaeus's system of taxonomy. *Proc. Linn. Soc. Lond.* 169:144–163.

Cain, A. J., and G. A. Harrison. 1960. Phyletic weighting. *Proc. Zool. Soc. Lond.* 135:1–31.

Camin, J. H., and R. R. Sokal. 1965. A method for deducing branching sequences in phylogeny. *Evolution* 19:311–326.

Colless, D. H. 1970. The phenogram as an estimate of phylogeny. *Syst. Zool.* 19:352–362.

———. 1980. Congruence between morphometric and allozyme data for the *Menidia* species: A reappraisal. *Syst. Zool.* 29:288–299.

Cracraft, J. 1983. The significance of phylogenetic classifications for systematic and evolutionary biology. Pages 1–17 in J. Felsenstein, ed. *Numerical taxonomy*. Springer-Verlag, Berlin.

DuPraw, E. J. 1964. Non-Linnean taxonomy. *Nature* (Lond.) 202:849–852.

Eldredge, N., and J. Cracraft. 1980. *Phylogenetic patterns and the evolutionary process*. Columbia University Press, New York.

Estabrook, G. F. 1978. Some concepts for the estimation of evolutionary relationships in systematic botany. *Syst. Bot.* 3:146–158.

Farris, J. S. 1970. Methods for computing Wagner trees. *Syst. Zool.* 19:83–92.

Farris, J. S., A. G. Kluge, and M. J. Eckardt. 1970. A numerical approach to phylogenetic systematics. *Syst. Zool.* 19:172–189.

Felsenstein, J. 1982. Numerical methods for inferring evolutionary trees. *Q. Rev. Biol.* 57:379–404.

Fiala, K. L., and R. R. Sokal. 1985. Factors determining the accuracy of cladogram estimation: Evaluation using computer simulation. *Evolution* 39:609–622.

Fitch, W. M., and E. Margoliash. 1968. The construction of phylogenetic trees. II. How well do they reflect past history? *Brookhaven Symp. Biol.* 21:217–242.

Gilmour, J. S. L. 1937. A taxonomic problem. *Nature* (Lond.) 139:1040–1042.

———. 1940. Taxonomy and philosophy. Pages 461–474 *in* J. Huxley, ed. *The new systematics.* Clarendon, Oxford.

———. 1951. The development of taxonomic theory since 1851. *Nature* (Lond.) 168:400–402.

———. 1961. Taxonomy. Pages 27–45 *in* A. M. MacLeod and L. S. Cobley, eds. *Contemporary botanical thought.* Oliver & Boyd, Edinburgh.

Kluge, A. G. 1985. Ontogeny and phylogenetic systematics. *Cladistics* 1:13–27.

Mayr, E. 1969. *Principles of systematic zoology.* McGraw-Hill, New York.

———. 1982. *The growth of biological thought.* Belknap, Cambridge, Mass.

Neel, J. Y. 1978. The population structure of an Amerindian tribe, the Yanomama. *Annu. Rev. Genet.* 12:365–413.

Platnick, N. I. 1985. Philosophy and the transformation of cladistics revisited. *Cladistics* 1:87–94.

Remane, A. 1956. *Die Grundlagen des natuerlichen Systems, der vergleichenden Anatomie und der Phylogenetik.* Theoretische Morphologie und Systematik. Vol. I. 2d ed. Akademische Verlagsgesellschaft Geest & Portig, Leipzig.

Riedl, R. 1976. *Die Strategie der Genesis.* Piper, Munich.

Rohlf, F. J. 1982. Consensus indices for comparing classifications. *Math. Biosci.* 59:131–144.

———. 1984. A note on minimum length trees. *Syst. Zool.* 33:341–343.

Rohlf, F. J., and R. R. Sokal. 1981. Comparing numerical taxonomic studies. *Syst. Zool.* 30:459–490.

Rohlf, F. J., D. H. Colless, and G. Hart. 1983. Taxonomic congruence—reexamined. *Syst. Zool.* 32:144-158.

Shao, K. T. 1983. Consensus methods in numerical taxonomy. Ph.D. diss., State University of New York, Stony Brook.

Smith, E. E., and D. L. Medin. 1981. *Categories and concepts.* Harvard University Press, Cambridge, Mass.

Sneath, P. H. A. 1962. The construction of taxonomic groups. *Symp. Soc. Gen. Microbiol.* 12:289–332.

Sneath, P. H. A., and R. R. Sokal. 1973. *Numerical taxonomy.* Freeman, San Francisco.

Sokal, R. R. 1974. Classification: purposes, principles, progress, prospects. *Science* (Wash., D.C.) 185:1115–1123.

———. 1983a. A phylogenetic analysis of the Caminalcules. I. The data base. *Syst. Zool.* 32:159–184.

———. 1983b. A phylogenetic analysis of the Caminalcules. II. Estimating the true cladogram. *Syst. Zool.* 32:185–201.

————. 1983c. A phylogenetic analysis of the Caminalcules. III. Fossils and classification. *Syst. Zool.* 32:248–258.

————. 1983d. A phylogenetic analysis of the Caminalcules. IV. Congruence and character stability. *Syst. Zool.* 32:259–275.

————. 1984. Die Caminalcules als taxonomische Lehrmeister. Pages 15–31 *in* H. H. Bock. ed., *Anwendungen der Klassifikation: Datenanalyse und numerische Klassifikation.* Indeks Verlag. Frankfurt.

Sokal, R. R., and F. J. Rohlf. 1970. The intelligent ignoramus, an experiment in numerical taxonomy. *Taxon* 19:305–319.

————. 1980. An experiment in taxonomic judgment. *Syst. Bot.* 5:341–365.

————. 1981. Taxonomic congruence in the Leptopodomorpha re-examined. *Syst. Zool.* 30:309–325.

Sokal, R. R., and K. T. Shao. 1985. Character stability in 39 data sets. *Syst. Zool.* 34:83–89.

Sokal, R. R., and P. H. A. Sneath. 1963. *Principles of numerical taxonomy.* Freeman, San Francisco.

Sokal, R. R., K. L., Fiala, and G. Hart. 1984. OTU stability and factors determining taxonomic stability: Examples from the Caminalcules and the Leptopodomorpha. *Syst. Zool.* 33:387–407.

Stinebrickner, R. 1984. An extension of intersection methods from trees to dendrograms. *Syst. Zool.* 33:381–386.

Tateno, Y., M. Nei, and F. Tajima. 1982. Accuracy of estimated phylogenetic trees from molecular data. I. Distantly related species. *J. Mol. Evol.* 18:387–404.

Wiley, E. O. 1981. *Phylogenetics.* Wiley, New York.

13 Phylogenetic Systematics

Willi Hennig

Since the advent of the theory of evolution, one of the tasks of biology has been to investigate the phylogenetic relationship between species. This task is especially important because all of the differences which exist between species, whether in morphology, physiology, or ecology, in ways of behavior or even in geographical distribution, have evolved, like the species themselves, in the course of phylogenesis. The present-day day multiplicity of species and the structure of the differences between them first becomes intelligible when it is recognized that the differences have evolved in the course of phylogenesis; in other words, when the phylogenetic relationship of the species is understood.

Investigation of the phylogenetic relationship between all existing species and the expression of the results of this research, in a form which cannot be misunderstood, is the task of the phylogenetic systematics.

The problems and methods of this important province of biology can be understood only if three fundamental questions are posed and answered: what is phylogenetic relationship, how is it established, and how is knowledge of it expressed so that misunderstandings are excluded?

The definition of the concept "phylogenetic relationship" is based on the fact that reproduction is bisexual in the majority of organisms, and that it usually takes place only within the framework of confined reproductive communities which are genetically isolated from each other. This is especially true for the insects, with which this chapter is mainly concerned. The reproductive communities which occur in nature we call species. New species originate exclusively because parts of existing reproductive communities have first become externally isolated from one another for such extended periods that genetic isolation mechanisms have developed which make reproductive relationships between these parts impossible when the external barriers which have led to their isolation are removed. Thus, all species (= reproductive communities) which exist together at a given time—e.g., the present—have originated by the splitting of older homogeneous reproductive communities.

The survey of the literature pertaining to this review was concluded in 1963.

From *Annual Review of Entomology*, 1965, 10:97–116.

On this fact is based the definition of the concept "phylogenetic relationship": under this concept, species B is more nearly related to species C than to another species, A, when B has at least one ancestral species source in common with species C which is not the ancestral source of species A (Hennig, 8).

"Phylogenetic relationship" is thus a relative concept. It is pointless (since it is self-evident) to say, as is often said, that a species or species-group is "phylogenetically related" to another. The question is rather one of knowing whether a species or species-group is more or less closely related to another than to a third. The measurement of the degree of phylogenetic relationship is, as the definition of the concept shows, "recency of common ancestry" (Bigelow, 1). A phylogenetic relationship of varying degree exists between all living species, whether we know of it or not. The aim of research on phylogenetic systematics is to discover the appropriate degrees of phylogenetic relationship within a given group of organisms.

The degree of phylogenetic relationship which exists between different species, and thus also the results of research on phylogenetic systematics, can be represented in a visual form which is not open to misinterpretation, as is a so-called phylogeny tree (dendrogram). To be able to discuss this, not only the species but also the monophyletic groups included in the diagram must be given names. "Monophyletic groups" are small or large species-groups whose member species can be considered to be more closely related to one another than to species which stand outside these groups (Hennig, 8). When a phylogeny diagram, conforming to this postulate, has been rendered suitable for discussion by the naming of all of the monophyletic groups, then the diagram can be discarded and its information may be expressed solely by ranking the names of the groups:

A. Myriopoda
B. Insecta
　B.1 Entognatha
　　B.1a Diplura
　　B.1b Ellipura
　　　B.1ba Protura
　　　B.1bb Collembola
　B.2 Ectognatha

Such arrangement of monophyletic groups of animals according to their degree of phylogenetic relationship is called, in the narrower sense, a phylogenetic system of the group in question. Such a system belongs to the type called a "hierarchical" system. Since "system" in the wider sense means every arrangement of elements according to a given principle, the phylogeny tree, too, can be termed a phylogenetic system. Phylogeny diagrams and arrangement of the names of monophyletic groups in a hierarchical sequence are merely different but closely comparable forms of presentation whose content is the same. Therefore, everything which can be said about the methods of phylogenetic systematics (see below) applies irrespective of whether the results sought by the use of these methods are expressed only as a phylogeny tree or as a phylogenetic system in the narrower sense, in a hierarchically arranged list of the names of monophyletic groups.

In some cases, a hierarchical arrangement of group names, that is, a phylogenetic system in the narrower sense, is to be preferred to a phylogeny tree. One can, for instance, in a catalogue or checklist of Nearctic Diptera, give expression to all that one thinks is known about the phylogenetic relationship of all Nearctic species of Diptera in a form which can in no way be misinterpreted, without using a single phylogeny tree.

However, considerable difficulties arise because systems of the hierarchical type have also been used in biology with intentions other than of expressing the phylogenetic relationship of species. Long before the advent of the theory of evolution, "systematics" existed as the branch of biological science which had adopted as its aim an orderly survey of the plurality of organisms. Naturally, the principle of classification in systematics could not then be the phylogenetic relationship of species, which was still unrecognized, but only a morphological resemblance between organisms. This morphological systematics also used the hierarchical type of system to express its results although Linnaeus already held the view that morphological resemblances between organisms corresponded to a multidimensional net. Numerous attempts have also been made to introduce other types of system, which differ from the hierarchical, into biological systematics (see Wilson and Doner, 21). But they have not been successful.

Today, there are still many authors who consider that the purpose of biological "systematics" is to classify organisms according to their morphological resemblance, and who use a system of the hierarchical type to this end. It is hardly surprising that misunderstandings and serious errors can be produced by this formal identity between morphological and phylogenetic systems.

The source of danger in the formal identity between systems based on such different principles of classification is that, in a hierarchical system, each group formation relates to a "beginner," which is linked in "one-many relations" with all of the members of that group and only those (Gregg, 3). In morphological systems the "beginner" that belongs to each group is a formal idealistic standard ("Archetype") whose connections with the other members of the group are likewise purely formal and idealistic. But in a phylogenetic system the "beginner" to which each group formation relates is a real reproductive community which has at some time in the past really existed as the ancestral species of the group in question, independently of the mind which conceives it, and which is linked by genealogical connections with the other members of the group and only with these. One could, without difficulty, adduce many examples from the literature in which the formal beginner ("Archetype") of a group, conceived according to the principles of morphological systematics, has been erroneously taken, with all of the consequences of such an error of logic, as the real beginner (ancestral species) of a monophyletic group.

This dangerous difference between a formal morphological (typological) hierarchical system and the equally hierarchical system of phylogenetic systematics, would not arise if the degree of morphological resemblance were an exact measurement of the degree of phylogenetic relationship. But this is not the case. Furthermore, there is yet no definition of the concept of morphological resemblance which is not open to theoretical objection, nor any method which can be accepted as the one and only method which achieves a satisfactory determination of more than the threshold of morphological resemblance, that is, the degree of resemblance between relatively similar species which agree in very many characters.

In these circumstances, the dangers which arise from the formal identity of phylogenetic and morphological systems will be avoided if agreement can be reached on whether or not the branch of biological science known simply as systematics will, in the future, always try to express the morphological resemblance of organisms or their phylogenetic relationship in the system in which it works.

It has often been stated, in defense of a system of morphological resemblance, that this has historical primacy over endeavors to express phylogenetic relationship in a system, because the morphological system had already existed as the aim of "systematics" before the advent of the theory of evolution. Even today, this reasoning is often augmented with the argument that the theory of evolution was established with the help, among other things, of the system of graduated morphological resemblances between organisms, and that therefore one is prescribing a circle if, in reverse, one wishes to take the theory of evolution and the notion of the phylogenetic relationship of organisms which follows from it as the theoretical starting-point of their classification in a system (Sokal, 17; Blackwelder, Alexander, and Blair, 2). This "ebenso haltwie heillose Einwand" (Günther, in discussing the work of Sokal, 12) has already been so often refuted that one can only attribute, to authors who persist in asserting it today, a lack of information.

It is certainly correct that the classification of organisms according to their morphological resemblance has led to the theory of evolution. This was possible only because the morphological differences between organisms are the result of a historical (phylogenetic) development and because, at least in rough terms, very similar organisms, are, in fact, generally more closely related than are very different ones. It was therefore inevitable that the classification of organisms according to their morphological resemblance, in association with certain features of their ontogenetic development and their geographical distribution, would sooner or later lead to the discovery of their successive degrees of phylogenetic relationship and thus to the theory of evolution.

However, there are historical origins not only of the morphological differences between organisms in the narrower sense, but also differences in their physiological functions, their ways of behavior and, in addition to these physical ("holomorphological") attributes, differences in their distribution in geographical and ecological space. Since it has been recognized and, moreover, become widely known that there are not the same degrees of agreement and difference in the various holomorphological and chorological resemblances which connect organisms, the way is open for establishing the phylogenetic relationship itself of organisms as the principle of classification, instead of successive degrees of resemblance in a single category of characters; for only from the phylogenetic relationship is it possible to establish direct connections with all other thinkable kinds of agreement and difference between organisms. The demand for a phylogenetic system is thus not so much a renunciation of pre-phylogenetic resemblance, systematics, as its consequential further development.

The claim of the phylogenetic system to elevation into the universal reference system of biology has a logical, even if not historical, foundation, and arises because few areas of research can be conceived which do not bear fruit and lead to more profound conclusions through a knowledge of the phylogenetic relationship of its objects, and which cannot, in turn, lead to the discovery of hitherto unknown relationships in the course of mutual exchange of information. This is not true to the same extent for any other system built on any other principle of classification. Other systems may also have their value as knowledge; but this value is, in each case, restricted to answering particular questions.

The logical primacy of the phylogenetic system also arises because it alone provides all parts of the field studied by biological systematics with a common theoretical foundation (Kiriakoff, 14). It is true that phylogenetic relationship exists only between different species, and species are not the simplest elements of biological systematics. These are not even the "individuals," but the individuals in given short periods of their lifetime ("semaphoronts"). The first and basic task of systematics is to establish that different individuals, or rather "semaphoronts," belong to particular species. The difficulty within this task rests in the fact that the species, which exist in nature as real phenomena independent of the men who perceive them, are units which are not morphologically but genetically defined. They are communities of reproduction, not resemblance. Of course, the morphological resemblance between members of a species is not unimportant for the practical establishment of specific limits. But it has only the significance of an auxiliary criterion whose capabilities of use are limited. This is because the definition of the phylogenetic relationship between species, as well as the definition of the species-concept, is deduced from the fact that the reproduction of species generally takes place within the framework of defined communities which cannot be unqualified communities of resemblance if, in the demand for a phylogenetic system, biological systematics has acquired for all its spheres of activity a common aim, that is, the discovery and recording of the "hologenetical" connections which exist between all organisms. In contrast with this, morphological resemblance-systematics, though not denying the modern genetic species-concept, employs different principles of classification above and below the specific level.

It would, of course, be meaningless to extol the need for a phylogenetic system, however well founded it might be theoretically, if this demand could not be put into practice. There is, in fact, a widespread notion that phylogenetic systematics, at least in those groups of animals for which no fossil finds are available, possesses no method of its own, but can only interpret the results of morphological systematics according to the principle that the degree of morphological resemblance equals the degree of phylogenetic relationship. This notion is false. The fundamental difference between the method of morphological and phylogenetic systematics is that the latter breaks up the simple concept of "resemblance" (figure 13.1).

It is a consequence of the theory of evolution that the differences between various organisms must have arisen through changes of characters in the course of a historical process. Therefore it is not the extent of resemblance or difference between various organisms that is of significance for research into phylogenetic relationship, but the connection of the agreeing or divergent characters with earlier conditions. It is valid to distinguish different categories of resemblance according to the nature of these connections.

The division of the concept of resemblance into various categories of resemblances probably began, in the history of systematics, with the introduction of the concept of convergence. Often this concept was linked with the distinction between analogous and homologous organs. Convergence is, in fact, commonly manifested by similar organs having arisen in adaptation to the same functions from different morphological foundations in different organisms. But there are also cases where virtually complete agreement in the form of homologous organs rests on convergence. "Convergence" means resemblance between the characters of different species which has evolved through the independent change of divergent earlier conditions of these characters. It shows how species which differed from one another are ancestors of species which have become similar to one another. If one associates in a group the species whose resemblance rests on convergence, then this is not a monophyletic but a polyphyletic group. There are few authors today who would specifically support the inclusion of demonstrably

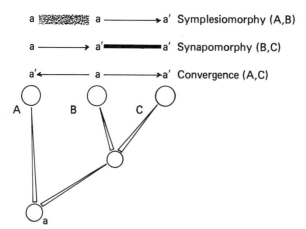

Figure 13.1 The three different categories of morphological resemblance a: plesiomorph. a': apomorph expression of the morphological character a. Agreement may rest on symplesiomorphy ($a - a$), symapomorphy ($a' - a'$), or convergence ($a' - a'$).

polyphyletic groups in a system. "Convergence" and "polyphyletic groups" are concepts which presuppose acceptance of the theory of evolution. Therefore, some systematists think they are already working with a "phylogenetic system" when, in their evaluation of morphological resemblance, they exclude convergence and thus polyphyletic groups from their system.

But even when purged of convergence, morphological resemblance is still not a satisfactory criterion for the degree of phylogenetic relationship between species. It still does not provide one with exclusively monophyletic groups, such as a phylogenetic system demands. This arises from the fact that characters can remain unchanged during a number of speciation processes. Therefore, it follows that the common possession of primitive ("plesiomorph") characters which remained unchanged cannot be evidence of the close relationship of their possessors.

Often, a given species can be phylogenetically more closely related to a species which possesses a particular character in a derivative ("apomorph") stage of expression than to species with which it agrees in the possession of the primitive ("plesiomorph") stage in the expression of this character. Therefore, a resemblance which rests on symplesiomorphy is of no more value in justifying a supposition of closer phylogenetic relationship than is a resemblance which has occurred through convergence. If, in a system, one associates in a group species whose agreement rests on convergence, a polyphyletic group is thereby formed, as has been established above and is generally recognized. If one associates species whose agreement rests on symplesiomorphy, then a paraphyletic group is formed (figure 13.2). Paraphyletic groups among insects are the "Apterygota" and Palaeoptilota (= Palaeoptera), if one considers the closer relationship of the Odonata with the Neoptera as established. Paraphyletic vertebrate groups are the "Pisces" and the "Reptilia."

The supposition that two or more species are more closely related to one another than to any other species, and that together they form a monophyletic group, can be confirmed only by demonstrating their common possession of derivative characters ("synapomorphy"). When such characters have been demonstrated, then the supposition has been confirmed that they have been inherited from an ancestral species common only to the species showing these characters.

It must be recognized as a principle of inquiry for the practice of systematics that agreement in characters must be interpreted as synapomorphy as long as there are no grounds for suspecting its origin to be symplesiomorphy or convergence.

The method of phylogenetic systematics as that part of biological science whose aim is to investigate the degree of phylogenetic relationship between species and to express this in the system which it has designed thus has the following basis: that morphological resemblance between species cannot be considered simply as a criterion of phylogenetic relationship, but that this concept should be divided into the concepts of symplesiomorphy, convergence, and synapomorphy, and that only the last-named category of resemblance can be used to establish states of relationship.

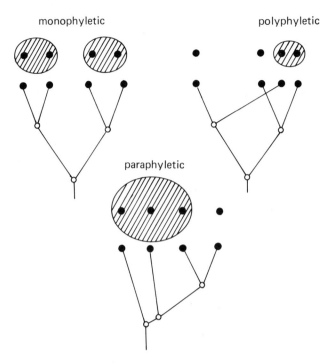

Figure 13.2 The three different categories of systematic group formations corresponding to the resemblance of their constituents resting on synapomorphy (monophyletic groups), convergence (polyphyletic groups), or symplesiomorphy (paraphyletic groups).

The differences between the phylogenetic system and all other systems which likewise classify species on the basis of their morphological resemblance are as follows: (a) Systems which employ the simple criterion of morphological resemblance. Such systems include polyphyletic, paraphyletic, and monophyletic groups. (b) Systems which employ the criterion of morphological resemblance but fail to consider characters whose agreement rests on convergence. In such systems, polyphyletic groups are excluded but paraphyletic as well as monophyletic groups are admitted. (c) Phylogenetic systems. Characters whose agreement rests on convergence or symplesiomorphy are not considered. Therefore, polyphyletic and paraphyletic groups are excluded and only monophyletic groups admitted.

The systems names under (b) have also often been termed phylogenetic systems in the literature (e.g., Stammer, 18; Verheyen, 20). But it is thereby overlooked that the paraphyletic groups admitted in these "pseudo-phylogenetic" or "cryptotypological" systems (Kiriakoff, 14) are similar in many respects to polyphyletic groups. No one would think of considering polyphyletic groups in studies concerned with the course and eventual rules of phylogenesis (zoogeographical studies, for instance, belong here), since they have no ancestors solely of their own and therefore no individual history. Exactly the same holds true, however, for paraphyletic groups. The sole common ancestors of all the so-called Apterygota, for instance, were also the ancestors of the Pterygota, and the beginning of the history of the Apterygota was not the beginning of an individual history of this group, but the beginning of the individual history of the Insecta, which were at first Apterygota in the morphological-typological sense. Also, the concept of "extinction" is different in paraphyletic and monophyletic groups. Only monophyletic groups can become "extinct" in the sense that from a particular point in time no physical progeny of any member of the group have existed. But if, however, one says that a paraphyletic group has become "extinct," this can only mean that after a particular point of time no bearers of the morphological characters of this group have existed. But physical progeny of many of its members may, with changed characters, continue to live. Monophyletic and paraphyletic groups thus cannot be compared with each other in any question concerning their history. Failure to take account of this fact and invalid uncritical comparison of paraphyletic ad monophyletic groups has led to some false conclusions in studies about the "Grossablauf der phylogenetischen Entwicklung" (Müller, 15) and the history of the distribution of animals.

From the premise that morphological agreement only confirms a supposition that the species concerned belong to a monophyletic group when it can be interpreted as synapomorphy, is derived for the practical work of the systematist, the "Argumentation plan of phylogenetic systematics" (figure 13.3). This plan shows that in a phylogenetic system which must contain only monophyletic groups, every group formation, irrespective of the rank to which it belongs, must be established by demonstration of derivative ("apomorph") characters in its ground plan. But it also shows clearly that in two monophyletic groups which together form a monophyletic group of higher rank and are therefore to be termed "sister-groups," one particular character must always occur in a more primitive (relatively plesiomorph) condition in one group than in its sister-group. For the latter, the same is true in respect to other characters. This mosaic-like distribution of relatively primitive and relatively derivative characters in related species and species-groups (Spezialisationskreuzungen, Heterobathmie der Merkmale: Takhtajan, 19) has long been known. But one still finds it occasionally mentioned in the literature as a special peculiarity of some groups of animals that the classification of their constituent groups cannot be achieved in a definite sequence, because there are no solely primitive and no solely derivative species or species-groups. In a phylogenetic system there can indeed be no solely primitive and no solely derivative groups. The possession of at least one derivative (relatively apomorph) ground-plan character is a precondition for a group to be recognized at all as a monophyletic group. But it also follows from this that this same character is the nearest related group must be present in a more primitive (relatively plesiomorph) stage of expression. The exclusive presence of relatively plesiomorph characters is indicative of paraphyletic groupings: these are to be found only in pseudo-phyletic (see above under *b*) and purely morphological systems (see above under *a*), but not in phylogenetic systems. Heterobathmy of characters is therefore a precondition for the establishment of the phylogenetic relationship of species and hence a phylogenetic system.

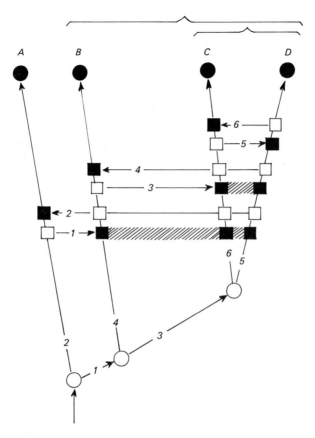

Figure 13.3 Argumentation plan of phylogenetic systematics. Empty box = plesiomorph, solid box = apomorph expression of characters. Equal numbers indicate how sister-group relations are established by the distribution of relatively plesiomorph (empty box) and relatively apomorph (solid box) characters ("heterobathmy of characters"). Adapted from Hennig (11).

It is sometimes said that the aims of phylogenetic systematics are not only practically but also theoretically unattainable, because the comparison of species living in a given time-horizon, such as the present, cannot in any way reveal their phylogenetic relationship, which refers to a completely different dimension. This view is false. Just as two stereoscopic views of a landscape, which themselves assume only a two-dimensional form, together contain exact information about the third spatial dimension, so the mosaic of heterobathmic characters in its distribution over a number of simultaneously living species contains reliable information about the sequence in which the species have evolved from common ancestors at different times. The study and use of the methods which serve to reveal this information needs, it is true, a far greater amount of knowledge and experience than some systematists are willing to employ. The theoretical foundation and refinement of these methods forms a special chapter in the theory of phylogenetic systematics which can only be touched upon in the present brief chapter.

It is sometimes alleged that consideration of as many characters as possible which have so far not been studied is a prerequisite for the progress of phylogenetic systematics. In particular, the restriction of entomological systematics to comparatively easily recognizable characters of the external skeleton which lie open to view is often not highly regarded. This has some justification. The phenomena of convergence (particularly in its variant known under the name "parallel development"), reversed development of characters, and paedomorphosis, which leads to pseudoplesiomorph conditions, make the establishment of true synapomorphy difficult. The more complex is the mosaic of heterobathmic characters which we have at our disposal in a chosen group of species, the more surely can their phylogenetic relationship be deduced from it.

Consideration of new and hitherto unobserved characters can, however, represent progress only if these are analyzed with the special methods of phylogenetic systematics. Thus, it is also necessary to distinguish between plesiomorph and apomorph expressions of characters in the internal anatomy and chemical structure, physiology, and serology and when considering different ways of behavior. Symplesiomorphy must be excluded just as much as convergence. If this is not observed, then consideration of however many characters leads, at best, only to a more precise determination of the overall similarity of the bearers of all these characters, but not to a more precise establishment of their degree of phylogenetic relationship.

This becomes particularly obvious in animal groups such as the insects, in which the life of the individual is subject to the phenomenon of metamorphosis. This is the cause of the incongruences which are so often discussed between larval, pupal, and imaginal classification in morphological and pseudo-phylogenetic systematics. A theoretically acceptable solution of such "incongruences" is possible only in phylogenetic systematics. It can indeed be the case that particular instances of synapomorphy, and therefore of monophyletic groups, can be recognized only in the larval or pupal stages and others only in the imaginal stage. But this is not a true incongruence, for the phylogenetic system does not try to classify organisms according to their degree of resemblances, but species according to their degree of phylogenetic relationship. It does not matter therefore which stage of development is used to establish relationship on the ground of synapomorphy. A monophyletic group remains such even if it can be established only with the characters of a single stage of development (for more detailed exposition see Hennig, 11).

The fact that not resemblance as such, but only agreement in a particular category of characters is significant for the study of phylogenetic relationship, also makes it possible for phylogenetic systematics to adduce for its purposes features other than physical (holomorphological) characters. Such nonholomorphological characters are the life history and geographical distribution of species. Phylogenetic systematics can, for instance, proceed from the plausible hypothesis that species which show a clearly derivative ("apooec") life history, and for which a certain relationship is probable on other grounds, form a monophyletic group. This is, for instance, often true with parasites. However, hypotheses of this kind must always be verified by close morphological studies, for it is particularly with similar life histories that adaptive convergence is common.

A particularly great importance for phylogenetic systematics is presently often ascribed to parasites and to monophagous and oligophagous plant-feeders which are to be equated with them from the standpoint of phylogenetic method. The theoretical justification for this is supplied by the so-called parasitophyletic rules. Particularly important among these is the so-called Fahrenholz rule, which supposes a marked parallelism between the phylogenetic development of parasitic groups of animals and their hosts in the majority of cases. If this is correct, then it might be concluded from the restriction of a monophyletic group of parasites to a particular group of host species that the latter, too, form a monophyletic group. but it can easily be shown that this conclusion would be correct only if one could assume that the ancestral species of the host group was attacked by one parasite species and that thereafter each process of speciation in the host group has been accompanied by one speciation process in the parasites. Clearly, this precondition is only rarely fulfilled, since the evolution of parasites often seems to be retarded in comparison with that of their hosts, in respect to both character changes and speciation. The result of this is that paraphyletic host groups can also be attacked by monophyletic groups of parasites. Moreover, it happens that parasites can transfer secondarily (without being passed from ancestors to progeny in the course of speciation) to host species which offer them similar conditions of life. This, too, is often seen as an indication of close phylogenetic relationship between host species which are exclusively attacked by particular parasite species or a monophyletic group of parasites. But this assumption would be valid only if one could assume that the "degree of resemblance" of different species and the "degree of their phylogenetic relationship" corresponded closely with each other. As has been shown, this is not the case. Resemblance can also be based, for instance, on symplesiomorphy, and this cannot be assumed to establish phylogenetic relationship. Since one cannot assume that parasites distinguish, in their choice of host range, the categories of resemblance connections (symplesiomorphy, synapomorphy, and convergence) whose differences are important for phylogenetic systematics, the greatest care is necessary in attempting to draw conclusions about the phylogenetic relationship of their hosts from the occurrence of monophyletic groups of parasites. The importance of parasitology for phylogenetic systematics is considerable. But on the grounds given it is not so great as is sometimes supposed. In particular there is still no really satisfactory clarification of this whole complex of questions.

The geographical distribution of organisms is also of restricted, though not to be underestimated, importance for phylogenetic systematics. This can often proceed from the hypothesis that parts of a group which are restricted to a defined, more or less separated, part of the total range, whose ancestors may be assumed to have arrived from other regions, form a monophyletic group. This is particularly valid for the fauna of the marginal continents (Australia and South America), whose ease of accessibility has been different at different periods of the earth's history, and for some islands (e.g., Madagascar, New Zealand). One can, for instance, proceed on the working hypothesis that the Marsupialia of Australia form a monophyletic group, and then seek either to sustain or refute this hypothesis with the morphological methods of phylogenetic systematics. With groups of animals with disjunctive distribution, one may proceed on the hypothesis that both parts of the range (Australia and South America in the case of pouched mammals) have been settled by monophyletic subgroups and that between these a sister-group relationship exists. Extensive investigation of the phylogenetic development of animal groups (e.g., Hofer on the Marsupialia) often in themselves remain fruitless, since they do not proceed from a working hypothesis of this kind and as a result contain no statements which serve to answer the questions which first come clearly to light in such a hypothesis. This is often of even greater importance in studies of the history of the settlement of geographical space. Discussions about the earlier existence of direct land connections between now separate regions (Madagascar and the Oriental Region: Günther, 5; New Zealand and South America: Hennig, 12) have somewhat the same significance as have attempts to sustain or refute hypotheses about the monophyletic, paraphyletic, or polyphyletic character of particular groups of animals. The inadequacy of morphological or pseudo-phyletic systems is shown here with particular clarity.

A special chapter in the theory of phylogenetic systematics which can only be touched upon here is the position of fossils in the system (Hennig, 9). Despite a widely held opinion, establishing the phylogenetic relationships of fossil animal forms is usually more difficult than that of recent species. The cause of this is that in fossil finds usually only a small, often extremely small, section is available from the character structure of the whole organism. But since the methods of phylogenetic systematics have a numerical character insofar as the certainty of their conclusions grows as the number of characters at their disposal increases (see above), it follows necessarily that the reliability with which relationships can be established cannot usually be as great with fossils as with recent species. In the sphere of the lower categories of the system, the species and their subunits, palaeosystematics is, in addition, at a decisive disadvantage because it can never observe its objects alive, and can therefore only solve its problems with the help of relatively unreliable morphological criteria. It is true that the systematics of recent organisms also satisfies itself mainly with morphological criteria to help it establish the limits of species. However, there is always the possibility, in principle, of testing in important cases that individuals of similar or different appearance actually belong to one or to different reproductive communities by observation of their life in nature or by breeding and crossing experiments. In species with seasonal and sexual dimorphism and those in which the life of the individual contains a metamorphosis, systematics depends upon such methods. But, in paleontology, they cannot be employed. Here systematics can establish the specific limits only with a much lower degree of accuracy than with recently known organisms. It would, however, be completely false to deduce from this, as is sometimes done, that paleontological systematics operates with other concepts (e.g., a different species-concept) and other methods. It differs from the systematics applicable to recent animal forms only in the lesser degree of certainty and accuracy with which it is able to apply itself.

This applies to inquiry into specific limits just as it does to establishing the degree of the phylogenetic relationship between species. If the purpose of systematics does not consist exclusively of conducting a survey of the animal forms which have existed on the earth at any time, then paleontology must also try to relate its objects to the phylogenetic system of recent organisms—that is, to include them in this system. But this can be meaningful and fruitful only if the limits of the knowledge it can supply are known very precisely and are clearly expressed in each particular case.

Subject to these conditions, the value of fossil finds lies in enabling one to interpret character agreements in recent species when this cannot be done solely from a knowledge of these recent forms. There are, in the recent fauna, monophyletic groups which agree in certainly derivative (apomorph) characters with other diverse groups which are just as surely monophyletic. Some of these agreements must therefore rest on convergence. But it is often impossible to decide with certainty which of these agreements are based on convergence and which are to be considered as true synapomorphy. The possibility of decision in such cases depends on a knowledge of the sequence in which the characters in question evolved. This is sometimes clarified by fossils. An example of this kind is supplied by the sea urchins (Echinoidea).

The Cidaroidea, which are shown to be a monophyletic group by their peculiar spine formation, agree completely with most other recent sea urchins in their possession of a rigid corona. The more primitive expression of this character, a flexible corona, is present only in the Echinothuridae. On the other hand, the Echinothuridae agree completely, in their possession of external gills, with the sea urchins which do not belong to the Cidaroidea. This is likewise a derivative character. This character distribution allows no decision on the question of whether the Cidaroidea or the Echinothuridae are more closely related to the bulk of recent sea urchins. One of the two derivative characters, the external gills or the rigid corona, must thus have evolved through convergence at least twice independently. The oldest fossil (Cidaroidea, which are shown to belong to this group by their spine formation, possess a flexible corona. This is decisive evidence that the ridigity of the corona in recent Cidaroidean and in the remaining recent sea urchins (except the Echinothuridae) has evolved through convergence. Concerning the external gills, there are no reasons to suggest convergent evolution. Their presence in recent sea urchins which do not belong to the Cidaroidea may therefore be regarded as synapomorphy. However, it must also be said that they have often been lost secondarily. In other cases, only fossil finds make it possible to establish which expression of a character should be regarded as plesiomorph in a group and which as apomorph.

The importance of fossils thus lies not so much in the fact that they reduce the morphological gap between different monophyletic groups of the recent fauna but in that they help to make it possible to decide the categories of resemblance (symplesiomorphy, synapomorphy, or convergence) to which particular agreements of character belong.

Still greater is the value of fossils for determining the age of animal groups. But in this context it should be realized that age determinations have a meaning only in monophyletic groups, since only they have a history of their own (see above). It can be difficult, however, to demonstrate the relationship of a fossil to a given monophyletic group of animals. As has been shown above, heterobathmy of characters is characteristic for nearly related monophyletic groups. Therefore, it often happens that one of two sister-groups can be established as a monophyletic group only by a few apomorph characters which are difficult to verify or only present at a particular stage of metamorphosis. For the distinction of the two groups and the identification of the species belonging to them, this has no significance, because plesiomorph characters can also be employed for diagnosis, though they must be left out of consideration in establishing the monophyly of a group. One can, for instance, recognize at once that a recent arthropod species belongs to the Myriopoda from its possession of homonomous body segmentation with jointed appendages on more than three of its trunk segments, although both are plesiomorph characters and cannot be used to justify the supposition that the Myriopoda are monophyletic. But this is not the case with fossils. One cannot assume without qualification that fossils, especially from the early Palaeozoic, belong to the Myriopoda if they possess a homonomous segmentation and jointed appendages on more than three trunk segments. Both are plesiomorph characters which must also have been present in the common ancestors of the Insecta and Myriopoda. To demonstrate that fossils in fact belong to the Myriopoda, one must demonstrate in them those apomorphy characters in the ground plan of the group which suggest its monophyly—that is, the absence of ocelli and compound eyes. Such demonstration is often very difficult, since these characters are not preserved for us in the fossils. If, in this case, one proceeds uncritically, and classifies fossils on the basis of plesiomorph characters which suffice as diagnostic characters for the certain recognition of all recent species of a monophyletic group, then it can happen that the group will become a paraphyletic group solely through its acquisition of fossils. This can then become the source of all the errors which necessarily arise if one compares monophyletic and paraphyletic groups with one another in phylogenetic studies (see above).

When, however, it has been firmly established that a fossil belongs to a given monophyletic group, that fossil can then be of importance not only for determining the minimum age of the group to which it belongs, but also for determining the minimum age of related groups, of which no fossil finds are available. The existence of *Rhyniella praecursor* in the Devonian not only proves that Collembola, the group to which *Rhyniella* belongs, already occurred then, but from our relatively certain knowledge of the phylogenetic relationships of the principal monophyletic groups of insects it follows that at the same period the Protura, Diplura, and Ectognatha must also have existed, although, of course, not in the form of their present-day progeny.

In determining the age of animal groups, another factor should be considered as well. In the history of a monophyletic group of animals, there are two points of time which are especially important (figure 13.4): one is at the time at which the group in question was separated from its sister-group by the splitting of their common ancestor (age of origin), and the other the time at which the last common ancestral species of all recent species of the group ceased to exist as a homogeneous reproductive community (age of division). The distinction between these two points of time is especially important in those groups whose recent species are distinguished from species of other groups by their agreement in a large number of derivative characters. One must assume that these characters were already present in the last common ancestral species to whose progeny they have been transmitted unchanged or, in part, further developed. These characters must have evolved in the period between the two named points of time.

Speculation upon the age of a particular group of animals can have three appropriate but different meanings. The following may be intended: (a) When did the last common ancestral species of all the recent species of this group which have inherited their derivative characters from it, live? (question about the group's age of division). (b) When was the group separated from its sister group? (question about the group's age of origin). (c) When, in the period

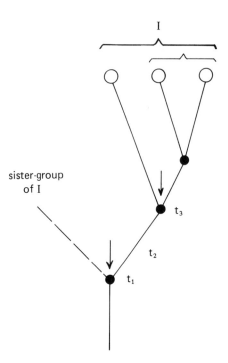

Figure 13.4 The three different meanings of questions about the "age" of an animal group. t_1: age of origin (separation of group I from its sister-group); t_2: first appearance of the "typical" characters of group I; t_3: age of division (last common ancestor of all recent species of group I).

between these two points of time, did species for the first time occur with the characters which justify their ascription to the "type" represented by the recent species?

It is seldom clear which of these three essentially different questions is intended when questions are asked about the age of fleas, lice, or other animal groups. This fact, in conjunction with the custom of seeing in phylogenesis mainly the emergence of particular "types" or "Baupläne" whose delimitation is dependent on subjective criteria, is the cause of endless and fruitless debate on the question of whether certain fossils should be considered "reptiles," "birds," "mammals," or "humans," and when these groups evolved.

It might seem that questions about the age of animal groups lie outside the field of systematics. But this is not the case. The examples quoted should have shown that answering these questions has the same significance as systematically classifying fossils in particular groups, and that the meaning of an answer depends on the classificatory principle used in forming them.

The age of animal groups also has yet another significance for phylogenetic systematics, under some circumstances. It has been said above that the phylogeny diagram and the hierarchical system are closely corresponding kinds of presentation whose content is one and the same. The phylogeny tree presents, as the most important factor, the time dimension in which the degree of phylogenetic relationship between species or monophyletic groups of species is expressed by the sequence in which they evolved from each common ancestral species (i.e., recency of common ancestry); in a hierarchical system this is shown by the sequence of subordination in the group categories. It is a justifiable aim to perfect the phylogeny diagram by giving, not only the relative sequence of origin of the monophyletic groups, but also the actual time of their origin. This detail of a perfected phylogeny diagram can also be reproduced in a hierarchical system by means of the absolute rank of its group categories. In a hierarchical system, not only are the names of the monophyletic groups quoted but they are also given a specific absolute rank (class, order, family, etc.). Some clear-sighted sighted authors (e.g., Simpson, 16) have quite correctly realized that the absolute rank which is attributed to a given group (e.g., family) does not generally mean that this group can be compared with any other of the same rank in any particular respect. Only within one and the same sequence of subordination is it true that the lower ranks show a higher degree of phylogenetic relationship than the higher. This situation can be accepted without injury to the basic principles of phylogenetic systematics. It could be changed, without injury to these principles, only if the absolute rank of categories was linked to their time of origin, just as in geology the sequence of strata in different continents is made comparable by its correlation with specific periods of the earth's history (e.g., Triassic, Jurassic, Cretaceous). Some authors (e.g., Stammer, 18) think that one must take into account, when according absolute rank to systematic groups, their different rates of evolution which have led to greater or lesser morphological "differentiation." But it needs little reflection to see that this is incompatible with the theoretical foundations of phylogenetic systematics and necessarily leads to pseudo-phylogenetic systems. This should already have been shown by the fact that sister-groups must have the same rank in a phylogenetic system, entirely without regard for the way in which this rank is established; for sister-groups can, of course, have morphologically unfolded (i.e., diverged from the form of their common ancestors) with completely different rates of evolution.

Biological systematics can no more do without a theoretical foundation for its work than can any other science. The theory of phylogenetic systematics is a comprehensive and complex edifice of thought, which here can only be touched upon lightly, even in its most important aspects. In this edifice there is, as always, a logical arrangement of individual problems. In critical expositions, this logical order must be observed. It is not permissible, as sometimes happens, to confuse the critique for answering logically subordinate questions with the critique concerning the principles of the phylogenetic system. From a thoroughgoing theory of phylogenetic systematics, there arise necessarily some unexpected demands on the practical work of the systematist. If the theory as such is accepted in principle it is not permissible to refuse these demands or leave them unconsidered merely because they conflict with certain customary methods obtaining at the time when systematics had no theory. There are many problems in biology whose solution presupposes knowledge of the phylogenetic relationship of one or many species—that is, a phylogenetic system of one or more groups of animals. To avoid false conclusions it is therefore especially important that every author of a system should make it easy to recognize whether, or rather to what extent, his system ought to meet the demands imposed by the theory of phylogenetic systematics. But even when these demands should be met in a system, according to the expressed wish of its author, there will always be differences of opinion over the actual relationships of some species or species groups. The person who requires a phylogenetic system as a premise for his own work will then have to decide on which side lie the better arguments; the criteria for this must again emerge from the theory of phylogenetic systematics. Differences of opinion on matters of fact are not, however, a special defect of phylogenetic systematics but the universal mark of every science.

It is impossible in a short chapter to treat even sketchily the extensive field of phylogenetic systematics with all of the questions of detail which are important for the practical work of the systematist. A more comprehensive account in Spanish and another in English with detailed bibliography, are in the course of preparation. Excellent introductions on its theoretical and methodological foundations with many critical comments on recent systematic works are given in the writings of Günther (4, 6). A valuable study on the philosophical foundations of biological systematics has very recently been published by Kiriakoff (14).

REFERENCES

1. Bigelow, R. S. 1956. Monophyletic classification and evolution. *System Zool.* 5:145–46.

2. Blackwelder, R. A., Alexander, R. D., and Blair, W. F. 1962. The data of classification, a symposium. *System. Zool.* 11:49–84 [Critical review by Günther, K., in *Ber. Wiss. Biol.*, 1963].

3. Gregg, J. R. 1954. *The Language of Taxonomy. An Application of Symbolic Logics to the Study of Classificatory Systems.* New York, Columbia University Press.

4. Günther, K. 1956. Systematik und Stammesgeschichte der Tiere 1939–1953. *Fortschr. Zool.* (N.F.) 10:33–278.

5. ———. 1959. Die Tetrigidae von Madagaskar, mit einer Erörterung ihrer zoogeographischen Beziehungen und ihrer phylogenetischen Verwandtschaften. *Abhandl. Ber. Staatl. Mus. Tierkde Dresden* 24:3–56.

6. ———. 1962. Systematik und Stammesgeschichte der Tiere 1954–1959. *Fortschr. Zool.* (N.F.) 14:268–547.

7. Hennig, W. 1950. *Grundzüge einer Theorie der Phylogenetischen Systematik.* Berlin, Deutscher Zentralverlag.

8. ———. 1953. Kritische Bemerkungen zum phylogenetischen System der Insekten. *Beitr. Entomol.* 3:1–85.

9. ———. 1954. Flügelgeäder und System der Dipteren, unter Berückschtigung der aus dem Mesozoikum beschriebenen Fossilien. *Beitr. Entomol.* 4: 245–388.

10. ———. 1955. Meinungsverschiedenheiten über das System der niederen Insketen. *Zool. Ans.* 155:21–30.

11. ———. 1957. Systematik und Phylogenese. *Ber. Hundertjahrs. Deut. Entomol. Ges.* (Berlin, 1956), 50–71.

12. ———. 1960. Dipterenfauna von Neuseeland als systematisches und tiergeographisches Problem. *Beitr. Entomol.* 10:221–329.

13. ———. 1962. Veränderungen am phylogenetischen System der Insekten seit 1953. *Ber. Wandervers. Deut. Entomol.* (Berlin, 1951) 9:29–42.

14. Kiriakoff, S. G. 1960. Les Fondaments philosophiques de la systématique biologique. *Natuurw. Tijdschr.* (Ghent) 42:35–57.

15. Müller, A. H. 1955. *Der Grossablaus der stammesgeschichtlichen Entwicklung* (Gustav Fisher, Jena).

16. Simpson, G. G. 1937. Supra-specific variation in nature and in classification from the view-point of paleontology. *Am. Naturalist* 71:236–67.

17. Sokal, R. R. 1963. Typology and empiricism in taxonomy *J. Theoret. Biol.* 3:230–67; critical review by Günther, K., in *Ber. Wiss. Biol.*, A, 191, 70.

18. Stammer, H. J. 1961. Neue Wege der Insektensystematik. *Verhandl. Intern. Kongr. Entomol.*, Wien, 1950, 1:1–7.

19. Takhtajan, A. 1959. *Die Evolution der Angiospermen* (Gustav Fischer, Jena).

20. Verheyen, R. A. 1961. A new classification for the non-passerine birds of the world. *Bull. Inst. Roy. Sci. Nat. Belg.* 37 (27).

21. Wilson, H. F., and Doner, M. H. 1937. The historical development of insect classification. (Planographed by John S. Swift Co., Inc., St. Louis, Chicago, New York, Indianapolis.)

14 Biological Classification: Toward a Synthesis of Opposing Methodologies

Ernst Mayr

Currently a controversy is raging as to which of three competing methodologies of biological classification is the best: phenetics, cladistics, or evolutionary classification. The merits and seeming deficiencies of the three approaches are analyzed. Since classifying is a multiple-step procedure, it is suggested that the best components of the three methods be used at each step. By such a synthetic approach, classifications can be constructed that are equally suited as the basis of generalizations and as an index to information storage and retrieval systems.

For nearly a century after the publication of Darwin's *Origin* (1) no well-defined schools of classifiers were recognizable. There were no competing methodologies. Taxonomists were unanimous in their endeavor to establish classifications that would reflect "degree of relationship." What differences there were among competing classifications concerned the number and kinds of characters that were used, whether or not an author accepted the principle of recapitulation, whether he attempted to "base his classification on phylogeny," and to what extent he used the fossil record (2). As a result of a lack of methodology, radically different classifications were sometimes proposed for the same group of organisms; also new classifications were introduced without any adequate justification except for the claim that they were "better." Dissatisfaction with such arbitrariness and seeming absence of any carefully thought out methodology led in the 1950s and 1960s to the establishment of two new schools of taxonomy, numerical phenetics and cladistics, and to a more explicit articulation of Darwin's methodology, now referred to as evolutionary classification.

THE MAJOR SCHOOLS OF TAXONOMY

Numerical Phenetics

From the earliest preliterary days, organisms were grouped into classes by their outward appearance, into grasses, birds, butterflies, snails, and others. Such grouping "by inspection" is the expressly stated or unspoken starting

From *Science*, 1981, 510–516.

point of virtually all systems of classification. Any classification incorporating the method of grouping taxa by similarity is, to that extent, phenetic.

In the 1950s to 1960s several investigators went one step further and suggested that classifications be based exclusively on "overall similarity." They also proposed, in order to make the method more objective, that every character be given equal weight, even though this would require the use of large numbers of characters (preferably well over a hundred). In order to reduce the values of so many characters to a single measure of "overall similarity," each character is to be recorded in numerical form. Finally, the clustering of species and their taxonomic distance from each other is to be calculated by the use of algorithms that operationally manipulate characters in certain ways, usually with the help of computers. The resulting diagram of relationship is called a phenogram. The calculated phenetic distances can be converted directly into a classification.

The fullest statement of this methodology and its underlying conceptualization was provided by Sokal and Sneath (3). They called their approach "numerical taxonomy," a somewhat misleading designation, since numerical methods, including numerical weighting, can be and have been applied to entirely different approaches to classification. The term "numerical phenetics" is now usually applied to this school. This has introduced some ambiguity since some authors have used the term "phenetic" broadly, applying it to any approach making use of the "similarity" of species and other taxa, while to the strict numerical pheneticists the term "phenetic" means the "theory-free" use of unweighted characters.

Cladistics (or Cladism)

This method of classification (4), the first comprehensive statement of which was published in 1950 by Hennig (5), bases classifications exclusively on genealogy, that is, on the branching pattern of phylogeny. For the cladist phylogeny consists of a sequence of dichotomies (6), each representing the splitting of a parental species into two daughter species; the ancestral species ceases to exist at the time of the dichotomy; sister groups must be given the same categorical rank; and the ancestral species together with all of its descendants must be included in a single "holophyletic" taxon.

Evolutionary Classification

Phenetics and cladistics were proposed in the endeavor to replace the methodology of classification that had prevailed ever since Darwin and that was variously designated as the "traditional" or the "evolutionary" school, which bases its classifications on observed similarities and differences among groups of organisms, evaluated in the light of their inferred evolutionary history (7). The evolutionary school includes in the analysis all available attributes of these organisms, their correlations, ecological stations, and patterns of distributions and attempts to reflect both of the major evolutionary processes, branching, and the subsequent diverging of the branches (clades). This school follows Darwin (and agrees in this point with the cladists) that classification must be based on genealogy and also agrees with Darwin (in contrast to the cladists) "that genealogy by itself does not give classification" (8).

The results of the evolutionary analysis are incorporated in a diagram, called a phylogram, which records both the branching points and the degrees of subsequent divergence. The method of inferring genealogical relationship with the help of taxonomic characters, as it was first carried out by Darwin, is an application of the hypothetico-deductive approach. Presumed relationships have to be tested again and again with the help of new characters, and the new evidence frequently leads to a revision of the influences on relationship. This method is not circular (9) as has sometimes been suggested.

IS THERE A BEST WAY TO CLASSIFY?

Each of the three approaches to classification—phenetics, cladistics, and evolutionary classification—has virtues and weaknesses. The ideal classification would be one that would meet best as many as possible of the generally acknowledged objectives of a classification.

A biological classification, like any other, must serve as the basis of a convenient information storage and retrieval system. Since all three theories produce hierarchical systems, containing nested sets of subordinated taxa, they permit the following of information up and down the phyletic tree. But this is where the agreement among the three methods ends. Purely phenetic systems, derived from a single set of arbitrarily chosen characters, sometimes provide only low retrieval capacity as soon as other sets of characters are used. The effectiveness of the phenetic method could be improved by careful choice of selected characters. However, the method would then no longer be "automatic," because any selection of characters amounts to weighting.

Cladists use only as much information for the construction of the classification as is contained in the cladogram. They convert cladograms, quite unaltered, into classifications, only when the cladograms are strictly dichotomous. Even though cladists lose much information by this simplistic approach, the information on lines of descent can be read off their classifications directly. However, a neglect of all ancestral-descendant information reduces the heuristic value of their classifications. By contrast, since evolutionary taxonomists incorporate a great deal more information in their classifications than do the cladists, they cannot express all of it directly in the names and ranking of the taxa in their classifications. Therefore, they consider a classification simply to be an ordered index that refers them to the information that is stored elsewhere (in the detailed taxonomic treatments).

A far more important function of a classification, even though largely compatible with the informational one, is that it establishes groupings about which generalizations can be made. To the extent that classifications are explicitly based on the theory of common descent with modification, they postulate that members of a taxon share a common heritage and thus will have many characteristics in common. Such classifications, therefore, have great heuristic value in all comparative studies. The validity of specific observations can be generalized by testing them against other taxa in the system or against other kinds of characters (10–12).

Pheneticists, as well as cladists, have claimed that their methods of constructing classifications are nonarbitrary, automatic, and repeatable. The criticisms of these methods over the last fifteen years (13) have shown, however, that these claims cannot be substantiated. It is becoming increasingly evident that a one-sided methodology cannot achieve all the above-listed objectives of a good classification.

The silent assumption in the methodologies of phenetics and cladistics is that classification is essentially a single-step procedure: clustering by similarity in phenetics and establishment of branching patterns in cladistics. Actually a classification follows a sequence of steps, and different methods and concepts are pertinent at each of the consecutive steps. It seems to me that we might arrive at a less vulnerable methodology by developing the best method for each step consecutively. Perhaps the steps could eventually be combined in a single algorithm. In the meantime, their separate discussion contributes to the clarification of the various aspects of the classifying process.

ESTABLISHMENT OF SIMILARITY CLASSES

The first step is the grouping of species and genera by "inspection." that is, by a phenetic procedure. (I use "phenetic" in the broadest sense, not in the narrow one of numerical phenetics.) All of classifying consists of, or at least begins with, the establishment of similarity classes, such as a preliminary grouping of plants into trees, shrubs, herbs, and grasses. The reason why the method is so often successful is simply that—other things being equal— descendants of a common ancestor tend to be more similar to each other than they are to species that do not share immediate common descent. The method is thus excellent in principle. Numerical phenetics has nevertheless proved to be largely unsuccessful because (1) claims, such as "results are objective and strictly repeatable," were not always justifiable since in practice different results are obtained when different characters are chosen or different programs of computation are used; (2) the method was inconsistent in its claim of objectivity since subjective biological criteria were used in the assigning of variants (for example, sexes, age classes, and morphs) to "operational taxonomic units" (OTUs); and, most important, the method insisted on the equal weighting of all characters.

It is now evident that no computing method exists that can determine "true similarity" from a set of arbitrarily chosen characters. So-called similarity is a complex phenomenon that is not necessarily closely correlated with common descent, since similarity is often due to convergence. Most major improvements in plant and animal classifications have been due to the discovery of such convergence (14).

Different types of characters—morphological characters, chromosomal differences, enzyme genes, regulatory genes, and DNA matching—may lead to rather different grouping. Different stages in the life cycle may result in different groupings.

The ideal of phenetics has always been to discover a measure of total (overall) similarity. Since it is now evident that this cannot be achieved on the basis of a set of arbitrarily chosen characters, the question has been asked whether there is not a method to measure degrees of difference of the genotype as a whole. Improvements in the method of DNA hybridization offer hope that this method might give realistic classifications on a phenetic basis, at least up to the level of orders (15). The larger the fraction of the nonhybridizing DNA, the less reliable this method is, because it cannot be determined whether the nonmatching DNA is only slightly or drastically different.

TESTING THE NATURALNESS OF TAXA

In the first step of the classifying procedure clusters of species were assembled that seemed to be more similar to each other than to species in other clusters. These clusters are the taxa we recognize tentatively (16). In order to make these clusters conform to evolutionary theory, two, operationally more or less inseparable, tests must be made: (1) determine for all species of a cluster (taxon) whether they are descendants of the nearest common ancestor and (2) connect the taxa by a branching tree of common descent, that is, construct a cladogram. An indispensable preliminary of this testing is an analysis of the characters used to establish the similarity clusters.

Character Analysis

A careful analysis shows almost invariably that some characters are better clues to relationship (have greater weight) than others. The fewer the number of available characters, the more carefully the weighting must be done. This weighting is one of the most controversial aspects of the classifying procedure. Investigators who come to systematics from the outside, say from mathematics, or who are beginners tend to demand objective or quantitative methods of weighting. There are such methods, principally ones based on the covariation of characters, but they are not nearly as informative as methods based on the biological evaluation of characters (17). But such an evaluation requires an understanding of many aspects of the to-be-classified group (that is, its life history, the inferred selection pressures to which it is exposed, and its evolutionary history) that may not be available to an outsider. This creates a genuine dilemma. If strictly taxometric methods were available that would produce satisfactory weighting, everyone would surely prefer them to weighting based on experience and biological knowledge. But so far such methods are still in their infancy.

The greatest difficulty for a purely phenetic method, indeed for any method of classification, is the discordance (noncongruence) of different sets of characters. Entirely different classifications may result from the use of characters of different stages of the life cycle as, for instance, larval versus adult characters. In a study of species of bees, Michener (18) obtained four different classifications when he sorted them into similarity classes on the basis of the characters of (1) larvae, (2) pupae, (3) the external morphology of the adults, and (4) male genitalic structures. Phenetic delimitation of taxa unavoidably necessitates a great deal of decision making on the use and weighting of characters. Often, when new sets of characters become available, their use may lead to a new delimitation of taxa or to a change in ranking.

Determination of the Genealogy

Each group (taxon) tentatively established by the phenetic method is, so to speak, a hypothesis as to common descent, the validity of which must be tested. Is the delimited taxon truly monophyletic (19)? Are the species included in this taxon nearest relatives (descendants of the nearest common ancestor)? Have all species been excluded that are only superficially or convergently similar?

Methods to answer these questions have been in use since the days of Darwin, particularly the testing of the homology of critical characteristics of the included species. However, Hennig (5) was the first to articulate such methods explicitly, and these have been modified by some of his followers. These methods can be designated as the cladistic analysis.

Such an analysis involves first the partitioning of the joint characters of a group into ancestral ("plesiomorph" in Hennig's terminology) characters and derived ("apomorph") characters, that is, characters restricted to the descendants of the putative nearest common ancestor (20). The joint possession of homologous derived characters proves the common ancestry of a given set of species. A character is derived in relation to the ancestral condition of the character. The end product of such a cladistic character analysis is a cladogram, that is, a diagram (dendrogram) of the branching points of the phylogeny.

Although this procedure sounds simple, numerous practical difficulties have been pointed out (21, 22). Very often the branching points are inferred by way of single or very few characters and are affected by all the weaknesses of single character classifications. More serious are two other difficulties.

1. *Polarity*. A derived character is often simpler or less specialized than the ancestral condition. For this reason it can be difficult to determine polarity in a transformation series of characters, that is, to determine which end of the series is ancestral. Tattersall and Eldredge (23) stressed that "in practice it is hard, even impossible, to marshall a strong, logical argument for a given polarity for many characters in a given group." Are they primitive (ancestral) or derived? Much of the controversy concerning the phylogeny of the invertebrates, for instance, is due to differences of opinion concerning polarity. Hennig tried to elaborate methods for determining polarity but, as others (24,

25) have shown, with rather indifferent success. Since characters come and go in phyletic lines and since there is much convergence, the problem of polarity can rarely be solved unequivocally. There are three best types of evidence for polarity reconstruction. First is the fossil record. Although primitiveness and apparent ancientness are not correlated in every case, nevertheless as Simpson (26) stressed, "for any group with even a fair fossil record there is seldom any doubt that characters usual or shared by older members are almost always more primitive than those of later members." Second is sequential constraints. Consecutive chromosomal inversions (as in *Drosophila*) or sets of amino acid replacements (and presumably certain other molecular events) form definite sequences. Which end of the sequence is the beginning can usually not be read off from the sequence itself, but additional information (polarity of other character chains, geographical distribution, and the like) often permits an un-equivocal determination of the polarity. Third is the reconstruction of the presumed evolutionary pathway. This can sometimes be done by studying evidence for adaptive shifts, the invasion of new competitors or the extinction of old ones, the behavior of correlated characters, and other biological evidence (see ref. 11, pp. 886–887; also ref. 24). Particular difficulties are posed when the polarity is reversed in the course of evolution, as documented in the fossil record.

2. *Kinds of derived characters.* Two taxa may resemble each other in a given character for one of three reasons: because the character existed already in the ancestry of the two groups before the evolution of the nearest common ancestor (symplesiomorphy in Hennig's terminology), because it originated in the common ancestor and is shared by all of his descendants (homologous apomorphy or synapomorphy), or because it originated independently by convergence in several descendant groups (nonhomologous or convergent apomorphy) (27). Since, according to the cladistic method, sister groups are recognized by the possession of synapomorphies, convergence poses a major problem. How are we to distinguish between homologous and convergent apomorphies? Hennig was fully aware of the critical importance of this problem, but it has been quietly ignored by many of his followers. Both grebes and loons, two orders of diving birds, have a prominent spur on the knee and were therefore called sister groups by one cladist. However, other anatomical and biochemical differences between the two taxa indicate that the shared derived feature was acquired by convergence. The reliability of the determination of mono-phyly of a group depends to a large extent on the care that is taken in discriminating between these two classes of shared apomorphy (11, pp. 880–890).

There is a third class of derived characters, so-called autapomorphies, which are characters that were acquired by and are restricted to a phyletic line after it branched off from its sister group.

The pheneticists do not undertake a character analysis. Cladists and evolutionary taxonomists agree with each other in principle on the importance of a careful character analysis. They disagree, however, fundamentally in how to use the findings of the character analysis in the construction of classifications, particularly the ranking procedure.

THE CONSTRUCTION OF A CLASSIFICATION

Cladistic Classification

Cladists convert the cladogram directly into a cladistic classification. In such a classification taxa are delimited exclusively by holophyly, that is, by the possession of a common ancestor, rather than by a combination of genealogy and degree of divergence (19). This results in such incongruous combinations as a taxon containing only crocodiles and birds, or one containing only lice and one family of Mallophaga.

Taxa based exclusively on genealogy are of limited use in most biological comparisons. Since, as Hull (28) pointed out, cladists really classify characters rather than organisms, they have to make the arbitrary assumption that new apomorph characters originate whenever a line branches from its sister line. This is unlikely in most cases. Surely the reptilian species that originated the avian lineage lacked any of the flight specializations characteristic of modern birds, except perhaps the feathers (29).

Two principles govern the conversion of a cladogram into a cladistic classification: 1) all branchings are bifurcations that give rise to two sister groups, and 2) branchings are usually connected with a change in categorical rank. Cladistic classifications are only representations of branching patterns, with complete disregard of evolutionary divergence, ancestor-descendant relationships, and the information content of autapomorph characters. Because these aspects of evolutionary change are neglected, the cladistic method of classification "either results in lumping very similar forms (parasites and their relatives) or in recognizing a multitude of taxa (perhaps also of other categories) regardless of the extreme similarity of some of them. Such simplistic procedures do violence to most biological attributes other than the pattern of the cladistic branching system, as well as to the function of a classification for convenient information transmittal and storage," as Michener remarked (18).

These objections show that the methodology of cladistic classification is not satisfactory. Anyone familiar with the history of taxonomy is strangely reminded of the principles of Aristotelian logical division when encountering cladistic classifications with their rigid dichotomies, the mandate that every taxon must have a sister group, and the principle of a straight-line hierarchy.

There has been much argument over the relationship between classification and phylogeny (30). Both cladists and evolutionary taxonomists agree that all members of a taxon must have a common ancestor. A phylogenetic analysis, and in particular a clear separation of homologous apomorphies from convergences, is a necessary component of the classifying procedure. Classificatory analysis often leads to new inferences on phylogeny, and new insights on phylogeny may necessitate changes in classification. These interactions are not in the least circular (9).

It is quite unnecessary in most cases to know the exact species that was the common ancestor of two diverging phyletic lines. An inability to specify such an ancestral species has rarely impeded paleontological research (31, 32). For instance, it is of little importance whether *Archaeopteryx* was the first real ancestor of modern birds or some other similar species or genus. What is important to know is whether birds evolved from lizard-like, crocodile-like, or dinosaur-like ancestors. If a reasonably good fossil record is available, it is usually possible, by the backward tracing of evolutionary trends and by the backward projection of divergent phyletic lines, to reconstruct a reasonably convincing facsimile of the representative of a phyletic line at an earlier time.

Simpson (32) has provided us with cogent arguments about why it is not permissible to reject information from the fossil record under the pretext that it fails to give the phylogenetic connections between fossil and recent taxa with absolute certainty. Hence, there is no merit in the suggestion to construct separate classifications for recent and for fossil organisms. After all, fossil species belong to the same tree of descent as living species. Indeed, enough evidence usually becomes available through a careful character analysis to permit relatively robust inferences on the most probable phylogeny. A number of recent endeavors have been made to develop a cladistic methodology that is quantitative and automatic. New methods in this area are published in rapid succession and it would seem too early to determine which is most successful and freest of possible flaws (33).

Evolutionary Classification

The taxonomic task of the cladist is completed with the cladistic character analysis. The genealogy gives him the classification directly, since for him classification is nothing but genealogy. The evolutionary taxonomist carries the analysis one step further. He is interested not only in branching, but, like Darwin, also in the subsequent fate of each branch. In particular, he undertakes a comparative study of the phyletic divergence of all evolutionary lineages, since the evolutionary history of sister groups is often strikingly different. Among two related groups, derived from the same nearest common ancestor, one may hardly differ from the ancestral group, while the other may have entered a new adaptive zone and evolved into a novel type. Even though they are sister groups in the terminology of cladistics, they may deserve different categorical rank, because their biological characteristics differ to such an extent as to affect any comparative study. The importance of this consideration was stated by Darwin (1, p. 420): "I believe that the *arrangement* of the groups within each class, in due subordination and relation to the other groups, must be strictly genealogical in order to be natural, but that the *amount* of difference in the several branches or groups, though allied in the same degree in blood to their common progenitor, may differ greatly, being due to the different degrees of modification which they have undergone, and this is expressed by the forms being ranked under different genera, families, sections or orders." Darwin refers then to a diagram of three Silurian genera that have modern descendants; one has not even changed generically, but the other two have become distinct orders, one with three and the other with two families.

The question as to what extent an analysis of degrees of divergence is possible, is still debated. The cladist makes only "horizontal" comparisons, cataloging the synapomorphies of sister groups. The evolutionary taxonomist, however, also makes use of derived characters that are restricted to a single line of descent, so-called autapomorph characters (figure 14.1), which are apomorph characters restricted to a single sister group. The importance of autapomorphy is well illustrated by a comparison of birds with their sister group (34). Birds originated from that branch of the reptiles, the Archosauria, which also gave rise to the pterodactyls, dinosaurs, and crocodilians. The crocodilians are the sister group of the birds among living organisms: a stem group of archosaurians represents the common ancestry of birds and crocodilians. Although birds and crocodilians share a number of synapomorphies that originated after the archosaurian line had branched off from the other reptilian lines, nevertheless crocodilians are on the whole very similar to other reptiles, that is, they have developed relatively few autapomorph characters. They represent the reptilian "grade," as many morphologists call it. Birds, by contrast, have acquired a vast array of new autapomorph characters in connection with their shift to aerial living. Whenever a clade (phyletic lineage) enters a new adaptive zone that leads to a drastic reorganization of the clade, greater taxonomic weight may have to be assigned to the resulting transformation than to the proximity of joint ancestry. The cladist virtually ignores this ecological component of evolution.

The main difference between cladists and evolutionary taxonomists, thus, is in the treatment of autapomorph characters. Instead of automatically giving sister groups the same rank, the evolutionary taxonomist ranks them by considering the relative weight of their autapomorphies (figure 14.1). For in-

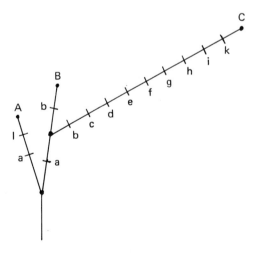

Figure 14.1 Cladogram of taxa A, B, and C. Cladists combine B and C into a single taxon because B and C share the synapomorphy character b. Evolutionary taxonomists separate C from A and B, which they combine, because C differs by many (c through k) autapomorph characters from A and B and shares only one (b) synapomorph character with B.

Ernst Mayr

stance, one of the striking autapomorphies of man (in comparison to his sister group, the chimpanzee) is the possession of Broca's center in the brain, a character that is closely correlated with man's speaking ability. This single character is for most taxonomists of greater weight than various synapomorphous similarities or even identities in man and the apes in certain macromolecules such as hemoglobins and cytochrome c. The particular importance of autapomorphies is that they reflect the occupation of new niches and new adaptive zones that may have greater biological significance than synapomorphies in some of the standard macromolecules.

I agree with Szalay (35) when he says: "The loss of biological knowledge when not using a scheme of ancestor-descendant relationship, I believe, is great. In fact, whereas a sister group relationship may ... tell us a little, a postulated and investigated ancestor-descendant relationship may help explain a previously inexplicable character in terms of its origin and transformation, and subsequently its functional (mechanical) significance." In other words, the analysis of the ancestor-descendant relationships adds a great deal of information that cannot be supplied by the analysis of sister group relationships.

It is sometimes claimed that the analysis of ancestor-descendant relationships lacks the precision of cladistic sister group comparisons. However, as was shown above and as is also emphasized by Hull (36), the cladistic analysis is actually full of uncertainties. The slight possible loss of precision, caused by the use of autapomorphies, is a minor disadvantage in comparison with the advantage of the large amount of additional information thus made available.

The information on autapomorphies permits the conversion of the cladogram into a phylogram. The phylogram differs from the cladogram by the placement of sister groups at different distances from the joint common ancestry (branching point) and by the expression of degree of divergence by different angles. Both of these topological devices can be translated into the respective categorical ranking of sister groups. These methods (37) generally attempt to discover the shortest possible "tree" that is compatible with the data. Yet, anyone familiar with the frequency of evolutionary reversals and of evolutionary opportunism realizes the improbability of the assumption that the tree constructed by this so-called parsimony method corresponds to the actual phylogenetic tree. "To regard [the shortest tree method] as parsimonious completely misconceives the intent and use of parsimony in science" (38).

It is not always immediately evident whether a tree construction algorithm is based on cladists' principles or on the methods of evolutionary classification. If the "special similarity" on which the trees are based is strictly synapomorphies, then the method is cladistic. If autapomorphies are also given strong weight, then the method falls under evolutionary classification.

The particular aspect of the method of evolutionary taxonomy found most unacceptable to cladists is the recognition of "paraphyletic" taxa. A paraphyletic taxon is a holophyletic group from which certain strikingly divergent members have been removed. For instance, the class Reptilia of the standard zoological literature is paraphyletic, because birds and mammals, two strikingly divergent descendants of the same common ancestor of all the Reptilia, are not included. Nevertheless, the traditional class Reptilia is monophyletic, because it consists exclusively of descendants from the common ancestor, even though it excludes birds and mammals owing to the high number of autapomorphies of these classes. The recognition of paraphyletic taxa is particularly useful whenever the recognition of definite grades of evolutionary change is important.

THE RANKING OF TAXA

Once species have been grouped into taxa the next step in the process of biological classification is the construction of a hierarchy of these taxa, the so-called Linnaean hierarchy. The hierarchy is constructed by assigning a definite rank such as family or order to each taxon, subordinating the lower categories to the higher ones. It is a basic weakness of cladistics that it lacks a sensitive method of ranking and simply gives a new rank after each branching point. The evolutionary taxonomist, following Darwin, ranks taxa by the degree of divergence from the common ancestor, often assigning a different rank to sister groups. Rank determination is one of the most difficult and subjective decision processes in classification. One aspect of evolution that causes difficulties is mosaic evolution (39). Rates of divergence of different characters are often drastically different. Conventionally taxa, such as those of vertebrates, are described and delimited on the basis of external morphology and of the skeleton, particularly the locomotory system. When other sets of morphological characters are used (for example, sense organs, reproductive system, central nervous system, or chromosomes), the evidence they provide is sometimes conflicting. The situation can become worse, if molecular characters are also used. The anthropoid genus *Pan* (chimpanzee), for instance, is very similar to *Homo* in molecular characters, but man differs so much from the anthropoid apes in traditional characters (central nervous system and its capacities) and occupation of a highly distinct adaptive zone that Julian Huxley even proposed to raise him to the rank of a separate kingdom—Psychozoa.

It has been suggested that different classifications should be constructed for each kind of character, or at least for morphological and molecular characters. Yet there is already much evidence that the acceptance of several classifications based on different characters would lead to insurmountable complications. But taking all available data into consideration simultaneously, a classification can usually be constructed that can serve conveniently as an all-purpose classification or, as Hennig (5) called it, "a general reference system."

It is usually possible to derive more than one classification from a phylogram, because higher taxa are usually composed of several end points of the phylogram, and different investigators differ by the degree to which they lump such terminal branches into a single higher taxon (40). An example is the phylogram of the higher ferns on which, as Wagner (41) has shown, six different classifications have been founded (figure 14.2), and many more are possible. The extent to which investigators "split" or "lump" higher taxa, thus, is of considerable influence on the classifications they produce.

COMPARISON OF THE THREE MAJOR SCHOOLS

Each school believes that its classification is the "best." Pheneticists as well as cladists claim that their respective methods have also the great merit of giving automatically nonarbitrary results. These claims cannot be substantiated. To be sure, grouping by phenetic characters and determination of holophyly by cladistic analysis are valuable components of the procedure of biological classification. The great deficiency of both phenetics and cladistics is the failure to reflect adequately the past evolutionary history of taxa.

What needs to be emphasized once more is the fact that groups of organisms are the product of evolution and that no classification can hope to be satisfactory that does not take this fact fully into consideration. Both pheneticists and cladists are ambiguous in their attitude toward the evolutionary

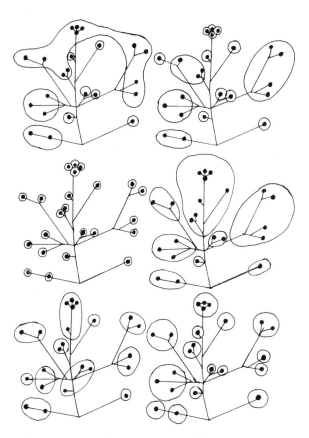

Figure 14.2 Six different possible classifications of ferns, based on the same dendogram. Each filled circle is a genus, and each open circle is a family. The differences are due to which and how many genera are combined to make up the families. (From W. H. Wagner [41, figure 7].)

theory. The pheneticists claim that their approach is completely theory-free, but they nevertheless assume that their method will produce a hierarchy of taxa that corresponds to descent with modification. On the basis of this assumption, they also claim to be "evolutionary taxonomists" (42), but the fact that different phenetic procedures may produce very different classifications and that their procedure is not influenced by evolutionary considerations refutes this assertion. The cladists exclude most of evolutionary theory (for example, inferences on selection pressures, shifts of adaptive zones, evolutionary rates, and rates of evolutionary divergence) from their consideration (43) and tend increasingly not to classify species and taxa, but only taxonomic characters (28) and their origin. The connection with evolutionary principles is exceedingly tenuous in many recent cladistic writings.

By contrast, the evolutionary taxonomists, as indicated by the name of their school and by well-articulated statements of some of its major representatives (7), expressly base their classifications on evolutionary theory. They aim to construct classifications that reflect both of the two major evolutionary processes, branching and divergence (cladogenesis and anagenesis). They make full use of information on shifts into new adaptive zones and rates of evolutionary change and believe that the resulting classifications are a key to a far richer information content.

Although the three schools still seem rather fundamentally in disagreement, as far as the basic principles of classification are concerned, the more moderate representatives have quietly incorporated some of the criteria of the opposing schools, so that the differences among them have been partially obliterated. For instance, Farris's (44) clustering of special similarities is a phenetic method based on the weighting of characters. The evolutionary school uses phenetic criteria to establish similarity classes and to construct a classification, and a cladistic criterion to test the naturalness of taxa. Comparing what McNeill (45) says in favor of phenetics (appropriately modified) and Farris (44) against it, we find that the gap has narrowed. I have no doubt that moderates will be able to develop an eclectic methodology, one that contains a proper balance of phenetics and cladistics that will produce far more "natural classifications" (16) than any one-sided approach that relies exclusively on a single criterion, whether it be overall similarity, parsimony of branching pattern, or what not. Evolutionary taxonomy, from Darwin on, has been characterized by the adoption of an eclectic approach that makes use of similarity, branching pattern, and degree of evolutionary divergence.

CLASSIFICATION AND INFORMATION RETRIEVAL

Biological classifications have two major objectives: to serve as the basis of biological generalizations in all sorts of comparative studies and to serve as the key to an information storage system. Up to this point, I have concentrated on those aspects of classifying that help to secure a sound basis for generalizations. This leaves unanswered the question of whether achievement of this first objective is, or is not, reconcilable with achievement of the second objective. Is the classification that is soundest as a basis of generalizations also most convenient for information retrieval? This, indeed, seems to have been true in most cases I have encountered. However, we can also look at this problem from another side.

It is possible at nearly each of the three major steps in the making of a classification to make a choice between several alternatives. These choices may be scientifically equivalent, but some may be more convenient in aiding information retrieval than others. If we choose one of them, it is not necessarily because the alternatives were "falsified," but rather because the chosen method is "more practical." In this respect, biological classifications are not unique. Scientific theories are nearly always judged by criteria additional to truth or falsity, for instance, by their simplicity or, in mathematics, by their "elegance." Therefore, it can be asserted that convenience in the use of a classification, including its function as key to information retrieval, is not necessarily in conflict with its more purely scientific objectives (46–48).

REFERENCES AND NOTES

1. C. Darwin. 1859. *On the Origin of Species*. London, Murray.

2. For an illuminating survey of the thinking of that period see F. A. Bather, *Proc. Geol. Soc. London*, 83, LXII (1927).

3. R. R. Sokal and P. H. A. Sneath. 1963. *Principles of Numerical Taxonomy*. San Francisco, Freeman. A drastically revised edition was published in 1973.

4. The method was first published under the misleading name "phylogenetic systematics," but since it is based on only a single one (branching) of the various processes of phylogeny, the terms "cladism" or "cladistics" have been substituted and are now widely accepted.

5. W. Hennig's original statement is *Grundzüge einer Theorie der Phylogenetischen Systematik*, Deutscher Zentralverlag, Berlin (1950). A greatly revised second edition (reprinted in 1979) is *Phylogenetic Systematics*, D. D. Davis and R. Zangerl, eds., University of Illinois Press, Urbana, 1966; see also W. Hennig (47). An independent phylogenetic analysis of characters was made by T. P. Maslin, *Syst. Zool.*, I (1952) 49. For an overview of the more significant recent literature see D. Hull (36) and J. S. Farris, *Syst. Zool.* 28 (1979) 483.

6. Some cladists in recent years have relaxed the requirements of strict dichotomy and have permitted tri- and polyfurcations or have quietly abandoned dichotomy by admitting empty internodes in the cladogram. Polyfurcations can be translated into several alternate bifurcations (see J. Felsenstein, *Syst. Zool.* 27 [1978] 27) and this makes the automatic conversion of the cladogram into a classification of sister groups impossible.

7. The classical statement of this theory is to be found in C. Darwin (1, pp. 411–434). G. G. Simpson, *Principles of Animal Taxonomy*, New York, Columbia University Press, 1961, and E. Mayr (48) provide comprehensive modern presentations of this theory. Several critical recent analyses are: W. Bock (11); C. D. Michener (18); *Syst. Zool.* 27 (1978) 112; P. D. Ashlock (12).

8. F. Darwin. 1887. *Life and Letters of Charles Darwin*. London, Murray, vol. 2, p. 247.

9. D. Hull, *Evolution* 21 (1967) 174; see also W. Bock (11).

10. F. E. Warburton. *Syst. Zool.* 16 (1967) 241; W. Bock, ibid. 22 (1973) 375.

11. W. Bock, 1977, in *Major Patterns in Vertebrate Evolution*, M. K. Hecht, P. C. Goody, B. M. Hecht, eds., NATO Advanced Study Institute Series, New York, Plenum, vol. 14:851–895.

12. P. D. Ashlock, *Syst. Zool.* 28 (1979) 441.

13. I shall not, at this time, recount the almost interminable controversies among the three schools. For critiques of phenetics see E. Mayr (48, pp. 203–211), L. A. S. Johnson, *Syst. Zool.* 19 (1970) 203, and D. Hull, *Annu. Rev. Ecol. Syst.* 1 (1970) 19. Some of the weaknesses pointed out by these early critics have been corrected in the 1973 edition of Sokal and Sneath (3) and by J. S. Farris (44). For critiques of cladistics see E. Mayr (21), R. R. Sokal (22), G. G. Simpson (32), D. Hull (36), P. D. Ashlock, *Annu. Rev. Ecol. Syst.* 5 (1974) 81; and L. van Valen (49).

14. A particularly illuminating example is the breaking up of the plant group *Amentiferae*, which has been shown to consist of taxa secondarily adapted for wind pollination (R. F. Thorne, *Brittonia* 25 [1973] 395). Examples among animals of radical reclassifications are the Rodentia, parasitic bees, certain beetle families, and the turbellarians.

15. C. G. Sibley, in preparation.

16. There have been arguments since before the days of Linnaeus about how to determine whether a system, a classification, is "natural." William Whewell, at a time before Darwin had proclaimed his theory of common descent, expressed the then prevailing pragmatic consensus, "The maxim by which all systems professing to be natural must be tested is this: that the arrangement obtained from one set of characters coincides with the arrangement obtained from another set" (W. Whewell, *Philos. Inductive Sci.* 1, 1840, 521). Interestingly, the covariance of characters is still perhaps the best practical test of the goodness of a classification. Since Darwin, of course, that classification is considered most natural that best reflects the inferred evolutionary history of the organisms involved.

17. For a tabulation and analysis of such qualitative methods of weighting, see E. Mayr (48, pp. 220–228).

18. C. D. Michener, *Syst. Zool.* 26 (1977) 32.

19. I use the word "monophyletic" in its traditional sense, as a qualifying adjective of a taxon. Various definitions of monophyletic have been proposed but all of them for the same concept, a qualifying statement concerning a taxon. A taxon is monophyletic if all of its members are derived from the nearest common ancestor (E. Haeckel, 1868, *Natürliche Schöpsungsgeschichte*, Berlin, Reimer). Cladists have attempted to turn the situation upside down by placing all descendants of an ancestor into a taxon. "Monophyletic" thus becomes a qualifying adjective for descent, and a taxon is not recognized by its characteristics but only by its descent. The transfer of such a well-established term as "monophyletic" to an entirely different concept is as unscientific and unacceptable as if someone were to "redefine" mass, energy, or gravity by attaching these terms to entirely new concepts. P. D. Ashlock, *Syst. Zool.* 20 (1971) 63, has proposed the term "holophyletic" for the assemblage of descendants of a common ancestor. See also P. D. Ashlock (12, p. 443).

20. Terms like "apomorph", "synapomorph," "derived", "ancestral", and so forth always refer to characters of taxa at all levels. A genus may have synapomorphies with another genus, and so may an order with another order. It is this applicability of the same criteria for taxa of all ranks that permits the construction of the Linnaean hierarchy.

21. E. Mayr, *Z. Zool. Syst. Evolutions-forsch.* 12 (1974) 94; reprinted in E. Mayr, *Evolution and the Diversity of Life*, Cambridge, Mass., Harvard University Press, 1976, pp. 433–478.

22. R. R. Sokal, *Syst. Zool.* 24 (1975) 257.

23. J. Tattersall and N. Eldredge, *Am. Sci.* 65 (1977) 204.

24. D. S. Peters and W. Gutmann, *Z. Zool. Syst. Evolutionsforsch.* 9 (1971) 237.

25. O. Schindewolf, *Acta Biotheor.* 18 (1968) 273; H. K. Erben, *Verh. Dtsch. Zool. Ges.* 79 (1979) 116.

26. G. G. Simpson (32); see also L. van Valen (49).

27. For a diagram of these three categories of morphological resemblance see Figure 1 in W. Hennig (47).

28. "Cladistic classifications do not represent the order of branching of sister groups, but the order of emergence of unique derived characters." See D. Hull (36).

29. G. G. Simpson, 1953, *The Major Features of Evolution*, New York, Columbia University Press, p. 348, discusses the fallacy of the cladist assumption.

30. Phylogeny is equated by cladists with cladogenesis (branching), while the evolutionary taxonomist subsumes both branching and evolutionary divergence (anagenesis) under phylogeny.

31. C. W. Harper, *J. Paleontol.* 50 (1976) 180.

32. G. G. Simpson, 1975, in *Phylogeny of the Primates*, W. Pluckett and F. S. Szalay, eds., New York, Plenum, pp. 3–19.

33. J. H. Camin and R. R. Sokal, *Evolution* 19 (1965) 311; W. M. Fitch and E. Margoliash, *Science* 155 (1967) 279; W. M. Fitch, in *Major Patterns of Vertebrate Evolution*, M. K. Hecht, P. C. Goody, B. M. Hecht, eds., NATO Advanced Study Institute Series, New York, Plenum, 1977, vol. 14:169–204.

34. There are literally hundreds of cases to illustrate this situation. I use again the classical case of birds and crocodilians because even a nonbiologist will understand the situation if such familiar animals are used. The holophyletic classification of the lice (Anoplura) derived from one of the suborders of the Mallophaga is another particularly instructive example. K. C. Kim and H. W. Ludwig, *Ann. Entomol. Soc. Am.* 71 (1978) 910.

35. F. S. Szalay, *Syst. Zool.*, 26 (1977) 12.

36. D. Hull, ibid., 28 (1979) 416.

37. J. W. Hardin, *Brittonia* 9 (1957) 145; W. H. Wagner, in *Plant Taxonomy: Methods and Principles*, L. Benson, ed., Ronald, New York (1962), pp. 415–417; A. G. Kluge and J. S. Farris, *Syst. Zool.* 18 (1969) 1; J. S. Farris, *Am. Nat.* 106 (1972) 645.

38. L. H. Throckmorton, in *Biosystematics in Agriculture*, J. A. Romberger, ed., New York, Wiley, 1978, p. 237. Others who have questioned the validity of the so-called parsimony principle are M. Ghiselin, *Syst. Zool.* 15 (1966) 214 and W. Bock (11).

39. Unequal rates of evolution for different structures or for any other components of phenotypes or genotypes are designated mosaic evolution.

40. See E. Mayr (48, pp. 238–241) on the differences between splitters and lumpers.

41. W. H. Wagner, 1969, The construction of a classification. In *Systematic Biology*, Publication 1692, National Academy of Sciences, Washington, D.C., pp. 67–90.

42. R. R. Sokal in (22). "I have yet to meet a nonevolutionary taxonomist."

43. Several leading cladists have recently published antiselectionist statements.

44. J. S. Farris, 1977, in *Major Patterns in Vertebrate Evolution*, M. K. Hecht, P. C. Goody, B. M. Hecht, eds., NATO Advanced Study Institute Series, New York, Plenum, vol. 14:823–850.

45. J. McNeill, *Syst. Zool.* 28 (1979) 468.

46. For criteria by which to judge the practical usefulness of biological classifications, see E. Mayr (48, pp. 229–242).

47. W. Hennig, *Annu. Rev. Entomol.* 10 (1965) 97.

48. E. Mayr, 1969, *Principles of Systematic Zoology*, New York, McGraw-Hill.

49. L. van Valen, Evol. Theory 3 (1978) 285.

50. Drafts were read by P. Ashlock, J. Beatty, W. Bock, W. Fink, C. G. Hempel, and D. Hull, to all of whom I am indebted for valuable suggestions and critical comments, not all of which was I able to accept.

15 Contemporary Systematic Philosophies

David L. Hull

During the past decade, taxonomists have been engaged in a controversy over the proper methods and foundations of biological classification. Although methodologically inclined taxonomists had been discussing these issues for years, the emergence of an energetic and vocal school of taxonomists, headed by Sokal and Sneath, increased the urgency of the dispute. This phenetic school of taxonomy had its origins in a series of papers in which several workers attempted to quantify the processes and procedures used by taxonomists to classify organisms. Of special interest was the process of weighting. These early papers give the impression that the primary motivation for the movement was the desire to make taxonomy sufficiently explicit and precise to permit quantification and, hence, the utilization of computers as aids in classification (22, 23, 41, 91, 106, 107, 111, 112). The initial conclusion that these authors seemed to come to was that taxonomy, as it was then being practiced, was too vague, intuitive, and diffuse to permit quantification. Hence, the procedures and foundations of biological classification had to be changed.

The central issue in this dispute, however, has not been quantification but the extremely empirical philosophy of taxonomy which the founders of phenetic taxonomy seemed to be propounding (54, 79). The pheneticists' position on these issues is not easy to characterize because it has undergone extensive development in the last few years. The words have remained the same. Pheneticists still maintain that organisms should be classified according to overall similarity without any a priori weighting. But the intent of these words has changed. However, one thing seems fairly certain. Pheneticists believed that there was something fundamentally wrong with taxonomy as it was being practiced, especially as set out by such evolutionists as Dobzhansky, Mayr, and Simpson. Later, a third group of taxonomists, led by Hennig, Brundin, and Kiriakoff, entered the dispute, appropriating the name "phylogenetic school" for themselves. The evolutionists and the phylogeneticists agree that evolutionary theory must play a central role in taxonomy and that biological classification must have a systematic relation to phylo-

From *Annual Review of Ecology and Systematics*, 1970, 1:19–53.

geny. They disagree only over the precise nature of this relation. For the purpose of this chapter, *evolutionary taxonomy* will refer to the views of the Dobzhansky-Mayr-Simpson school and *phylogenetic taxonomy* will refer to the views of Hennig, Brundin, and Kiriakoff. Together, these two schools will be referred to as *phyleticists* in contrast to the pheneticists.

Although the emphasis of this chapter will be on contemporary systematic philosophies and not on the role of quantification in taxonomy, some of the resistance which phenetic taxonomy met was due to a blanket distaste on the part of some taxonomists for mathematical techniques as such and, in particular, for the pheneticists' attempt to quantify taxonomic judgment (104, 105). When Huxley called for "more measurement" in the *New Systematics* (68), he did not have in mind the processes by which taxonomists judge affinity. It is easy to sympathize with both sides, with the biologists who were less than elated over the prospect of learning all the new, high-powered notations and techniques that were beginning to flood the literature and with the pheneticists whose work was rejected on occasion, not because the particular mathematical techniques suggested were inadequate, but because they were mathematical. Happily, this aspect of the conflict has largely abated, although pockets of resistance still remain. The question is no longer whether to quantify but which are the best methods for quantifying.[1]

Recognition should also be made of the majority of taxonomists who, though they consider themselves mildly evolutionary in outlook, feel that all such disputes over foundations and methodology are idle chatter. Taxonomy is not the kind of thing one has to talk about. One just does it. The closest approximation to a spokesman for this group is R. E. Blackwelder, but he is atypical of the majority for which he speaks since he still advocates essentialism in almost its pristine, Aristotelian form (7–13, 15, 121–123). The inadequacy of essentialism as a philosophical foundation for biological classification has been discussed so extensively that nothing more needs to be said here (63, 65, 83, 86).

Not only will this chapter be limited to the philosophical aspects of the phenetic-phyletic controversy, but also, of the various issues which have been raised, it will deal with only two—the relation of phylogeny to classification and the species problem. Many of the objections raised against evolutionary taxonomy are actually criticisms of the synthetic theory of evolution, rather than of the classifications built upon it. Nor are these criticisms of recent origin. Every objection raised by the pheneticists to evolutionary theory and evolutionary taxonomy can be found in the work of earlier biologists, usually in the writings of the evolutionists themselves. The difference is that the evolutionists are optimistic about the eventual resolution of these difficulties, whereas the pheneticists, in the early years of the school, believed that they were insoluble. When viewed in the context of the development of biology during the past thirty years, phenetic taxonomy does not appear so much a recent insurrection as the culmination of long-standing grievances.

David L. Hull

Soon after the turn of the century, both taxonomy and evolutionary theory had reached a low ebb in the esteem of the rest of the scientific community. Taxonomists seemed to be engaged in a frenzy of splitting and were viewed as nit-picking, skin-sorters, more as quarrelsome old librarians than scientists. Evolutionists had indulged themselves in reconstructing phylogenies in far greater detail and scope than the data and theory warranted and were looked upon as uncritical speculators, more authors of science fiction than science. Among evolutionists themselves, there were controversies. Were the laws of macroevolution different from those of microevolution? Was there such a thing as orthogenesis and, if so, what were the mechanisms for it? At this critical period, Mendel's laws were rediscovered, but instead of clarifying the situation, the birth of modern genetics confused it even further. A whole series of prejudices, conceptual confusions, and peculiarly pernicious terminologies made it seem as if the new genetics conflicted with evolutionary theory. Adding to the intensity of the controversy was the fact that evolutionists tended to be museum and field workers, whereas geneticists were, by and large, experimentalists at home in the laboratory. It was in this setting that the synthetic theory of evolution and the New Systematics had their inception.

The initial impetus for the rebirth of evolutionary theory was Fisher's *The Genetical Theory of Natural Selection* (49), followed by similar works by Haldane (58) and Wright (130, 131). In these works it was shown that a mathematical model of evolutionary theory could be constructed in which the genetic mechanisms of Mendelian genetics meshed perfectly with the selective mechanisms of evolutionary theory. Evolutionary theory and, hence, evolutionary taxonomy had become respectable again. However, the models supplied by Fisher, Haldane, and Wright were highly restrictive and very far removed from any situation a naturalist was likely to encounter in nature. Using the techniques of idealization which had proved so successful in physics, they showed that in certain overly simple, ideal cases, natural selection working on mutations which obeyed the laws of Mendelian genetics could result in the gradual evolution and splitting of species. The task still remained of showing how the insights gained in these idealizations could be applied to real situations in nature.[2] The classic works on this are those by Dobzhansky (37), Mayr (80, 85, 86), Huxley (68, 69), Simpson (99, 100, 102, 103), Rensch (94, 95), Stebbins (120), Hennig (60, 61), and Remane (93). In the following discussion, the earliest works of these authorities will be cited as freely as their later works because the basic features of the synthetic theory of evolution and evolutionary taxonomy have changed very little during this period.

PHYLOGENY AND CLASSIFICATION

One of the most persistent problems in biology has been the quest for a natural classification. Prior to Darwin a natural classification was one based on the essential natures of the organisms under study. Of the possible patterns that could be recognized in nature, a taxonomist would settle on one, partly because of his own peculiar psychological make-up and partly because of the scientific theories he held. Of course, another taxonomist with a different psychological make-up, perhaps holding different theoretical views, frequently recognized a different pattern. The controversies that ensued were usually settled by force of authority. The case of Cuvier and his disciples Owen and Agassiz is typical in this respect (83). There are four basic plans in the animal kingdom, no more, no less!

Evolutionary theory promised to put an end to all this dogmatic haggling. After Darwin a natural classification would be one that was genealogical. No longer would biologists have to search fruitlessly for some ideal plan but would need only to discover the genealogical relationships among the organisms being studied and record this information in their classifications. The alacrity with which many biologists adopted Darwin's suggestion stemmed in part from two illegitimate sources—an inherent vagueness in the proposal and a misconception of the relation which any system of indented, discontinuous words can have to something as continuous and complex as phylogeny. As Darwin (34) observed of Naudin's simile of a tree and classification, "He cannot, I think, have reflected much on the subject, otherwise he would see that genealogy by itself does not give classification." Nearly a century later Gilmour (56) was still forced to remark that he doubted "whether the real significance of the term 'phylogenetic relationship' is yet fully understood."

The purpose of this section will be to investigate the relationships which phylogeny can have to biological classification—assuming that phylogeny can be known with sufficient certainty. The major criticism of evolutionary taxonomy by pheneticists has been that such reconstructions are too often impossible to make. Discussion of this criticism will be postponed until the next section.

No term in taxonomy seems immune to ambiguity and misunderstanding; this includes the term "classification." Mayr (86) has already pointed out the process-product distinction between the process of classifying and the end product of this enterprise—a classification. But even the words "a classification" are open to misunderstanding. At one extreme, a classification is nothing but a list of taxa names indented to indicate category levels. Others would also include all the characters and the taxonomic principles used to construct a classification as part of the classification. At the other extreme, some authors use the words "a classification" to refer to the entire taxonomic monograph. Unless otherwise stipulated, "a classification" in [this chapter] will be used in the first, restricted sense.

The simplest view of the relation of a biological classification to phylogeny is that, given a classification, one can infer the phylogeny from which it was derived. One source of this misconception is a naive yet pervasive misconstrual of the relation between a hierarchical classification and a dendritic representation of phylogeny. According to this mistaken view, the classification of Order I sketched below

David L. Hull

Order I
  Family A
    Genus 1
      species a
    Genus 2
      species b
      species c
    Genus 3
      species d
      species e
      species f
  Family B
    Genus 4
      species g
      species h
    Genus 5
      species i
      species j

corresponds to the phylogenetic tree in figure 15.1. However, figure 15.1 is not a dendritic representation of a possible phylogeny. Rather it is merely a representation of the hierarchic indentations of the classification in a dendritic form. A true dendrogram of the possible phylogenetic development of the organisms involved would consist only of the species listed in the classification. One possible phylogeny from which the classification of Order I could have been derived is shown in figure 15.2.

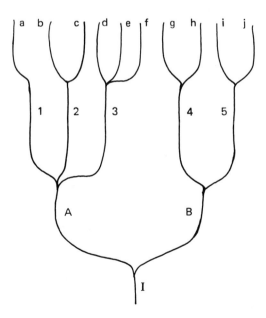

Figure 15.1   A dendritic representation of a hierarchical classification.

In this section we assume the phylogenetic development of the groups under discussion to be completely known. Hence, all the ancestral species are included in the classification along with extant species. In actual classifications, of course, not all ancestral species are known, but at least some are. At least sometimes, biological classifications contain reference to extinct forms. Hence, the interpretation of I in figure 15.1 as the unknown stem species which gave rise to Order I, of A and B as the unknown stem species which gave rise to Families A and B, respectively, and so on, cannot be carried through consistently. On occasion, at least, ancestral species will be known and will be included in the classification. The mistake is to confuse the inclusion relations in the taxonomic hierarchy with species splitting (103). An order does not split into genera nor genera into species.

A second impediment to seeing clearly the relation between phylogeny and classification has been a failure to distinguish cladistic from patristic relations (2, 5, 23, 75, 76, 84, 89, 115, 116). The primary difference between the phylogenetic school of Hennig, Kiriakoff, and Brundin and the evolutionary school of Dobzhansky, Mayr, and Simpson is that the former want classification to reflect only cladistic affinity, whereas the latter feel that the classification should also reflect such factors as degree of divergence, amount of diversification, or in general, patristic affinity.

Hennig's principles of classification are extremely straightforward (60, 61). The stem species of every single higher taxon must be included in that taxon and must be indicated as the stem species by not being included in any of the other subgroups of that taxon. Splitting is the only mechanism of species for-

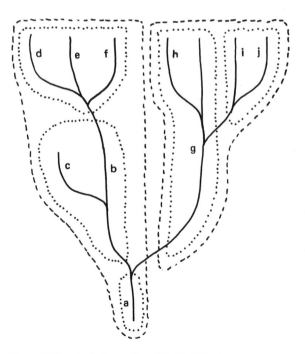

Figure 15.2   A phylogenetic subdivided into taxa.

mation that is recognized. Even though a group may evolve progressively until later members are extremely divergent from their ancestors, if no splitting has taken place, all the individuals are considered members of the same species. Upon splitting, the parental species is always considered to be extinct, even though individuals may persist which are morphologically identical to members of the parent species. As far as ranking is concerned, sister groups must always be given coordinate ranks. In addition, Hennig is predisposed to Bigelow's (2–5) observation that in a truly phylogenetic classification recency of common ancestry must be considered a criterion for ranking. Taxa that evolved earlier should be given a higher taxonomic rank than those that evolved later.

The major consequences of the adoption of these principles of classification is precisely the one intended by Hennig. Given a strictly phylogenetic classification, cladistic development can be read off directly. That is, given a classification and Hennig's principles, a dendrogram could be constructed which would accurately represent the cladistic relations of the groups classified. Hennig's principles of classification have something aesthetically satisfying about them. They are straightforward and exceptionless. But this satisfaction is purchased at a price higher than many biologists are willing to pay. Early groups, even if they immediately became extinct without leaving descendants, would have to be recognized as separate phyla, equivalent to highly diversified, persistent groups. Hence, if it could be shown that a species split off in the Precambrian but gave rise to no other species, it nevertheless would have to be classed as a phylum. The resulting classification would be exceedingly monotypic. Our increasing ignorance as phylogeny is traced further back through the geological strata saves phylogeneticists from actually having to introduce such extreme asymmetries into their classification, but even so, enough is known so that classifications erected on the purely cladistic principles of the Hennig school would be much more asymmetrical than those now commonly accepted. Evolutionists also complain that the Hennig school is too narrow since it limits itself just to cladistic affinity. Patristic affinity is also important. Thus, for both practical and theoretical reasons, the evolutionists feel that Hennig's solution to the problem of the relation between phylogeny and classification is unacceptable.

When we turn our attention to evolutionary taxonomy, the situation is not so straightforward. The principles of evolutionary taxonomy are extremely fluid and intricate. As Simpson (103) has said, the practice of evolutionary taxonomy requires a certain flair. There is an art to taxonomy. Vagueness as to the actual relation which evolutionary classification is to have to phylogeny can be discerned in the earliest statements on the subject. The main purpose of Dobzhansky's *Genetics and the Origin of Species* (37) was to reconcile the differences between naturalists and geneticists: to convince the naturalists that the geneticists' experimental findings in the laboratory were relevant to their work in the field and museum, and to convince the geneticists that their understanding of evolutionary theory was grossly inadequate. Dobzhansky is not a systematist and is not especially interested in the problems of systematics. The little that Dobzhansky (37) had to say about classification can be quoted in its entirety.

A knowledge of the position of an organism in an ideal natural system would permit the formation of a sufficient number of deductive propositions for its complete description. Hence, a system based on the empirically existing discontinuities in the materials to be classified, and following the hierarchical order of the discontinuous arrays, approaches most closely to the ideal natural one. Every subdivision made in such a system conveys to the student the greatest possible amount of information pertaining to the objects before him. The modern classification of organisms uses the principles on which an ideal system could be built, although it would be an exaggeration to think that the two are consubstantial.

On the other hand, since the time of Darwin and his immediate followers the term "natural classification" has meant in biology one based on the hypothetical common descent of organisms. The forms united together in a species, genus, or phylum were supposed to have descended from a single common ancestor, or from a group of very similar ancestors. The lines of separation between the systematic categories were, hence, adjusted, at least in theory, not so much to the discontinuities in the observed variations as to the branching of real or assumed phylogenetic trees. And yet the classification has continued to be based chiefly on morphological studies of the existing organisms rather than of the phylogenetic series of fossils. The logical difficulty thus incurred is circumvented with the aid of a hypothesis according to which the similarity between the organisms is a function of their descent. In other words, it is believed that one may safely base the classification on studies on the structures and functions of the organisms existing at one time level, in the assurance that if such studies are made complete enough, a picture of the phylogeny will emerge automatically. This comfortably complacent theory has received some rude shocks from certain palaeontological data that cast a grave doubt on the proposition that similarity is always a function of descent. Now, if similar organisms may, however rarely, develop from dissimilar ancestors, a phylogenetic classification must sometimes unite dissimilar, and separate similar, forms. The resulting system will be, at least in some of its parts, neither natural in the sense defined above nor convenient for practical purposes.

Fortunately, the difficulty just stated is more abstract than real. The fact is that the classification of organisms that existed before the advent of evolutionary theory has undergone surprisingly little change in the times following it, and whatever changes have been made depended only to a trifling extent on the elucidation of the actual phylogenetic relationships through palaeontological evidence. The phylogenetic interpretation has been simply superimposed on the existing classifications; a rejection of the former fails to do any violence to the latter. The subdivisions of the animal and plant kingdoms established by Linnaeus are, with few exceptions, retained in the modern classification, and this despite the enormous number of new forms discovered since then. These new forms were either included in the Linnaean groups, or else new groups were created to accommodate them. There has been no necessity for a basic change in the classification. This fact is taken for granted by most systematists, and all too frequently overlooked by the representatives of other biological disciplines. Its connotations are worth considering. For the only inference that can be drawn from it is that the classification now adopted is not an arbitrary but a natural one, reflecting the objective state of things.[3]

To begin with, the position of an organism in a hierarchical classification permits the inference of numerous propositions about it only if the characters used to classify the organisms are also listed. For example, knowledge that an organism is a chordate in conjunction with the defining characters of Chordata permits the inference that at some time in its ontogenetic development it has gill slits, a dorsal, hollow nerve cord, and probably a notochord. Knowledge that it is a vertebrate in conjunction with the defining characters of Vertebrata and the fact that Vertebrata is included in Chordata permits additional inferences and so on. A claim frequently made in the recent literature is that the best classification is the one with the highest information content; that is, the one which permits the greatest number of inferences. Colless (31), for example, says:

The current conflict between the "phenetic" and "phylogenetic" approaches to taxonomy thus boils down to whether a classification should in some fashion act as a storage-and-retrieval system for information about the distribution of attributes over organisms, and thus as a theory that predicts unexamined parts of that distribution, or whether it should reflect, as closely as possible, the historical course of evolution of the organisms concerned.

What is being blurred in the preceding quotation is that a biological classification as such (whether phenetic or phyletic) permits little in the way of inferences. Only a classification in conjunction with the principles and characters used to construct it is sufficient to permit any extensive inferences about the organisms being classified. For example, given the phenon levels of a phenetic classification, it is possible to infer that members of two taxa at a particular phenon level share a certain percentage of their characteristics, but it is not possible to infer which these may be. Similarly, from an evolutionary point of view, it would be reasonable to infer that two organisms classed together at the 40 phenon level are likely to have a more recent common ancestor than two organisms which are not classed together until the 10 phenon level—if one were given this information.[4] Only when the characteristics used to partition the organisms into taxa are included can specific predictions be made about which organisms are likely to have which characters. But with the addition of such information, we are rapidly approaching the point at which the word "classification" has become expanded to include the entire monograph. Classifications in the narrow sense are incapable of storing much in the way of specific information. Rather than being storage-and-retrieval systems themselves, they serve as indexes to such storage-and-retrieval systems. The information resides in the monograph, not in the classification (128). The classification merely provides a nested set of names which can be used to refer to the relevant taxa in as felicitous a manner as possible.

A second basic misunderstanding concerning the relation between a classification and a phylogeny has contributed to the belief that phylogeny can be inferred from an evolutionary classification. One commonly meets the assertion that proximity of names in a classification implies propinquity of descent. A glance back at figure 15.2 shows that this belief is mistaken. If our knowledge of phylogeny were reasonably complete, every single higher taxon would contain at least one species which would be as closely related to a species in another taxon of the same rank as it is to its closest relative in its own taxon. For example, in the classification sketched on p. 299, species *a* is twice removed from species *b* (species *b* appears two lines below species *a*) and nine places removed from species *g* (species *g* appears nine lines below species *a*)—and yet both of these species are directly descended from *a* (see figure 15.2).

When stated so baldly, the claim that inferences concerning propinquity of descent can actually be made from an evolutionary classification seems incredible; yet such a view is implicit in the writings of many phyleticists. From an evolutionary classification, even in conjunction with the stated criteria of classification, implications of cladistic relations are not possible. With a reconstructed phylogeny, indefinitely many classifications are possible. With any one of these classifications, an indefinite number of phylogenies are compatible. As Mayr has observed, "Even if we had a perfect understanding of phylogeny, it would be possible to convert it into many different classifications" (86; see also 14, 62, 103). Of course, one way to falsify these claims is to expand the meaning of "a classification" to include phylogenetic dendrograms. Then, in a trivial sense, phylogeny can be inferred from an evolutionary classification.

Note that not all classifications are acceptable to an evolutionist. For example, all taxa must be "monophyletic." Each taxon can "contain only the descendants of a common ancestor." Early in the history of evolutionary theory, this meant that all the members of a taxon had to be descended from at most a single individual or pair of individuals in an immediately ancestral taxon. As the emphasis in evolutionary theory shifted from individuals to populations and species, the principle of monophyly was expanded so that descent from a single immediately ancestral species was all that was necessary for a taxon to be monophyletic. Hennig and the phylogenetic school still retain this rather stringent notion of monophyly. Unfortunately, if this principle is adhered to, many well-known and easily recognizable taxa such as the mammals (à la Simpson) become polyphyletic. The compromise suggested by Simpson (101, 103), and Gilmour (56) before him, is that all the members of a taxon may be descended not from a single immediately ancestral species but from a single, immediately ancestral taxon of the same or lower rank. (For opposing views, see 32.) Thus, since all the species which contributed to the class Mammalia were in all likelihood therapsid reptiles, Mammalia is minimally monophyletic. As reasonable as this decision seems from the point of view of retaining well-marked groups and reflecting degree of divergence, its adoption further weakens the relation between classification and phylogeny. Not all classifications are compatible with a given phylogeny, but too many to permit any precise inferences.

Numerous authors before and after Dobzhansky (37) have observed that, from their classifications alone, "it is practically impossible to tell whether zoologists of the middle decades of the nineteenth century were evolutionists or not." Evolutionists have taken this fact to imply that preevolutionary taxonomists had been reflecting evolution in their classifications all along, though unwittingly. Pheneticists have argued for an additional factor. Classifications, before and after the introduction of evolutionary theory, are basically phenetic. Evolutionary theory, for all intents and purposes, is irrelevant to biological classification. It has been argued in this section that a third factor is actually responsible for the similarity between pre- and postevolutionary classifications. Hierarchical classifications, in the absence of a rigid adherence to principles of classifications like those of Hennig, do not permit any extensive inferences—whether phyletic or phenetic. Hennig says that hierarchic classifications are completely adequate to indicate phylogeny because he has incorporated the requirements of hierarchic classification into his principles of classification. From a strictly phylogenetic classification (just a list of indented names of taxa) and Hennig's principles, cladistic relations can be deduced. To the extent that this is not done, to that extent the number and variety of phylogenetic inferences which can be drawn from a classification will be diminished (28, 62, 125, 126).

Thus, biologists who maintain that biological classifications should be genealogical are presented with a dilemma. If they adopted a system like Hennig's, in which cladistic development is inferable from a classification, they would have to put up with the loss of information about patristic affinities and the cumbersome classifications that would result. If they retained the more tractable classifications that result from the more pliant principles of evolutionary taxonomy, they would have to abandon the ideal that classifications imply anything very precise about phylogenetic development. Evolution and evolutionary theory would still influence evolutionary classifications, but mainly in decisions as to homologies and the basic units of classification. The way in which evolutionary theory influences estimations of homologies will be discussed in the next section. The relation between evolutionary theory and the basic units of classification will be treated in the last section.

PHYLETIC INFERENCES AND PHENETIC TAXONOMY

In the preceding section, certain formal difficulties inherent in any attempt to establish a systematic relationship between classification and phylogeny were pointed out. The main thrust of the pheneticists' objections to evolutionary taxonomy, however, has been against permitting phylogeny to influence biological classification in the first place. The chief reasons that the pheneticists have given for excluding evolutionary considerations from biological classifications are as follows: (1) We cannot make use of phylogeny in classification since, in the vast majority of cases, phylogenies are unknown (3, 4, 6, 43, 119). (2) The methods which evolutionists use to reconstruct phylogeny, when not blatantly fallacious, are not sufficiently explicit and quantitative (6, 45, 115, 119). (3) With the help of techniques being developed by the pheneticists, it eventually may be possible to reconstruct reasonably accurate phylogenies for certain groups of organisms, but since phylogeny cannot be known with sufficient certainty for all groups, it should not be used in those few cases in which we do have good reconstructions (6, 24, 115, 119). (4) Even if the necessary evidence were available for all groups and the methods of reconstructing phylogeny were reformulated to make them completely acceptable, the resulting evolutionary classification would still be a special purpose classification and inadequate for biology as a whole; a general purpose classification would still be needed (55–57, 108–110, 119).

Like other criticisms of evolutionary taxonomy, these are not new. As early as 1874, Huxley (70), hardly an enemy of evolution, can be found saying, "Valuable and important as phylogenetic speculations are, as guides to, and suggestions of, investigation, they are pure hypotheses incapable of any objective test; and there is no little danger of introducing confusion into science by mixing up such hypotheses with Taxonomy, which should be a precise and logical arrangement of verifiable facts."

There is little that a philosopher can say about the first two objections to evolutionary taxonomy. After obvious inconsistencies have been removed and warnings about the type of certainty possible in empirical science duly entoned, the controversy becomes largely an empirical matter to be decided by scientists, not philosophers (64, 66). If extensive fossil evidence for a group is necessary for reconstructing the phylogeny of that group, then the phylogenetic development of a majority of plants and animals will never be known, but many biologists think that various laws (or rules of thumb, if you prefer) can be used to reconstruct tentative phylogenies even in the absence of more direct evidence (85, 103, 124).

An interesting development in the phyletic-phenetic controversy is that some numerically minded biologists are beginning to set out formalisms for inferring phylogeny which they feel fulfill the various criteria of objectivity, etc., which more traditional methods are reputed to lack (22–24, 27, 40, 47, 48, 77, 91, 112, 118, 126). Implicit in this endeavor is the conviction that attempts to reconstruct phylogeny even in the absence of fossil evidence are not inherently fallacious. Perhaps the practice of some evolutionists has been slipshod and certain reconstructions of the methods by which phylogenies are inferred have been mistaken, but the phyletic enterprise as such is not hopeless. For example, Colless (30) says, "I must stress at the outset that I am *not* denying that we can, and do, have available a body of reasonably credible phylogenies, which are probably fair reproductions of historical fact. I do, however, assert that some influential taxonomists have an erroneous view of the process by which such phylogenies are inferred; and, if my view is correct, such a situation clearly invites faulty inferences and sterile controversy."

Initially, phenetic and evolutionary taxonomy were treated by all those concerned as if they were in opposition to each other (84, 90, 108, 115, 116). Pheneticists argued that evolutionary classifications, based on a priori weighting, were limited in their uses because they were biased toward a single scientific theory. Phenetic classifications, on the other hand, were general purpose classification, based on the total number of unweighted or equally weighted characters, and were equally useful to all scientists because they were biased toward no scientific theory whatsoever. Pheneticists like Cain (18–21) attributed the mistakes which early taxonomists like Aristotle, Linnaeus, and Cuvier made to their letting theoretical and philosophical beliefs affect their classifications. Evolutionists had carried on in this misbegotten tradition. To eliminate such errors, pheneticists argued that no theoretical considerations should enter into the initial stages of a purely phenetic classification. A pheneticist must classify as if he were completely ignorant of all the scientific achievements (and failures) which preceded him. Characters must be delineated, homologies established, and clusters derived without recourse to any preconceived ideas whatsoever. No character could be weighted more heavily than another because it proved to be a "good" character in previous studies (unless those studies themselves were phenetic) or because the studies were theoretically important according to current scientific theories. There must be no a priori weighting! Later, after several such purely phenetic studies had been run, certain characters would be found that tended to covary. They then could be weighted a posteriori. This a posteriori weighting would be, however, purely a function of the observed covariations of the characters being studied, not of any theoretical considerations. Finally, evolutionary interpretations could be placed on these purely phenetic classifications which would transform them into special purpose evolutionary classifications. In short, phenetic taxonomy was just look, see, code, cluster.

This initial sharp contrast between evolutionary and phenetic classification has been modified considerably in recent years. In their latest utterances, pheneticists tend to view phenetic taxonomy somewhat differently. Doubts are raised as to whether any pheneticist ever held the views described above. Purely phenetic studies are still considered necessary preliminaries to scientific endeavors of any kind, including the construction of evolutionary classifications, but these phenetic studies are no longer thought of as being performed in isolation from all scientific theories—just from evolutionary theory. Homologies are not established just by observation, but are inferred via relevant genetic, embryological, physiological, and other scientific theories. Prior to any phenetic study, decisions are made as to which characters are to be considered the same, and in what sense they are to be so considered. For example, two organs which are structurally very similar in adult forms might be considered different organs because they have decidedly different embryological developments. Phenetic taxonomy is a matter of look, see, infer, code, and cluster. The resulting phenetic classifications are general purpose classifications because they have been constructed using all available knowledge, including all well-established scientific theories—except evolutionary theory. Finally, evolutionary interpretations can be placed on these phenetic classifications, but if the phenetic classification is properly constructed to begin with, it will actually be an evolutionary classification. Hence, phenetic and evolutionary classification, when properly constructed, are equivalent to each other and are equally general purpose classifications.

It will be the purpose of this section to trace the change in phenetic taxonomy from its early, antitheory stage to its current state and to point out the fallacies in the early phenetic position which made it seem attractive and the reasons for changing it. It will be argued that purely phenetic classifications, as they were originally explicated, are impossible and that even if they were possible, they would be undesirable. To the question "Theory now or theory later?" only one answer is possible. The two processes of constructing classifications and of discovering scientific laws and formulating scientific theories must be carried on together. Neither can outstrip the other very far without engendering mutually injurious effects. The idea that an extensive and elaborate classification can be constructed in isolation from all scientific theories and then transformed only later into a theoretically significant classification is purely illusory. A priori weighting of the theoretical kind is not only desirable in taxonomy, it is necessary. The price one pays for theoretical significance is, obviously, that any change or abandonment of the theories which give rise to the classification will necessitate corresponding changes in the classification (52, 53, 67, 68, 69).

There is less to criticize in the latest versions of the phenetic position. One still must question why, of all scientific theories, evolutionary theory must be scrupulously excluded from the process of biological classification. There may be reasons for such a rejection, but the pheneticists have not been very articulate in stating them. Most of the objections which they have raised against evolutionary theory would count equally against any scientific theory and must be interpreted as utterances stemming from their early, antitheory stage of development. Now that pheneticists are willing to accept the role of theory in science, it would be helpful if they were to spell out exactly what faults they still find with evolutionary theory. A final question must also be asked before we turn to a detailed analysis of the evolution of phenetic taxonomy. What was all the controversy about? Except for a greater emphasis on making taxonomic practice explicit and perhaps even quantitative, how does phenetic taxonomy differ from evolutionary taxonomy? If patristic affinity is equivalent to some function of phenetics, chronistics, and cladistics, why all the acrimony? Has the phenetic-phyletic controversy been just one extended terminological confusion?

That the pheneticists actually held the early views attributed to them can easily be demonstrated. For example, as late as 1965, Sokal, Camin, Rohlf, and Sneath (116) can be found saying, "Numerical taxonomists *do not disparage* interpretation or speculation or the inductive-deductive method in science. They simply feel that the process of constructing classification should be as free from such inferences as possible." (See also 18–23, 29, 30, 39, 115, 116.) According to Colless (30), phenetic taxonomy makes reference "only to the observed properties of such entities, without any reference to inferences that may be drawn *a posteriori* from the patterns displayed. Such a classification can, and, to be strictly phenetic *must*, provide nothing more than a summary of observed facts." Even in their most recent publications, pheneticists can still be found making such extremely empirical claims; for example, Sokal (114) says that taxonomy is "the grouping of like organisms based on direct observation."

The key notion in the empiricist philosophy is the claim that, ideally, a priori weighting is to be completely expunged from taxonomic practice. What pheneticists have intended by such interdictions has been extremely equivocal. At one extreme they claim that homologies must be established on the basis of pure observation (as if there were such a thing). Two instances of a character are instances of the same character if they look, smell, taste, sound, and feel the same; otherwise not. Systematics "is a pure science of relation, unconcerned with time, space, or cause" (15). All operational homologies are observational homologies.

So far no pheneticist has produced anything like a strict phenetic classification as described above. Pheneticists make reference to things like wings, antennae, anal gills, dorsal nerve cords, enzymes, and nucleotides. These are hardly pure observation terms. They presuppose all sorts of previous knowledge of a highly theoretical kind. For example, a taxonomist working on brachiopods today describes his specimens and forgets that at one time considerable effort was expended to decide whether brachiopod valves were front and back, dorsal and ventral, or right and left and that the eventual decision reached was based on various theoretical beliefs concerning their ontogenetic and phylogenetic development (35). As Sneath (108) has observed, "Many taxonomic problems start part of the way along the classificatory process, and one is apt to forget what previous knowledge is assumed."

Pheneticists take this to be a fault with traditional taxonomy rather than a characteristic of all scientific undertakings, including their own. They think that, ideally, a purely descriptive, nontheoretical classification must be possible. The source of the persuasiveness of this view can be found in empiricist epistemology, according to which all empirical knowledge stems from sense impressions. Hence, all knowledge must be reducible to pure observation statements. Empiricists themselves have shown that such a reduction is impossible and, specifically, that scientific theories are not replaceable by sets of observation statements (59). There remains the metaphysical compulsion to believe that such a reduction must be possible, and with it, the notion of a purely phenetic classification.

At times pheneticists are a little more liberal in their interpretation of what is to count as a priori weighting. For example, Colless (29) says, "Of course, the simple act of observation of 'existing' entities involves inferences, but they are of a primitive nature and, I believe, can be clearly distinguished from those which I am concerned to exclude." But how primitive is primitive enough? What criteria does Colless have for making this distinction? And why are primitive a priori weightings acceptable but sophisticated ones illegitimate?

There is a continuum between terms that are largely observations, like white precipitate, flammable fluids, and red appendage, and those that are more theoretical, like inertia, unit charge, and selection pressure. The reason why pheneticists want classifications to be constructed using those terms nearer the observational end of the scale is too apparent. Time and again Cain (18–21) has argued that the greatest source of error in early classifications is their reliance on scientific theories which we now know to be erroneous.

Wouldn't the safest procedure be to classify neutrally? That way theories would come and go and the classification, nevertheless, remain unchanged. Such a procedure would assuredly be safe, but in the extreme it is impossible to accomplish and in moderation undesirable.

The basic fallacy underlying the phenetic position on a priori weighting is the confusion of the logical order of epistemological reconstructions with the temporal order in actual scientific investigations (50). Perhaps an analogous example from a different discipline will help to bring this fallacy into sharper focus. In the epistemological approach advocated by Sneath (108), a classification of inorganic substances must begin with purely phenetic studies in which samples are collected of a wide variety of inorganic substances, purely observational homologies established, and various clustering techniques used to group these substances into OTUs. Certain characters might then turn out to be good indicators of certain clusters and weighted more heavily for future runs. Eventually, a classification would emerge which would be equally useful for all purposes. Later, if one wished, this general purpose classification could be transformed into a special purpose classification by introducing atomic theory and weighting atomic number more heavily than all other characters put together.

The actual history of the construction of the periodic table does not, of course, read anything like this epistemological reconstruction. For example, gold was orginally recognized and defined in terms of color, malleability, weight, and so on—a characterization inadequate to distinguish gold from various alloys. Thus, Archimedes was presented with the problem of discovering a more important characteristic of gold. He hit upon specific density. What we tend to forget is that his selection of specific density rather than a host of other characters was his acceptance of the physics of his day in which the four elements were fire, air, earth, and water! Later, as physical theory developed, atomic weight replaced specific density as the key character in distinguishing inorganic substances. In the interim a new concept of element had evolved in the context of atomic theory. Not until atomic number replaced atomic weight could elements, in this new sense, be distinguished from each other and from compounds.

The analogy to the development of evolutionary theory and the species concept is obvious. The point is that a priori considerations were not after-the-fact interpretations but necessary factors in every step of the formation of the periodic table. Inorganic elements are distinguished from compounds and from each other on largely theoretical grounds. Incidentally, some very rough clusters of observable characters also accompany this theoretically significant classification. Atomic number, even if considered a phenetic character, was not treated as of equal weight to all other characters. Nor was its weight established a posteriori by discovering that numerous other characters tended to covary with it. The correlation between atomic number and the overall similarity of physical elements is about on the same order of magnitude as that observed by Dobzhansky between breeding habits and the overall similarity of living organisms.

Pheneticists might reply that perhaps this is how the periodic table was constructed, but it should not have been. It should have been constructed by purely phenetic means, and to be justified it must be. This contention has yet to be proved. To do so, pheneticists would have to sample all inorganic substances. They could not limit themselves to just the elements, because that would presuppose that they knew which inorganic substances were elements, a blatant instance of a priori weighting. After establishing homologies purely on the basis of observation, pheneticists would then have to erect various alternative phenetic classifications. Atomic number could hardly appear as one of these phenetic characters, since electrons are observable in only the widest sense of the word. If electrons are observable, so is evolutionary development! If one of these phenetic classifications can distinguish between elements and compounds and can order the elements as they are ordered on the periodic table, then the pheneticists will have proved their case. If recourse to atomic theory is permitted in the early stages of the investigation and atomic number weighted more heavily than all other phenetic characters put together, then phenetic taxonomy, as it was originally explicated and as it is still propounded by many, has been abandoned. If it is to be abandoned, then the original criticisms of evolutionary classifications need to be reevaluated.

Pheneticists seem to have gradually come to realize that the notion of a theoretically neutral phenetic classification is an illusion and have modified their position accordingly. Operational homologies are established utilizing any respectable scientific theory except evolutionary theory. The reasons given for permitting morphological, behavioral, physiological, serological, and DNA homologies, but forbidding evolutionary homologies, have all depended on repeated equivocations on the terms "phenetic character" and "operational homology." Pheneticists claim that operational homologies are observed, whereas evolutionary homologies must be inferred. In the first place, only characters are observed. That two instances of a character are instances of the same character (i.e., that they are operationally homologous) must be inferred. Only if operational homologies are limited to observational homologies (i.e., if they both look blue, then they are blue) will these inferences be made solely on the basis of observation. All other types of inferences to operational homologies will make essential reference to a particular scientific theory, and with the introduction of theory the overly simplistic notion of observational homology must be abandoned. One cannot observe that two nucleotides are operationally homologous. Both the existence of the nucleotides and which of the nucleotides are homologous must be inferred from extremely indirect evidence in the context of current biochemical theories.

Colless (30) claims that there is a "phylogenetic fallacy"—the view that "in reconstructing phylogenies, we can employ something more than the observed attributes of individual specimens, plus some concept of 'overall resemblance' and some concept of 'attribute' of a set or class of such specimens." Scientists in general, not just evolutionists, do employ something more than observed attributes and some concept of overall resemblance. This something more is scientific theory. As Colless (30) himself says, "The codon elements thus employed as attributes must, surely, be the ultimate approximation to our notion of 'unit attributes'...." What is or is not a codon is determined in large measure by biochemical theory. Codons are certainly not observable. In this instance, pheneticists and not phylogeneticists are guilty of reasoning fallaciously. The phenetic fallacy is the belief that in reconstructing phylogenies, we employ anything less than all the data and all the scientific theories at our disposal. For example, even a theory as far removed from biology as quantum theory is used in the proces of carbon dating.

Each of the various kinds of homology has its own special problems. For example, behavioral homologies cannot be obtained very readily for extinct species, nor are the results obtained for extant species by controlled experiments in the laboratory very reliable. Thus, the argument that evolutionary homologies should not be used for any group because we cannot obtain them for all groups cannot be cogent, since, if it were, it would count against all types of homologies. Even morphological homology, the most pervasive type of homology used in classification, has limited applicability. For example, individual viruses and bacteria have few morphological characters which can be used in classifying them. The likelihood of obtaining information about DNA homologies for more than an infinitesimally small percentage of species (and these all extant) is very slim, and yet no one would want to argue that this information should not be used when we do have it. Sokal and Camin (115) say, "Because phenetic classifications require only description, they are possible for all groups and are more likely to be obtained as a first stage in the taxonomic process." The preceding claim is true only if operational homologies are limited to observational homologies. If not, then phenetic classifications require more than description. They require the establishment of theoretically significant operational homologies.

However, the abandonment of the distinction between a priori and a posteriori weighting has certain ramifications for the notions of overall similarity and a general purpose classification. If it is admitted that the establishment of homologies presupposes various scientific theories, then the idea of a single parameter which might be termed *overall similarity* loses much of its plausibility and all classifications become special purpose classifications. As Edwards and Cavalli-Sforza (40) observed, "To say that the purpose of a classification is 'general' is, in our view, too vague to be of use in its construction." The idea of a general purpose classification is still another phenetic illusion. Pheneticists themselves have come to realize that too many parameters exist which have equal right to be termed measures of overall similarity and, hence, that there is no such thing as a general purpose classification. As the Ehrlichs (44) have said recently, "Theoretical considerations make it seem unlikely that the idea of 'overall similarity' has any validity.... *All* classifications are inherently special." They quickly add, however, that "no special classification is any more or less 'correct' than any other." (See also 45, 50–53, 73, 110, 115, 116.)

All actual biological classifications are mixed classifications; that is to say, they are affected to a greater or lesser degree by all current biological theories. No classification is purely evolutionary, purely embryological, and certainly none is purely phenetic. The justifications for this irregular mixing of these various considerations in a single classification are both practical and theoretical. In the current state of these theories, evolutionary considerations could no more be untwined from all other considerations and excluded from classification than could embryological or physiological considerations. They are too interconnected. They are interconnected because the theories from which they are partially derived are themselves partially interdependent. Of course, this situation need not be permanent. These various theories may gradually become more carefully and completely formulated, and the relevant derivations more distinct. When this happens, the ideal of providing a straightforward reconstruction of the inferences involved in biological classification can be more closely approximated. We must resist at all costs the tendency to superimpose a false simplicity on the exterior of science to hide incompletely formulated theoretical foundations.

THE BIOLOGICAL SPECIES CONCEPT

Although Dobzhansky (36–38) first emphasized the biological species concept, it has received its most extensive development at the hands of Ernst Mayr. From his earliest to his most recent writings, Mayr (80–83, 85–87) has set himself the task of demolishing the typological species concept and replacing it with a species concept adequate for its role in evolutionary theory. According to the typological species concept, each species is distinguished by one set of essential characteristics. The possession of each essential character is necessary for membership in the species, and the possession of all the essential characters sufficient. On this view, either a character is essential or it is not. There is nothing intermediate. If a character is essential, it is all-important. If it is accidental, then it is of no importance (63, 65).

In taxonomy, the essentialist position is known as *typology*, a word with decidedly bad connotations, In the recent literature, every school of taxonomy has been called typological at one time or another. The phylogeneticists term the evolutionists typologists because they let degree of divergence take precedence over recency of common ancestry in their classifications (75, 76). The pheneticists call both evolutionists and phylogeneticists typologists because they claim to use criteria which are rarely tested and may not actually obtain (113, 116, 117). The pheneticists in turn are called typologists because their classifications are intended to reflect overall similarity (71, 98, 103). The pheneticists reply that they are typologists but without types and of a statistical variety (113). Their opponents reply that this is not typology but nominalism (84–86)! To put a nice edge on the dispute, some taxonomists openly claim the honor of being called typologists. "Now the great object of classification everywhere is the same. It is to group the objects of study in accordance with their essential natures." (See also 13, 15, 97, 121–123, 127.)

The key feature of essentialism is the claim that natural kinds have real essences which can be defined by a set of properties which are severally necessary and jointly sufficient for membership. Hence, strictly speaking, there can be no such things as statistical typology. Biologists were always aware that the characters which they used to distinguish species did not always universally covary, as the essentialist metaphysics which they tacitly assumed entailed, but not until evolutionary theory were they forced to admit that such variation was not an accidental feature of the organic world, but intrinsic to it. After evolutionary theory was accepted, variation was acknowledged as the rule, not the exception (63, 65). Instead of ignoring it, taxonomists had to take variation into account by describing it statistically. No one specimen could possibly be typical in any but a statistical sense. Species could no longer be viewed as homogeneous groups of individuals, but as polytypic groups, often with significant subdivisions. Polythetic definitions, in terms of statistically covarying properties, replaced essentialist definitions in terms of a single character or several universally covarying characters (1, 26, 33, 63, 92).

One of the accompanying characteristics of essentialism was the gradual insinuation of metaphysical properties and entities into taxonomy. Whenever naturalists attempted to define natural kinds in terms of observable attributes of the organisms being studied, exceptions always turned up. One way to reconcile this apparent contradiction was to dismiss all exceptions as monsters. Another way was to define the names of natural kinds in terms of unobservable attributes. However, two kinds of unobservable must be distinguished at this juncture—metaphysical entities and theoretical entities which, in the context of a particular scientific theory, are indirectly observable. The entities and attributes postulated by classical essentialists tended to be of the former type. The genetic criteria of the biological definition of species may be tested very rarely, but they are testable and, hence, are not metaphysical. What the pheneticists have in common with typologists is a belief in the existence of natural units of overall similarity. They differ in that these units can be defined only polythetically.

Recognizing the existence of variation among contemporary forms as a necessary consequence of the synthetic theory of evolution is one thing; formulating a methodology in taxonomy sufficient to handle such variation is another. The history of the biological species concept is a story of successive attempts to define species so that the resulting groups are significant units in evolution, or in Simpson's (101, 103) words, an evolutionary species is an "ancestral-descendant sequence of populations ... evolving separately from others and with its own unitary evolutionary role and tendencies." Dobzhansky (36, 37) began by defining a species as that stage of the evolutionary process "at which the once actually or potentially interbreeding array of forms becomes segregated in two or more separate arrays which are physiologically incapable of interbreeding," and he emphasized the necessity of geographic isolation in species formation. "Species formation without isolation is impossible." Mayr concurred with Dobzhansky and distinguished with him between various isolating mechanisms, as such, and geographic and ecological isolation, since these latter are temporary and are readily removed. The species level is reached "when the process of speciation has become irreversible, even if some of the (component) isolating mechanisms have not yet reached prefection" (85). The classic formulation of the biological species definition is as follows:

A species consists of a group of populations which replace each other geographically or ecologically and of which the neighboring ones intergrade or interbreed wherever they are in contact or which are potentially capable of doing so (with one or more of the populations) in those cases where contact is prevented by geographical or ecological barriers.

Or it may be defined more briefly:

Species are groups of actually or potentially interbreeding natural populations, which are reproductively isolated from other such groups. (80)

Special attention in the preceding definition must be paid to the fact that it is populations which are said to be actually or potentially interbreeding, reproductively isolated, and so on, not individuals. In ordinary discourse, the same terms are applied both to individuals and to groups of individuals—like populations. For example, both individuals and populations are frequently said to interbreed. In most cases, the use of two distinct senses of interbreed causes no confusion, especially since the notion of populations interbreeding is defined in terms of individuals interbreeding. Similarly, Mayr (85) says of isolating mechanisms that they are "biological properties of individuals that prevent the interbreeding of populations that are actually or potentially sympatric." By their very nature, claims about populations interbreeding, etc., are statistical notions derived from the corresponding actions and properties of individuals. Thus, complaints that evolutionists continue to consider two groups as separate species even though members of these groups occasionally cross and produce fertile offspring are misplaced. It is the amount of crossing and the degree of viability and fertility of the offspring that matter. Complaints that values for these variables are too often difficult or impossible to specify are obviously relevant.

Since there is a definitional interdependence between species and population, charges of circularity must be allayed before we proceed further. *Species* is defined in terms of interbreeding, potential interbreeding, and reproductive isolation. Populations are included in species. Hence, populations must at least fulfill all the requirements for species. Additional requirements are added for populations. *Populations* are defined in terms of geographic distribution, ecological continuity, and genetic exchange. A population is "the total sum of conspecific individuals of a particular locality comprising a single potential interbreeding unit" (85). The members of a population must not be separated from each other by ecological or geographic barriers. They must be actually interbreeding among themselves. As a unit, they are potentially interbreeding with other such units.

Throughout his long career, Mayr has continually opposed the typological species concept and essentialism, and yet on some interpretations, the biological species concept has itself been treated typologically, as if it provided both necessary and sufficient conditions for species status. Dobzhansky (37, 82), for example, has argued that individuals which never reproduce by interbreeding can form neither populations nor species because potential interbreeding is a necessary condition for the correct application of these terms. He even goes so far as to say that the terminal populations of a *Rasenkreis*, if intersterile, are to be included in separate species, even though these populations are exchanging genes through intermediary populations! Dobzhansky seems to be confusing the importance of a particular species criterion with the importance of the species concept. The crucial issue is not whether some one character is possessed, but whether the units functions in evolution as species. As Mayr (87) has said, "*Species are the real units of evolution*, they are the entities which specialize, which become adapted, or which shift their adaption." Do asexual "species" specialize, become adapted, split, diverge, become extinct, invade new ecological niches, compete, etc.? If so, then from the point of view of evolutionary theory, they form species and criteria must be found to delimit them.

The three elements in the biological species definition are actual interbreeding, potential interbreeding, and reproductive isolation. As succinct as Mayr's shorter version of the biological species definition is, it nevertheless contains redundancies. Two or more populations are reproductively isolated from each other if, and only if, they are neither actually nor potentially interbreeding with each other. Thus, one or the other side of the equivalence could be omitted with no loss of assertive content. Species are groups of natural populations which are not reproductively isolated from each other but which are reproductively isolated from other such groups. In his most recent publication, Mayr himself omits reference to potential interbreeding in his revised version of the biological species definition: "Species are groups of interbreeding natural populations that are reproductively isolated from other such groups" (86).

In his new biological species definition Mayr still retains reference to interbreeding to indicate that the definition is applicable only to populations whose members reproduce by interbreeding and because successful interbreeding is the most directly observable criterion for species status. Reference to potential interbreeding is omitted because interbreeding is omitted because anything that can be said in terms of potential interbreeding can be said in terms of reproductive isolation. Neither morphological similarity nor time is mentioned in any of the formulations of the biological species definition. Among synchronous populations, morphological similarity and difference are of no significance, as far as species status is concerned. Questions of inferring species status aside, they function only in distinguishing phena of the same population, subspecies, sibling species, etc. (See Mayr's [86] discrimination grid.)

The omission of any temporal dimension from the biological species definition is of greater significance. The application of the biological species definition successively in time would lead to the recognition of a series of biological species with minimal temporal dimensions. What is to integrate these successive time species? The answer, as Simpson (103) pointed out earlier, is descent. If species are to be significant evolutionary units, some

David L. Hull

reference to descent eventually must be made. It is also implicit in any definition of population, since males, females, young and adults, workers and asexual castes are all to be included in the same populations. Morphological similarity won't do, because the types of individuals listed are often morphologically quite dissimilar. However, once a temporal dimension is introduced into the species concept and speciation without splitting is permitted (contra Hennig), an additional criterion must be introduced to divide gradually evolving phyletic lineages into species. The only candidate for such a criterion is degree of divergence, as indicated by morphological and physiological similarity and difference. Thus in the discernment of biological species, morphological similarity and difference play a dual role, in most cases as the evidence by which the fulfillment of the other criteria is inferred and in some instances as criteria themselves. By now it should be readily apparent that any adequate definition of species as evolutionary units can no more be typological in form than can any definition of any theoretically significant term in science. As Julian Huxley (68) observed quite early in the development of the synthetic theory of evolution, "Species and other taxonomic categories may be of very different types and significance in different groups; and also ... there is no single criterion of species."

The objections, however, which have been made most frequently by the pheneticists against the biological species concept are not those just enumerated, but the following: (1) As important as biological species may be in evolutionary theory, such theoretical considerations should not be allowed to intrude into biological classification, both because they are theoretical and because the presence or absence of reproductive isolation can seldom be inferred with sufficient certainty. (2) There may be fairly pervasive evolutionary units in nature, but reproductive isolation does not mark them. (3) There are no pervasive evolutionary units in nature, regardless of the criteria used to discern them.

As in the case of inferring phylogeny, the commonest complaint raised by extreme empiricists in general, and the pheneticists in particular, against the biological species concept is that too often reproductive isolation cannot be inferred with sufficient certainty to warrant its intrusion into classification. As early as the New Systematics (68) Hogben objected that biological species could not be determined often enough, and recently Mayr (85) has said that to "determine whether or not an incipient species has reached the point of irreversibility is often impossible." The problem is not distinguishing one taxon from another but deciding when one or more taxa have reached the level of evolutionary unity and distinctness required of species. If two groups are reproductively isolated from each other, then they are included in separate species; but how often and with what degree of certainty can the presence or absence of reproductive isolation be determined?

If just the two factors space and time are taken into account, four possible situations confront the taxonomist: In the ideal case, two populations, for a while at least, are synchronous and partially overlap. Here, in principle, it is possible to confirm species status by observation. In practice, the situation is not so ideal because the making of such observations is expensive, time consuming, and difficult—not to mention that decisions have to be made regarding the frequency of crossing, the degree of viability and fertility of the offspring, etc. In most cases, even under such optimal conditions, taxonomists depend heavily on inferences from morphological similarity to aid them in their decisions. In cases of synchronic but allopatric populations, the presence or absence of reproductive isolation must be inferred. The advantage here is that on occasion such inferences can be checked, both indirectly by fertility tests in the laboratory and directly, if the populations happen to meet in nature. Usually, of course, species status is inferred via morphological similarity and difference. When two populations are separated by appreciable durations of time, inferences of species status are even more circumstantial and can never be checked by any of the more direct means. "Hence, while the definition of the BSC [biological species concept] does not involve phenetics, the actual determination of a biological species always will do so, even in the optimal case" (117).

Pheneticists have objected both to the failure of evolutionists to give phenetics its just due in the application of the biological species concept and to the deficiencies of phenetic similarity as an indicator of reproductive isolation. Since phenetics plays such a predominant role in species determination anyway and since inferences from phenetic similarity to interbreeding status are very shaky at best, they ask why one should not abandon oneself to phenetic taxonomy right from the start. The problem in replying to this question is in deciding precisely what phenetic taxonomy is. By a rigid interpretation, phenetic taxonomy, as it was originally set out, is something radically new, but by this interpretation it can be shown that there can be no such thing as phenetic taxonomy. By a more reasonable interpretation, phenetic taxonomy loses its originality, since it becomes by and large what traditional taxonomists have been doing all along. The jargon of phenetic taxonomy is different, and greater emphasis is placed on mathematical techniques of evaluation, but with such an interpretation phenetic taxonomy is not very revolutionary.

Sokal and Crovello (117) complain that since the words "potential interbreeding" have "never really been defined, let alone defined operationally, . . . it appears to us that the only possible answer one could get from the question whether or not two samples are potentially interbreeding is 'don't know.'" In the first place, potential interbreeding has been defined. If two populations are kept from interbreeding only by geographical or ecological barriers, then they are potentially interbreeding; otherwise not. It is another story, of course, whether or not ecologists and population biologists are in a position to make reasonable inferences on these matters. Sometimes, however, detailed analyses of particular situations have been provided and biologists are in a position to say more than "don't know." With equal justification, an evolutionist could say that since the words "phenetic similarity" have never really been defined, let alone defined operationally, the only possible answer one could get from the question whether or not two samples are potentially similar is "don't know." Of course, for specific studies, when the OTUs, characters, and clustering method are specified, more specific decisions can be made, but the same is true for potential interbreeding claims. In both disciplines loose and specific questions can be asked.

As unflattering as the appellation may sound, "phenetic" has been a weasel word in phenetic taxonomy. Its meaning changes as the occasion demands. When the principles of other schools of taxonomy are being criticized, it is given a strict interpretation. Phenetic taxonomy is look, see, code, cluster. A methodologically sophisticated ignoramus could do it. But when the pheneticists turn to the elaboration of the methods and procedures of phenetic taxonomy, it takes on a whole spectrum of more significant meanings, heedless of the fact that under these various interpretations the original criticisms of other taxonomic schools lose much of their decisiveness.

For example, in the flow chart designed by Sokal and Crovello (117) for the recognition of biological species, they begin by grouping individuals into rough-and-ready samples. "In the initial stages of the study it may be that sufficient estimations of phenetic similarity can be determined by visual inspection of the specimens." But they go on to admit that such groupings are not the result of mindless look-see. "Knowledge of the biology of the organisms involved may be invoked." Throughout this flow chart, *phenetically homogenous sets* must include all stages in the life cycle of the organism, various castes in social insects, males and females, etc., regardless of the polymorphisms involved (11, 16, 85, 88, 95, 96). They see this as a practical difficulty, when it is plainly a theoretical difficulty. The admission of such theoretical considerations in the initial stages of a phenetic study means that the pheneticists themselves are practicing a priori weighting, a practice which they have roundly condemned in others. Decisions to include males and females in the same taxon do not stem from earlier phenetic clustering but from previously accepted biological theories. Evolutionists emphasize reproductive isolation because they feel that it is of extreme importance in the phylogenetic development of species. They don't want to see evolutionary units broken up and scattered throughout the nomenclatural system. Similarly, biologists emphasize cellular continuity as a criterion for individuality because they feel that it is of extreme importance in the embryological development of the individual. They don't want to see embryological units broken up and scattered throughout the nomenclatural system. The theory of the individual, as Hennig calls it, may be so fundamental that it has become commonplace, but a biological theory does not cease to be a theory just because it has been around for a long time. As was argued earlier in the section on inferring phylogeny, pheneticists themselves admit theoretical (i.e., a priori) considerations in the initial stages of their studies—as well they should. The point in making this observation is not that pheneticists should be more rigorous in purging their procedures of such theoretical considerations—which are absolutely necessary—but that pheneticists should recognize them for what they are and modify their criticisms of evolutionary taxonomy accordingly.

What is a phenetic property, a phenetic classification, phenetic similarity? If a phenetic property is to be some minimal attribute analyzed in the absence of all scientific theories, regardless of how rudimentary, such characters will certainly be useless in any attempt to construct a scientifically meaningful classification. Arguments have even been set out that, in principle, such an analysis is impossible. If a phenetic property is to be some minimal unit analyzed in the context of some but not all scientific theories (and, in particular, not of evolutionary theory), then the criteria for deciding which scientific theories are legitimate and which illegitimate must be stated explicitly and defended. If some scientific theories are to be admitted even at the initial stages of a phenetic study, then the criticisms of comparable admissions of evolutionary theory must be reevaluated. The establishment of evolutionary homologies on the basis of evolutionary theory may still be illegitimate, but not just because it is a scientific theory entering into the initial stages of a taxonomic study.

Sokal and Crovello (117) say that phenetic taxonomy is closely related to what Blackwelder calls practical taxonomy—"the straight-forward description of the patterns of variation in nature for the purpose of ordering knowledge." As efforts of the pheneticists have ably proved, there are indefinitely many ways of describing the patterns of variation in nature, and in each way there are indefinitely many patterns to be recognized. The problem is not so much that there is nothing which might be called overall phenetic similarity, but that there are too many things which might answer to this title. The question is whether or not some of these possible ways of ordering knowledge are perhaps more significant than others. The whole course of science attests to the reply that there are some preferable orderings—those which are most compatible with current scientific theories.

Evolutionists claim that their classifications, though they may be constructed in part by intuitive means, are objective, real, nonarbitrary, and so on, because they reflect something which really exists in nature. Pheneticists reply that character covariation also really exists in nature. As might be expected, this sort of exchange has done little to clarify the issues. The difference between evolutionary and phenetic taxonomy in this respect is that evolutionists have biologically significant reasons for making one decision rather than another while, by a strict interpretation, pheneticists do not. On purely phenetic criteria, any group of organisms can be arranged in indefinitely many OTUs with coefficients of similarity ranging from zero to unity. In contrast, evolutionists contend that biological species are important units in nature, more important than numerous other units which might be discernible. They are functioning as evolutionary units in evolution. Hence, from the point of view of evolutionary theory, there is good reason to pay special attention to these units and not to others.

If science were a theoretically neutral exercise, all decisions would be on a par. There would be no difference between the claims that it rains a lot in San Francisco and that all bodies attract each other with a force equal to the product of their masses divided by the square of their distances. As soon as scientific theory is allowed to intrude, certain alternatives are closed, certain decisions are preferable. This is the important sense of natural which has lurked behind the distinction between natural and artificial classifications from the beginning.

In the absence of any scientific theory, the only difference between a natural and an artificial classification is the number of characters used. A natural classification is constructed using a large number of characters, while an artificial classification is constructed using only a few (22, 55–57, 78, 115, 128). Biologists have tended to object to this characterization because it seemed to leave something out, but they have not been too articulate in describing this something. They have argued that a natural biological classification is one based on biologically relevant attributes—as many as possible. An artificial classification is one based on biologically irrelevant attributes—regardless of how many. The controversy has surrounded the sense in which attributes can be biologically relevant or irrelevant.

Taxonomists have tended to term an attribute relevant or taxonomically useful if it has served to cluster organisms into reasonably discrete groups. Thus, for future runs on a group, it would be given greater weight a posteriori. Pheneticists are in full agreement with this usage. But taxonomists also wish to extend their taxonomically useful attributes to cover additional, unstudied groups. This is the a priori weighting to which the pheneticists raised such vocal objections. The justification for such an extension, when it is justified, rests on the second and more important sense of biologically relevant. Certain concepts are central to biological theories; others are not. For example, canalization, geographic isolation, crossing over, epistatic interaction, and gene flow are important concepts in contemporary biological theory. Hence, a classification in which they were central would be natural in the above sense. Of course, gene flow is not used to define the name of a particular taxon, but it does serve two other functions. It plays an important part in the definition of species, and this definition, in turn, determines which taxa are classed at the species level and which are not. In addition, it might play a part in justifying the claim that an attribute which was taxonomically useful in group *A* should also prove to be taxonomically useful in group *B*. To the extent that such claims are justified, they must be backed up by appropriate scientific laws.

An empiricist might object that all attributes of organisms are equally real. This is certainly true. The broken setae of an insect are as real as a mutation which permits it to produce double the number of offspring, but they hardly are equally important. Just as physical elements are classified on the basis of their atomic number—an attribute selected because of its theoretical significance—evolutionary elements are classified on the basis of their reproductive habits and for the same reasons. Evolutionists contend that if all the data were available, a high percentage of organisms which reproduce by interbreeding could be grouped for long periods of their duration into phylogenetically significant units by the biological species definition.

The pheneticists have attacked this contention on two fronts. First, they have argued that biological species, like phenetic species, are arbitrary units and, second, that biological species, even if they could be determined, would not form pervasive, significant units in evolution. At the heart of the first criticism is the evaluative term "arbitrary." Claiming to use "arbitrary" in Simpson's sense, Sokal and Crovello (117) say, "Our study of the operations necessary to delimit a biological species revealed considerable arbitrariness in the application of the concept. This is in direct conflict with the claim of nonarbitrariness by proponents of the BSC.... The degree of sterility required in any given cross, the number of fertile crosses between members of populations, not to mention the necessarily arbitrary decisions proper to the hidden phenetic components of the BSC, make this concept no less arbitrary than a purely phenetic species concept, and perhaps even more so, since phenetics is one of its components."

Simpson's definition of "arbitrary" is hardly relevant to the issues at hand. According to Simpson (101, 103), when there is a criterion of classification and a classification, groups in this classification are nonarbitrary to the extent that they have actually been classified according to the criterion. For example, if species A is defined in terms of property f, then the species is nonarbitrary if all of its members have f; otherwise, it is not. Simpson's definition is extraneous to this discussion since it assumes precisely what is at issue.

What then do Sokal and Crovello mean by "arbitrary"? Since they repeatedly designate decisions in phenetic taxonomy as arbitrary and since they are advocates of phenetic taxonomy, one might reasonably infer that they do not take it to be a term of condemnation. Yet in one place they talk of arbitratiness as being a drawback to various species definitions. "Arbitrary" is used in ordinary discourse in a host of different senses, and the pheneticists, in a manner not confined to themselves, seem to switch casually from one to another in their criticisms of evolutionary taxonomy. At one extreme, a decision is arbitrary if more than one choice is possible. This is unfortunate because in science more than one reasonable decision is always possible. Hence, all scientific decisions become arbitrary, and the term ceases to make a distinction. For example, should physicists retain Euclidean geometry and complicate their physical laws, or should they retain the simplicity of their laws and treat space as non-Euclidean? Either choice is possible, but physicists' decision for the latter is hardly arbitrary.

A more reasonable use of "arbitrary" is in the division of continua into segments. Biologists of all persuasions commonly admit that whenever an even gradation exists, any classificatory decision automatically becomes arbitrary (17, 103, 119). Here there are not just two or a few possible choices, but many, perhaps infinitely many. Hidden in this line of reasoning is the essentialist prejudice that the only distinctions that exist are sharp distinctions. Unless there is a complete, abrupt break in the distribution of the characters being used for classification, no meaningful decisions can be made. This prejudice was one of the primary motives for philosophers' refusing to countenance even the possibility of evolution by gradual variation and for many philosophers' and biologists' opting for evolution by saltation (65). But this prejudice runs counter to both the very nature of modern science and the methods being introduced by the pheneticists. Various statistical means exist for clustering elements, even when at least one element exists at every point in the distributional space. For example, there are reasons for dividing a bimodal curve at some points rather than at others. Darwin argued that species as well as varieties intergraded insensibly. He concluded, therefore, that they were equally arbitrary. Owing to the mathematical and philosophical prejudices of his day, Darwin's conclusion is understandable. There is no excuse for similar prejudices still persisting (25, 72).

All decisions in phenetic taxonomy are hardly arbitrary in any meaningful sense. If they were, then all the techniques of phenetic taxonomy could be replaced by the single expedient of flipping a coin. Similarly, all decisions as to the degree of crossing, the number of fertile offspring and their viability, etc., sufficient to ensure the presence or absence of reproductive isolation are hardly arbitrary in any meaningful sense of the term. From all indications, various thresholds exist in the empirical world. The temperature of water can be varied continuously, but it does not follow thereby that the attendant physical phenomena also vary continuously. At the boiling point, at the freezing point, and near absolute zero, a change of a single degree is accompanied by extremely discontinuous changes in the attendant physical phenomena. Similary, for example, Simpson refers to quantum evolution, the burst of proliferation that follows a population managing to make its way through an adaptive valley to invade a new ecological niche (88, 101, 103; see also Lewontin's "The Units of Selection," *Annual Rev. Ecol. Syst.* 1970, 1:1–14.

There seems to be no question that such significant thresholds exist in evolution. Recently, however, pheneticists have contended that the biological species concept does not mark such a threshold (42, 45, 115). Of all the criticisms leveled at evolutionary taxonomy in the last ten years, this is the most serious. Most of the other criticisms have been largely methodological, resting uneasily on certain dubious philosophical positions, but this criticism is empirical. In a recent study by Ehrlich and Raven (46), evidence was adduced to show that selection is so overwhelmingly important in speciation that the occasional effects of gene flow can safely be ignored in the general evolutionary picture. If this contention is borne out by additional investigation, then the role of the biological definition of species will have been fatally undermined and the synthetic theory of evolution will have to be modified accordingly.

Sokal and Crovello (117), concurring with the position of Ehrlich and Raven, observe that "possibly concepts such as the BSC are more of a burden that a help in understanding evolution." They go on to conclude, however, that "the phenetic species as normally described and whose definition may be improved by numerical taxonomy is the desirable appropriate concept to be associated with the category, species, while the local population may be the most useful unit for evolutionary study." If it can be shown that biological species are not significant units in evolution, then from the point of view of evolutionary taxonomy, the role of the biological species has been fatally undermined. It does not follow, therefore, that the phenetic species, as normally described, should automatically replace them in biological classification, if for no other reason than that no description has been provided yet for the phenetic species. Instead, there are literally an infinite number of phenetic units, all of which have an equal right, on the principles of numerical taxonomy, to be called species.

CONCLUSION

Numerous distinctions have been drawn in the preceding pages [of this chapter], but little notice has been taken of the most important distinction underlying the phenetic-phyletic controversy—the difference between explicit and implicit or intuitive taxonomy. Simpson (103) has argued that taxonomy, like many other sciences, is a combination of science and art. For example, tempering vertical with horizontal classification, dividing a gradually evolving lineage into species, deciding how much interbreeding is permissible before two populations are included in the same species, the assignment of category rank above the species level, choices between alternative ways of classifying the same phylogeny, balancing splitting and lumping tendencies, and the inductive inferences by which phylogenies are inferred are all to some extent part of the art of taxonomy. The question is whether the intuitive element in taxonomy should be decreased and, if so, at what cost.

It has been assumed in this chapter that decreasing the amount of art in taxonomy is desirable. Taxonomists can be trained to produce quite excellent classifications without being able to enunciate the principles by which they are classifying, just as pigeons can be trained to use the first-order functional calculus in logic. Human beings can be trained to be quite efficient classifying machines. They can scan complex and subtle data and produce estimates of similarity with an accuracy which far exceeds the capacity of current techniques of multivariate analysis. Taxonomists as classifying machines, however, have several undesirable qualities. Although taxonomists, once trained, tend to produce consistent, accurate classifications, the programs by which they are producing these classifications are unknown to other taxonomists and vary from worker to worker. In addition, just when a taxonomist is reaching the peak of his abilities, he tends to die. Only recently one of the most accomplished taxonomists passed away and with her, all the experience which she had accumulated during decades of doing taxonomy.

The resistance to making taxonomic practice and procedures explicit seems to have stemmed from two sources: one, an obscurantist obsession with the ultimate mystery of the human intellect; the other, a concern over how much theoretical significance one must sacrifice in order to make biological classification explicit. With respect to the first reservation, Kaplan (74) has distinguished between reconstructed logic and logic-in-use. Frequently, during the course of development of formal and empirical science, empirical scientists use certain modes of inference which are beyond the current formal reconstructions. There is the tendency to dismiss these modes of inference by attributing them to genius, imagination, and unanalyzable, fortuitous guesswork. Kaplan (74) views the intuition of great scientists, not as lucky guesswork, but as currently unreconstructed logic-in-use. Intuition is any logic-in-use which is preconscious and outside the inference schemata for which we have readily available reconstructions. "We speak of intuition, in short, when neither we nor the discoverer himself knows quite how he arrived at his discoveries, while the frequency or pattern of their occurrence makes us reluctant to ascribe them merely to chance."

The second reservation which taxonomists have had about making taxonomy less intuitive and more explicit is less subtle, but equally important. In the early days of phenetic taxonomy, pheneticists seemed willing to dismiss the theoretical side of biological classification, since it seemed to make straightforward reconstructions extremely difficult, if not impossible. They tended to conflate the complexity of taxonomic inferences with taxonomists being muddle-headed. Certainly some of the complexity of traditional taxonomy may well have been due just to sloppy thinking, but instead of this evaluation being the immediate, initial response, it should have been the last resort. Traditional taxonomists and computer taxonomists are going to have to adapt to each other, but this adaptation cannot be purchased at the expense of the purposes of scientific investigation. These ends are better characterized by the words "theoretical significance" than by "usefulness". An extremely accurate scientific theory of great scope will certainly be useful, but there are many things which are useful, though of little theoretical significance.

NOTES

I wish to thank Donald H. Colless, Theodore J. Crovello, Michael T. Ghiselin, Ernst Mayr, and Robert R. Sokal for reading and criticizing this paper. The preparation of this paper was supported in part by NSF grant GS-1971.

1. For those interested in a review of the numerical aspects of the phenetic-phyletic controversy, I recommend Johnson (73).

2. For a more realistic, formal axiomatization of evolutionary theory, see Williams (129).

3. Dobzhansky condenses the preceding discussion to about half its length in the third edition of his work, and Mayr quotes the final paragraph in his *Systematics and the Origin of Species* (80).

4. One of the most persistent problems in taxonomy has been the explication of the notion of "similarity," which is to be some function of descent. An analysis of this concept must be postponed until the next section.

REFERENCES

1. Beckner, M. 1959. *The Biological Way of Thought*. New York, Columbia University Press.

2. Bigelow, R. S. 1956. Monophyletic classification and evolution. *Syst. Zool.* 5:145–46.

3. ———. 1958. Classification and phylogeny. *Syst. Zool.* 7:49–59.

4. ———. 1959. Similarity, ancestry, and scientific principles. *Syst. Zool.* 8:165–68.

5. ———. 1961. Higher categories and phylogeny. *Syst. Zool.* 10:86–91.

6. Birch, L. C., and Ehrlich, P. R. 1967. Evolutionary history and population biology. *Nature* 214:349–52.

7. Blackwelder, R. E. 1959. The present status of systematic zoology. *Syst. Zool.* 8:69–75.

8. ———. 1959. The functions and limitations of classification. *Syst. Zool.* 8:202–11.

9. ———. 1962. Animal taxonomy and the new systematics. *Surv. Biol. Progr.* 4:1–57.

10. ———. 1964. Phyletic and phenetic *versus* omnispective classification. In *Phenetic and Phylogenetic Classification*, ed. V. H. Heywood and J. McNeill, 17–28, London, Systematics Assoc.

11. ———. 1967. A critique of numerical taxonomy. *Syst. Zool.* 16:64–72.

12. ———. 1967. *Taxonomy*, New York, Wiley.

13. Blackwelder, R. E., and Boyden, A. 1952. The nature of systematics. *Syst. Zool.* 1:26–33.

14. Bock, W. J. 1963. Evolution and phylogeny in morphologically uniform groups. *Am. Natur.* 97:265–85.

15. Borgmeier, T. 1957. Basic questions of systematics. *Syst. Zool.* 6:53–69.

16. Boyce, A. J. The value of some methods of numerical taxonomy with reference to hominoid classification. See Ref. 10:47–65.

17. Burma, B. H. 1949. The species concept: A semantic review. *Evolution* 3:369–70.

18. Cain, A. J. 1958. Logic and memory in Linnaeus's system of taxonomy. *Proc. Linn. Soc. London* 169:144–63.

19. ———. 1959. Deductive and inductive methods in post-Linnaen taxonomy. *Proc. Linn. Soc. London* 170:185–217.

20. ———. 1959. Taxonomic concepts. *Ibis* 101:302–18.

21. ———. 1962. Zoological classification. *Aslib Proc.* 14:226–30.

22. Cain, A. J., and Harrison, G. A. 1958. An analysis of the taxonomist's judgment of affinity. *Proc. Zool. Soc., London* 131:85–98.

23. ———. 1960. Phyletic weighting. *Proc. Zool. Soc. London* 135:1–31.

24. Camin, J. H., and Sokal, R. R. 1965. A method for deducing branching sequences in phylogeny. *Evolution* 19:311–26.

25. Cargile, J. 1969. The sorites paradox. *Brit. J. Phil. Sci.* 20:193–202.

26. Carmichael, J. W., George, J. A., and Julius, R. S. 1968. Finding natural clusters. *Syst. Zool.* 17:144–50.

27. Cavalli-Sforza, L. L., and Edwards, A. W. F. 1967. Phylogenetic analysis: models and estimation procedures. *Evolution* 21:550–70.

28. Clark, R. B. 1956. Species and systematics. *Syst. Zool.* 5:1–10.

29. Colless, D. H. 1967. An examination of certain concepts in phenetic taxonomy. *Syst. Zool.* 16:6–27.

30. ———. 1967. The phylogenetic fallacy. *Syst. Zool.* 16:289–95.

31. ———. 1970. The relationship of evolutionary theory to phenetic taxonomy. *Evolution*.

32. Crowson, R. A. 1965. Classification, statistics and phylogeny. *Syst. Zool.* 14:144–48.

33. Daly, H. V. 1961. Phenetic classification and typology. *Syst. Zool.* 10:176–79.

34. Darwin, F. 1959. *The Life and Letters of Charles Darwin*. New York, Basic Books, 2 vols.

35. Dexter, R. W. 1966. Historical aspects of studies on the Brachipoda by E. E. Morse, *Syst. Zool.* 15:241–43.

36. Dobzhansky, T. 1935. A critique of the species concept in biology. *Phil. Sci.* 2:344–55.

37. ———. 1937. *Genetics and the Origin of Species*, New York, Columbia University Press.

38. ———. 1940. Speciation as a stage in evolutionary divergence. *Am. Natur.* 74:312–21.

39. DuPraw, E. J. 1964. Non-Linnaean taxonomy. *Nature* 202:849–52.

40. Edwards, A. W. F., and Cavalli-Sforza, L. L. 1964. Reconstruction of evolutionary trees. See Ref. 10, 67–76.

41. Ehrlich, P. R. 1958. Problems of higher classification. *Syst. Zool.* 7:180–84.

42. ———. 1961. Has the biological species concept outlived its usefulness? *Syst. Zool.* 10:167–76.

43. ———. 1964. Some axioms of taxonomy. *Syst. Zool.* 13:109–23.

44. Ehrlich, P. R., and Ehrlich, A. H. 1967. The phenetic relationships of the butterflies. *Syst. Zool.* 16:301–27.

45. Ehrlich, P. R., and Holm, R. W. 1962. Patterns and populations. *Science* 137:652–57.

46. Ehrlich, P. R., and Raven, P. H. 1969. Differentation of populations. *Science* 165:1228–31.

47. Farris, J. S. 1967. The meaning of relationship and taxonomic procedure. *Syst. Zool.* 16:44–51.

48. ———. 1968. Categorical rank and evolutionary taxa in numerical taxonomy. *Syst. Zool.* 17:151–59.

49. Fisher, R. A. 1930. *The Genetical Theory of Natural Selection*. Oxford: Clarendon.

50. Ghiselin, M. T. 1966. On psychologism in the logic of taxonomic principles. *Syst. Zool.* 15:207–15.

51. ———. 1967. Further remarks on logical errors in systematic theory. *Syst. Zool.* 16:347–48.

52. ———. 1969. *The Triumph of the Darwinian Method*. Berkeley, University of California Press.

53. ———. 1969. The principles and concepts of systematic biology. In *Systematic Biology*, Publ. 1962. Nat. Acad. Sci., ed. C. G. Sibley, 45–55.

54. Gilmartein A. J. 1967. Numerical taxonomy—an eclectic viewpoint. *Taxon* 16:8–12.

55. Gilmour, J. S. L. 1937. A taxonomic problem. *Nature* 139:1040–47.

56. ———. 1940. Taxonomy and philosophy. In *The New Systematics*, ed. J. Huxley, 461–74. London, Oxford University Press.

57. Gilmour, J. S. L., and Walters, S. M. 1964. Philosophy and classification. *Vistas Bot.* 4:1–22.

58. Haldane, J. B. S. 1932. *The Causes of Evolution*. London, Harpers.

59. Hempel, C. G. 1965. *Aspects of Scientific Explanation*. New York, Free Press.

60. Hennig, W. 1950. *Grundzüge einer Theorie der phylogenetischen Systematik*. Berlin, Deut. Zentralverlag.

61. ———. 1966. *Phylogenetic Systematics*. Chicago, University of Illinois Press.

62. Hull, D. L. 1964. Consistency and monophyly. *Syst. Zool.* 13:1–11.

63. ———. 1965. The effect of essentialism on taxonomy. *Brit. J. Phil. Sci.* 15:314–26, 16:1–18.

64. ———. 1967. Certainty and circularity in evolutionary taxonomy. *Evolution* 2:174–89.

65. ———. 1967. The metaphysics of evolution. *Brit. J. Hist. Sci.* 3:309–37.

66. ———. 1968. The operational imperative—sense and nonsense in operationism. *Syst. Zool.* 16:438–57.

67. ———. 1969. The natural system and the species problem. In *Systematic Biology*, Publ. 1962, Nat. Acad. Sci., ed. C. G. Sibley, 56–61.

68. Huxley, J., ed. 1940. *The New Systematics*, London, Oxford University Press.

69. ———. 1942. *Evolution: The Modern Synthesis.* London, Allen & Unwin.

70. Huxley, T. H. 1874. On the classification of the animal kingdom. *Nature* 11:101–2.

71. Inger, R. R. 1958. Comments on the definition of genera. *Evolution* 12:370–84.

72. James, M. T. 1963. Numerical vs. phylogenetic taxonomy. *Syst. Zool.* 12:91–93.

73. Johnson, L. A. S. 1968. Rainbow's end: the quest for an optimal taxonomy. *Proc. Linn. Soc. N.S.W.* 93:8–45.

74. Kaplan, A. 1964. *The Conduct of Inquiry.* San Francisco, Chandler, 428 pp.

75. Kiriakoff, S. G. 1959. Phylogenetic systematics versus typology. *Syst. Zool.* 8:117–18.

76. ———. 1965. Cladism and phylogeny. *Syst. Zool.* 15:91–93.

77. Kluge, A. G., and Farris, J. S. 1969. Quantitative phyletics and the evolution of Anurans. *Syst. Zool.* 18:1–32.

78. Lorch, J. 1961. The natural system in biology. *Phil. Sci.* 28:282–95.

79. Mackin, J. H. 1963. Rational and empirical methods of investigation in geology. In *The Fabric of Geology*, ed. C. C. Albritton, 135–63, New York: Addison-Wesley.

80. Mayr, E. 1942. *Systematics and the Origin of Species.* New York, Columbia University Press.

81. ———. 1949. The species concept. *Evolution* 3:371–72.

82. ———. 1957. *The Species Problem.* AAAS Publ. N. 50, Washington.

83. ———. 1959. Agassiz, Darwin, and evolution. *Harvard Libr. Bull.* 13:165–94.

84. ———. 1963. *Animal Species and Evolution.* Cambridge, Harvard University Press.

85. ———. 1965. Numerical phenetics and taxonomic theory. *Syst. Zool.* 14:73–97.

86. ———. 1969. *Principles of Systematic Zoology.* New York, McGraw-Hill.

87. ———. 1969. The biological meaning of species. *Biol. J. Linn. Soc.* 1:311–20.

88. Megletsch, P. A. 1954. On the nature of the species. *Syst. Zool.* 3:49–65.

89. Michener, C. D. 1957. Some bases for higher categories in classification. *Syst. Zool.* 6:160–73.

90. ———. 1963. Some future developments in taxonomy. *Syst. Zool.* 12:151–72.

91. Michener, C. D., and Sokal, R. R. 1957. A quantitiative approach to a problem in classification. *Evolution* 11:130–62.

92. Minkoff, E. C. 1964. The present state of numerical taxonomy. *Syst. Zool.* 13:98–100.

93. Remane, A. 1952. *Die Grundlagen des natürlichen Systems, der vergleichenden Anatomie und der Phylogenetik.* Leipzig, Geest & Portig.

94. Rensch, B. 1929. *Das Prinzip geographischer Rassenkreise und das Problem der Artbildung.* Berlin, Borntraeger.

95. ———. 1947. *Neure Probleme der Abstammungslehre.* Stuttgart, Enke.

96. Rohlf, F. J. 1963. The consequence of larval and adult classification in Aedes. *Syst. Zool.* 12:97–117.

97. Sattler, R. 1963. Methodological problems in taxonomy. *Syst. Zool.* 13:19–27.

98. ———. 1963. Phenetic contra phyletic systems. *Syst. Zool.* 12:94–95.

99. Simpson, G. G. 1944. *Tempo and Mode in Evolution.*, New York, Columbia University Press.

100. ———. 1945. The principles of classification and a classification of mammals. *Bull. Am. Mus. Natur. Hist.* 85:1–350.

101. ———. 1951. The species concept. *Evolution* 5:285–98.

102. ———. 1953. *The Major Features of Evolution.* New York, Columbia University Press.

103. ———. 1961. *Principles of Animal Taxonomy.* New York, Columbia University Press.

104. ———. 1964. Numerical taxonomy and biological classification. *Science* 144:712–13.

105. Simpson, G. G., Roe, A., and Lewontin, R. C., 1960, *Quantitative Zoology.* New York, Harcourt, Brace & World.

106. Sneath, P. H. A. 1957. The application of computers to taxonomy. *J. Gen. Microbiol.* 17:201–26.

107. ———. 1958. Some aspects of Adansonian classification and of the taxonomic theory of correlated features. *Ann. Microbiol. Enzimol.* 8:261–68.

108. ———. 1961. Recent developments in theoretical and quantitative taxonomy *Syst. Zool.* 10:118–39.

109. ———. 1964. Introduction. See Ref. 10, 43–45.

110. ———. 1968. International conference on numerical taxonomy. *Syst. Zool.* 17:88–92.

111. Sokal, R. R. 1959. Comments on quantitative systematics. *Evolution* 13:420–23.

112. ———. 1961. Distance as a measure of taxonomic similarity. *Syst. Zool.* 10:70–79.

113. ———. 1962. Typology and empiricism in taxonomy. *J. Theor. Biol.* 3:230–67.

114. ———. 1969. Review of Mayr's *Principles of Systematic Zoology. Quart. Rev. Biol.* 44:209–11.

115. Sokal, R. R., and Camin, J. H. 1965. The two taxonomies: Areas of agreement and conflict. *Syst. Zool.* 14:176–95.

116. Sokal, R. R., Camin, J. H., Rohlf, F. J., and Sneath, P. H. A. 1965. Numerical taxonomy: Some points of view. *Syst. Zool.* 14:237–43.

117. Sokal, R. R., and Crovello, T. J. 1970. The biological species concept: A critical evaluation. *Am. Natur.* 104:127–53.

118. Sokal, R. R., and Michener, C. D. 1958. A statistical method for evaluating systematic relationships. *Univ. Kansas Sci. bull.* 38:1409–38.

119. Sokal, R. R., and Sneath, P. H. A. 1963. *The Principles of Numerical Taxonomy.* San Francisco, Freeman.

120. Stebbins, G. L. 1950. *Variation and Evolution in Plants.* New York, Columbia University Press.

121. Thompson, W. R. 1952. The philosophical foundations of systematics. *Can. Entomol.* 84:1–16.

122. ———. 1960. Systematics: The ideal and the reality. *Studio Entomol.* 3:493–99.

123. ———. 1962. Evolution and taxonomy. *Studio Entomol.* 5:549–70.

124. Thorne, R. F. 1963. Some problems and guiding principles of Angiosperm phylogeny. *Am. Natur.* 97:287–305.

125. Throckmorton, L. H. 1965. Similarity *versus* relationship in *Drosophila. Syst. Zool.* 14:221–36.

126. ———. 1968. Concordance and discordance of taxonomic characters in *Drosophila* classification. *Syst. Zool.* 17:355–87.

127. Troll, W. 1944. Urbild und Ursache in der Biologie, *Bot. Arch.* 45:396–416.

128. Warburton, F. E. 1967. The purposes of classification. *Syst. Zool.* 16:241–45.

129. Williams M. B. 1970, Deducing the consequences of evolution: A mathematical model. *J. Theor. Biol.* 29:343–385.

130. Wright, S. 1931. Evolution in Mendelian populations. *Genetics* 16:97–159.

131. ———. 1931. Statistical theory of evolution. *Am. Statist. J.* March suppl., 201–8.

VIII Phylogenetic Inference

16 The Logical Basis of Phylogenetic Analysis

James Farris

Phylogeneticists hold that the study of phylogeny ought to be an empirical science, that putative synapomorphies provide evidence on genealogical relationship, and that (aside possibly from direct observation of descent) those synapomorphies constitute the only available evidence on genealogy. Opponents of phylogenetic systematics maintain variously that genealogies cannot (aside from direct observation) be studied empirically, that synapomorphies are not evidence of kinship because of the possibility of homoplasy, or that raw similarities also provide evidence on genealogy. Most phylogeneticists recognize that inferring genealogy rests on the principle of parsimony—that is, choosing genealogical hypotheses so as to minimize requirements for ad hoc hypotheses of homoplasy. But other criteria as well have been proposed for phylogenetic analysis, and some workers believe that parsimony is unnecessary for that purpose. Others contend that that principle is not truly "parsimonious," or that its application depends crucially on the false supposition that homoplasy is rare in evolution. Authors of all these criticisms have in common the view that phylogenetic systematics as it is now practiced may be dismissed as futile or at best defective. Phylogeneticists must refute that view, but accomplishing that goal seems complicated both by the apparent multiplicity of phylogenetic methods and by the diversity of the objections. I shall show here that the complexity of this problem is superficial. An analysis of parsimony will not only provide a resolution of the objections to that criterion, but will supply as well an understanding of the relationship of genealogical hypotheses to evidence, and with it a means of deciding among methods of phylogenetic inference.

AD HOC HYPOTHESES

I share Popper's disdain for arguing definitions as such, but is it important to make intended meanings clear, and so I shall first dismiss terminological objections to the parsimony criterion. These all come to the idea that parsimonious

From N. Platnick and V. Funk (eds.), *Advances in Cladistics: Proceedings of the Second Meeting of the Willi Hennig Society*, Columbia University Press, 1982, 7–36.

phylogenetic reconstructions are so primarily by misnomer: the word might equally well refer to any of several other qualities. The meanings of "parsimony" would surely take volumes to discuss, but doing so would be quite pointless. Whether the word is used in the same way by all has no bearing on whether the phylogenetic usage names a desirable quality. I shall use the term in the sense I have already mentioned: most parsimonious genealogical hypotheses are those that minimize requirements for ad hoc hypotheses of homoplasy. If minimizing ad hoc hypotheses is not the only connotation of "parsimony" in general usage, at least it is scarcely novel. Both Hennig (1966) and Wiley (1975) have advanced ideas closely related to my usage. Hennig defends phylogenetic analysis on grounds of his auxiliary principle, which states that homology should be presumed in the absence of evidence to the contrary. This amounts to the precept that homoplasy should not be postulated beyond necessity, that is to say, parsimony. Wiley discusses parsimony in a Popperian context, characterizing most parsimonious genealogies as those that are least falsified on available evidence. In his treatment, contradictory character distributions provide putative falsifiers of genealogies. As I shall discuss below, any such falsifier engenders a requirement for an ad hoc hypothesis of homoplasy to defend the genealogy. Wiley's concept is then equivalent to mine.

Cartmill (1981) has effectively objected to that last equivalence, claiming that neither phylogenetic analysis nor parsimony can be scientific in Popper's sense. His argument is superficially technical, but his principal conclusion is in fact based on a terminological confusion, and so I shall discuss his ideas here.

Cartmill cites Gaffney (1979) to the effect that character distributions are falsifiers of genealogical hypotheses, and that it is possible that every conceivable genealogy will be falsified at least once. From the first of these admissions he "deduces" that Gaffney must have relied on the "theorem" that any genealogy contradicted by a character distribution is false. Cartmill then reasons: Some genealogy must be true. "Gaffney's" theorem, together with a falsifier for every genealogy, implies that every genealogy is false. Therefore, Gaffney's claim that character distributions are falsifiers is false.

Cartmill's argument rests directly and entirely on a misrepresentation of the Popperian meaning of "falsifier": a test statement that, if true, allows a hypothesis to be rejected. There is a great difference between "falsify" in Popper's sense and "prove false." The relationship between a theory and its falsifiers is purely logical; Popper never claimed that proof of falsity could literally be achieved empirically. "Observing" a falsifier of a theory does not prove that the theory is false; it simply implies that either the theory or the observation is erroneous. It is then seen that the only implication that can be derived from falsification of every genealogy is that some of the falsifiers are errors—homoplasies. It is thus seen as well that Cartmill's "syllogism" is nothing other than an equivocation.

So much for the claim that characters cannot be Popperian falsifiers, but is phylogenetic parsimony Popperian? Cartmill admits that phylogeneticists hold that the least falsified genealogy is to be preferred. The reason for this preference is that each falsifier of any accepted genealogy imposes a requirement for an ad hoc hypothesis to dispose of the falsifier. According to Popper—as Cartmill also cites—ad hoc hypotheses must be minimized in scientific investigation. Cartmill never attempts to argue that conflicts between characters and genealogies do not require hypotheses of homoplasy, and so none of his claims can serve to question the connection between parsimony and Popper's ideas.

PARSIMONY AND SYNAPOMORPHY

The objection that parsimony requires rarity of homoplasy in evolution is usually taken to be just that: a criticism of parsimony. It might seem that the problem posed by that objection could be avoided simply by using some other criterion for phylogenetic analysis. Some quite nonphylogenetic proposals, such as grouping according to raw similarity, have been made along those lines, and I shall discuss those eventually. Of more immediate interest is the question whether grouping by putative synapomorphy can do without the parsimony criterion.

Watrous and Wheeler (1981) suggest that parsimony is needed only when characters conflict, with the implication that a set of congruent characters can be analyzed without invoking ad hoc hypotheses of homoplasy. A similar idea would appear to underlie advocacy by Estabrook and others (reviewed by Farris and Kluge 1979) of techniques ("clique" methods) that "resolve" character conflicts by discarding as many characters as necessary so that those surviving (the clique) are mutually congruent. The surviving characters are then used to construct a tree. Proponents of such methods maintain that the tree so arrived at rests on a basis different from parsimony.

The character selection process itself may well have a distinctive premise, a possibility that I shall discuss below. To claim that the interpretation of the characters selected rests on a basis other than parsimony, however, seems not to be defensible. The tree constructed from a suite of congruent characters by a clique method is chosen to avoid homoplasy in any of those characters, the possibility of doing so being assured by the selection. (Selection aside, Watrous and Wheeler proceed likewise.) It seems accurate, then, to describe that construction as minimizing requirements for ad hoc hypotheses of homoplasy for the characters within the congruent suite, but, more particularly, there seems to be no other sensible rationale for the construction. No one seems to have suggested any such principle, aside from the obvious: that if the characters were free of homoplasy (were "true" as it is often put), then the tree would follow. But the characters comprising a congruent suite are hardly observed to be free of homoplasy. At the most it might be said that the selected characters seem to suggest no genealogy other than the obvious one.

Of course it is what data suggest, or how they do it, that is at issue. If a suite of congruent characters is interpreted by avoiding unnecessary postulates of homoplasy, then the interpretation embodies parsimony. But the only apparent motivation for concentrating just on congruent characters is to avoid reliance on parsimony. That avoidance would seem sensible only on the supposition that parsimony is ill founded, and the only apparent reason for that supposition is the charge that parsimony depends crucially on unrealistic assumptions about nature. If that charge means anything at all, it must mean that taking conditions of nature realistically into account would lead to preference for a less parsimonious arrangement over a more parsimonious one. But if that charge were correct, then it would be—to say the least—less than obvious why the implications of those natural conditions would be expected to change simply because any characters incongruent with those chosen had been ignored.

If avoiding ad hoc hypotheses of homoplasy is unjustified, then neither Watrous and Wheeler nor clique advocates are entitled to the inferences on phylogeny that they draw, but the significance of parsimony for Hennigian methods is much more general than that. Watrous and Wheeler probably thought that they had no need for anything so questionable as parsimony, because they were simply applying Hennig's well-established principle of grouping according to synapomorphy. Just how did that principle come to be well established? It is usually explained by taking note of the logical relationship between monophyletic groups and true synapomorphies, but that leaves open the question of how genealogies are related to observed features. It might well be questioned whether the logical construct can legitimately be extended into a principle to guide interpretation of available characters. That question has in fact often been raised, and almost always in the form of the suggestion that putative synapomorphies are not evidence of kinship because they might well be homoplasies. Hennig's (1966) own reply to that objection was his auxiliary principle, which, as I have already observed, is a formulation of the parsimony criterion.

Hennig's defense of the synapomorphy principle by recourse to parsimony is not accidental, but necessary. The analytic relationship of correct synapomorphies to phylogeny is just that a property that evolved once and is never lost must characterize a monophyletic group. Synapomorphies are converted into a genealogy, that is, by identifying the tree that allows a unique origin for each derived condition. A phylogeny based on observed features is parsimonious to the degree that it avoids requirements for homoplasies—multiple origins of like features. Secondary plesiomorphies aside, a plesiomorphic trait will already have a single origin at the root of the putative tree, so that the effect of parsimony is precisely to provide unique derivations wherever possible. (Secondary plesiomorphies, being a kind of apomorphy, are treated likewise.) Grouping by synapomorphy would thus have to behave like parsimony, but further, the latter applies to actual traits, whereas the logic of true synapomorphies does not. Superficially, the use of the synapomorphy principle in phylogenetic inference seems to be just a consequence of the logical connection between true synapomorphies and genealogies, but it cannot be just that, as the condition of that logic—that the traits are indeed synapomorphies—need not be met. Grouping by putative synapomorphy is instead a consequence of the parsimony criterion.

ABUNDANCE OF HOMOPLASY

There are two main varieties of the position that use of the parsimony criterion depends crucially on the supposition that homoplasy is rare in evolution. In the first, the observation that requirements for homoplasy are minimized is taken as prima facie evidence that the supposition is needed. In the second, the claim is advanced in conjunction with some more elaborate, often statistical, argument. The conclusion from the first kind of reasoning is quite general, while that from the second is necessarily limited by the premises of the argument employed. If the first kind of criticism were correct, there would be little point to considering arguments of the second sort. I shall thus first point out why the first type of objection rests on a fallacy.

To evaluate the claim that an inference procedure that minimizes something must ipso facto presuppose that the quality minimized is rare, it is useful to consider a common application of statistics. In normal regression analysis, a regression line is calculated from a sample of points so as to minimize residual variation around the line, and the residual variation is then used to estimate the parametric residual variance. Plainly the choice of line has the effect of minimizing the estimate of the residual variance, but one rarely hears this procedure criticized as presupposing that the parametric residual variance is small. Indeed, it is known from normal statistical theory that the least squares line is the best point estimate of the parametric regression line, whether the residual variance is small or not. The argument that the parsimony criterion must presume rarity of homoplasy just because it minimizes required homoplasy is thus at best incomplete. That reasoning presumes a general connection between minimization and supposition of minimality, but it is now plain that no such general connection exists. Any successful criticism of phylogenetic parsimony would have to include more specific premises.

The same conclusions can readily be reached in a specifically phylogenetic context. Suppose that for three terminal taxa A, B, C, there are ten putative synapomorphies of A + B and one putative apomorphy shared by B and C. We assume for simplicity of discussion that the characters are independent and all of equal weight, and that attempts to find evidence to support changes in the data have already failed. Parsimony then leads to the preference for ((A, B), C) over alternative groupings. We will be interested in whether abundance of homoplasy leads to preference for some other grouping. If it does not, then the claim that parsimony presupposes rarity of homoplasy is at best not generally true.

It is plain that the grouping ((B, C), A) is genealogically correct if the one B + C character is, in fact, a synapomorphy, and that ((A, B), C) is instead correct if the A + B characters are synapomorphies. Truth of the latter grouping does not require, however, that all ten of the putative synapomorphies of A + B be accurate homologies. If just one of those characters were truly a synapomorphy while all the other characters in the data were in fact parallelisms, the genealogy would necessarily be ((A, B), C). That A and B share a common ancestor unique to them, in other words, does not logically require that every feature shared by A and B was inherited from that ancestor. In the extreme, if all the characters were parallelisms, this would not imply that ((A, B), C) is genealogically false. Under those circumstances the data would simply leave the question of the truth of that (or any other) grouping entirely open.

The relationship between characters and genealogies thus shows a kind of asymmetry. Genealogy ((A, B), C) requires that the B + C character be homoplasious, but requires nothing at all concerning the A + B characters. The genealogy can be true whether the conforming characters are homoplasious or not. One kind of objection to phylogenetic parsimony runs that ad hoc hypotheses are indeed to be minimized, but this does not mean minimizing homoplasies, because a genealogy also requires ad hoc hypotheses of homology concerning the characters that conform to it. It is seen that such is not the case. Only characters conflicting with a genealogy lead to requirements for ad hoc hypotheses, and so the only ad hoc hypotheses needed to defend a genealogy are hypotheses of homoplasy.

The sensitivity of inference by parsimony to rarity of homoplasy is readily deduced from these observations. If homoplasy is indeed rare, it is quite likely with these characters that ((A, B), C) is the correct genealogy. In order for that grouping to be false, it would be required at least that all ten of the A + B characters be homoplasious. As these characters are supposed to be independent, the coincidental occurrence of homoplasy in all ten should be quite unlikely. Suppose, then, that homoplasy is so abundant that only one of the characters escapes its effects. That one character might equally well be any of the eleven in the data, and if it is any one of the ten A + B characters the parsimonious grouping is correct. That grouping is thus a much better bet than is ((B, C), A). At the extreme, as has already been seen, if homoplasy is universal, the characters imply nothing about the genealogy. In that case the parsimonious grouping is no better founded than is any other, but then neither is it any worse founded.

It seems that no degree of abundance of homoplasy is by itself sufficient to defend choice of a less parsimonious genealogy over a more parsimonious one. That abundance can diminish only the strength of preference for the parsimonious arrangement; it can never shift the preference to a different scheme. In this the relationship of abundance of homoplasy to choice of genealogical hypothesis is quite like that between residual variance and choice of regression line. Large residual variance expands the confidence interval about the line, or weakens the degree to which the least squares line is to be preferred over nearby lines, but it cannot by itself lead to selection of some other line that fits the data even worse.

STOCHASTIC MODELS

The supposition of abundance of homoplasy by itself offers no grounds for preferring unparsimonious arrangements, but it is easy enough to arrive at that preference by resorting to other premises. Felsenstein (1973, 1978, 1979) objects to parsimony on statistical grounds. He suggests (as others have) that genealogies ought to be inferred by statistical estimation procedures. In his approach he devises stochastic models of evolution, then applies the principle of maximum likelihood, choosing the genealogy that would assign highest probability to the observed data if the model were true. With the models that he has investigated, it develops that the maximum likelihood tree is most parsimonious when rates of change of characters are very small, under which circumstances the models would also predict very little homoplasy. He concludes from this that parsimony requires rarity of homoplasy. In his 1978 paper he discusses a model according to which both parsimony and clique methods would be certain to yield an incorrect genealogy if a large enough random sample of independent characters were obtained. He contends that maximum likelihood estimation under the same conditions would yield the correct tree, as that estimate possesses the statistical property termed consistency. That last is the logical property that if an indefinitely great number of independent characters were sampled at random from the distribution specified by the model, then the estimate would converge to the parameter of the model, the hypothetical true tree.

Felsenstein does not try to defend his models as realistic. his attitude on their purpose seems to be instead that "if a method behaves poorly in this simple model framework, this calls into question its use on real characters" (1981, p. 184), or perhaps (1978, p. 409), "If phylogenetic inference is to be a science, we must consider its methods guilty until proven innocent." The first is preposterous except on the supposition that reasoning from false premises cannot lead to false conclusions. As for the second: to the extent that these models are intended seriously, they comprise empirical claims on evolution. If science required proof concerning empirical claims in order to draw conclusions, no kind of science would be possible.

Felsenstein nonetheless apparently believed that he had demonstrated that practical application of parsimony requires rarity of homoplasy, but in fact such is hardly the case. The dependence of parsimony on rarity of homoplasy is in Felsenstein's analysis a consequence of his models. These models, as he is well aware, comprise "strong assumptions about the biological situation" (1978, p. 403). If those assumptions do not apply to real cases, then so far as Felsenstein can show, the criticism of parsimony need not apply to real cases either. But, again, Felsenstein does not maintain that the assumptions of his models are realistic. He has not shown that abundance of homoplasy implies preference for unparsimonious genealogies. Instead he has shown at most that if homoplasy were abundant, and if in addition the conditions of his models prevailed in nature, then one should prefer unparsimonious schemes. We have already seen that abundance of homoplasy by itself does not justify departure from parsimony. If Felsenstein's argument offers any reason for that departure, then that reason would have to rely on the supposition that his models apply to nature. An ironic result indeed. The original criticism of parsimony was that it required an unrealistic assumption about nature. It now seems instead that unparsimonious methods require such assumptions, whereas parsimony does not.

Felsenstein's arguments from consistency and maximum likelihood have a related drawback. Consistency is a logical relationship between an estimation method and a probability model. In the hypothetical case imagined by Felsenstein, his method would have obtained the right answer, but whether the method would work in practice depends on whether the model is accurate. If it is not, then the consistency of the estimator under the model implies nothing about the accuracy of the inferred tree. The status of a procedure as a maximum likelihood estimator is also bound to the probability model. If the model is false, the ability of a procedure to find the most likely tree under the model implies nothing of how likely the chosen tree might actually be. Likewise the conclusion that parsimony would arrive at the wrong tree depends on the model, and so the hypothetical analysis implies little about the practical accuracy of parsimony.

One might say, of course, that the model illustrates a potential weakness of parsimony: That criterion will fail if the conditions of the model should happen to be met. And how are we to know that this will not happen? This seems in fact to be the intended substance of Felsenstein's remarks. While admitting that his premises are unrealistic, he rejects realism as a criticism of his attack on parsimony, claiming that an objection based on realism "amounts to a confession of ignorance rather than a validation of the inference method in question" (1978, p. 408). A derivation that implies nothing about reality is not much of an improvement on ignorance, of course. To the extent that Felsenstein has a point, then, it seems to be just that parsimony is invalid because we cannot be certain that it will not lead to errors of inference. But there is nothing distinctive about Felsenstein's model in that regard. One may always concoct fantastic circumstances under which scientific conclusions might prove incorrect. It is hardly necessary to resort to mathematical manipulations in order to produce such fears. One need only imagine that his characters have evolved in just the right way to lead him to a false conclusion. Or, with Descartes, that his perceptions and reason have been systematically and maliciously distorted by a demon. None of these possibilities can be disproved, but it hardly matters. There is likewise no reason for accepting any of them, and so collectively they amount to no more than the abstract possibility that a conclusion might be wrong. No phylogeneticist—or any scientist—would dispute that anyway, and so such "objections" are entirely empty. That thinking provides no means of improving either conclusions or methods, but instead offers, if anything, a rejection of all conclusions that cannot be established with certainty. If that attitude were taken seriously, no scientific conclusions whatever could be drawn.

EXPLANATORY POWER

A number of authors, myself among them (Farris, 1973, 1977, 1978), have used statistical models to defend parsimony, using, of course, different models from Felsenstein's. Felsenstein has objected to such derivations on grounds of statistical consistency: as before, parsimonious reconstructions are not consistent under his models. That is no more than an equivocation, as the models differ, and consistency is a relationship between method and model. But I do not mean by this to defend those favorable derivations, for my own models, if perhaps not quite so fantastic as Felsenstein's, are nonetheless like the latter in comprising uncorroborated (and no doubt false) claims on evolution. If reasoning from unsubstantiated suppositions cannot legitimately question parsimony, then neither can it properly bolster that criterion. The modeling approach to phylogenetic inference was wrong from the start, for it rests on the idea that to study phylogeny at all, one must first know in great detail how evolution has proceeded. That cannot very well be the way in which scientific knowledge is obtained. What we know of evolution must have been learned by other means. Those means, I suggest, can be no other than that phylogenetic theories are chosen, just as any scientific theory, for their ability to explain available observation. I shall thus concentrate on evaluating proposed methods of phylogenetic analysis on that basis.

That ad hoc hypotheses are to be avoided whenever possible in scientific investigation is, so far as I am aware, not seriously controversial. That course is explicitly recommended by Popper, For example. No one seems inclined to maintain that ad hoc hypotheses are desirable in themselves; at most they are by-products of conclusions held worthy on other grounds. Nonetheless, I suspect that much of the criticism of the phylogenetic parsimony criterion arises from a failure to appreciate the reasons why ad hoc hypotheses must be avoided. Avoiding them is no less than essential to science itself. Science requires that choice among theories be decided by evidence, and since the effect of an ad hoc hypothesis is precisely to dispose of an observation freely, there could be no effective connection between theory and observation, and the concept of evidence would be meaningless. The requirement that a hypothesis of kinship minimize ad hoc hypotheses of homoplasy is thus no more escapable than the general requirement that any theory should conform to observation; indeed, the one derives from the other.

There are a number of properties commonly held to characterize a theory that gives a satisfactory account of observation. The theory must first of all provide a description of what is known, else it would serve little purpose. As Sober (1975) puts it, theories serve to make experience redundant. But not all descriptions are equally useful. Good theories describe in terms of a coherent framework, so that experience becomes comprehensible; in short they are explanatory. Explanations in turn may be judged on their ability to cover observations with few boundary conditions, that is, with little extrinsic information. Sober has characterized theories satisfying this goal as most informative, or simplest. All these criteria are interrelated in the case of phylogenetic inference, so that they effectively yield a single criterion of analysis. These connections have already been recognized, and I shall summarize them only briefly.

I have elsewhere (Farris 1979, 1980, 1982) already analyzed the descriptive power of hierarchic schemes. I showed that most parsimonious classifications are descriptively most informative in that they allow character data to be summarized as efficiently as possible. That conclusion has aroused some opposition, as syncretistic taxonomists had been inclined to suppose that grouping according to (possibly weighted) raw similarity gave hierarchies of greatest descriptive power. There seems to be no reason for taking that view seriously, however, as no attempt has been made to derive clustering by raw similarity from the aim of effective description of character information.

In my treatment I found that a hierarchic classification provides an informative or efficient description of the distribution of a feature to the degree that the feature need occur in the diagnoses of few taxa. The utility of efficient descriptions is precisely that they minimize redundancy. As I have observed before (particularly Farris, 1980), the presence of a feature in the diagnosis of a taxon corresponds to the evolutionary interpretation that the feature arose in the stem species of that taxon. There is thus a direct equivalence between the descriptive utility of a phylogenetic taxon and the genealogical explanation of the common possession of features by members of that group. Sober (1975) has stressed the importance of informativeness of theories, and has developed a characterization of informativeness in terms of simplicity. It is no surprise to find that simplicity is related to parsimony, and Beatty and Fink (1979) have lucidly discussed the connection in terms of Sober's ideas. Sober (1982) has likewise concluded that phylogenetic parsimony corresponds to simplicity (efficiency, informativeness) of explanation.

In choosing among theories of relationship on the basis of explanatory power, we wish naturally to identify the genealogy that explains as much of available observation as possible. In general, deciding the relative explanatory power of competing theories can be a complex task, but it is simplified in the present case by the fact that genealogies provide only a single kind of explanation. A genealogy does not explain by itself why one group acquires a new feature while its sister group retains the ancestral trait, nor does it offer any explanation of why seemingly identical features arise independently in distantly related lineages. (Either sort of phenomenon might, of course, be explained by a more complex evolutionary theory.) A genealogy is able to explain observed points of similarity among organisms just when it can account for them as identical by virtue of inheritance from a common ancestor. Any feature shared by organisms is so by reason of common descent, or else it is a homoplasy. The explanatory power of a genealogy is consequently measured by the degree to which it can avoid postulating homoplasies.

It is necessary in applying that last observation to distinguish between homoplasies postulated by the genealogy and those concluded for other reasons. A structure common to two organisms might be thought to be a homoplasy on grounds extrinsic to the genealogy. Such a conclusion would amount to specifying that the structure is not a point of heritable similarity. A genealogy would not explain such a similarity, but that would be no grounds for criticizing the genealogy. Rather, the extrinsic conclusion would make the feature irrelevant to evaluating genealogies by effectively stipulating that there is nothing to be explained. The same would hold true for any trait that is known not to be heritable, such as purely phyenotypic variations. The explanatory power of a genealogy is consequently diminished only when the hypothesis of kinship requires ad hoc hypotheses of homoplasy.

By analogy with the abundance of homoplasy argument, it might be objected that seeking a genealogical explanation of similarities is pointless, inasmuch as most similarities are likely to be homoplasies anyway. If homoplasy were universal, that point might well hold. It seems unlikely, however, that homoplasy is universal. It is seldom maintained that segmented appendages have arisen independently in each species of insect. Universality of homoplasy would imply in the extreme that organisms do not generally resemble their parents, a proposition that seems at best contrary to experience. That the character distributions of organisms generally correspond to a hierarchic pattern, furthermore, seems comprehensible only on the view that the character patterns reflect a hierarchy of inheritance. Indeed, the recognized organic hierarchy was one of the chief line of evidence for Darwin's theory of descent with modification. The idea that homoplasy is abundant is not usually intended in such extreme form, of course. Usually it is meant just to suggest that there is room for doubt concerning whether a shared feature is a homology or a homoplasy. Under those circumstances, however, genealogies retain explanatory power. More to the point, the explanatory power of alternative genealogies is still related to their requirements for homoplasies. Suppose that one genealogy can explain a particular point of similarity in terms of inheritance, while a second hypothesis of kinship cannot do so. If that point of similarity is, in fact, a homoplasy, the similarity is irrelevant to evaluating genealogical hypotheses, as has already been seen. If the similarity is, instead, a homology, then only the first genealogy can explain it. If there is any chance that the similarity is homologous, the first genealogy is to be preferred.

There is nothing unusual in the relationship of genealogical hypotheses to characters: scientific theories are generally chosen to conform to data. But it is seldom possible to guarantee that observations are free of errors, and it is no criticism of a theory if it turns out that some of the observations that conform to it are susceptible to error. If a theory does not conform to some observation, however, then the mere suspicion that the observation might be erroneous is not logically adequate to save the theory. Instead the data must be dismissed outright by recourse to an ad hoc hypothesis. Establishing that an observation is erroneous, on the other hand, simply makes it irrelevant to evaluating the theory. The relationship between the explanatory power of genealogies and their requirements for ad hoc hypotheses is likewise characteristic of theories in general. Any observation relevant to evaluating a theory will either conform—and so be explained—or fail to do so, in which case an ad hoc hypothesis is needed to defend the theory. It is generally true that a theory explains relevant observations to the degree that it can be defended against them without recourse to ad hoc hypotheses.

INDEPENDENCE OF HYPOTHESES

Identifying the more explanatory of two alternative hypotheses of kinship is accomplished by finding the total of ad hoc hypotheses of homoplasy required by each. Reckoning those totals will generally involve summing over both separate characters and over observed similarities within characters. Only required ad hoc hypotheses diminish the explanatory power of a putative genealogy. It is thus important to ensure that the homoplasies combined in such totals are logically independent, since otherwise their number need not reflect required ad hoc hypotheses. If two characters were logically or functionally related so that homoplasy in one would imply homoplasy in the other, then homoplasy in both would be implied by a single ad hoc hypothesis. The "other" homoplasy does not require a further hypothesis, as it is subsumed by the relationship between the characters. This is the principle underlying such common observations as that only independent lines of evidence should be used in evaluating genealogies, and that there is no point to using both number of tarsal segments and twice that number as characters. Phylogeneticists seldom attempt to use logically related characters as separate sources of evidence (although an example of this mistake is discussed by Riggins and Farris 1983), and so it seems unnecessary to discuss this point further here.

A different sort of interdependence among homoplasies may arise in considering similarities within a single character. Suppose that 20 of the terminal taxa considered show a feature X, and that a putative genealogy distributes these taxa into two distantly related groups A and B of 10 terminals each. There are 100 distinct two-taxon comparisons of members of A with members of B, and each of those similarities in X considered in isolation comprises a (pairwise) homoplasy. Those homoplasies do not constitute independent required hypotheses, however. The genealogy does not require that similarities in X within either group be homoplasies; it is consistent with identity by descent of X within each group. If X is identical by descent in any two members of A, and also in any two members of B, then the A-B similarities are all homoplasies if any one of them is. The genealogy thus requires but a single ad hoc hypothesis of homoplasy. Of course the numbers in the groups do not matter; the same conclusion would follow if they were 15 and 5, or 19 and 1.

Similar reasoning can be extended to more complex examples, but the problem can be analyzed more simply. If a genealogy is consistent with a single origin of a feature, then it can explain all similarities in that feature as identical by descent. A point of similarity in a feature is then required to be a homoplasy only when the feature is required to originate more than once on the genealogy. A hypothesis of homoplasy logically independent of others is thus required precisely when a genealogy requires an additional origin of a feature. The number of logically independent ad hoc hypotheses of homoplasy in a feature required by a genealogy is then just one less than the number of times the feature is required to originate independently. (The lack of a structure might, of course, be a feature; "origins" should be interpreted broadly, to include losses.)

LENGTHS

That last observation reduces to the rule that genealogies with greatest explanatory power are just those that minimize the (possibly weighted) total of required independent origins of known features. There is another way of putting that characterization, in terms of length. Each required origin of a feature can be assigned (although not necessarily uniquely) to a particular branch of a putative tree, and the weighted total of the origins in a branch can be regarded as that branch's length. If such lengths are summed over branches of the tree, the result is the total of required origins, or the length of the tree. Early work on automatic techniques of parsimony analysis (particularly the Wagner method formulation of Kluge and Farris 1969) used the length conception of parsimony. That formulation has turned out to be technically very useful and has facilitated considerable progress in methods of analysis. (Basic principles are described by Farris 1970; for some applications of greatly improved procedures see Mickevich 1978, 1980; Schuh and Farris 1981; Mickevich and Farris 1981). Nonetheless, its use was in a way unfortunate, for the length terminology has probably caused more misunderstanding than has any other single aspect of parsimony methods.

The length measure used by Kluge and Farris is coincidentally a familiar mathematical measure of distance in abstract spaces, the Manhattan metric. Once ideas have been reduced to formulas, it is easy to forget where the formulas came from, and to devise new methods with no logical basis simply by modifying formulas directly. Phylogenetic reconstructions typically infer the features of hypothesized ancestors, so that the length of a branch lying between two nodes of a tree can be regarded as the distance between two points in the space of possible combinations of features. If one notes only that length is to be minimized, then he might just as well seek trees of minimum Euclidean length, or indeed of minimum length in any of the other uncountably many possible measures of abstract distance. But even that does not exhaust the possibilities. Numerical values—lengths of a sort—can be calculated for branches without regard to the possible features of nodes, by fitting the tree directly to a matrix of pairwise distances between terminal taxa (such methods are reviewed by Farris 1981). Such trees, too, might be selected to minimize "length," and this might be done for any of the huge number of ways of arriving at a matrix of pairwise distances.

The analogy through length has allowed methods such as this to become confused with parsimony analysis, and that confusion has played a role in specious criticisms of phylogenetic methods. Felsenstein (1981)—one of the main proponents of the idea that "parsimony" might mean almost anything—for example, attributes "parsimony" to one such method that had been used by Edwards and Cavalli-Sforza (1964). Rohlf and Sokal (1981) used a procedure for fitting branch lengths to a distance matrix to analyze the data of

The Logical Basis of Phylogenetic Analysis

Schuh and Polhemus (1980), then criticized the parsimony analysis of the latter authors on the grounds that the distance-fitted tree is "shorter." As Schuh and Farris (1981) pointed out, the length that Rohlf and Sokal attribute to their tree is quite meaningless, inasmuch as it is smaller than the number of origins of features required to account for the data (for related discussion see Farris 1981).

That will serve as a general commentary on this class of methods, which are too numerous in their possibilities to discuss here individually. The lengths arrived at by such calculations are generally incapable of any interpretation in terms of origins of features, and the evaluation of trees by such lengths consequently has nothing to do with the phylogenetic parsimony criterion. What is worse, the trees produced by these methods frequently differ in their grouping from parsimonious genealogies, and to that extent the use of these procedures amounts to throwing away explanatory power.

PAIRWISE HOMOPLASIES

The situation is somewhat different with some types of comparative data, such as matrices of immunological distance, in which no characters are directly observed. I have emphasized before (Farris 1981) that the parsimony criterion cannot be directly applied to such cases, and so I shall not consider them here. (The paper just cited offers other bases for evaluating methods of distance analysis.) Some analogies with distance analytic methods, however, can be related to the present discussion.

In fitting a tree to distances, branch lengths are used to determine a matrix of pairwise tree-derived distances between terminal taxa. The derived distance between a pair is just the sum of the lengths of the branches that lie on the path connecting the two taxa on the tree. Evaluation of the fit of the tree to observed distances is based on conformity of the derived to the observed distance values, this being measured by, say, the sum of the unsigned differences between the corresponding elements of the matrices (other measures are discussed by Farris 1981). Parsimony analyses can also be used to produce derived distances, the patristic differences of Farris (1967). I had earlier (Farris 1967) termed the departure of patristic from observed differences (pairwise) homoplasies, and from this, as well as by analogy with distance analytic procedures, one might be tempted to evaluate genealogies according to the total of those pairwise homoplasies (such a suggestion has been made by D. Swofford). The drawback of doing so is already clear from earlier discussion: the pairwise homoplasies are not independent.

A more extreme problem of interpretation of pairwise homoplasies arises in some methods for analyzing electrophoretic data. Suppose that each of three terminal taxa A, B, and C is fixed for a different allele at some locus, and that these taxa are related through an unresolved tree with common ancestor X. There are a number of ways of calculating distance between gene frequency distributions (see Farris 1981). To fix ideas, suppose that the Manhattan distance on frequencies is used. The distance between any two terminals is then

2. The ancestor X might plausibly be assigned frequency 1/3 for each observed allele. In that case the distance between X and any terminal is 4/3, the patristic difference between any two terminals is 8/3, and the corresponding pairwise homoplasy is 2/3. That implies that there is homoplasious similarity between any two terminals, but the conclusion is nonsense, inasmuch as there is no similarity between them at all. The three terminals simply have three entirely different conditions of the locus.

The details of that example depend on how gene frequencies are assigned to X, but no assignment can bring all the pairwise "homoplasies" to 0 simultaneously. In part this observation reflects the difficulties inherent in any attempt to utilize distances between gene frequency distributions as evidence in phylogenetic analysis (discussed in further detail by Farris 1981). Of greater interest for present purposes is what the example reveals about alleles as characters. The algebraic reason for the existence of those spurious homoplasies is that the distance coefficient treats shared 0 frequencies as points of similarity. Two taxa are assessed as similar in that both lack some allele, whereas in fact they simply possess different alleles. It is clear that those shared absences offer no independent assessment of the resemblances among the taxa, as the 0 frequency in any one allele is a necessary consequence of the fixation of any other. This problem then results from treating dependent quantities as if they were independent.

That difficulty is not limited to analysis of frequency data. Mickevich and Johnson (1976) introduced a method in which frequencies are transformed into a two-state coding: any frequency above a cutting point is coded as 1 (presence), any other as 0 (absence). The standard Wagner method is then used to find a tree minimizing required origins of states for the coded data. This procedure obviates many of the difficulties of analyzing frequency data through distance measures, but it still suffers from dependence of variables. Fixation of one allele will necessarily control the codes of others at the same locus. The number of state origins for the coding thus need not indicate the number of independent hypotheses of homoplasy for a genealogy, and this procedure should not then be regarded as a parsimony method. The problem of interdependence can, however, be avoided by choosing a better means of coding. Mickevich and Mitter (1981) and Mitter and Mickevich (1982) have made impressive progress in developing coding methods for analyzing electrophoretic data.

COVERING ASSUMPTIONS

Inasmuch as the aim is to minimize ad hoc hypotheses, it might seem that one could do better still by posing single hypotheses to cover several separate cases of homoplasy. Any putative genealogy might on that reasoning be defended against any character by concocting some premise to imply that all similarities in that character are homoplasies—or against any set of characters by dismissing evidence in general. I shall refer to such mass dismissals of evidence as covering assumptions.

The Logical Basis of Phylogenetic Analysis

The danger of using covering assumptions can be readily seen through a consideration of usual scientific practice. Suppose that an experiment is designed to evaluate a theory on the basis of readings from several instruments, and that some of the readings do not conform to the theory. If the nonconforming observations are only a few of the many readings made, the theory may seem to offer a generally satisfactory explanation; it is less so to the degree that such observations are abundant. Even then attempts may be made to salvage the theory. If the offending readings all come from the same instrument and so are logically related, they might be dismissed through the premise that the instrument is defective. (If it is found to be defective, so much the better.) But if no connection can be found among the nonconforming readings, the claim that they are coincidentally erroneous would have to be viewed with suspicion. Even the best theories seldom conform to every relevant observation, and so theories are well founded to the degree that nonconforming observations are rare. If contradictory observations could be dismissed as uninformative without regard to their abundance, the link between theory and observation would be tenuous at best.

Of course this is generally recognized, and attempts to defend theories by doing away with entire masses of evidence are typically rationalized by postulating mechanisms to account for what would otherwise be coincidental departures of observation from expectation. The legitimacy of that procedure depends crucially on validity of the postulates used. If the postulated mechanisms can themselves be corroborated by other sources of evidence, their use to defend the original theory is justified, and indeed they constitute improvements or extensions of the original theory. But if such mechanisms cannot be defended on extrinsic grounds, then they amount to no more than ad hoc excuses for the failure of the theory. Logically (albeit not rhetorically), they have no more force than the flat assertion that all nonconforming observations must be erroneous because the theory is true. Covering assumptions must be forbidden in scientific study, not only because they are ad hoc, but more particularly because they provide false license to dismiss any amount of evidence whatever.

The reason for prohibiting covering assumptions might be encapsulated by the observation that their use would allow theories to be chosen without regard to explanatory power. This effect can be seen directly in phylogenetic application. If twenty terminals share a particular feature, a genealogy consistent with a single origin of that feature explains those similarities fully. A hypothesis of kinship that broke those terminals into two separate groups of ten would not explain all the similarities among taxa, but it would still explain similarities within those groups. A tree that divided the same terminals into four separate groups of five would explain still less of observed similarities, but would still retain some explanatory power, while a scheme that required twenty separate origins would leave the observed similarities entirely unexplained. Some ad hoc rationale might be used to combine three or nineteen logically independent hypotheses of homoplasy into a "single" hypothesis. The possibility of that combination might be interpreted to mean that all these genealogies but the first conform equally well with observation. If such a course were followed, the differences in explanatory power among the last three hypotheses of kinship would play no role in choosing among them.

Almost any method that led to departure from parsimony might be suspected of involving a covering assumption. One might presume that the various length measures discussed before arise from some underlying premises that would amount to assumptions about the nature of evolution. But inasmuch as those premises, supposing that they exist, have never been made explicit, there is no real possibility of evaluating them as theories, and it is more immediately useful to view those methods as resulting simply from misunderstanding of the explanatory relationship between genealogies and characters.

Felsenstein's maximum likelihood methods offer fine examples of reliance on covering assumptions. The stochastic models would—if they were realistic—explain why seemingly independent characters would depart systematically from a parsimonious arrangement, hence would justify preference for unparsimonious schemes. Likewise, that neither Felsenstein nor anyone else maintains that those models can be corroborated makes it clear that in practice that justification would be entirely specious. But most of these methods have never been advocated for practical application, anyway. Felsenstein's own recent efforts center on likelihood interpretations of procedures that had already been advocated on other grounds, as I shall discuss later. It is of more practical interest to analyze methods that have been proposed more or less seriously.

IRREVERSIBILITY

Some techniques have been proposed as restricted "parsimony" methods. In these the number of origins of features is minimized, subject to the condition that some kinds of origins be rare or forbidden. Commonly, methods of this sort embody some version of the idea that evolution of individual characters is irreversible. In the method of Camin and Sokal (1965) secondary plesiomorphies are supposed not to occur, and so are excluded from reconstruction, the tree being chosen to minimize parallelisms. In the "Dollo" method of Farris (1977), origins of structures are supposed to be unique—structures once lost cannot be regained—and the tree is chosen to minimize secondary plesiomorphies.

Any of these methods might yield the same genealogy as would be obtained without the restriction, but none of them needs to do so, and in general applying the restriction will increase the number of hypotheses of homoplasy needed to defend the conclusion. Since there is no particular limit to that increase, using the restriction amounts simply to dismissing en mass any evidence that might otherwise seem to vitiate the conclusion. The motivation for doing so seems often to be more a matter of technical convenience than of conviction of the propriety of the restriction. That seems particularly to apply to the Camin-Sokal method, as it was one of the earliest techniques to be implemented as a computer algorithm. The reason for my own (Farris 1977) development of the Dollo method likewise had little to do with the realism of the assumption. That study was intended primarily to show logical flaws in Le Quesne's (1974) earlier attempt to analyze the same problem.

In a serious study, defending conclusions that depended crucially on use of a restricted method would require defending the restriction itself. I would not claim that the supposition of irreversible character evolution could never be supported by extrinsic evidence. I would suggest, however, that what acceptance that idea has gained has been based mostly on generalizations derived from hypotheses of kinship. The common notion that evolution generally proceeds from many, similar, parts to fewer, differentiated, parts, for example, seems to have been arrived at by induction from putative lineages. If the putative phylogenies used to draw such conclusions had been arrived at by presupposing irreversibility, then the conclusion would have no legitimate empirical support. If the idea of irreversibility is supported at all, then, it must have been derived from analyses that did not depend crucially on its truth. The evidence for a directed evolutionary trend, then, would be that the postulated trend conforms to a pattern of kinship that is in turn supported by other evidence—that is, that itself conforms to other characters. If it were known that evolution is irreversible, application of that knowledge might lead to genealogical inferences that otherwise might seem unparsimonious. But in fact no such thing is known, and the attempt to apply an empirically supported claim of irreversibility as a criticism of parsimony leads to a peculiar difficulty. Any body of characters might be made to appear consistent with the postulate of irreversibility. It is always technically possible to construct a tree so that all homoplasies take the form of parallelisms. It might seem from this that character information could never challenge the theory. But if the evidence for irreversibility was originally based on character distributions, then it would be quite unwarranted to analyze further cases so as to force them into conformity with irreversibility. The effect of doing so would be precisely to confer on irreversibility the status of an empirical conclusion that cannot be questioned by evidence—a contradiction in terms. In order to avoid that fallacy, it is necessary to allow that character information may support a conclusion of reversal. Whenever a putative reversal offers a more complete (that is, as already seen, more parsimonious) explanation of observed similarities than does a reconstruction enforcing irreversibility, irreversibility must be discounted. (In that particular case trends might still be accepted as rough descriptive generalizations.)

A proponent of irreversibility might nevertheless insist that when an analysis that does not presuppose irreversibility gives a different result from another that does use that premise, then the conclusion of the former depends crucially on the supposition that reversal is possible. Moreover, as the procedure just outlined will always discount irreversibility when parsimony requires, there is no way of rejecting the possibility of reversal. That possibility might seem, then, to be an ad hoc hypothesis, so that a conclusion of reversal actually requires more ad hoc hypotheses than would be suggested just by counting independent origins. But even if possibility of reversal did constitute an ad hoc hypothesis, it would certainly not be an additional independent hypothesis, for it is entailed by the particular hypothesis of reversal postulated. That observation, in fact, contains the key to the defect of the whole objection. If a particular conclusion of reversal could be legitimately criticized as presupposing the possibility of reversal, then any scientific conclusion whatever could be dismissed as requiring the supposition of its own possibility. The argument outlined is seen in that light to be simply another rationalization for discarding evidence.

It is clear that the reasoning outlined effectively views irreversibility and the possibility of reversal as competing theories. The charge that possibility of reversal cannot be rejected by parsimony analysis would be pertinent only as a criticism of a way of testing an empirical claim. But that view is itself suspicious. Irreversibility is certainly an empirical claim, and, furthermore, it is plainly testable in principle, inasmuch as it prohibits something, namely reversals. The possibility of reversal, on the other hand, can hardly be by itself an empirical claim in the same sense (although the claim that particular reversals have occurred might be), as it does not prohibit anything. One might think that admitting that reversal might occur, if it is not itself directly an empirical contention, nonetheless implies one, in that using a method that can discard irreversibility for parsimony would necessarily yield conclusions of reversal. But in fact it is quite possible for a parsimonious reconstruction to lack requirements for reversal. (The contrary, of course, is also possible, and is often observed. But that is a consequence of the idea in conjunction with particular observations, not of the idea itself.) While irreversibility and the possibility of reversal seem superficially to be simply alternative theories, then, they are in fact not the same kind of idea. The first is a theory that forbids conclusions that might otherwise seem supported by observation, and, when confronted with such cases, can be saved only by ad hoc supposition. The second is simply an attitude. The possibility that irreversibility (or any theory) is false must be considered in order to test the theory. No kind of empirical science would be possible without such attitudes.

POLYMORPHISM

Because of their reliance on covering assumptions to justify otherwise unnecessary ad hoc hypotheses of homoplasy, the Camin-Sokal and Dollo techniques should not be regarded as proper parsimony methods, prior usage notwithstanding. The situation may be different, however, for another restricted procedure. In the chromosome inversion model of Farris (1978), it is presumed that each of two alternative inversion types originated uniquely. Inversion types may nonetheless show incongruence with a genealogy through independent fixations from polymorphic ancestral populations, and the tree is chosen to minimize such fixations. The accuracy of the premise of unique origin might, of course, be questioned, but the idea is accepted by specialists on grounds extrinsic to genealogical hypotheses, and I shall not attempt to dispute its validity here. A further observation in this connection, however, seems worthwhile.

As this model presumes a unique origin for each inversion type, it might seem that similarity between organisms would on this premise be due to inheritance regardless of the genealogy postulated, so that the relationship between parsimony and genealogical explanation would no longer hold. The inherited similarity covered by the premise, however, holds only for chromosomes of individuals. Resemblance between populations fixed for the same inversion may still be explained by inheritance, or else the coincidental result of independent fixations. As it is populations that are grouped in postulating a genealogy, it is still possible to compare alternative genealogical hypotheses on explanatory power. There is in fact nothing unusual in this conclusion. It is generally true that features used to arrange taxa are characteristics of populations, rather than of individuals. The observation that deer have antlers is just a contracted way of stating that normal, adult, male deer in breeding condition possess those structures. The females, young, and deformed are not given a separate place in the system by reason of lacking the characteristic. The same principle underlies Hennig's emphasis of the idea that holomorphs rather than specimens are classified. Mickevich and Mitter (1981) arrive at the same concept in developing their greatly improved methods for analyzing electrophoretic data. They concentrate on recognizing suites of alleles as features of populations, rather than attempting to use single alleles—traits of individuals—as characters.

PHENETIC CLUSTERING

Clustering by raw similarity (phenetic clustering) has sometimes been advocated as a means of making genealogical inferences, typically with the justifying assumption that rates of evolutionary change (or divergence) are nearly enough constant so that degree of raw similarity reflects recency of common ancestry. The method is most often used with comparative biochemical data, but it has been recommended for morphological data as well (for example, by Colless 1970).

Constancy of rate is rather a different theory from irreversibility of evolution, but many of the comments made earlier apply here as well. Phenetic clustering might coincidentally produce a parsimonious scheme, but it certainly need not do so, and again there is no limit in principle to the number of otherwise unnecessary requirements for hypotheses of homoplasy that this method might impose. The assumption is certainly an empirical claim, and advocates of the method usually defend it by producing evidence for rate constancy. (Colless is an exception; he shows no inclination to resort to evidence.) That evidence typically takes the form of correlations between observed raw similarities and putative recency of common ancestry. Those last naturally depend on hypotheses of kinship, and this raises the familiar dilemma. If the genealogies used as evidence depended crucially on rate constancy, there would be no evidence. Supposing, then, that they do not, the evidence must consist of agreement between the theory and arrangements that conform to character distributions. Just as before, if the premise of rate constancy is used to justify unparsimonious conclusions, the effect is to consider rate constancy as empirical and irrefutable at one. Likewise parsimony analysis might be accused of presupposing that rates can vary, but discussion of that idea would precisely parallel what has already been said in connection with irreversibility.

A molecular evolutionist is quite happy with the generally good correlation that is observed between raw similarity and putative recency of common ancestry; for him it substantiates the molecular "clock." But as I have emphasized before (Farris 1981), such correlations are not enough to justify clustering by raw similarity. The correlations reported show considerable scatter. The implication of this, accepting the usual interpretation of the general correlation, is that rates of divergence vary somewhat. Even if it is often true, then, that genealogically most closely related taxa are also mutually most similar, there are evidently exceptions. Those exceptions could not be identified if genealogy were inferred by presupposing that raw similarity reflects kinship. To make accurate inferences in such cases—to discover what the cases are— it is necessary to use a method that can discount raw similarity as indicative of kinship if the data seem to require doing so—if doing so is required to achieve a more complete explanation of observed features. By analogy with the discussion of irreversibility, the same conclusion would be reached just by requiring that the relationship of raw similarity to kinship be vulnerable to evidence. It seems, then, that a correlation between raw similarity and kinship —even if it often holds—can provide no legitimate grounds for accepting unparsimonious inferences.

I commented before on the distinction between an ad hoc covering assumption and a corroborated improved theory able to account systematically for observations that would otherwise seem coincidental departures from its predecessor. This distinction suggests a further defect in the attempt to defend phenetic clustering on grounds of a correlation between raw similarity and kinship. In a legitimate extension of theory, the old coincidences are not dismissed as such, but explained by the extension. The process, that is, expands explanatory power, rather than discarding it. Suppose that clustering by raw similarity in some case requires otherwise unnecessary hypotheses of homoplasy, and that the conclusion is defended on grounds of a theoretical relationship between raw similarity and recency of common ancestry. If this is not ad hoc, then the theory must offer an explanation of the putative homoplasies. It is far from clear, however, that it can do so. Homoplasies, as already observed, are not explained by the inferred genealogy, from the standpoint of which the shared features that they represent are so only coincidentally. Inasmuch as raw similarities are calculated from features, it seems curious that they could either explain or be explained by a scheme that left the feature themselves unaccounted for. In order for a relationship between raw similarity and kinship to explain homoplasies, furthermore, it would seem necessary to suppose that that relationship rests on some real mechanism. That mechanism would have to have the property that organisms would come to possess features in common for reasons other than inheritance, and in just such a way as to maintain the correlation between raw similarity and recency of common ancestry. As no known natural process appears to have this property, it would seem that use of a postulated correlation between raw similarity and kinship to defend clustering by raw similarity rests necessarily on an ad hoc covering assumption.

A related conclusion can be reached by another route. Phenetic clustering ignores considerations of parsimony and so effectively proceeds by freely introducing whatever hypotheses of homoplasy are needed to derive a result conforming to the rate constancy premise. The procedure would be highly questionable on statistical grounds alone, then, if homoplasy were supposed to be rare. The method then requires the assumption that homoplasy is abundant, and indeed its proponents are prominent in criticizing parsimony as requiring rarity of homoplasy. The premise that homoplasy is abundant, however, poses a problem for clustering by raw similarity as well. That method infers recency of common ancestry of two taxa from the fraction of characters in which the two are similar. If homoplasy were rare (and rates constant), that would be superficially reasonable. Similarity between two lineages would decrease in clocklike fashion as ancestral similarities were lost. But if homoplasy is abundant, many of the similarities between two taxa are likely to be homoplasious, in which case they need indicate nothing about how recently the pair diverged. Two populations having only a remote common origin, and so (if rates were constant) very little homologous similarity, might have many recently acquired homoplasies, and so be judged to be of recent common ancestry. It is easy enough to identify conditions under which inferences might still be valid. If pairwise homoplasies were all the same, or nearly so, homoplasy would not alter the relative degrees of raw similarity among taxa, and then (if rates of change were constant) the method would still work. Phenetic clustering effectively presumes, then, that the variance of pairwise homoplasies is small. Keeping that variance small would be the task of the hypothetical mechanism just discussed.

While phenetic clustering does not consider homoplasies as such, it does select a tree by finding a constant-rate (ultrametric) model that conforms to observed raw similarities as closely as possible. If rates of evolution were constant, homologous similarities would conform to the constant rate model, so that departure from the model would be due to variation in pairwise homoplasies. The phenetic clustering procedure most commonly applied for genealogical inference, UPGMA, has precisely the effect of minimizing the variance of pairwise departures of observed from ultrametric similarities (Farris 1969b). Phenetic clustering and parsimony analysis are similar, then, in the sense that each minimizes a criterion. But whereas abundance of homoplasy need not imply error by parsimonious inference, large variations in pairwise homoplasies would certainly vitiate the conclusion of phenetic clustering. Phenetic clustering, unlike parsimony, depends crucially on minimality in nature of the quantity that it minimizes. Clustering by raw similarity possesses the very sort of defect that its proponents had incorrectly claimed as a weakness of phylogenetic analysis.

CLIQUES

Clique methods rely on parsimony to interpret suites of congruent characters, but their trees require homoplasy for characters outside the selected clique, and often the clique tree will be quite unparsimonious for those characters. In practice the excluded characters are often numerous, so that basing the inferred genealogy just on the clique imposes a considerable loss of explanatory power. These methods are then prime suspects for reliance on a covering assumption, but for a long time it was not clear from the clique literature what that assumption was supposed to be.

Le Quesne (1972) offered an approximate method (later made exact by Meacham 1981) for finding the probability that a suite of characters would all be congruent if features were distributed independently and at random among taxa. He suggested selecting the clique with the lowest such probability, and other proponents also commonly refer to cliques as "least likely." It is possible that this idea is intended as a justification of clique methods. If so, the justifying reasoning amounts to no more than misunderstanding of statistics. If a clique were evaluated just on its probability under a random model, the evaluation would be bound to the model. In that case the covering assumption of cliques would be that characters—being randomly distributed—have no relationship to genealogy. Perhaps it was intended that low probability under a null model would lend credence to an alternative, genealogical interpretation of the clique, but that idea, too, rests on a fallacy. Observing that a clique (or anything else) has low probability under a model might provide statistical grounds for rejecting the model, but it does not by itself offer any basis for choosing any particular alternative hypothesis. Once a model has been rejected, the probabilities it assigns to events necessarily become irrelevant. In this case rejecting the null model is uninformative, as no one interested in making phylogenetic inferences would have taken it seriously anyway. The statistical reason for accepting a new hypothesis is that it assigns much higher probability to observation than does the old. In normal statistics, a large enough difference between sample means serves as grounds for rejecting the hypothesis that the two samples were drawn from populations with the same parametric mean. If an alternative hypothesis is chosen so that it assigns maximum probability to the observed difference, the new theory conforms best with observation. But one hardly proceeds by choosing observations so as to minimize their nominal probability under the original hypothesis, let alone using such observations as the basis for choosing a new theory. Making statistical genealogical inferences from characters that had been used to reject the hypothesis of randomness would likewise require choosing a genealogy that would assign maximum probability to available characters—a maximum likelihood tree. I have already commented on the difficulties of applying that approach in practice, but this case is far worse. No model other than the rejected one of randomness is provided, and so neither are any grounds whatever for accepting the tree from the "least likely" clique as a genealogical inference best conforming to observation. (Felsenstein has made much the same point.)

As none of the ideas just discussed provides any legitimate rational for clique methods, those procedures must rest on an undisclosed assumption, if indeed they rest on anything at all. It is not difficult to discern what that assumption would have to be. Cliques are usually chosen to comprise as many mutually congruent characters as possible, and any characters that must be discarded to achieve this are simply counted as excluded. If the genealogy corresponding (by parsimony) to the clique is accepted, each of the excluded characters will require at least one hypothesis of homoplasy, but the number required may well vary among those characters. As characters are counted just as excluded or not, the number of hypotheses of homoplasy required by excluded characters plays no role in the analysis: similarities in those characters are dismissed en masse. The covering assumption involved is thus like the archetypical one discussed before. Ad hoc hypotheses of a sort are counted, but the counts do not reflect simple hypotheses of homoplasy. Instead any and all similarities in each excluded character are discounted by recourse to a "single" covering assumption. Excluding a character amounts to treating all similarities in it as irrelevant to assessing kinship. Those similarities could all be logically irrelevant only if they were all homoplasies. The covering assumption utilized is, then, that excluding a character—concluding that it shows some homoplasy—implies that all points of similarity in that character are homoplasies.

As discussed before, the collective dismissal of similarities in a character would be justified if the multiple required origins of features were not logically independent. It is readily seen, however, that such is not the case. The conclusion that endothermy has evolved independently in mammals and in birds does not imply that each species of bird or mammal has independently achieved that condition. Such being the case, it is likewise clear, from earlier discussion, that use of such a covering assumption leads to loss of explanatory power. As before, a single requirement of homoplasy may leave many of the similarities in a character explained, while a large enough number of required homoplasies will leave the same similarities entirely unexplained. Counting characters as simply excluded or not produces an evaluation oblivious to that distinction.

An attempt might be made to defend clique methods by advancing their covering assumption as an empirical claim on evolution, although of course doing so would raise the same sort of difficulties already discussed for irreversibility and rate constancy. Clique advocates have not tried to take that course—perhaps for fear of inviting ridicule. Unlike superficially tenable premises such as rate constancy, the clique assumption implies a theory that no one would take seriously as a realistic possibility. A different approach to rationalizing cliques has been taken by Felsenstein, though. He has proposed two stochastic models (Felsenstein 1979, 1981) under which he derives cliques as maximum likelihood estimates of genealogy. Both of these operate, just as would be expected from the clique assumption, by supplying principles that would excuse dismissing characters as units. In the 1979 paper this effect is achieved by introducing the possibility of a carefully selected type of error. Any character incongruent with the accepted tree is characterized as erroneous, and this is taken to mean that the character has been so completely misinterpreted that it is uninformative on genealogy. In the other model, incongruent characters are instead regarded as having changed so frequently in evolution as to be unrelated to genealogical grouping. Felsenstein himself emphasized that his models are inconsistent with the observed frequency of incongruence among characters. Realism thus plays no role in these justifications, which seem aimed instead at defending the clique assumption just by translating it into statistical terminology. Both rationalizations, consequently, have the same faults as the clique assumption itself.

The error idea rests on a misrepresentation of how systematists recognize characters. At one time it was believed that the eyes of octopi and of vertebrates were the same. That was certainly an error, for the two organs differ in both structure and ontogeny. But that the mistake was made does not mean that the sameness attributed to the eyes of rats and of mice is likewise a misrepresentation. Concluding a homoplasy in a feature may well invite renewed inspection and possible reinterpretation. But that reinspection certainly need not lead to dismissing all the agreements between taxa that the original feature had been intended to summarize. While it is reasonable to attribute some homoplasies to errors, it does not follow from this that those errors will turn out to have universal effects. It is seen from this that Felsenstein's use of the idea of wholesale error as a defense of the clique assumption amounts to no more than stating the desired conclusion as a premise. The defect of cliques is just that they treat every conclusion of homoplasy as if it implied universal homoplasy. Felsenstein attributes homomplasies to errors, but bolsters cliques only by supposing that any conclusion of error implies universal error. Neither implication is valid, and so either is merely an ad hoc rationalization for dismissing relevant evidence.

Much the same applies to Felsenstein's second argument. Dismissing an incongruent character on the grounds that it must have changed very frequently clearly depends on discounting the possibility that it changed only a few times. As Felsenstein (1981, p. 183) puts it, the clique method is suitable when "it is known that a few characters have very high rates of change, and the rest very low rates, but it is not known which characters are the ones having high rates." He does not disclose, however, how one comes to know the rate of change of a character without a prior phylogenetic analysis. Nor does he explain how one would apply that undisclosed method to gain the knowledge that his method calls for, without in the process incidentally learning which characters had the high rates. Nor, again, does he offer any pretense of a reason why rates should restrict themselves to be either very high or else very low—or why rapidly changing characters ought to be "few." In the absence of such explanation, it is seen that the covering assumption that one conclusion of homoplasy implies universal homoplasy has once again been "defended" simply by restating it as the entirely equivalent—and equally unsubstantiated—premise that any feature that originates more than once must have done so a very great number of times.

CONCLUSION

Advocacy of nonphylogenetic methods has consistently been based on the charge that parsimony depends on unrealistic assumptions. That allegation has never been supported by substantial argument. It has been instead motivated by the dependency of other approaches on false suppositions: Proponents of other views have tried to bolster their position through the pretense that no means of phylogenetic analysis can be realistic.

Parsimony analysis is realistic, but not because it makes just the right suppositions on the course of evolution. Rather, it consists exactly of avoiding uncorroborated suppositions whenever possible. To a devotee of supposition, to be sure, parsimony seems to presume very much indeed: that evolution is not irreversible, that rates of evolution are not constant, that all characters do not evolve according to identical stochastic processes, that one conclusion of homoplasy does not imply others. But parsimony does not suppose in advance that those possibilities are false—only that they are not already established. The use of parsimony depends just on the view that the truth of those—and any other—theories of evolution is an open question, subject to empirical investigation.

The dichotomy between parsimony and supposition is just that; parsimony offers no barrier to evolutionary theories as such. Rate constancy—or any other supposition—seems to be in conflict with parsimony in the abstract, as it seems to offer a different basis for making genealogical inferences. But it would conflict with parsimony in application only in conjunction with observation, if maintaining the supposition required discarding a parsimonious—explanatory—interpretation of evidence. In that case, however, the same evidence would serve to question the supposition, which could then be defended only by presupposing its truth, or—entirely equivalently—simply dismissing the evidence. But if parsimonious interpretation of evidence did not refute the supposition, then the latter would become a corroborated theory. Parsimony does not require that no such theories will be corroborated, but offers a means for that corroboration, provided evidence allows it. Unlike prior supposition, empirically supported evolutionary theories can offer no criticism of parsimony, for those theories could have become corroborated just to the extent that they require few dismissals of evidence. The insistence by proponents of suppositions that parsimony is unrealistic, it is then seen, is merely a subterfuge. Ostensibly the objection is to parsimony, but in fact the complaint is that some cherished idea does not conform to evidence.

I return finally to the questions raised at the beginning. Phylogenetic analysis is necessarily based on parsimony, both because it is precisely the criterion that leads to grouping according to putative synapomorphy, and because empirical investigation is impossible without avoiding ad hoc hypotheses. Only synapomorphy provides evidence of kinship, for the attempt to use raw similarity as evidence would necessarily either rest on uncorroborated—and so nonevidential—supposition, or else could lead to no conclusion conflicting with synapomorphy. And phylogenetic analysis is most certainly empirical, for in applying the parsimony criterion, it chooses among alternative hypotheses of relationship on the basis of nothing other than their explanatory power. Differing as it thus does from all other approaches, phylogenetic systematics alone provide a logical basis for the empirical study of the relationships among organisms.

REFERENCES

Beatty, J., and W. L. Fink. 1979. Review of *Simplicity*. *Syst. Zool.* 28:643–651.

Camin, J. H., and R. R. Sokal. 1965. A method for deducing branching sequences in phylogeny. *Evolution* 19:311–326.

Cartmill, M. 1981. Hypothesis testing and phylogeny reconstruction. *Zeit. Zool. Syst. Evolut.-forsch.* 19:73–95.

Colless, D. H. 1970. The phenogram as an estimate of phylogeny. *Syst. Zool.* 19:352–362.

Edwards, A. W. F., and L. L. Cavalli-Sforza. 1964. Reconstruction of evolutionary trees. In V. H. Heywood and J. McNeill, eds., *Phenetic and Phylogenetic Classification*. London, Systematics Association, pp. 67–76.

Farris, J. S. 1966. Estimation of conservatism of characters by constancy within biological populations. *Evolution* 20:587–591.

————. 1967. The meaning of relationship and taxonomic procedure. *Syst. Zool.* 16:44–51.

————. 1969a. On the cophenetic correlation coefficient. *Syst. Zool.* 18:279–285.

————. 1969b. A successive approximations approach to character weighting. *Syst. Zool.* 18:374–385.

————. 1970. Methods for computing Wagner trees. *Syst. Zool.* 19:83–92.

————. 1973. On the use of the parsimony criterion for inferring evolutionary trees. *Syst. Zool.* 22:250–256.

————. 1977. Phylogenetic analysis under Dollo's law. *Syst. Zool.* 26:77–88.

————. 1978. Inferring phylogenetic trees from chromosome inversion data. *Syst. Zool.* 27:275–284.

————. 1979. The information content of the phylogenetic system. *Syst. Zool.* 28:483–519.

————. 1980. The efficient diagnoses of the phylogenetic system. *Syst. Zool.* 29:386–401.

————. 1981. Distance data in phylogenetic analysis. In *Advances in Cladistics: Proceedings of the First Meeting of the Willi Hennig Society*, ed. V. A. Funk and D. R. Brooks. New York Botanical Garden.

————. 1982. Simplicity and informativeness in systematics and phylogeny. *Syst. Zool.* 31:413–444.

Farris, J. S., and A. G. Kluge. 1979. A botanical clique. *Syst. Zool.* 28:400–411.

Farris, J. S, A. G. Kluge, and M. F. Mickevich. 1982. Phylogenetic analysis, the monothetic group method, and myobatrachid frogs. *Syst. Zool.* 31:317–327.

Felsenstein, J. 1973. Maximum likelihood and minimum steps methods for estimating evolutionary trees from data on discrete characters. *Syst. Zool.* 22:240–249.

———. 1978. Cases in which parsimony or compatibility methods will be positively misleading. *Syst. Zool.* 27:401–410.

———. 1979. Alternative methods of phylogenetic inference and their interrelationship. *Syst. Zool.* 28:49–62.

———. 1981. A likelihood approach to character weighting and what it tells us about parsimony and compatibility. *Biol. Jour. Linn. Soc.* 16:183–196.

Funk, V. A., and D. R. Brooks. 1981. National Science Foundation workshop on the theory and application of cladistic methodology. Organized by T. Duncan and T. Stuessy. University of California, Berkeley, March 22–28, 1981. *Syst. Zool.* 30:491–498.

Gaffney, E. S. 1979. An introduction to the logic of phylogeny reconstruction. In J. Cracraft and N. Eldredge, eds., *Phylogenetic Analysis and Paleontology*. New York, Columbia University Press, pp. 79–112.

Hennig, W. 1966. *Phylogenetic Systematics*, Urbana, Ill., University of Illinois Press.

Kluge, A. G., and J. S. Farris. 1969. Quantitative phyletics and the evolution of anurans. *Syst. Zool.* 18:1–32.

Le Quesne, W. J. 1969. A method of selection of characters in numerical taxonomy. *Syst. Zool.* 18:201–205.

———. 1972. Further studies based on the uniquely derived character concept. *Syst. Zool.* 21:281–288.

———. 1974. The uniquely evolved character concept and its cladistic application. *Syst. Zool.* 23:513–517.

Meacham, C. A. 1981. A probability measure for character compatibility. *Math. Biosci.* 57:1–8.

Mickevich, M. F. 1978. Taxonomic congruence. *Syst. Zool.* 27:143–158.

———. 1980. Taxonomic congruence: Rohlf and Sokal's misunderstanding. *Syst. Zool.* 29:162–176.

Mickevich, M. F., and M. S. Johnson. 1976. Congruence between morphological and allozyme data in evolutionary inference and character evolution. *Syst. Zool.* 25:260–270.

Mickevich, M. F., and C. Mitter. 1981. Treating polymorphic characters in systematics: A phylogenetic treatment of electrophoretic data. In V. A. Funk and D. R. Brooks, eds., *Advances in cladistics: Proceedings of the First Meeting of the Willi Hennig Society*. New York Botanical Garden, pp. 45–60.

Mickevich, M. F., and J. S. Farris. 1981. The implications of congruence in *Menidia*. *Syst. Zool.* 30:351–370.

Nei, M. 1972. Genetic distance between populations. *Amer. Nat.* 106:283–292.

Riggins, R., and J. S. Farris. 1983. Cladistics and the roots of angiosperms. *Syst. Bot.* 8:96–101.

Rohlf, F., and Sokal, R. 1981. Comparing numerical taxonomic studies. *Syst. Zool.* 30:459–485.

Schuh, R. T., and J. T. Polhemus. 1980. Analysis of taxonomic congruence among morphological, ecological, and biogeographic data sets for the Leptopodomorpha (Hemiptera). *Syst. Zool.* 29:1–26.

Schuh, R. T., and J. S. Farris. 1981. Methods for investigating taxonomic congruence and their application to the Leptopodomorpha. *Syst. Zool.* 30:331–351.

Sober, E. 1975. *Simplicity*, London, Oxford University Press.

————. 1982. Parsimony methods in systematics. *Hennig Society II.* New York, Columbia University Press.

Watrous, L. E., and Q. D. Wheeler. 1981. The out-group comparison method of character analysis. *Syst. Zool.* 30:1–11.

Wiley, E. O. 1975. Karl R. Popper, systematics, and classification: A reply to Walter Bock and the evolutionary taxonomists. *Syst. Zool.* 24:233–243.

————. 1981. *Phylogenetics: The Theory and Practice of Phylogenetic Systematics.* New York, John Wiley and Sons.

17　The Detection of Phylogeny

Joseph Felsenstein

After an extraordinarily acrimonious start, debates in systematics over the proper logical framework for inferring phylogenies have become centred on two positions, the hypothetico-deductivist and the statistical. Formal statistical methods such as maximum likelihood have been applied to gene frequencies and nucleic acid sequences. It is possible to use statistical criteria to ask when existing non-statistical approaches such as parsimony methods are also those recommended by statistical criteria; the only circumstance in which this has so far been shown to happen is when rates of evolution are assumed to be low, so that little change is expected. The statistical properties of parsimony, compatibility, and distance matrix methods can be investigated in a larger range of cases, and it can sometimes be shown that they misbehave. We still lack statistically robust methods of inferring phylogenies. Effective computer programs for inferring phylogenies have been widely distributed. Perhaps the greatest single barrier to further development of phylogenetic methods is the lack of communication between microevolutionists and systematists. This is reflected in inadequate methods for treating measurable quantitative characters and gene frequencies. Future developments will concentrate on the development of robust nonparametric methods, methods for a posteriori character weighting, and integration of morphological and molecular data. Phylogenetic methods can only become more effective by relating processes which happen on micro- and macro-evolutionary time scales.

INTRODUCTION

Twenty-three years ago in 1964, when the Systematics Association was barely half as old as it is now [1988], it published a volume containing a paper by Edwards and Cavalli-Sforza (1964) which was a landmark. It described for the first time a parsimony method for reconstructing phylogenies; the paper was the first to propose a statistical approach to inferring phylogenies, and presented the first phylogeny inferred from gene frequencies. This was an extremely influential paper, but one which is, at the same time, little known and even less read. Therein lies a tale, which has mostly to do with the way the way numerical phylogenetic methods have spread in systematics.

　　The development of numerical and algorithmic methods for inferring phylogenies has three roots: the phenetic clustering methods of Sneath (1957) and Sokal (Michener and Sokal 1957), the writing of algorithms to analyze

From D. Hawksworth (ed.), *Prospects in Systematics*. Systematics Association, Clarendon Press, 1988, 112–127.

molecular sequence data (Eck and Dayhoff 1966; Fitch and Margoliash 1967), and the analysis by the German entomologist Hennig (1950, 1966) of the logic of inferring phylogenies and making classifications.

One might think that the story of systematics from 1950 to the present would then be the tale of a conflict between two camps, arrayed along an axis of enthusiasm for numerical methods. When Hennig's work was popularized in the 1970s, it was taken up by young, but traditionally trained morphological systematists who tended, on the whole, to be uncomfortable with both numerical and molecular methods. They found Hennig's qualitative discussion more accessible than the numerical work which had been slowly spreading for the previous decade. In addition, the people who had developed numerical phylogenetic methods were also either advocates of phenetic classification or of using molecular data, both of which were seen as unacceptable.

A strange and acrimonious three-way split in systematics then occurred, particularly acute in the United States, between traditional systematists, Hennigian phylogenetic systematists, and people developing numerical methods. The Hennigian phylogenetic systematists, struggling to make an impact on systematic practice, adopted an evangelical and at times messianic style. The reader who thinks these adjectives excessive is invited to scan the debates in the pages of *Systematic Zoology* and *Cladistics* in recent years, and to consider what must have gone on in unpublished personal interactions.

Edwards and Cavalli-Sforza's (1964) use of a minimum-evolution method for gene frequencies influenced the work of Camin and Sokal (1965), who stated the first discrete-characters parsimony method. Although it was widely noticed, that paper in turn tends not to be cited by phylogenetic systematists, who usually prefer to cite those of Kluge and Farris (1969), and Farris (1970). Camin and Sokal's parsimony method did not allow reversion from the derived to the ancestral state; Farris's "Wagner parsimony" did allow change in both directions, and phylogenetic systematists often regard their methods as derived from this more recent ancestor. A Wagner parsimony criterion had previously been used by Eck and Dayhoff (1966), but the algorithms to make it practically useable were first given by Kluge and Farris (1969) and by Farris (1970).

Thus, the extreme ferocity of conflict in systematics has tended to obscure the contribution of Edwards and Cavalli-Sforza's (1964) paper to the use of parsimony methods. Their paper was also remarkable for another reason. As students of R. A. Fisher, both authors regarded their algorithms as statistical methods. To reconcile their differences (Cavalli-Sforza had proposed a pioneering distance matrix method, Edwards the parsimony method) they tried to see whether a statistical method, maximum likelihood, would lead to either of these approaches. Although they could not, for purely technical reasons, get maximum likelihood to work, it was apparent that it would not lead to either the parsimony or the distance method. Thus, the same paper that introduced the first explicit parsimony method also started the use of statistical methods as a framework for phylogenetic inference. It very nearly introduced the first distance matrix method as well, but that was only published three years later (Cavalli-Sforza and Edwards 1967) at the same time as the method of Fitch and Margoliash (1967).

TWO FRAMEWORKS FOR PHYLOGENETIC INFERENCES

A decade of hostilities is now coming to an end, with signs that tempers are cooling and normal scientific communication between advocates of different approaches is resuming. As the smoke clears, there appear to be two major logical frameworks for making inferences about phylogenies. One is hypothetico-deductive and the other statistical. In the hypothetico-deductive approach characters are considered as falsifying candidates for the true phylogeny, and parsimony is preferred as a method of inferring phylogenies because it chooses "the phylogenetic hypothesis which has been rejected the least number of times" (Wiley 1975). In the statistical approach, conflicting evidence is reconciled by reference to probability models of the evolutionary process.

Both of these approaches have limitations. The obvious one in the hypothetico-deductive approach is that falsification cannot be regarded as absolute, in the sense that an experiment in physics is said to reject a hypothesis. The levels of "noise" in systematics are great enough that we often find data sets which falsify every possible phylogeny. No matter what the phylogeny, one character or another must have evolved with more than the minimum possible number of steps. We can regard such a case as one in which the data have falsified the statements made by a character, namely the character which required extra steps.

At this point there is a substantial ambiguity. Is one to regard each extra step as an additional falsification? Or is one to regard each character having extra steps as a falsification? In the first case we should seek to minimize the number of extra steps, in the second case the number of noisy characters. The first is a parsimony method, the second a compatibility method. There has recently been a lengthy and hard-fought exchange (Duncan et al. 1980; Churchill et al. 1985; Duncan 1984, 1986; Farris and Kluge 1985, 1986) over which of those two methods of reconciling conflict in the data are supported by the principles of Hennig. There is also the sticky question of why, and whether, we should count all falsifications as of equal value, without regard to the relative probabilities of homoplasy in different characters.

The statistical approach suffers from the assumption that we can model the processes of evolution accurately enough to employ probabilistic models. The models we make, of independent change in different characters, each following the same probabilistic process, are manifestly inadequate to the uncertainty, complexity, and just plain messiness of biological systems. It is therefore evident that any statistical statement about the uncertainty in an estimate of the phylogeny must be increased by a substantial and unknown factor for the uncertainty as to whether the probability model employed was near the truth.

Each position lends itself to abuses. The hypothetico-deductivist is apt to convince onlookers that all but the most parsimonious phylogenies have been falsified, and thus need not be considered. The statistician is apt to forget to notify the consumer that the model itself is to some extent wrong. Unfortunately, there exists no well-defined intermediate position, nor is it obvious that a better position *would* be intermediate between these two views.

My own efforts of recent years have been devoted to developing the statistical inference approach and seeing what it says about the assumptions implicit in other methods. It is possible, if one has a probabilistic model of the process of evolution, to derive from it a statistical method for estimating the phylogeny from a given type of data. Although there are a number of different statistical frameworks which could be used, maximum likelihood has been the one most often applied. After Edwards and Cavalli-Sforza (1964) had technical difficulties, Thompson (1975) and myself (Felsenstein 1973a, 1981b) were able to find practical methods for making maximum likelihood estimates of phylogenies from gene frequency data under the presumption that the populations involved had diverged exclusively by genetic drift. Thompson's method was a maximum likelihood method under a model of constant rates of drift in all lineages; mine was a REML (reduced maximum likelihood) method under a model in which each lineage could have a different effective population size.

The other major class of data to which maximum likelihood methods can be applied is nucleic acid sequences. Here one must have a model of random change in nucleotides (a Markov process involving change among four states, A, C, G, and T or U). The overall model assumes that change in a sequence involves the same process operating independently at each site. Neyman (1971) was the first to discuss how to make maximum likelihood estimates, for the case of three species. Felsenstein (1981c) showed how to compute likelihoods for larger trees and to estimate branch lengths iteratively. The computer program for doing so has now been generalized to allow for different rates of transitions and transversions. Hasegawa et al. (1984a,b 1985a,b) have made extensive use of maximum likelihood on various bodies of sequence data.

Ahead is the difficult problem of generalizing the method to allow for unequal rates of evolution at different sites, either when it is known which sites are the ones evolving quickly, or when it is not. An order of magnitude more difficult are the problems of allowing for dependence of change in adjacent sites, and carrying out the alignment of sequences as one makes the tree. Bishop and Thompson (1986) have made a start on this. It is currently fashionable for molecular algorithmists to construct multiple sequence alignment algorithms; it will ultimately be realized that the processes of alignment and tree construction should be combined.

Another type of data for which maximum likelihood methods are under active development are restriction site data (e.g., Debry and Slade 1985; Nei and Tajima 1985; Li 1986).

STATISTICAL PROPERTIES OF OTHER METHODS

Even though other methods are not stated in explicitly statistical terms, we can use statistical criteria to ask what are their implicit assumptions, and we can ask what are their statistical properties. In several papers (Felsenstein 1973b, 1979, 1981a) I have asked under what assumptions we can show that existing parsimony and compatibility methods themselves make a maximum likelihood estimate of the phylogeny. When rates of evolution were sufficiently small, and sufficiently equal among lineages, it proved possible to show that these methods would make a maximum likelihood estimate. Exactly which of the various parsimony methods (Camin-Sokal parsimony, Wagner parsimony, Dollo parsimony, and polymorphism parsimony) or compatibility methods were recommended depended on the relative rates of different kinds of events; forward change, reversion, gain and loss of polymorphism, and misinterpretation of the data.

The pattern found makes a great deal of intuitive sense. When it is believed to be improbable that we should have seen even one occurrence of some particular event in evolution, then that phylogeny which requires us to assume as few as possible of these implausible events is most credible. When rates of evolution are not small, there seems to be no general correspondence between maximum likelihood and parsimony (or compatibility) methods. This does not, however, mean that parsimony methods can necessarily be dismissed. They might have acceptable statistical properties even if they were not making maximum likelihood estimates.

Cavender (1978) and Felsenstein (1978) investigated the statistical properties of parsimony and compatibility methods in a simple evolutionary model with four species. We both found the same counterexample to the use of parsimony. When we assumed that evolutionary rates varied greatly enough among lineages, it turned out to be more probable that a shared derived state was the result of parallel changes in two branches of the tree that had high evolutionary rates, than that they were the result of a single change in a branch having low evolutionary rate. In these cases parsimony and compatibility methods fail to have the statistical property of consistency: they fail to converge to the true tree as we accumulate characters, but instead converge with greater and greater certainty to the tree which mistakes parallelism for common ancestry. When rates of change in lineages are made low, the requisite inequality of rates becomes greater and greater for parsimony and compatibility to misbehave.

This counterexample to the use of parsimony and compatibility has no force if evolutionary rates are sufficiently equal among lineages. Even a very rough molecular clock should be enough to prevent statistical inconsistency. Inferences from morphological data, in which great rate differences among lineages are common, would be more prone to statistical inconsistency. Nevertheless, I do not believe that this problem ought to cause abandonment of parsimony and compatibility methods, just a realization that they too have assumptions, and that these include either a low rate of evolution or sufficiently equal rates in different lineages.

Sober (in Felsenstein and Sober 1986) called me to task, with some justification, for overstating what is known by saying flatly that parsimony assumes these conditions. However, in the only models which can be investigated, those do turn out to be the critical assumptions and any fuller exposition of the conditions must include those cases.

Sober (1983, 1984, 1985) has presented a proof that, under general conditions, parsimony and likelihood always recommend the same tree. He then argues that a maximum likelihood estimate is to be preferred, even if not statistically consistent. I have taken the contrary view, that consistency is a fundamental property and that likelihood methods that fail to be consistent are not acceptable. This is a familiar dispute in statistics, one which is not going to be settled by anything Sober or I say. In any case, Sober (in Felsenstein and Sober 1986) has more recently retracted his proof as erroneous. At present we have no general proof of correspondence between parsimony and likelihood, whether or not such a proof would settle the issue.

Farris (1983) has taken a different tack, dismissing the inconsistency of parsimony as irrelevant because it was demonstrated in an oversimplified model. The inconsistency result was intended as a counterexample to claims that the parsimony method makes no substantive assumptions about evolutionary process. The inconsistency cannot be exorcised by claiming that the model is special, unless it can be shown that the same behavior would not occur in more complex models. Does the inconsistency arise as a result of the particular special symmetry assumptions of the model, or is it a more general phenomenon? I think the latter.

Another controversy on which a statistical approach sheds light is that over the use of distance matrix methods (such as that of Fitch and Margoliash 1967). Distance matrix methods are frequently held to be "phenetic" rather than "cladistic" and thus to discard the information essential to make phylogenetic inferences. The issue arises because of the increasing use of DNA hybridization infer phylogenies. If there is in principle no logically defensible way of analyzing such data, then their collection is called into question. Farris (1981, 1985, 1986) has argued that distance matrix methods are flawed. I have argued (Felsenstein 1984, 1986) that, once a statistical viewpoint is adopted, Farris's criticisms can be seen to lack force, and that although there are assumptions about linearity and independence of the individual distance values, there is no inherent flaw in distance matrix methods.

PROGRESS, PROGRAMS, AND PROBLEMS

From small beginnings twenty years ago, the use of explicit algorithms for inferring phylogenies has grown to become the standard approach today. The acceptance of statistical approaches is increasing, and so has the availability of computer programs to carry out various methods. Farris's FORTRAN program WAGNER-78 was the first phylogenetic program to achieve any currency. It has been supplanted by several larger and more sophisticated packages. My program package PHYLIP was first distributed in 1980, and has since been provided, free of charge, to over 700 installations. Farris's package PHYSYS was released in 1982 and is distributed for a substantial charge. Swofford's widely distributed program PAUP was released in 1983, and is inexpensive. A particularly interesting program is Maddison's MacClade, a program for the Apple MacIntosh which allows the user to interactively manipulate the tree and watch the resulting changes in the distribution of character state changes.

In spite of the rapid acceptance and spread of algorithmic and numerical methods, and in spite of the increasing communication between Hennigian phylogenetic systematists and other exponents of numerical methods, there are still some major obstacles to progress in phylogenetic inference, and these are reflected in obvious inadequacies of treatment of certain kinds of data.

Anyone trained in population genetics, as I was, notices immediately how little communication there is between microevolutionists and systematists. Until recently, population geneticists tended not to extend their attention beyond the species boundary. Systematists, on their part, have tended to be innumerate, and have paid relatively little attention to within-population variation. The result was that there has been little communication or understanding between the two groups.

This is beginning to change as systematists find numerical methods more relevant, and as population geneticists confront molecular data, which extend across species. Even a brief consideration of the treatment of gene frequencies and of quantitative characters in contemporary phylogenetic inference should be enough to persuade anyone that more communication is needed. Systematists tend to want to reduce all kinds of characters to discrete alternatives and to ignore within-population variation. This shows up in the treatment of gene frequencies in the popularity of all-or-none coding methods (see Buth 1984). A similar concern pervades the "character coding problem" of reducing continuous quantitative characters to discrete states.

Population geneticists have used gene frequencies in the methods they have developed for inferring times of divergence between populations. They have used two models, pure genetic drift (Edwards and Cavalli-Sforza 1964) and genetic drift with infinite-alleles neutral mutation (Nei 1972). This is appropriate for work at that time scale, but ignores information about the presence or absence of alleles in different populations (see, however, the use of presence-absence considerations in the technically sophisticated work of Griffiths 1979). Systematists, on their part, have been eager to discard frequency information and code a population in terms of the presence or absence of alleles (Mickevich and Johnson 1976; Mickevich and Mitter 1981, 1983).

There seems to be something gained and something lost each way. If the time scale of change is such that it is plausible that an allele could have arisen since the origin of the group, and if an allele is found only in one branch of a tree, this should be counted as supporting that tree topology, and doing so more strongly than calculation of the usual genetic distance statistics would indicate. Systematists are right to yearn for the introduction of such considerations into the inference of phylogenies from gene frequencies. They are not right when they assume that frequencies are all but irrelevant. Presence-absence methods of coding have obvious limitations, which cannot be made to disappear by appeals to William of Ockham.

For example, there is the obvious problem of sample size. Small samples are common in systematics. An allele may be absent because the sample has simply not been large enough to detect its presence. This might be avoided by using a cutoff frequency such as 5 or 10 percent, below which we treat the allele as absent. That in turn suffers the problem that when sample size is large, we find ourselves coding as greatly different populations with gene frequencies of (say) 0.04 and 0.06. Felsenstein has (1985d) shown that a distance measure based on presence-absence considerations can behave very strangely, declining with increasing divergence time, and this is due to the frequency information being ignored. Clearly, something important is being lost.

What is needed is for systematists and population geneticists to look closely into the behavior of gene frequencies in models of genetic drift and mutation, or of randomly varying selection, over long periods of time. This might lead to improved methods that took both frequency and presence/absence information into account in a statistically valid way.

QUANTITATIVE CHARACTERS

The treatment of quantitative characters in systematics is even worse. Most characters collected by systematists are what a population geneticist would recognize as quantitative characters, although this sometimes disguised by taking a length and coding it as "long" or "short." Morphometric methods are spreading rapidly, and with the availability of microcomputers and digitizers (and in the future automatic image processing) the data are beginning to be collected automatically. How will this increasing volume of quantitative data be analyzed? At the moment it is analyzed by discretizing the quantitative scale into two or a few states, and resolutely ignoring any genetic or environmental correlation between the characters, unless that correlation is complete.

Of course much information is lost by doing things this way. Having the data in the form of discrete states certainly helps if you want to apply the hypothetico-deductive framework and think of homoplasy as falsifying a tree. However, where possession of a complex and unusual structure may be a state for which parallelism is improbable, this cannot be said for possession of "long" versus "short." There are only two ways one can change on a linear scale, so that parallelism may not be improbable.

Arguments to the effect that parsimony methods make no assumptions about the correlation between characters are no doubt reassuring to someone who knows that the measurements that underlie their discrete states are correlated. In a statistical framework life is harder: when change in two characters is positively correlated then the observation of two changes, one in each character, is less support for a grouping than would be the case if the characters are not correlated. It then becomes important to know about genetic correlations, environmental correlations, and correlations of the selection pressures acting on the characters. The first two of these are rarely available to the systematist; the last almost never.

Quantitative genetic considerations are relevant even to discrete characters. Suppose that the discrete character is the result of a developmental threshold applied to an underlying polygenic quantitative character, a model well known in quantitative genetics. If we see a change of state $0 \rightarrow 1$ in one lineage, the mean of the distribution on the underlying scale has passed the threshold. Does this mean that a parallel $0 \rightarrow 1$ change is to be regarded as equally likely in a sister lineage as in one far removed on the tree? The quantitative genetics immediately suggests that on the sister lineage the population mean is also near the threshold, and hence that a parallel change is more likely there.

The treatment of quantitative characters should not be regarded as simply a matter of finding the right way to solve the "character coding problem." One also has to ask whether discrete coding is the best approach (I think not), and how we are to deal with the possibility of known and unknown sources of correlation in change of different characters. All of these questions will have to be faced as the flood of digitized data rises.

FUTURE DIRECTIONS

I have already suggested some of the directions in which future work will go, or at any rate ought to. We need much more development of statistical methods for inferring phylogenies. Parametric methods such as maximum likelihood need to be developed for more realistic models of change in morphological, molecular, and gene frequency characters. Nonparametric methods, which make fewer assumptions, are, of course, of great importance, in view of our lack of knowledge of the detailed genetics and detailed patterns of natural selection acting on the characters. Nonparametric methods are scarcely developed at all, or are they? Can we regard methods like parsimony and compatibility as nonparametric statistical methods? We will never find out unless we discuss them in statistical terms.

As people come to think more statistically, the issue of how and whether we can place confidence intervals on phylogenies will be addressed more often. Parametric statistical methods such as least squares and maximum likelihood do allow us to make statements about the variability of our estimates, and allow us to test alternative hypotheses. Cavender (1978, 1981) and Felsenstein (1985c) have also addressed the question of by how many steps one tree must be worse than another to be significantly worse, in some special cases involving three or four species. Templeton (1983) has suggested another nonparametric test, whose statistical properties are not yet well known. Felsenstein (1985b) applied the resampling technique called the bootstrap to placing confidence limits on phylogenies in larger cases. Cavender (Cavender and Felsenstein 1987) has found a quadratic function of observed character pattern frequencies with properties better than parsimony counts; Lake (1987) has found a linear function with similar properties in the case of DNA sequences. There needs to be much more work on resampling methods such as the bootstrap and jack-knife, on how we can use parsimony statistics and clique sizes to reject one tree in favor of another, and on the parametric tests such as likelihood ratio tests.

Character weighting seems to be another area of likely future progress. It can be shown that the more frequent change is expected to be, the less we ought to weight a character in parsimony methods, if we want to make a maximum likelihood estimate under low rates of evolution (Felsenstein 1981a). This suggests strongly that when we have two trees, and the number of steps on each in two different characters is:

Tree:	I	II
Character:		
# 1	2	3
# 2	39	37
Total:	41	40

that we might consider the first character more reliable than the second, as it changes less frequently. If so, then it is possible that the first character is sufficiently more reliable as a guide to the phylogeny that it would lead us to prefer the first tree, even though the total number of steps says otherwise. Farris (1969) suggested some families of iterative weighting methods that could be used to carry out such an assessment, and I showed (Felsenstein 1981a) that a particular threshold weighting scheme was preferred in certain cases. However, there is much more that can be done to develop weighting methods. At present they are under an anathema from which they have not yet recovered, as many phylogenetic systematists refuse to allow weighting, possibly because they confuse a priori with a posteriori weighting.

The rapid increase in molecular data, and the difficulties carrying out a proper statistical inference of phylogenies on morphological data, will create pressure toward using the molecular data to infer the phylogeny, followed by use of the estimated phylogeny to make inferences about the patterns of change of the morphological data. There are at the moment only a few studies where both types of data are collected, but their number is bound to increase, particularly as the advantages become apparent. Felsenstein outlined (1985a) a method of extracting independent contrasts from a phylogeny which could be used for estimating correlations of changes in characters. The method is dependent on a rather idealized Brownian motion model of character change, and assumes that the tree is known in detail. Both of these assumptions being rather unrealistic, there is much room for further work. Note that in this case the objective is to find out how the characters have evolved, and, although it cannot be ignored, the phylogeny is of secondary importance.

In molecular evolution studies, the collection of population samples is not yet common, but will inevitably become so. When we want to make inferences about population phylogenies or about rates of change in different parts of a sequence, the detailed genealogy of the genes is not of primary interest, but it too cannot be ignored. In principle the likelihood of the observed sample for a given population phylogeny is the sum of the probability of obtaining the observed data, summed over all possible detailed genealogies that could have led to the samples. There is no computationally effective way to compute this. When recombination is allowed for, the genealogy will contain loops and the problem becomes correspondingly worse. Work such as that of Hudson and Kaplan (1985) is a start, but much more is needed.

With a bit of luck and much hard work, we will find ourselves in an era in which there is a much more frutiful interaction between systematists and population geneticists. Ancient prejudices of each group against the other will have to be modified. We will need, on the way, to set aside the hypothetico-deductive model of phylogenetic inference and adopt a statistical one. Whether we can treat statistical models in phylogenetic inference with the appropriate mixture of reverence and scepticism remains to be seen.

ACKNOWLEDGMENT

[The] work [for this chapter] was supported by grant BSR-8614807 from the US National Science Foundation.

REFERENCES

Bishop, M. J., and Thompson, E. A. 1986. Maximum likelihood alignment of DNA sequences. *J. Mol. Biol.* 190:159–65.

Buth, D. G. 1984. The application of electrophoretic data in systematic studies. *A. Rev. Ecol. Syst.* 15:501–22.

Camin, J. H., and Sokal, R. R. 1965. A method for deducing branching sequences in phylogeny. *Evolution* 19:311–26.

Cavalli-Sforza, L. L., and F. Edwards, A. W. 1967. Phylogenetic analysis: Models and estimation procedures. *Evolution* 32:550–70; *Am. J. Hum. Genet.* 19:233–57.

Cavender, J. A. 1978. Taxonomy with confidence. *Math. Biosci.* 40:271–80. [Erratum, 44, 308, 1979.]

———. 1981. Tests of phylogenetic hypotheses under generalized models. *Math. Biosci.* 54:217–29.

Cavender, J. A., and Felsenstein, J. 1987. Invariants of phylogenesis in a simple case with discrete states. *J. Classification* 4:57–71.

Churchill, S. P., Wiley, E. O., and Hauser, L. A. 1985. Biological realities and the proper methodology: A reply to Duncan. *Taxon* 34:124–30.

Debry, R. W., and Slade, N. A. 1985. Cladistic analysis of restriction endonuclease cleavage maps within a maximum-likelihood framework. *Syst. Zool.* 34:21–34.

Duncan, T. 1984. Willi Hennig, character compatibility, Wagner parsimony, and the "Dendrogrammaceae" revisited. *Taxon* 33:698–704.

———. 1986. Editorial: Semantic fencing—a final riposte with a Hennigian crutch. *Taxon* 35:110–7.

Duncan, T., Philips, R. B., and Wagner, W. H. 1980. A comparison of branching diagrams derived by various phenetic and cladistic methods. *Syst. Bot.* 5:264–93.

Eck, R. V., and Dayhoff, M. O. 1966. *Atlas of Protein Sequence and Structure 1966*. National Biomedical Research Foundation, Silver Spring, Maryland.

Edwards, A. W. F., and Cavalli-Sforza, L. L. 1964. Reconstruction of evolutionary trees. In *Phenetic and Phylogenetic Classification* (ed. V. H. Heywood and J. McNeill), pp. 67–76. Systematics Association, London.

Farris, J. S. 1969. A successive approximations approach to character weighting. *Syst. Zool.* 18:375–85.

———. 1970. Methods for computing Wagner trees. *Syst. Zool.* 19:83–92.

———. 1981. Distance data in phylogenetic analysis. In *Advances in Cladistics* (ed. V. A. Funk and D. Brooks), pp. 3–23. New York Botanical Garden, New York.

———. 1983. The logical basis of phylogenetic analysis. In *Advances in Cladistics* (ed. N. I. Platnick and V. A. Funk), vol. 2, pp. 7–36. Columbia University Press, New York.

———. 1985. Distance data revisited. *Cladistics* 1:67–85.

———. 1986. Distances and statistics. *Cladistics* 2:144–57.

———. 1986. Editorial: Synapomorphy, parsimony, and evidence. *Taxon* 35:298–306.

Farris, J. S., and Kluge, A. G. 1985. Parsimony, synapomorphy, and explanatory power: A reply to Duncan. *Taxon* 34:130–35.

Felsenstein, J. 1973a. Maximum likelihood estimation of evolutionary trees from continuous characters. *Am. J. Hum. Genet.* 25:471–92.

———. 1973b. Maximum-likelihood and minimum-steps methods for evolutionary trees from data on discrete characters. *Syst. Zool.* 22:240–49.

———. 1978. Cases in which parsimony and compatibility methods will be positively misleading. *Syat. Zool.* 27:401–10.

———. 1979. Alternative methods of phylogenetic inference and their interrelationship. *Syst. Zool.* 28:49–62.

———. 1981a. A likelihood approach to character weighting and what it tells us about parsimony and compatibility. *Biol. J. Linn. Soc.* 16:183–96.

———. 1981b. Evolutionary trees from gene frequencies and quantitative characters: Finding maximum likelihood estimates. *Evolution* 35:1229–42.

———. 1981c. Evolutionary trees from DNA sequences: A maximum likelihood approach. *J. mol. Evol.* 17:368–76.

———. 1984. Distance methods for inferring phylogenies: A justification. *Evolution* 38:16–24.

———. 1985a. Phylogenies and the comparative method. *Am. Nat.* 125:1–15.

———. 1985b. Confidence limits on phylogenies: An approach using the bootstrap. *Evolution* 39:783–91.

———. 1985c. Confidence limits on phylogenies with a molecular clock. *Syst. Zool.* 34:152–61.

———. 1985d. Phylogenies from gene frequencies: A statistical problem. *Syst. Zool.* 34:300–11.

———. 1986. Distance methods: Reply to Farris. *Cladistics* 2:130–43.

Felsenstein, J., and Sober, E. 1986. Parsimony and likelihood: An exchange. *Syst. Zool.* 35:617–26.

Fitch, W. M., and Margoliash, E. 1967. Construction of phylogenetic trees. *Science, N.Y.* 155:279–84.

Griffiths, R. C. 1979. A transition density expansion for a multi-allele diffusion model. *Adv. Appl. Probability* 11:310–25.

Hasegawa, M., and Yano, T. 1984a. Phylogeny and classification of Hominoidea as inferred from DNA sequence data. *Proc. Jap. Acad.* 60B:389–92.

———. 1984b. Maximum likelihood method of phylogenetic inference from DNA sequence data. *Bull. Biomet. Soc. Jap.* 5:1–7.

Hasegawa, M., Iida, Y., Yano, T., Takaiwa, F., and Iwabuchi, M. 1985a. Phylogenetic relationships among eukaryotic kingdoms inferred by ribosomal RNA sequences. *J. Mol. Evol.* 22:32–38.

Hasegawa, M., Kishino, H., and Yano, T. 1985b. Dating of the human-ape splitting by a molecular clock of mitchondrial DNA. *J. Mol. Evol.* 22:160–74.

Hennig, W. 1950. *Grundzüge einer Theorie der Phylogenetischen Systematik.* Deutshcer Zentral-verlag, Berlin.

———. 1966. *Phylogenetic Systematics.* University of Illinois Press, Urbana.

Hudson, R. R., and Kaplan, N. L. 1985. Statistical properties of the number of recombination events in the history of a sample of DNA sequences. *Genetics* 111:147–64.

Kluge, A. R., and Farris, J. S. 1969. Quantitative phyletics and the evolution of anurans. *Syst. Zool.* 18:1–32.

Lake, J. A. 1987. A rate-independent technique for anlysis of nucleic acid sequences: Evolutionary parsimony. *Mol. Biol. Evol.* 42:167–91.

Li, W. H. 1986. Evolutionary change of restriction cleavage sites and phylogenetic inference. *Genetics* 113:187–213.

Michener, C. D., and Sokal, R. R. 1957. A quantitative approach to a problem in classification. *Evolution* 11:130–62.

Mickevich, M. F., and Johnson, M. S. 1976. Congruence between morphological and allozyme data in evolutionary inference. *Syst. Zool.* 25:260–70.

Mickevich, M. F., and Mitter, C. 1981. Treating polymorphic characters in systematics: A phylogenetic treatment of electrophoretic data. In *Advances in Cladistics* (ed. V. A. Funk and D. Brooks), pp. 45–58. New York Botanical Garden, New York.

———. 1983. Evolutionary patterns in allozyme data: A systematic approach. In *Advances in Cladistics* (ed. N. I. Platnick and V. A. Funk), vol. 2, pp. 169–76. Columbia University Press, New York.

Nei, M. 1972. Genetic distance between population. *Am. Nat.* 106:283–92.

Nei, M., and Tajima, F. 1985. Evolutionary change of restriction cleavage sites and phylogenetic inference for man and apes. *Mol. Biol. Evol.* 2:189–205.

Neyman, J. 1971. Molecular studies of evolution: A source of novel statistical problems. In *Statistical Decision Theory and Related Topics* (ed. S. S. Gupta and J. Yackel), pp. 1–27. Academic Press, New York.

Sneath, P. H. A. 1957. The application of computers to taxonomy. *J. Genet Microbiol.* 17:201–26.

Sober, E. 1983. Parsimony in systematics: Philosophical issues. *Rev. Ecol. Syst.* 14:335–57.

———. 1984. Common cause explanation. *Phil. Sci.* 51:212–41.

———. 1985. A likelihood justification of parsimony. *Cladistics* 1:209–33.

Templeton, A. R. 1983. Phylogenetic inference from restriction endonuclease cleavage site maps with particular reference to the evolution of humans and the apes. *Evolution* 37:221–24.

Thompson, E. A. 1975. *Human Evolutionary Trees.* Cambridge University Press, Cambridge.

Wiley, E. O. 1975. Karl R. Popper, systematics, and classification: A reply to Walter Bock and other evolutionary taxonomists. *Syst. Zool.* 24:233–43.

IX Reduction of Mendelian Genetics to Molecular Biology

18   1953 and All That: A Tale of Two Sciences

Philip Kitcher

Must we geneticists become bacteriologists, physiological chemists and physicists, simultaneously with being zoologists and botanists? Let us hope so.
—H. J. Muller, 1922[1]

1   THE PROBLEM

Toward the end of their paper announcing the molecular structure of DNA, James Watson and Francis Crick remark, somewhat laconically, that their proposed structure might illuminate some central questions of genetics.[2] Thirty years have passed since Watson and Crick published their famous discovery. Molecular biology has indeed transformed our understanding of heredity. The recognition of the structure of DNA, the understanding of gene replication, transcription and translation, the cracking of the genetic code, the study of gene regulation, these and other breakthroughs have combined to answer many of the questions that baffled classical geneticists. Muller's hope—expressed in the early days of classical genetics—has been amply fulfilled.

Yet the success of molecular biology and the transformation of classical genetics into molecular genetics bequeath a philosophical problem. There are two recent theories which have addressed the phenomena of heredity. One, *classical genetics*, stemming from the studies of T. H. Morgan, his colleagues and students, is the successful outgrowth of the Mendelian theory of heredity rediscovered at the beginning of this century. The other, *molecular genetics*, descends from the work of Watson and Crick. What is the relationship between these two theories? How does the molecular theory illuminate the classical theory? How exactly has Muller's hope been fulfilled?

There used to be a popular philosophical answer to the problem posed in these three connected questions: classical genetics has been reduced to molecular genetics. Philosophers of biology inherited the notion of reduction from general discussions in philosophy of science, discussions which usually center on examples from physics. Unfortunately attempts to apply this notion in the case of genetics have been vulnerable to cogent criticism. Even after considerable tinkering with the concept of reduction, one cannot claim that classical

From *Philosophical Review*, 1984, 93:335–373.

genetics has been (or is being) reduced to molecular genetics.[3] However, the antireductionist point is typically negative. It denies the adequacy of a particular solution to the problem of characterizing the relation between classical genetics and molecular genetics. It does not offer an alternative solution.

My aim in this chapter is to offer a different perspective on intertheoretic relations. The plan is to invert the usual strategy. Instead of trying to force the case of genetics into a mold, which is alleged to capture important features of examples in physics, or resting content with denying that the material can be forced, I shall try to arrive at a view of the theories involved and the relations between them that will account for the almost universal idea that molecular biology has done something important for classical genetics. In so doing, I hope to shed some light on the general questions of the structure of scientific theories and the relations which may hold between successive theories. Since my positive account presupposes that something is wrong with the reductionist treatment of the case of genetics, I shall begin with a diagnosis of the foibles of reductionism.

2 WHAT'S WRONG WITH REDUCTIONISM?

Ernest Nagel's classic treatment of reduction[4] can be simplified for our purposes. Scientific theories are regarded as sets of statements.[5] To reduce a theory T_2 to a theory T_1, is to deduce the statements of T_2 from the statements of T_1. If there are nonlogical expressions which appear in the statements of T_2, but do not appear in the statements of T_1, then we are allowed to supplement the statements of T_1 with some extra premises connecting the vocabulary of T_1 with the distinctive vocabulary of T_2 (so-called *bridge principles*). Intertheoretic reduction is taken to be important because the statements which are deduced from the reducing theory are supposed to be explained by this deduction.

Yet, as everyone who has struggled with the paradigm cases from physics knows all too well, the reductions of Galileo's law to Newtonian mechanics and of the ideal gas laws to the kinetic theory do not exactly fit Nagel's model. Study of these examples suggests that, to reduce a theory T_2 to a theory T_1, it suffices to deduce the laws of T_2 from a suitably modified version of T_1, possibly augmented with appropriate extra premises. Plainly, this sufficient condition is dangerously vague. I shall tolerate its vagueness, proposing that we understand the issue of reduction in genetics by using the examples from physics as paradigms of what "suitable modifications" and "appropriate extra premises" are like. Reductionists claim that the relation between classical genetics and molecular biology is sufficiently similar to the intertheoretical relations exemplified in the examples from physics to count as the same type of thing: to wit, as intertheoretical reduction.

It may seem that the reductionist thesis has now become so amorphous that it will be immune to refutation. But this is incorrect. Even when we have amended the classical model of reduction so that it can accommodate the

Philip Kitcher

examples that originally motivated it, the reductionist claim about genetics requires us to accept three theses:

R1: Classical genetics contains general laws about the transmission of genes which can serve as the conclusions of reductive derivations.

R2: The distinctive vocabulary of classical genetics (predicates like "① is a gene," "① is dominant with respect to ②") can linked to the vocabulary of molecular biology by bridge principles.

R3: A derivation of general principles about the transmission of genes from principles of molecular biology would explain why the laws of gene transmission hold (to the extent that they do).

I shall argue that each of the theses is false, offering this as my diagnosis of the ills of reductionism. . . .

Philosophers often identify theories as small sets of general laws. However, in the case of classical genetics, the identification is difficult and those who debate the reducibility of classical genetics to molecular biology often proceed differently. David Hull uses a characterization drawn from Dobzhansky: classical genetics is "concerned with gene differences; the operation employed to discover a gene is hybridization: parents differing in some trait are crossed and the distribution of the trait in hybrid progeny is observed."[6] This is not unusual in discussions of reduction in genetics. It is much easier to identify classical genetics by referring to the subject matter and to the methods of investigation, than it is to provide a few sentences that encapsulate the content of the theory.

Why is this? Because when we read the major papers of the great classical geneticists or when we read the textbooks in which their work is summarized, we find it hard to pick out *any* laws about genes. These documents are full of informative statements. Together, they tell us an enormous amount about the chromosomal arrangement of particular genes in particular organisms, about the effect on the phenotype of various mutations, about frequencies of recombination, and so forth. In some cases, we might explain the absence of formulations of general laws about genes (and even of reference to such laws) by suggesting that these things are common knowledge. Yet that hardly accounts for the nature of the textbooks or of the papers that forged the tools of classical genetics.

If we look back to the pre-Morgan era, we do find two general statements about genes, namely Mendel's Laws (or "Rules"). Mendel's second law states that, in a diploid organism which produces haploid gametes, genes at different loci will be transmitted independently; so, for example, if A, a, and B, b are pairs of alleles at different loci, and if an organism is heterozygous at both loci, then the probabilities that a gamete will receive any of the four possible genetic combinations, AB, Ab, aB, ab, are all equal.[7] Once it was recognized that genes are (mostly) chromosomal segments (as biologists discovered soon after the rediscovery of Mendel's laws), we understand that the law will not hold in general: alleles which are on the same chromosome (or, more exactly, close together on the same chromosome) will tend to be transmitted together because (ignoring recombination) one member of each homologous pair is distributed to a gamete.

Now it might seem that this is not very important. We could surely find a correct substitute for Mendel's second law by restricting the law so that it only talks about genes on nonhomologous chromosomes. Unfortunately, this will not quite do. There can be interference with normal cytological processes so that segregation of nonhomologous chromosomes need not be independent. However, my complaint about Mendel's second law is not that it is incorrect: many sciences use laws that are clearly recognized as approximations. Mendel's second law, amended or unamended, simply becomes irrelevant to subsequent research in classical genetics.

We envisaged amending Mendel's second law by using elementary principles of cytology, together with the identification of genes as chromosomal segments, to correct what was faulty in the unamended law. It is the fact that the application is so easy and that it can be carried out far more generally that makes the "law" it generates irrelevant. We can understand the transmission of genes by analyzing the cases that interest us from a cytological perspective —by proceeding from "first principles," as it were. Moreover, we can adopt this approach whether the organism is haploid, diploid, or polyploid, whether it reproduces sexually or asexually, whether the genes with which we are concerned are or are not on homologous chromosomes, whether or not there is distortion of independent chromosomal segregation at meiosis. Cytology not only teaches us that the second law is false; is also tells us how to tackle the problem at which the second law was directed (the problem of determining frequencies for pairs of genes in gametes). The amended second law is a restricted statement of results obtainable using a general technique. What figures largely in genetics after Morgan is the technique, and this is hardly surprising when we realize that one of the major research problems of classical genetics has been the problem of discovering the distribution of genes *on the same chromosome*, a problem which is beyond the scope of the amended law.

Let us now turn from R1 to R2, assuming, contrary to what has just been argued, that we can identify the content of classical genetics with general principles about gene transmission. (Let us even suppose, for the sake of concreteness, that the principles in question are Mendel's laws—amended in whatever way the reductionist prefers.) To derive these principles from molecular biology, we need a bridge principle. I shall consider first statements of the form

(*) (x) (x is a gene $\leftrightarrow$ Mx),

where Mx is an open sentence (possibly complex) in the language of molecular biology. Molecular biologists do not offer any appropriate statement. Nor do they seem interested in providing one. I claim that no appropriate bridge principle can be found.

Most genes are segments of DNA. (There are some organisms—viruses—whose genetic material is RNA; I shall henceforth ignore them.) Thanks to Watson and Crick, we know the molecular structure of DNA. Hence the problem of providing a statement of the above form becomes that of saying, in molecular terms, which segments of DNA count as genes.

Genes come in different sizes, and, for any given size, we can find segments of DNA of that size that are not genes. Therefore genes cannot be identified as segments of DNA containing a particular number of nucleotide pairs. Nor will it do to give a molecular characterization of those codons (triplets of nucleotides) that initiate and terminate transcription, and take a gene to be a segment of DNA between successive initiating and terminating codons. In the first place, mutation might produce a *single* allele containing within it codons for stopping and restarting transcription. Second, and much more important, the criterion is not general since not every gene is transcribed on mRNA.

The latter point is worth developing. Molecular geneticists recognize regulatory genes as well as structural genes. To cite a classic example, the operator region in the *lac* operon of *E. coli* serves as a site for the attachment of protein molecules, thereby inhibiting transcription of mRNA and regulating enzyme production. Moreover, it is becoming increasingly obvious that genes are not always transcribed, but play a variety of roles in the economy of the cell.

At this point, the reductionist may try to produce a bridge principle by brute force. Trivially, there are only a finite number of terrestrial organisms (past, present, and future) and only a finite number of genes. Each gene is a segment of DNA with a particular structure and it would be possible, in principle, to provide a detailed molecular description of that structure. We can now give a molecular specification of the gene by enumerating the genes and disjoining the molecular descriptions. The point made above, that the segments which we count as genes do not share any structural property, can now be put more precisely: any instantiation of (*) which replaces M by a structural predicate from the language of molecular biology will insert a predicate that is essentially disjunctive.

Why does this matter? Let us imagine a reductionist using the enumerative strategy to deduce a general principle about gene transmission. After great labor, it is revealed that all actual genes satisfy the principle. I claim that more than this is needed to reduce a *law* about gene transmission. We envisage laws as sustaining counterfactuals, as applying to examples that might have been but which did not actually arise. To reduce the law it is necessary to show how possible but nonactual genes would have satisfied it. Nor can we achieve the reductionist's goal by adding further disjuncts to the envisaged bridge principle. For although there are only finitely many *actual* genes, there are indefinitely many genes which *might* have arisen.

At this point, the reductionist may protest that the deck has been stacked. There is no need to produce a bridge principle of the form (*). Recall that we are trying to derive a general law about the transmission of genes, whose paradigm is Mendel's second law. Now the gross logical form of Mendel's second law is:

$$(x) (y) ((Gx \& Gy) \rightarrow Axy). \tag{1}$$

We might hope to obtain this from statements of the forms

$$(x) (Gx \rightarrow Mx) \tag{2}$$

$$(x) (y) ((Mx \& My) \rightarrow Axy) \tag{3}$$

where Mx is an open sentence in the language of molecular biology. Now there will certainly be true statements of the form (2): for example, we can take Mx as "x is composed of DNA v x is composed of RNA." The question is whether we can combine some such statement with other appropriate premises—for example, some instance of (3)—so as to derive, and thereby explain (1). No geneticist or molecular biologist has advanced any suitable premises, and with good reason. We discover true statements of the form (2) by hunting for weak necessary conditions on genes, conditions which have to be met by genes but which are met by hordes of other biological entities as well. We can only hope to obtain *weak* necessary conditions because of the phenomenon that occupied us previously: from the molecular standpoint, genes are not distinguished by any common structure. Trouble will now arise when we try to show that the weak necessary condition is jointly sufficient for the satisfaction of the property (independent assortment at meiosis) that we ascribe to genes. The difficulty is illustrated by the example given above. If we take Mx to be "x is composed of DNA v x is composed of RNA" then the challenge will be to find a general law governing the distribution of all segments of DNA and RNA!

I conclude that R2 is false. Reductionists cannot find the bridge principles they need, and the tactic of abandoning the form (*) for something weaker is of no avail. I shall now consider R3. Let us concede both of the points that I have denied, allowing that there are general laws about the transmission of genes and that bridge principles are forthcoming. I claim that exhibiting derivations of the transmission laws from principles of molecular biology and bridge principles would not explain the laws, and, therefore, would not fulfill the major goal of reduction.

As an illustration, I shall use the envisaged amended version of Mendel's second law. Why do genes on nonhomologous chromosomes assort independently? Cytology provides the answer. At meiosis, chromosomes line up with their homologues. It is then possible for homologous chromosomes to exchange some genetic material, producing pairs of recombinant chromosomes. In the meiotic division, one member of each recombinant pair goes to each gamete, and the assignment of one member of one pair to a gamete is probabilistically independent of the assignment of a member of another pair to that gamete. Genes which occur close on the same chromosome are likely to be transmitted together (recombination is not likely to occur between them), but genes on nonhomologous chromosomes will assort independently.

This account is a perfectly satisfactory explanation of why our envisaged law is true to the extent that it is. (We recognize how the law could fail if there were some unusual mechanism linking particular nonhomologous chromosomes.) To emphasize the adequacy of the explanation is not to deny that it could be extended in certain ways. For example, we might want to know more about the mechanics of the process by which the chromosomes are passed on to the gametes. In fact, cytology provides such information. However, appeal to molecular biology would not deepen our understanding of the transmission law. Imagine a successful derivation of the law from principles of chemistry and a bridge principle of the form (*). In charting the details of the molecular rearrangements the derivation would only blur the outline of a simple cytological story, adding a welter of irrelevant detail. Genes on nonhomologous chromosomes assort independently because nonhomologous chromosomes are transmitted independently at meiosis, and, so long as we recognize this, we do not need to know what the chromosomes are made of.

In explaining a scientific law, L, one often provides a deduction of L from other principles. Sometimes it is possible to explain some of the principles used in the deduction by deducing them, in turn, from further laws. Recognizing the possibility of a sequence of deductions tempts us to suppose that we could produce a better explanation of L by combining them, producing a more elaborate derivation in the language of our ultimate premises. But this is incorrect. What is relevant for the purposes of giving one explanation may be quite different from what is relevant for the purposes of explaining a law used in giving that original explanation. This general point is illustrated by the case at hand....

There is a natural reductionist response. The considerations of the last paragraphs presuppose far too subjective a view of scientific explanation. After all, even if *we* become lost in the molecular details, beings who are cognitively more powerful than we could surely recognize the explanatory force of the envisaged molecular derivation. However, this response misses a crucial point. The molecular derivation forfeits something important.

Recall the original cytological explanation. It accounted for the transmission of genes by identifying meiosis as a process of a particular kind: a process in which paired entities (in this case, homologous chromosomes) are separated by a force so that one member of each pair is assigned to a descendant entity (in this case, a gamete). Let us call processes of this kind *PS-processes*. I claim first that explaining the transmission law requires identifying PS-processes as forming a natural kind to which processes of meiosis belong, and second that PS-processes cannot be identified as a kind from the molecular point of view.

If we adopt the familiar covering law account of explanation, then we shall view the cytological narrative as invoking a law to the effect that processes of meiosis are PS-processes and as applying elementary principles of probability to compute the distribution of genes to gametes from the laws that govern PS-processes. If the illumination provided by the narrative is to be preserved in a molecular derivation, then we shall have to be able to express the relevant laws as laws in the language of molecular biology, and this will require that we be able to characterize PS-processes as a natural kind from the molecular point of view. The same conclusion, to wit that the explanatory power of the cytological account can be preserved only if we can identify PS-processes as a natural kind in molecular terms, can be reached in analogous ways if we adopt quite different approaches to scientific explanation—for example, if we conceive of explanation as specifying causally relevant properties or as fitting phenomena into a unified account of nature.

However, PS-processes are heterogeneous from the molecular point of view. There are no constraints on the molecular structures of the entities which are paired or on the ways in which the fundamental forces combine to pair them and to separate them. The bonds can be forged and broken in innumerable ways: all that matters is that there be bonds that initially pair the entities in question and that are subsequently (somehow) broken. In some cases, bonds may be formed directly between constituent molecules of the entities in question; in others, hordes of accessory molecules may be involved. In some cases, the separation may occur because of the action of electromagnetic forces or even of nuclear forces; but it is easy to think of examples in which the separation is effected by the action of gravity. I claim, therefore, that PS-processes are realized in a motley of molecular ways. (I should note explicitly that this conclusion is independent of the issue of whether the reductionist can find bridge principles for the concepts of classical genetics.)

We thus obtain a reply to the reductionist charge that we reject the explanatory power of the molecular derivation simply because we anticipate that our brains will prove too feeble to cope with its complexities. The molecular account objectively fails to explain because it cannot bring out that feature of the situation which is highlighted in the cytological story. It cannot show us that genes are transmitted in the ways that we find them to be because meiosis is a PS-process and because any PS-process would give rise to analogous distributions. Thus R3—like R1 and R2—is false.

3   THE ROOT OF THE TROUBLE

Where did we go wrong? Here is a natural suggestion. The most fundamental failure of reductionism is the falsity of R1. Lacking an account of theories which could readily be applied to the cases of classical genetics and molecular genetics, the attempt to chart the relations between these theories was doomed from the start. If we are to do better, we must begin by asking a preliminary question: what is the structure of classical genetics?

I shall follow this natural suggestion, endeavoring to present a picture of the structure of classical genetics which can be used to understand the intertheoretic relations between classical and molecular genetics. As we have seen, the main difficulty in trying to axiomatize classical genetics is to decide what body of statements one is attempting to axiomatize. The history of genetics makes it clear that Morgan, Muller, Sturtevant, Beadle, McClintock, and others have made important contributions to genetic theory. But the statements occurring in the writings of these workers seem to be far too specific to serve as parts of a general theory. They concern the genes of particular kinds of organisms—primarily paradigm organisms, like fruit flies, bread molds, and maize. The idea that classical genetics is simply a heterogeneous set of statements about dominance, recessiveness, position effect, nondisjunction, and so forth, in *Drosophila, Zea mays, E. coli, Neurospora,* etc. flies in the face of our intuitions. The statements advanced by the great classical geneticists seem more like *illustrations* of the theory than *components* of it. (To know classical genetics it is not necessary to know the genetics of any particular organism, not even *Drosophila melanogaster.*) But the only alternative seems to be to suppose that there are general laws in genetics, never enunciated by geneticists but reconstructible by philosophers. At the very least, this supposition should induce the worry that the founders of the field, and those who write the textbooks of today, do a singularly bad job.

Our predicament provokes two main questions. First, if we focus on a particular time in the history of classical genetics, it appears that there will be a set of statements about inheritance in particular organisms, which constitutes the corpus which geneticists of that time accept: what is the relationship between this corpus and the version of classical genetic theory in force at the time? (In posing this question, I assume, contrary to fact, that the community of geneticists was always distinguished by unusual harmony of opinion; it is not hard to relax this simplifying assumption.) Second, we think of genetic theory as something that persisted through various versions: what is the relation among the versions of classical genetic theory accepted at different times (the versions of 1910, 1930, and 1950, for example) which makes us want to count them as versions of the same theory?

We can answer these questions by amending a prevalent conception of the way in which we should characterize the state of a science at a time. The corpus of statements about the inheritance of characteristics accepted at a given time is only one component of a much more complicated entity that I shall call the *practice* of classical genetics at that time. There is a common language used to talk about hereditary phenomena, a set of accepted statements in that language (the corpus of beliefs about inheritance mentioned above), a set of questions taken to be the appropriate questions to ask about hereditary phenomena, and a set of patterns of reasoning which are instantiated in answering some of the accepted questions; (also: sets of experimental procedures and methodological rules, both designed for use in evaluating proposed answers; these may be ignored for present purposes). The practice of classical genetics at a time is completely specified by identifying each of the components just listed.[8]

A pattern of reasoning is a sequence of *schematic sentences*, that is sentences in which certain items of nonlogical vocabulary have been replaced by dummy letters, together with a set of *filling instructions* which specify how substitutions are to be made in the schemata to produce reasoning which instantiates the pattern. This notion of pattern is intended to explicate the idea of the common structure that underlies a group of problem-solutions.

The foregoing definitions enable us to answer the two main questions I posed above. Beliefs about the particular genetic features of particular organisms illustrate or exemplify the version of genetic theory in force at the time in the sense that these beliefs figure in particular problem-solutions generated by the current practice. Certain patterns of reasoning are applied to give the answers to accepted questions, and, in making the application, one puts forward claims about inheritance in particular organisms. Classical genetics persists as a single theory with different versions at different times in the sense that different practices are linked by a chain of practices along which there are relatively small modifications in language, in accepted questions, and in the patterns for answering questions. In addition to this condition of historical connection, versions of classical genetic theory are bound by a common structure: each version uses certain expressions to characterize hereditary phenomena, accepts as important questions of a particular form, and offers a general style of reasoning for answering those questions. Specifically, throughout the career of classical genetics, the theory is directed toward answering questions about the distribution of characteristics in successive generations of a genealogy, and it proposes to answer those questions by using the probabilities of chromosome distribution to compute the probabilities of descendant genotypes.

The approach to classical genetics embodied in these answers is supported by reflection on what beginning students learn. Neophytes are not taught (and never have been taught) a few fundamental theoretical laws from which genetic "theorems" are to be deduced. They are introduced to some technical terminology, which is used to advance a large amount of information about special organisms. Certain questions about heredity in these organisms are posed and answered. Those who understand the theory are those who know what questions are to be asked about hitherto unstudied examples, who know how to apply the technical language to the organisms involved in these examples, and who can apply the patterns of reasoning which are to be instantiated in constructing answers. More simply, successful students grasp general patterns of reasoning which they can use to resolve new cases.

I shall now add some detail to my sketch of the structure of classical genetics, and thereby prepare the way for an investigation of the relations between classical genetics and molecular genetics. The initial family of problems in classical genetics, the family from which the field began, is the family of *pedigree problems*. Such problems arise when we confront several generations of organisms, related by specified connections of descent, with a given distribution of one or more characteristics. The question that arises may be to understand the given distribution of phenotypes, or to predict the distribution of phenotypes in the next generation, or to specify the probability that a particular phenotype will result from a particular mating. In general, classical genetic theory answers such questions by making hypotheses about the relevant genes, their phenotypic effects, and their distribution among the individuals in the pedigree. Each version of classical genetic theory contains one or more problem-solving patterns exemplifying this general idea, but the detailed character of the pattern is refined in later versions, so that previously recalcitrant cases of the problem can be accommodated.

Each case of a pedigree problem can be characterized by a set of *data*, a set of *constraints*, and a question. In any example, the data are statements describ-

Philip Kitcher

ing the distribution of phenotypes among the organisms in a particular pedigree, or a diagram conveying the same information. The level of detail in the data may vary widely: at one extreme we may be given a full description of the interrelationships among all individuals and the sexes of all those involved; or the data may only provide the numbers of individuals with specific phenotypes in each generation; or, with minimal detail, we may simply be told that from crosses among individuals with specified phenotypes a certain range of phenotypes is found.

The constraints on the problem consist of general cytological information and descriptions of the chromosomal constitution of members of the species. The former will include the thesis that genes are (almost always) chromosomal segments and the principles that govern meiosis. The latter may contain a variety of statements. It may be pertinent to know how the species under study reproduces, how sexual dimorphism is reflected at the chromosomal level, the chromosome number typical of the species, what loci are linked, what the recombination frequencies are, and so forth. As in the case of the data, the level of detail (and thus of stringency) in the constraints can very widely.

Lastly, each problem contains a question that refers to the organisms described in the data. The question may take several forms: "What is the expected distribution of phenotypes from a cross between a and b?" (where a, b are specified individuals belonging to the pedigree described by the data), "What is the probability that a cross between a and b will produce an individual having P?" (where a, b are specified individuals of the pedigree described by the data and P is a phenotypic property manifested in this pedigree), "Why do we find the distribution of phenotypes described in the data?" and others.

Pedigree problems are solved by advancing pieces of reasoning that instantiate a small number of related patterns. In all cases the reasoning begins from a *genetic hypothesis*. The function of a genetic hypothesis is to specify the alleles that are relevant, their phenotypic expression, and their transmission through the pedigree. From that part of the genetic hypothesis that specifies the genotypes of the parents in any mating that occurs in the pedigree, together with the constraints on the problem, one computes the expected distribution of genotypes among the offspring. Finally, for any mating occurring in the pedigree, one shows that the expected distribution of genotypes among the offspring is consistent with the assignment of genotypes given by the genetic hypothesis.

The form of the reasoning can easily be recognized in examples—examples that are familiar to anyone who has ever looked at a textbook or a research report in genetics. What interests me is the style of reasoning itself. The reasoning begins with a genetic hypothesis that offers four kinds of information: (1) Specification of the number of relevant loci and the number of alleles at each locus; (2) specification of the relationships between genotypes and phenotypes; (3) specification of the relations between genes and chromosomes, of facts about the transmission of chromosomes to gametes (for example, resolution of the question whether there is disruption of normal segregation) and about the details of zygote formation; (4) assignment of genotypes to individuals in the pedigree. After showing that the genetic hypothesis is consistent with the data and constraints of the problem, the principles of cytology and the laws of probability are used to compute expected distributions of genotypes from crosses. The expected distributions are then compared with those assigned in part (4) of the genetic hypothesis.

Throughout the career of classical genetics, pedigree problems are addressed and solved by carrying out reasoning of the general type just indicated. Each version of classical genetic theory contains a pattern for solving pedigree problems with a method for computing expected genotypes which is adjusted to reflect the particular form of the genetic hypotheses that it sanctions. Thus one way to focus the differences among successive versions of classical genetic theory is to compare their conceptions of the possibilities for genetic hypotheses. As genetic theory develops, there is a changing set of conditions on admissible genetic hypotheses. Prior to the discovery of polygeny and pleiotropy (for example), part (1) of any adequate genetic hypothesis was viewed as governed by the requirement that there would be a one-one correspondence between loci and phenotypic traits.[9] After the discovery of incomplete dominance and epistasis, it was recognized that part (2) of an adequate hypothesis might take a form that had not previously been allowed: one is not compelled to assign to the heterozygote a phenotype assigned to one of the homozygotes, and one is also permitted to relativize the phenotypic effect of a gene to its genetic environment.[10] Similarly, the appreciation of phenomena of linkage, recombination, nondisjunction, segregation distortion, meiotic drive, unequal crossing over, and crossover suppression, modify conditions previously imposed on part (3) of any genetic hypothesis. In general, we can take each version of classical genetic theory to be associated with a set of conditions (usually not formulated explicitly) which govern admissible genetic hypotheses. While a general form of reasoning persists through the development of classical genetics, the patterns of reasoning used to resolve cases of the pedigree problem are constantly fine-tuned as geneticists modify their views about what forms of genetic hypothesis are allowable.

So far I have concentrated exclusively on classical genetic theory as a family of related patterns of reasoning for solving the pedigree problem. It is natural to ask if versions of the theory contain patterns of reasoning for addressing other questions. I believe that they do. The heart of the theory is the theory of *gene transmission*, the family of reasoning patterns directed at the pedigree problem. Out of this theory grow other subtheories. The theory of *gene mapping* offers a pattern of reasoning which addresses questions about the relative positions of loci on chromosomes. It is a direct result of Sturtevant's insight that one can systematically investigate the set of pedigree problems associated with a particular species. In turn, the theory of gene mapping raises the question of how to identify mutations, issues which are to be tackled by the *theory of mutation*. Thus we can think of classical genetics as having a central theory, the theory of gene transmission, which develops in the ways I have described above, surrounded by a number of satellite theories that are directed at questions arising from the pursuit of the central theory. Some of these satellite theories (for example, the theory of gene mapping) develop in the same continuous fashion. Others, like the theory of mutation, are subject to rather dramatic shifts in approach.

4 MOLECULAR GENETICS AND CLASSICAL GENETICS

Armed with some understanding of the structure and evolution of classical genetics, we can finally return to the question with which we began. What is the relation between classical genetics and molecular genetics? When we look at textbook presentations and the pioneering research articles that they cite, it is not hard to discern major ways in which molecular biology has advanced our understanding of hereditary phenomena. We can readily identify particular molecular explanations which illuminate issues that were treated incompletely, if at all, from the classical perspective. What proves puzzling is the connection of these explanations to the theory of classical genetics. I hope that the account of the last section will enable us to make the connection.

I shall consider three of the most celebrated achievements of molecular genetics. Consider first the question of *replication*. Classical geneticists believed that genes can replicate themselves. Even before the experimental demonstration that all genes are transmitted to all the somatic cells of a developing embryo, geneticists agreed that normal processes of mitosis and meiosis must involve gene replication. Muller's suggestion that the central problem of genetics is to understand how mutant alleles, incapable of performing wild-type functions in producing the phenotype, are nonetheless able to replicate themselves, embodies this consensus. Yet classical genetics had no account of gene replication. A molecular account was an almost immediate dividend of the Watson-Crick model of DNA.

Watson and Crick suggested that the two strands of the double helix unwind and each strand serves as the template for the formation of a complementary strand. Because of the specificity of the pairing of nucleotides, reconstruction of DNA can be unambiguously directed by a single strand. This suggestion has been confirmed and articulated by subsequent research in molecular biology.[11] The details are more intricate than Watson and Crick may originally have believed, but the outline of their story stands.

A second major illumination produced by molecular genetics concerns the characterization of mutation. When we understand the gene as a segment of DNA we recognize the ways in which mutant alleles can be produced. "Copying errors" during replication can cause nucleotides to be added, deleted, or substituted. These changes will often lead to alleles that code for different proteins, and which are readily recognizable as mutants through their production of deviant phenotypes. However, molecular biology makes it clear that there can be *hidden* mutations, mutations that arise through nucleotide substitutions that do not change the protein produced by a structural gene (the genetic code is redundant) or through substitutions that alter the form of the protein in trivial ways. The molecular perspective provides us with a general answer to the question, "What is a mutation?" namely, that a mutation is the modification of a gene through insertion, deletion, or substitution of nucleotides. This general answer yields a basic method for tackling (in principle) questions of form, "Is *a* a mutant allele?" namely, a demonstration that *a* arose through nucleotide changes from alleles that persist in the present population. The method is frequently used in studies of the genetics of bacteria and bacteriophage, and can sometimes be employed even in inquiries about more complicated organisms. So, for example, there is good biochemical evidence for believing that some alleles which produce resistance to pesticides in various species of insects arose through nucleotide changes in the alleles naturally predominating in the population.[12]

I have indicated two general ways in which molecular biology answers questions that were not adequately resolved by classical genetics. Equally obvious are a large number of more specific achievements. Identification of the molecular structures of particular genes in particular organisms has enabled us to understand why those genes combine to produce the phenotypes they do....

The claim that genes can replicate does not have the status of a central law of classical genetic theory. It is not something that figures prominently in the explanations provided by the theory (as, for example, the Boyle-Charles law is a prominent premise in some of the explanations yielded by phenomenological thermodynamics). Rather, it is a claim that classical geneticists took for granted, a claim presupposed by explanations, rather than an explicit part of them. Prior to the development of molecular genetics that claim had come to seem increasingly problematic. If genes can replicate, how do they manage to do it? Molecular genetics answered the worrying question. It provided a theoretical demonstration of the possibility of an antecedently problematic presupposition of classical genetics.

We can say that a theory presupposes a statement *p* if there is some problem-solving pattern of the theory, such that every instantiation of the pattern contains statements that jointly imply the truth of *p*. Suppose that, at a given stage in the development of a theory, scientists recognize an argument from otherwise acceptable premises which concludes that it is impossible that *p*. Then the presupposition *p* is problematic for those scientists. What they would like would be an argument showing that it is possible that *p* and explaining what is wrong with the line of reasoning which appears to threaten the possibility of *p*. If a new theory generates an argument of this sort, then we can say that the new theory gives a theoretical demonstration of the possibility of an antecedently problematic presupposition of the old theory....

Because theoretical demonstrations of the possibility of antecedently problematic presuppositions involve derivation of conclusions of one theory from the premises supplied by a background theory, it is easy to assimilate them to the classical notion of reduction. However, on the account I have offered, there are two important differences. First, there is no commitment to the thesis that genetic theory can be formulated as (the deductive closure of) a conjunc-

Philip Kitcher

tion of laws. Second, it is not assumed that all general statements about genes are equally in need of molecular derivation. Instead, one particular thesis, a thesis that underlies all the explanations provided by classical genetic theory, is seen as especially problematic, and the molecular derivation is viewed as addressing a specific problem that classical geneticists had already perceived. Where the reductionist identifies a general benefit in deriving all the axioms of the reduced theory, I focus on a particular derivation of a claim that has no title as an axiom of classical genetics, a derivation which responds to a particular explanatory difficulty of which classical geneticists were acutely aware. The reductionist's global relation between theories does not obtain between classical and molecular genetics, but something akin to it does hold between special fragments of these theories.

The second principal achievement of molecular genetics, the account of mutation, involves a conceptual refinement of prior theory. Later theories can be said to provide conceptual refinements of earlier theories when the later theory yields a specification of entities that belong to the extensions of predicates in the language of the earlier theory, with the result that the ways in which the referents of these predicates are fixed are altered in accordance with the new specifications. Conceptual refinement may occur in a number of ways. A new theory may supply a descriptive characterization of the extension of a predicate for which no descriptive characterization was previously available; or it may offer a new description which makes it reasonable to amend characterizations that had previously been accepted. . . .[13]

Finally, let us consider the use of molecular genetics to illuminate the action of particular genes. Here we again seem to find a relationship that initially appears close to the reductionist's ideal. Statements that are invoked as premises in particular problem-solutions—statements that ascribe particular phenotypes to particular genotypes—are derived from molecular characterizations of the alleles involved. On the account of classical genetics offered in section 3, each version of classical genetic theory includes in its schema for genetic hypotheses a clause which relates genotypes to phenotypes. . . . [W]e might hope to discover a pattern of reasoning within molecular genetics that would generate as its conclusion the schema for assigning phenotypes to genotypes.

It is not hard to characterize the relation just envisioned. Let us say that a theory T' provides an *explanatory extension* of a theory T just in case there is some problem-solving pattern of T one of whose schematic premises can be generated as the conclusion of a problem-solving pattern of T'. When a new theory provides an explanatory extension of an old theory, then particular premises occurring in explanatory derivations given by the old theory can themselves be explained by using arguments furnished by the new theory. However, it does not follow that the explanations provided by the old theory can be improved by replacing the premises in question with the pertinent derivations. What is relevant for the purposes of explaining some statement S may not be relevant for the purposes of explaining a statement S' which figures in an explanatory derivation of S.

Even though reductionism fails, it may appear that we can capture part of the spirit of reductionism by deploying the notion of explanatory extension. The thesis that molecular genetics provides an explanatory extension of classical genetics embodies the idea of a global relationship between the two theories, while avoiding two of the three troubles that were found to beset reductionism. That thesis does not simply assert that some specific presupposition of classical genetics (for example, the claim that genes are able to replicate) can be derived as the conclusion of a molecular argument, but offers a general connection between premises of explanatory derivations in classical genetics and explanatory arguments from molecular genetics. It is formulated so as to accommodate the failure of R1 and to honor the picture of classical genetics developed in section 3. Moreover, the failure of R2 does not affect it. . . .

Nevertheless, even born-again reductionism is doomed to fall short of salvation. Although it is true that molecular genetics belongs to a cluster of theories which, taken together, provide an explanatory extension of classical genetics, molecular genetics, on its own, cannot deliver the goods. There are some cases in which the ancillary theories do not contribute to the explanation of a classical claim about gene action. In such cases, the classical claim can be derived and explained by instantiating a pattern drawn from molecular genetics. The example of human hemoglobin provides one such case. [Individuals who are homozygous for a mutant allele for the synthesis of human hemoglobin develop sickle-cell anemia, a phenomenon that can be explained at the molecular level (ed.).] But this example is atypical.

Consider the way in which the hemoglobin example works. Specification of the molecular structures of the normal and mutant alleles, together with a description of the genetic code, enables us to derive the composition of normal and mutant hemoglobin. Application of chemistry then yields descriptions of the interactions of the proteins. With the aid of some facts about human blood cells, one can then deduce that the sickling effect will occur in abnormal cells, and, given some facts about human physiology, it is possible to derive the descriptions of the phenotypes. There is a clear analogy here with some cases from physics. The assumptions about blood cells and physiological needs seem to play the same role as the boundary conditions about shapes, relative positions, and velocities of planets that occur in Newtonian derivations of Kepler's laws. In the Newtonian explanation we can see the application of a general pattern of reasoning—the derivation of explicit equations of motion from specifications of the forces acting—which yields the general result that a body under the influence of a centrally directed inverse square force will travel in a conic section; the general result is then applied to the motions of the planets by incorporating pieces of astronomical information. Similarly, the derivation of the classical claims about the action of the normal and mutant hemoglobin genes can be seen as a purely chemical derivation of the generation of certain molecular structures and of the interactions among them. The chemical conclusions are then applied to the biological system under consideration by introducing three "boundary conditions"; first, the claim that the altered molecular structures only affect development to the extent of substituting a different molecule in the erythrocytes (the blood cells that transport hemoglobin); second, a description of the chemical conditions in the capillaries; and third, a description of the effects upon the organism of capillary blockage.

The example is able to lend comfort to reductionism precisely because of an atypical feature. In effect, one concentrates on the *differences* among the phenotypes, takes for granted the fact that in all cases development will proceed normally to the extent of manufacturing erythrocytes—which are, to all intents and purposes, simply sacks for containing hemoglobin molecules—and compares the difference in chemical effect of the cases in which the erythrocytes contain different molecules. *The details of the process of development can be ignored.* However, it is rare for the effect of a mutation to be so simple. Most structural genes code for molecules whose presence or absence make subtle differences. Thus, typically, a mutation will affect the distribution of chemicals in the cells of a developing embryo. A likely result is a change in the timing of intracellular reactions, a change that may, in turn, alter the shape of the cell. Because of the change of shape, the geometry of the embryonic cells may be modified. Cells that usually come into contact may fail to touch. Because of this, some cells may not receive the molecules necessary to switch on certain batteries of genes. Hence the chemical composition of these cells will be altered. And so it goes.

Quite evidently, in examples like this (which include most of the cases in which molecular considerations can be introduced into embryology) the reasoning that leads us to a description of the phenotype associated with a genotype will be much more complicated than that found in the hemoglobin case. It will not simply consist in a chemical derivation adapted with the help of a few boundary conditions furnished by biology. Instead, we shall encounter a sequence of subarguments: molecular descriptions lead to specifications of cellular properties, from these specifications we draw conclusions about cellular interactions, and from these conclusions we arrive at further molecular descriptions. There is clearly a pattern of reasoning here which involves molecular biology and which extends the explanations furnished by classical genetics by showing how phenotypes depend upon genotypes—but I think it would be folly to suggest that the extension is provided by molecular genetics alone.

In section 2, we discovered that the traditional answer to the philosophical question of understanding the relation that holds between molecular genetics and classical genetics, the reductionist's answer, will not do. Section 3 attempted to build on the diagnosis of the ills of reductionism, offering an account of the structure and evolution of classical genetics that would improve on the picture offered by those who favor traditional approaches to the nature of scientific theories. In the present section, I have tried to use the framework of section 3 to understand the relations between molecular genetics and classical genetics. Molecular genetics has done something important for classical genetics, and its achievements can be recognized by seeing them as instances of the intertheoretic relations that I have characterized. Thus I claim that the problem from which we began is solved.

So what? Do we have here simply a study of a particular case—a case which has, to be sure, proved puzzling for the usual accounts of scientific theories and scientific change? I hope not. Although the traditional approaches may have proved helpful in understanding some of the well-worn examples that have been the stock-in-trade of twentieth-century philosophy of science, I believe that the notion of scientific practice sketched in section 3 and the intertheoretic relations briefly characterized here will both prove helpful in analyzing the structure of science and the growth of scientific knowledge *even in those areas of science where traditional views have seemed most successful*. Hence the tale of two sciences which I have been telling is not merely intended as a piece of local history that fills a small but troublesome gap in the orthodox chronicles. I hope that it introduces concepts of general significance in the project of understanding the growth of science.

5 ANTIREDUCTIONISM AND THE ORGANIZATION OF NATURE

One loose thread remains. The history of biology is marked by continuing opposition between reductionists and antireductionists. Reductionism thrives on exploiting the charge that it provides the only alternative to the mushy incomprehensibility of vitalism. Antireductionists reply that their opponents have ignored the organismic complexity of nature. Given the picture painted above, where does this traditional dispute now stand?

I suggest that the account of genetics which I have offered will enable reductionists to provide a more exact account of what they claim, and will thereby enable antireductionists to be more specific about what they are denying. Reductionists and antireductionists agree in a certain minimal physicalism. To my knowledge, there are no major figures in contemporary biology who dispute the claim that each biological event, state, or process is a complex physical event, state, or process. The most intricate part of ontogeny or phylogeny involves countless changes of physical state. What antireductionists emphasize is the organization of nature and the "interactions among phenomena at different levels." The appeal to organization takes two different forms. When the subject of controversy is the proper form of evolutionary theory, then antireductionists contend that it is impossible to regard all selection as operating at the level of the gene.[14] What concerns me here is not this area of conflict between reductionists and their adversaries, but the attempt to block claims for the hegemony of molecular studies in understanding the physiology, genetics, and development of organisms.

A sophisticated reductionist ought to allow that, in the current practice of biology, nature is divided into levels which form the proper provinces of areas of biological study: molecular biology, cytology, histology, physiology, and so forth. Each of these sciences can be thought of as using certain language to formulate the questions it deems important and as supplying patterns of reasoning for resolving those questions. Reductionists can now set forth one of two main claims. The stronger thesis is that the explanations provided by any biological theories can be reformulated in the language of molecular biology and be recast so as to instantiate the patterns of reasoning supplied by molecular biology. The weaker thesis is that molecular biology provides explanatory extension of the other biological sciences.

Strong reductionism falls victim to the considerations that were advanced against R3. The distribution of genes to gametes is to be explained, not by rehearsing the gory details of the reshuffling of the molecules, but through the observation that chromosomes are aligned in pairs just prior to the meiotic division, and that one chromosome from each matched pair is transmitted to each gamete. We may formulate this point in the biologists' preferred idiom by saying that the assortment of alleles is to be understood at the cytological level. What is meant by this description is that there is a pattern of reasoning which is applied to derive the description of the assortment of alleles and which involves predicates that characterize cells and their large-scale internal structures. That pattern of reasoning is to be objectively preferred to the molecular pattern which would be instantiated by the derivation that charts the complicated rearrangement of individual molecules because it can be applied across a range of cases which would look heterogeneous from a molecular perspective. Intuitively, the cytological pattern makes connections which are lost at the molecular level, and it is thus to be preferred.

So far, antireductionism emerges as the thesis that there are *autonomous levels of biological explanation*. Antireductionism construes the current division of biology not simply as a temporary feature of our science stemming from our cognitive imperfections but as the reflection of levels of organization in nature. Explanatory patterns that deploy the concepts of cytology will endure in our science because we would foreswear significant unification (or fail to employ the relevant laws, or fail to identify the causally relevant properties) by attempting to derive the conclusions to which they are applied using the vocabulary and reasoning patterns of molecular biology. But the autonomy thesis is only the beginning of antireductionism. A stronger doctrine can be generated by opposing the weaker version of sophisticated reductionism.

In section 4, I raised the possibility that molecular genetics may be viewed as providing an explanatory extension of classical genetics through deriving the schematic sentence that assigns phenotypes to genotypes from a molecular pattern of reasoning. This apparent possibility fails in an instructive way. Antireductionists are not only able to contend that there are autonomous levels of biological explanation. They can also resist the weaker reductionist view that explanation always flows from the molecular level up. Even if reductionists retreat to the modest claim that, while there are autonomous levels of explanation, descriptions of cells and their constituents are always explained in terms of descriptions about genes, descriptions of tissue geometry are always explained in terms of descriptions of cells, and so forth, antireductionists can resist the picture of a unidirectional flow of explanation. Understanding the phenotypic manifestation of a gene, they will maintain, requires constant shifting back and forth across levels. Because developmental processes are complex and because changes in the timing of embryological events may produce a cascade of effects at several different levels, one sometimes uses descriptions at higher levels to explain what goes on at a more fundamental level. . . .

It would be premature to claim that I have shown how to reformulate the antireductionist appeals to the organization of nature in a completely precise way. My conclusion is that, to the extent that we can make sense of the present explanatory structure within biology—that division of the field into subfields corresponding to levels of organization in nature—we can also understand the antireductionist doctrine. In its minimal form, it is the claim that the commitment to several explanatory levels does not simply reflect our cognitive limitations; in its stronger form, it is the thesis that some explanations oppose the direction of preferred reductionistic explanation. Reductionists should not dismiss these doctrines as incomprehensible mush unless they are prepared to reject as unintelligible the biological strategy of dividing the field (a strategy which seems to me well understood, even if unanalyzed).

The examples I have given seem to support both antireductionist doctrines. To clinch the case, further analysis is needed. The notion of explanatory levels obviously cries out for explication, and it would be illuminating to replace the informal argument that the unification of our beliefs is best achieved by preserving multiple explanatory levels with an argument based on a more exact criterion for unification. Nevertheless, I hope that I have said enough to make plausible the view that, despite the immense value of the molecular biology that Watson and Crick launched in 1953, molecular studies cannot cannibalize the rest of biology. Even if geneticists must become "physiological chemists" they should not give up being embryologists, physiologists, and cytologists.

NOTES

Earlier versions of this chapter were read at Johns Hopkins University and at the University of Minnesota, and I am very grateful to a number of people for comments and suggestions. In particular, I would like to thank Peter Achinstein, John Beatty, Barbara Horan, Patricia Kitcher, Richard Lewontin, Kenneth Schaffner, William Wimsatt, an anonymous reader, and the editors of the *Philosophical Review*, all of whom have had an important influence on the final version. Needless to say, these people should not be held responsible for residual errors. I am also grateful to the American Council of Learned Societies and the Museum of Comparative Zoology at Harvard University for support and hospitality while I was engaged in research on the topics of this paper.
[This article has been abridged. (ed.).]

1. "Variation due to Change in the Individual Gene," reprinted in J. A. Peters ed., *Classic Papers in Genetics* (Englewood Cliffs, N.J.: Prentice-Hall, 1959), pp. 104–116. Citation from p. 115.

2. "Molecular Structure of Nucleic Acids," *Nature* 171 (1953), pp. 737–738; reprinted in Peters, op. cit., pp. 241–243. Watson and Crick amplified their suggestion in "Genetic Implications of the Structure of Deoxyribonucleic Acid" *Nature* 171 (1953), pp. 934–937.

3. The most sophisticated attempts to work out a defensible version of reductionism occur in articles by Kenneth Schaffner.

4. E. Nagel, *The Structure of Science* (New York: Harcourt Brace, 1961), Chapter 11. A simplified presentation can be found in Chapter 8 of C. G. Hempel, *Philosophy of Natural Science* (Englewood Cliffs, N.J.: Prentice-Hall, 1966).

5. Quite evidently, this is a weak version of what was once the "received view" of scientific theories, articulated in the works of Nagel and Hempel cited in the previous note.

6. Hull, *Philosophy of Biological Science* (Englewood Cliffs, N.J.: Prentice-Hall, 1974), p. 23, adapted from Theodosius Dobzhansky, *Genetics of the Evolutionary Process* (New York: Columbia University Press, 1970), p. 167.

7. A *locus* is the place on a chromosome occupied by a gene. Different genes which can occur at the same locus are said to be *alleles*. In diploid organisms, chromosomes line up in pairs just before the meiotic division that gives rise to gametes. The matched pairs are pairs of *homologous chromosomes*. If different alleles occur at corresponding loci on a pair of homologous chromosomes, the organism is said to be *heterozygous* at these loci.

8. My notion of a practice owes much to some neglected ideas of Sylvain Bromberger and Thomas Kuhn. See, in particular, Bromberger, "A Theory about the Theory of Theory and about the Theory of Theories" (W. L. Reese ed., *Philosophy of Science, The Delaware Seminar*, New York, 1963); and "Questions" (*Journal of Philosophy* 63 [1966], pp. 597–606); and Kuhn, *The Structure of Scientific Revolutions* (Chicago: University of Chicago Press, 1962) Chapters II–V. The relation between the notion of a practice and Kuhn's conception of a paradigm is discussed in Chapter 7 of my book *The Nature of Mathematical Knowledge* (New York: Oxford University Press, 1983).

9. *Polygeny* occurs when many genes affect one characteristic; *pleiotropy* occurs when one gene affects more than one characteristic.

10. *Incomplete dominance* occurs when the phenotype of the heterozygote is intermediate between that of the homozygotes; *epistasis* occurs when the effect of a particular combination of alleles at one locus depends on what alleles are present at another locus.

11. See Watson, *Molecular Biology of the Gene* (Menlo Park, CA: W. A. Benjamin, 1976), Chapter 9; and Arthur Kornberg, *DNA Synthesis* (San Francisco: W. H. Freeman, 1974).

12. See. G. P. Georghiou, "The Evolution of Resistance to Pesticides," *Annual Review of Ecology and Systematics* 3 (1972), pp. 133–168.

13. There are numerous examples of such modifications from the history of chemistry. I try to do justice to this type of case in "Theories, Theorists, and Theoretical Change," *Philosophical Review* 87 (1978), pp. 519–547 and in "Genes," *British Journal for the Philosophy of Science* 33 (1982), pp. 337–359.

14. The extreme version of reductionism is defended by Richard Dawkins in *The Selfish Gene* (New York: Oxford University Press, 1976) and *The Extended Phenotype* (San Francisco: W. H. Freeman, 1982). For an excellent critique, see Elliott Sober and Richard C. Lewontin, "Artifact, Cause, and Genic Selection," *Philosophy of Science* 49 (1982), pp. 157–180.

19 Why the Antireductionist Consensus Won't Survive the Case of Classical Mendelian Genetics

C. Kenneth Waters

Philosophers now treat the relationship between Classical Mendelian Genetics and molecular biology as a paradigm of nonreduction and this example is playing an increasingly prominent role in debates about the reducibility of theories ranging from macrosocial science to folk psychology. Patricia Churchland (1986), for example, draws an analogy between the alleged elimination of the "causal mainstay" of classical genetics and her view that today's psychological theory will be eliminated by neuroscience. Patricia Kitcher takes an autonomous rather than eliminativist view of the reported nonreduction in genetics and reasons that psychology will retain a similar autonomy from lower level sciences (1980, 1982). Although Churchland and Kitcher offer different interpretations of the apparent failure of molecular biology to reduce classical genetics, they agree that this failure will help illuminate theoretical relations between psychology and lower level sciences. The appearance of the Mendelian example alongside the usual ones from physics and chemistry marks a turning point in philosophy of science. Philosophers now look to biology in general, and the case of genetics in particular, for insights into the nature of theoretical relations. If I am correct, however, the current antireductionist consensus about genetics is mistaken and threatens to misguide our attempt to understand relations between other scientific theories. My aim is to defuse the arguments offered in support of the antireductionist consensus. Although the question of whether molecular biology is reducing Classical Mendelian Genetics will not be settled in any single chapter, my critical analysis will reveal the signs of a significant theoretical reduction and uncover issues relevant to gaining a better understanding of what is now happening in genetics and of what we might expect to occur in other sciences.

The current consensus among philosophers is that, despite the appearances, Classical Mendelian Genetics (hereafter called CMG) is not being reduced to molecular biology, at least *not in the spirit of Nagel's (1961) postpositivist conception of theoretical reduction*[1] (Hull 1972 and 1974. Wimsatt 1976, Maull 1977, Darden and Maull 1977, Hooker 1981, Kitcher 1984, and Rosenberg 1985, but Schaffner 1969 and 1976, Ruse 1976, and Richardson 1979 and 1982 disagree). There are important differences within the consensus view, but according to the general antireductionist thrust, the relations between the levels of organization represented by the classical and molecular theories are too complex to be connected in the systematic way essential for a successful theoretical reduction. Antireductionists support this view by arguing that the gene concepts of the respective theories cannot be linked in an appropriate way. If the concepts cannot be linked, the reasoning goes, neither can the theoretical claims couched in terms of them. Hence, reduction will never be achieved. Before considering the antireductionists' arguments in greater detail, I will briefly describe the conception of reduction at issue and review CMG and the molecular theory of the gene.

PRELIMINARIES

The Spirit of Postpositivist Reduction

The consensus against reductionism in genetics has focused on Nagel's (1961) formal analysis of theoretical reduction. One of the two formal requirements set out by Nagel was that the laws of the reduced theory must be derivable from the laws and associated coordinating definitions of the reducing theory. The second formal requirement was that all terms of the reduced theory must either be contained within or be appropriately connected to the reducing theory by way of "additional assumptions." It is this condition of connectability that proponents of the consensus think cannot be satisfied in the case of genetics because of the contrasting gene concepts in the classical and molecular theories. A difficulty of relying on this formal conception is that it is couched within an account of theories discarded by most philosophers of biology.[2] In order to render the antireductionist consensus nontrivial, the spirit behind Nagel's conception of theoretical reduction will need to be separated from his formal analysis.

Nagel's discussion of nonformal conditions for reduction provides an opening for freeing his conception of theoretical reduction from his outmoded account of theories. In a section on these conditions, Nagel admitted, "The two formal conditions for reduction discussed in the previous section [connectability and derivability] do not suffice to distinguish trivial from noteworthy scientific achievements" (p. 358). He identified two sets of nonformal considerations to explain why the reduction of thermodynamics was a significant achievement. The first set concerned the establishment of new experimental laws that were in better agreement with a broader range of facts than were the original ones. The second set involved the discovery of surprising connections between various experimental laws.

Nagel's reliance on nonformal conditions indicates that he had an unarticulated notion of theoretical reduction which he failed to capture in his formal account. I would like to suggest, therefore, that his underlying conception of theoretical reduction can be separated from his formal treatment and in fact

C. Kenneth Waters

reformulated with respect to an updated account of theories. When I say that intertheoretical relations satisfy *the spirit of postpositivist reduction*, I simply mean that they would satisfy conditions set out in such a reformulation. Since postpositivists tended to view both explanation and reduction as special kinds of derivation, it is natural to suppose that their conception of theoretical reduction centered on the idea that reducing theories explain the success of reduced theories. Hence, the fundamental question for us is whether CMG is being explained by molecular biology. According to the antireductionist consensus, CMG is not and will not be systematically explained by molecular biology.

Classical Mendelian Genetics (CMG)

The consensus view concerns the reducibility of the theory of Classical Mendelian Genetics (CMG), not the reducibility of Mendel's theory. CMG was developed during the first decades of this century, in large part by Thomas Hunt Morgan and his graduate students who worked on the genetics of *Drosophila*. According to the classical theory, patterns of inheritance can be explained by postulating the existence of genes. Differences in outward appearances (or phenotypes) of organisms are explained as the result of organisms' inheriting different genes (or genotypes). Genes in *Drosophila* come in twos on corresponding pairs of linear chains. Each gene of a given pair has 50 percent chance of having a copy distributed to a particular gamete (an egg or sperm). Genes located on different (nonpaired) chains assort independently from one another. Genes located on the same chain tend to be assorted together, but are sometimes distributed separately because paired chains occasionally exchange segments. The relative positions of genes can be determined by the frequency of such exchanges (on the assumption that genes located farther apart from one another are assorted separately more often than genes located closer together). CMG concerns a wide range of gene behavior including, but not limited to, mutation, expression, interaction, recombination, and distribution.

The classical account of gene expression is complicated. In the simplest kind of system, two alleles with complete dominance, there are two contrasting phenotypic traits and two kinds of genes, one of which is dominant. Each trait is associated with one kind of gene and every organisms has two genes. If an organism has two copies of the same gene, it exhibits the trait associated with the matching genes. If an organism has a pair of contrasting genes, it exhibits the characteristic associated with the dominant gene. This is but the simplest model of gene expression; classical geneticists have constructed models to represent systems of much greater complexity.

This abstract theory has a cytological interpretation. Gene chains are identified as chromosomes. Meiosis, the process in which chromosomes are distributed to gametes, offers an explanation of segregation and assortment. During the first division of this process, homologous chromosomes pair and then separate as two daughter cells are produced. The lack of complete linkage of genes located on the same chromosome is explained in terms of the crossing over (the exchange) of chromosomal segment.[3]

The Molecular Theory

The molecular theory of the gene is based on the Watson and Crick Model of DNA. According to molecular theory, a gene is a relatively short segment of a DNA molecule, which consists of two very long chains of nucleotides held together by hydrogen bonds. The genetic information is encoded in the linear sequence of nucleotides making up individual genes. On the basis of this model and empirical studies, molecular biologists soon succeeded in explaining a number of important genetic phenomena including: gene replication; the multistep process by which the information encoded in structural genes eventually gets translated during polypeptide synthesis; and mechanisms of gene regulation. Polypeptides are the constituents of proteins and the regulation of biosynthetic pathways is for the most part directed by enzymatic proteins. Hence, the molecular explanation of how genes direct polypeptide synthesis offered an abstract picture of the biochemistry of gene expression.[4]

These successes led Kenneth Schaffner (1969) to conclude that CMG was being reduced to molecular biology. But enthusiasm for reductionism soon waned (at least among philosophers) when Michael Ruse (1971) and David Hull (1972) criticized Schaffner's specific account of the apparent reduction. Since then, these rather narrowly focused criticisms have been generalized into self-contained arguments against the general idea that CMG is being reduced. I now turn to these antireductionist objections.

DEFUSING THE ANTIREDUCTIONIST OBJECTIONS

Arguments against the idea that CMG is being reduced (in the spirit of postpositivist reduction) fall into two general categories. The most prominent arguments are those aimed at showing that there are unbridgeable conceptual gaps between CMG and molecular biology. According to these arguments, subtle differences in the meaning of parallel terms from the classical and molecular theories obstruct reduction. The second category consists of arguments which conclude that molecular theory cannot deliver the explanatory power that reductionism requires. These arguments allegedly show that the explanatory relations between the classical and molecular theories are incomplete and that if a fuller explanation of Mendelian genetics is possible, it will come from a variety of biological fields, not just from molecular genetics as reductionism seems to imply. My critical analysis of these objections will not only show that the relationship between CMG and molecular biology is misunderstood, it will also reveal signs of a successful theoretical reduction in progress.

The Unconnectability Objection

The unconnectability objection can be traced to David Hull's seminal works (1972 and 1974) where he proposed and defended the then-heretical notion that Mendelian genetics is not being reduced by molecular biology, at least not according to Nagel's conception of theoretical reduction. The most rigorous formulation of this objection can be found in Alexander Rosenberg's provocative text (1985).

Rosenberg's opposition to reductionism in genetics rests on an alleged conceptual gap between the classical and molecular theories of genetics. He argues that relations between the gene concepts of the two theories are hopelessly complicated "many-many relations" that will forever frustrate any attempt to systematically connect the two theories. Rosenberg begins his analysis by pointing out that in CMG, genes are always identified by way of their phenotypic effects. Classical geneticists identified the gene for red eye color in *Drosophila*, for example, by following the distribution of red and white phenotypes in successive generations of a laboratory population. The reason CMG will never be reduced to molecular biology, Rosenberg argues, is that there is no manageable connection between the concept of a Mendelian phenotype and that of a molecular gene. The relation between them is complicated by the fact that scores of Mendelian phenotypes are potentially affected by an individual molecular gene and that a vast array of molecular genes are responsible for the production of any given Mendelian phenotype. Rosenberg explains the problem as follows:

Suppose we have set out to explain the inheritance of normal red eye color in *Drosophila* over several generations. The pathway to red eye pigment production begins at many distinct molecular genes and proceeds through several alternative branched pathways. Some of the genes from which it begins are redundant, in that even if they are prevented from functioning the pigment will be produced. Others are interdependent, so that if one is blocked the other will not produce any product. Still others are "ambiguous"—belonging to several distinct pathways to different phenotypes. The pathway from the genes also contains redundant, ambiguous, and interdependent paths. If we give a biochemical characterization of the gene for red eye color either by appeal to the parts of its pathway of synthesis, or by appeal to the segments of DNA that it begins with, our molecular description of this gene will be too intricate to be of any practical explanatory upshot. (Rosenberg 1985, p. 101)

Rosenberg reasons that since Mendelian genes are identified through their phenotypes, and since the relation between molecular genes and Mendelian phenotypes is exceedingly complex, the connection between the molecular and Mendelian gene concepts must also be exceedingly complex. Hence, he concludes, CMG will forever remain beyond the reductive grasp of molecular biology. Rosenberg does not deny that molecular biologists will occasionally furnish individual accounts of various Mendelian phenomena on a piecemeal basis (as they have done with the genetics of sickle-cell anemia). He insists, however, that the unmanageable complex relations between the gene concepts of the two theories will prevent any systematic, reductive explanation of CMG in terms of molecular theory.

What Rosenberg's persuasive argument does not take into consideration is the relationship between the Mendelian gene and the Mendelian phenotype. According to the classical theory, one gene can affect different phenotypic traits and each phenotypic trait can be affected by different (nonallelic) genes. I will argue that the relationship between the Mendelian gene and the Mendelian phenotype exhibits the same complexity that Rosenberg discusses from the molecular perspective. My argument will not depend upon historical hindsight. Alfred H. Sturtevant, one of the architects of CMG, discussed the complex relation between the Mendelian gene and phenotype in his Ph.D. thesis (1916), which he wrote under T. H. Morgan. Ironically, he illustrated the point with the very same example that Rosenberg considers:

The difference between normal red eyes and colorless (white) ones in *Drosophila* is due to a difference in a single gene. Yet red is a very complex color, requiring the interaction of at least five (and probably of very many more) different genes for its production. And these genes are quite independent, each chromosome bearing some of them. Moreover, eye-color is indirectly dependent upon a large number of other genes, such as those on which the life of the fly depends. We can then, in no sense identify a given gene with the red color of the eye, even though there is a single gene differentiating it from the colorless eye. So it is for all characters—as Wilson (1912) has put it "... the entire germinal complex is directly or indirectly involved in the production of every character.[5]"

The parallel between Sturtevant's and Rosenberg's accounts of the complex relationship between Mendelian phenotypes and Mendelian genes (Sturtevant's) and between Mendelian phenotypes and molecular genes (Rosenberg's) is striking. Both identify a web of relations too complex for the kind of explanation that Rosenberg seeks. My claim is that the molecular perspective offers a reductive interpretation of the complex picture offered by the classical theory. Our understanding of the biosynthetic pathways explains why there should be many-many relations between classical genes and Mendelian phenotypes.

The problem with Rosenberg's antireductionist line of reasoning is that it assumes that the existence of a particular gene can explain the presence of particular traits in an individual when in fact *genes can only explain* phenotypic *differences* and only *in given populations*. The presence of a gene for red eye-color on the X chromosome explains why the red-eyed *Drosophila* in a certain population have red eyes instead of white ones. The reason why classical geneticists found manageably simple relations between genes and phenotypic differences is because the genetic backgrounds against which particular genes produced differences were sufficiently uniform from one organism to another in the laboratory populations (of highly related individuals) under study. This can be explained from the molecular perspective in terms of a uniformity in relevant portions of the DNA, which in turn provided a uniform potential for bringing about certain results within the complex web of biosynthetic reactions.

Rosenberg's is but one of several lines of reasoning against the idea that the concepts of CMG and molecular biology can be systematically connected. Others focus on the problem of specifying a precise biochemical definition of a Mendelian gene. If the behavior of Mendelian genes can be explained in terms of molecular biology, some critics reason, then the central concepts of Mendelian theory must be defined in purely biochemical terms. The attempt to define a gene as a relatively short stretch of DNA won't do, the anti-reductionists point out, because not all relatively short stretches are genes. Furthermore, the attempt to define the gene in terms of a finer structure associated with a specific molecular mechanism will not work because of the diversity of molecular ways in which genes produce their effects. For example, Mendelian genes cannot be identified with reading frames (sections of DNA that are transcribed into RNA) because regulatory genes function without being transcribed. Such considerations reveal that a simple molecular definition of a Mendelian gene is not forthcoming.

The obvious response for the reductionist is simply to hold out for a disjunctive connection.[6] As we learn more about the molecular nature of Mendelian genes, we have discovered that they do not all function by way of the same mechanism. Some genes function by being transcribed into segments of RNA which code for polypeptides. Others function by regulating the transcription of neighboring genes. Furthermore, although all Mendelian genes are relatively short segments of DNA (or perhaps RNA), their finer structure varies with their role. Hence, any definition of Mendelian gene in terms of fine molecular structure will be disjunctive.

While I am not prepared to insist that molecular biology already provides the means for completing a disjunctive definition *in terms of molecular structure*, I do think the elements for such a definition are falling in place. For the time being, I believe it suffices to point out that the behavior of specific Mendelian genes has been explained by identifying them with relatively short segments of DNA which function as units to influence the course of chemical reactions within a biochemical system. The fact that such a characterization has been sufficient for the development of molecular models of a variety of Mendelian phenomena leads me to think that the philosophers' attempt to formulate precise syntactical connections (in the form of explicit and detailed definitions) has been counterproductive. The focus on formal aspects of the postpositivist conception of reduction has led to too much haggling over syntax and not enough analysis of whether genetics exhibits the sort of semantic and pragmatic features that motivated the formal account in the first place.

The Mendelian gene can be specified in molecular biology as a relatively short segment of DNA that functions as a biochemical unit. This specification provides an appropriate interpretation of the many-many relation between a Mendelian gene and phenotype. In addition, it provides a general statement of the precise connections that practicing molecular biologists have drawn between genes and phenotypes in individual cases. Most important, however, it has proved to be tremendously fruitful in research. For it has enabled molecular biologists to apply traditional strategies from classical genetics to uncover the biochemistry underlying many life processes. I conclude that the antireductionist thesis that there is some unbridgeable conceptual gap lurking between CMG and its molecular interpretation is wrong.

The Explanatory Incompleteness Objection

The idea that CMG is being reduced to molecular biology has also been opposed on the grounds that molecular biology will never explain, and hence will never reduce, the classical theory of genetics. Since the postpositivist account of theoretical reduction is centered on the idea that the reducing theory explains the reduced one, this complaint strikes at the very heart of the claim that CMG is being reduced in the spirit of postpositivism. Although this kind of objection can be found interspersed throughout the antireductionistic literature and seems to be an important element motivating the consensus against reductionism in genetics, it is seldom put forth as rigorously as the unconnectability objection. Nevertheless, I will reconstruct and defuse two separate arguments falling under this category.[7]

The Gory Details Argument Antireductionists have argued that knowledge of the molecular makeup of genes does not enhance our understanding of their classical Mendelian behavior. For example, Philip Kitcher (1984), in his brilliant essay which marks the culmination of the antireductionist literature, argues that the assortment of genes is best understood at the cytological level: "The distribution of genes to gametes is to be explained, not by rehearsing the gory details of the reshuffling of the molecules, but through the observation that chromosomes are aligned in pairs just prior to the meiotic division, and that one chromosome from each matched pair is transmitted to each gamete" (Kitcher 1984, p. 370). He goes on to argue that the cytological pattern of explanation is objectively preferable because it can uniformly account for a wide range of cases that would look heterogeneous from a molecular perspective.

Kitcher does not describe a diversity of molecular processes responsible for the segregation of genes during meiosis. Instead, he offers an abstract account of the cytological explanation of gene distribution. According to his account, the distribution of genes is explained by identifying meiosis as belonging to the natural kind of "pair-separation processes." This natural kind of process, he says, is heterogeneous from the molecular perspective because different kinds of forces are responsible for bringing together and pulling apart different paired "entities." The separation of paired entities, he claims, "may occur because of the action of electromagnetic forces or even of nuclear forces; but it is easy to think of examples in which the separation is effected by the action of gravity" (Kitcher 1984, p. 350). Kitcher, I think, is *not* making the claim that some paired chromosomes are pulled apart by nuclear forces and others by the force of gravity (such a claim would be completely at odds with today's evidence). Rather, when he is discussing the multiple realizations of pair-separation processes he seems to be conceiving of a natural kind that includes processes quite unlike anything that occurs during meiosis. Hence, his reasoning only suggests that at some high level of abstraction, it is possible to draw an analogy between the process of meiosis and (yet to be specified) processes that have quite different molecular mechanisms. This is a far cry from showing that cytological theory offers a uniform explanation of a range of cases that would appear heterogeneous at the molecular level.

Although meiosis appears to be an unpromising candidate, there are other phenomena that are explained uniformly by CMG, but which are caused by a variety of molecular mechanisms. Phenomena of gene expression provide obvious examples. CMG, for instance, lumps together different kinds of gene expression under the category of dominance. This Mendelian category includes genes that code for structural proteins, genes which code for enzymes, and even regulatory genes. The molecular mechanisms by which these different kinds of genes are eventually expressed are quite different. Yet, when examining concrete cases where CMG offers a more uniform perspective, it is difficult to accept the antireductionist judgment that the shallow explanations of CMG are objectively preferable to the deeper accounts provided by molecular theory.

The idea that the uniformity provided by CMG gives it some sort of explanatory edge over the less uniform molecular account seems plausible only when our attention is called away from the actual biology. But even if uniformity of explanation did provide a potentially decisive advantage, there would be no reason to suppose that the uniformity represented by CMG could not also be captured within the molecular perspective through the familiar scientific practices of abstraction and idealization. The reductionists' view is not that the pictures offered by the reduced and reducing theory are the same, but that they can be connected by auxiliary assumptions such that the reducing theory stands in an explanatory relation to the reduced one. The fact that the reducing theory, when not accompanied by such auxiliary assumptions, more accurately represents the true diversity of mechanisms responsible for various processes should not be held against it.

Antireductionists, of course, do not deny the fact that molecular biology has greatly improved our understanding of genetics. Kitcher (1984), for example, provides an interesting discussion of various ways that molecular genetics has advanced our understanding. But they seem pessimistic when it comes to the issue of whether molecular theory will help us understand what (they think) are the essentials of CMG: the processes by which genes are distributed to gametes. The phenomenon of independent assortment of nonlinked genes, it is claimed, depends only on the pairwise separation of chromosomes. The classical theory apparently tells us all we need to know: nonlinked genes are located on separate nonhomologous chromosomes and nonhomologous chromosomes segregate independently. The identification of genes as segments of a molecular double helix allegedly adds nothing to this account.

This antireductionist argument is problematic for two reasons: first, it becomes less plausible when we flesh it out within CMG (as opposed to Mendel's genetics) and second, it seems unduly pessimistic. To flesh the argument out within CMG, we need to consider not just the independent assortment of nonlinked genes, but also the distribution of linked ones. Recall that of central importance to the classical theory was the fact that linkage is incomplete because of the process of crossing over. At the cytological level, not much can be said about this process except that homologous chromosomes sometimes wrap around each other and swap segments during cellular division. Shortly after the double helical structure of the genetic chains was understood, however, molecular models of crossing over were proposed. The basic Holliday Model (Holliday 1964), illustrated in figure 19.1, has been especially fruitful. Since then, laboratory studies have led to a more detailed, though admittedly tentative, biochemical understanding of the individual

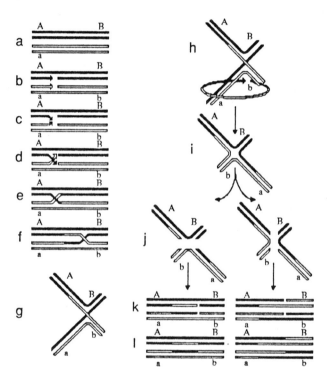

Figure 19.1 The Holliday Model for genetic recombination. (*a*) Two homologous double helices are aligned. (*b*) The two + or − strands are cut. (*c*) The free ends leave the complementary strands to which they have been hydrogen bonded. (*d*) The free ends become associated with the complementary strands in the homologous double helix. (*e*) Ligation creates partially heteroduplex double helices. (*f*) Migration of the branch point occurs by continuing strand transfer by two polynucleotide chains involved in a crossover. (*g*) The Holliday structure shown in extended form. (*h*) The rotation of the structure shown in (*g*) can yield the form depicted in (*i*). Resolution of the structure shown in (*i*) can proceed in two ways, depending on the points of enzymatic cleavage, yielding the structures shown in (*j*), which can be depicted as shown in (*k*), and repaired to the forms shown in (*l*). Figure from Potter and Dressler (1979, p. 970). Explanation quoted from Suzuki et al. (1986, p. 360).

C. Kenneth Waters

steps outlined in this model (see Potter and Dressler 1988). Our understanding of the exchange of segments between paired chains of genes is being greatly enhanced by our knowledge of the molecular structure of those chains. The biochemistry of genetic recombination is a tremendously active area of research and will bring our understanding of the classical Mendelian process of crossing over to the molecular level.

Antireductionists might respond by insisting that although the molecular perspective will contribute to our understanding of this bit of CMG, reductionism is a global thesis and requires that it contribute to all bits of the theory. "What about the independent assortment of nonlinked genes?" they might ask. "How do the molecular details improve the cytological explanation according to which nonlinked genes segregate independently because they are located on different chromosomes, which have been observed (via the microscope) to segregate independently?" This is the point at which I think the gory details objection becomes unduly pessimistic. Surely, the conjugation and separation of homologous chromosomes depends upon molecular mechanisms. While our understanding of why homologous chromosomes pair, why nonhomologous chromosomes don't pair, why separately paired chromosomes segregate independently, and so forth is not well developed, antireductionists haven't offered sufficient reason for thinking these questions won't eventually be answered. The answers to these questions will be given from the molecular perspective and will enhance our understanding of why nonlinked genes assort independently.

Research in the general area of genetic recombination has already displayed signs identified by Nagel as the distinguishing features of an important theoretical reduction. One sign is the discovery of surprising connections between seemingly unrelated processes. Recent biochemical research has revealed unexpected connections among the processes of recombination, replication, and repair (see Low 1988). Another sign of a significant reduction is the establishment of new experimental laws that are in better agreement with the facts. Recent lines of biochemical research hold promise for explaining why recombination is not entirely random and for helping us discover the finer patterns of genetic recombination (e.g., patterns of interference in closely spaced exchanges). Hence, even with respect to the Mendelian phenomena for which molecular explanations have tended to lag (i.e., transmission phenomena), the relation between CMG and molecular theory is beginning to exhibit characteristics corresponding to the two nonformal conditions set out by Nagel (1961) in his classic account of theoretical reduction.

The claim that the gory details of molecular biology do not enhance our understanding of key processes underlying CMG is quickly becoming outdated. There is no question that molecular theory has greatly improved our understanding of gene replication, expression, mutation, and recombination. Furthermore, it is just a matter of time before it accounts for the pair-wise coupling and separation of chromosomes during meiosis. Antireductionists need to justify their pessimism and explain why we should not expect molecular biology to continue on its path toward explaining CMG in accordance with the spirit of postpositivist reduction.

The Splintering Argument   The antireductionist literature contains hints of a way to dodge reductionism without denying the impending molecular explanation of CMG. Antireductionists might argue that even if a molecular explanation is imminent, the explanation will not come from molecular genetics; instead, it will come from a multitude of theories or fields of molecular biology.[8] Following Hull (1974), antireductionists have typically classified CMG as a theory of transmission genetics and molecular genetics as a theory of development. Presupposing this taxonomy, it might be argued that the classical and molecular theories of genetics explain different aspects of heredity. Hence, antireductionists might argue that even if transmission is explained at the molecular level, it will not be explained by molecular genetics.

It is tempting to dismiss such an antireductionist response as a case of sour grapes. "After volumes of denial," the reductionist might complain, "when the antireductionists are finally forced to admit that molecular biology systematically improves our understanding of classical genetics, they turn around and say that the explanation does not count because it comes from the wrong parts of molecular biology." While tempting, such a reply might miss the crux of the antireductionist complaint.

The issue at stake is whether molecular theory will offer a reasonably coherent explanation of CMG. The possible complaint is that molecular explanations will splinter into numerous fields. Instead of a case of one theory reducing another, one might envision a number of distinct theories explaining bits or pieces of the higher-level theory. If unification is taken to be the hallmark of scientific explanation, the splintering of explanatory paths might appear to clinch the case against reductionism.

While such reasoning sounds plausible in the abstract, it depends on a number of slippery points in need of careful examination. The conceptual division between transmission and developmental genetics, for instance, though widely adopted in the philosophical literature and introductory chapters of genetic texts, has never been carefully analyzed and provides a weak footing for antireductionism. The chief reason offered in favor of this division, that CMG was developed on the basis of transmission studies, applies to molecular genetics as well (transmission studies have played and will continue to play an important role in the development of molecular genetics). Furthermore, the history of classical genetics supports the idea that the scope of CMG encompasses more than transmission. Debates about the presence and absence hypothesis and the position effect, to take just two examples, clearly went beyond issues of transmission.

Perhaps the most serious obstacle to developing the splintering argument is it rests on the idea that there are significant divisions between theories of molecular biology when in fact molecular theory seems to have a diffuse structure. It is far from clear that molecular biology contains a separate theory of molecular genetics. Perhaps molecular biology consists of numerous molecular models of various phenomena, which are not organized into more discrete theories, but are loosely unified by their grounding in a set of common biochemical and biophysical principles. If this is indeed the case, the molecular explanation of CMG will not splinter into a number of different theories at the molecular level.

Developing the splintering objection would also entail substantiating premises about the structure of CMG. Antireductionists minimize the explanatory fit between CMG and molecular theory by deemphasizing the parts of CMG that can be elegantly explained at the molecular level. The explanatory relations between CMG and molecular theory appear fractured, for example, when Kitcher characterizes the principle of gene replication as a "presupposition," as opposed to a "central law" of CMG (1984, p. 361). Such structural accounts of CMG depend on controversial philosophical views about the structure of scientific theories, which I believe are poorly motivated.[9] In any case, they should not be taken for granted.

The prospects for developing the splintering objection appear dim. The objection entails controversial philosophical views about the structure of theories and the nature of explanation as well as highly questionable assumptions about the taxonomy of genetics and the makeup of CMG and molecular biology.

CONCLUSION

The major objections to the view that CMG is being reduced by molecular biology have not withstood rigorous scrutiny. Perhaps the most surprising result is that the unconnectability objection was found to be so seriously flawed. In retrospect, however, the claimed unconnectability seems unlikely. After all, researchers are successfully identifying the molecular constitutents and pinpointing the exact locations of genes contributing to many classically characterized traits (e.g., Duchenne muscular dystrophy). With sufficient experimental ingenuity, the molecular constituents and locations of the *Drosophila* genes mapped by Morgan et al. could also be identified and pinpointed.[10] As a matter of fact, researchers have just determined the molecular identity of the first Mendelian gene ever discovered, the gene for wrinkled-seed character in pea plants (Bhattacharyya et al. 1990). While molecular biologists have had to conquer many obstacles in their search for the molecular identity of Mendelian genes, the alleged conceptual gap between gene concepts was not one of them.

My examination of the arguments aimed at showing that molecular theory will never explain (and hence never reduce) classical genetics provides a partial explanation of why philosophers and molecular biologists disagree about the reduction of Mendelian genetics. In each case, the antireductionist arguments were based on admittedly brilliant philosophical analyses that appeared plausible in the abstract. But when scrutinized with respect to the details of the actual science, the arguments were found to rest on undue pessimism, on implausible judgments of comparative explanatory value, and on highly questionable assumptions about the structure of CMG and molecular biology. Practicing geneticists believe that the classical theory can be systematically explained at the molecular level, I suggest, because they have a firm grasp of the explanatory power and structure of molecular biology.

Practicing geneticists are also well aware of the achievements that I have identified as signs of a significant theoretical reduction. These achievements include the discovery of unexpected molecular connections among several different genetic processes and promises to improve the precision of our generalizations in genetics. In addition, genetics has provided tremendously fruitful strategies for biochemical research. While such signs indicate that a theoretical reduction is in the making, I have not offered an account of that reduction. I will conclude by briefly anticipating some of the philosophical work that lies ahead.

The main philosophical task will involve reformulating the postpositivist conception of theoretical reduction. Reformulating the postpositivist conception will require an explicit account of explanation as well as an updated account of theories.[11] The antireductionist arguments are tacitly or explicitly linked to accounts of explanation that place a very high premium on unification. This premium is associated with the idea that theoretical reduction requires unification. While the postpositivist view assumes that reduction is accompanied by unification, it is not clear whether the view takes unification to be an essential ingredient or just an expected dividend. Unification is essential for reduction just in case it is essential for explanation. If, as I have hinted, the unificationist criterion for explanation is implausible when invoked within the nitty-gritty details of genetics, there will be strong incentive to treat unification as a valued bonus, rather than a necessary requirement in the reformulated account of theoretical reduction. The unificationist accounts of explanation and reduction, I suggest, should be assessed from the perspective of molecular biology rather than the other way around.

The reformulation of theoretical reduction will have to be carried out in terms of an explicit account of theories. Most philosophers of biology accept something akin to the semantic view, a view which holds some promise for helping us capture the spirit of postpositivist reduction. One advantage of the semantic view is that it can reportedly help us avoid the logical empiricists' preoccupation with syntactical matters, a preoccupation which plays a role in some antireductionist analyses. Another advantage is that its picture of piecemeal theorizing should enable us to formulate not just a conception of completed reduction, but also the conception of reduction in progress. A shortcoming of the original formulation is that it does not offer a dynamic picture of theoretical reduction. This is especially problematic with respect to genetics where the reduction is still being worked out.

Different philosophical views on the structure of theories and the nature of explanation will undoubtedly lead to different conceptions of theoretical reduction and different pictures of the theoretical relations between classical genetics and molecular biology. These, in turn, can be assessed on the basis of how well they illuminate the actual science. The question of whether CMG is

C. Kenneth Waters

being reduced deserves to be reconsidered, not just because we have good reason to suspect that the antireductionistic consensus is wrong, but also because it provides the opportunity to advance philosophical debates about the structure of theories and the nature of scientific explanation and theoretical reduction.[12]

NOTES

I thank Bob Knox for stimulating discussions which influenced my thinking on this subject. Earlier versions of this chapter were presented at the University of Pittsburgh and Indiana University where audiences provided helpful feedback. The National Science Foundation funded this research (Grant No. DIR 89-12221) and the Center for Philosophy of Science at the University of Pittsburgh provided additional support and hospitality while I worked on this chapter.

1. "Not in the spirit of the postpositivist conception of theoretical reduction" is emphasized because some critics acknowledge that there are important theoretical relations between CMG and molecular biology, but insist that these relations cannot be understood in terms of the postpositivist conception of reduction. Wimsatt (1976), for example, attacks Nagel's conception and offers his own functional account of the activities related to "explanatory reduction." The more recent literature (e.g., Kitcher 1984 and Rosenberg 1985), which heavily borrows from the earlier works, is less ambiguous and clearly denies that molecular biology will ever reduce CMG in any significant sense of "reduction." I suspect that some of the earlier papers will appear less antireductionist when antireductionism is no longer taken for granted.

2. Rosenberg, however, clings to the old account of theories. See Waters (1990).

3. A good primary source of CMG is Morgan (1928). Carlson (1966) offers a provocative historical account and Hull (1974) gives a succinct and clear presentation of the theory.

4. More detailed accounts of molecular theory can be found in practically any contemporary genetics text.

5. Quoted from Carlson (1988, p. 69).

6. Although this is the obvious response, another is available. For, as some antireductionists have admitted (Hull 1974, Kitcher 1984), the derivation (or explanation) of the principles of CMG does not require the formulation of a set of necessary and sufficient molecular conditions for the terms of CMG. Necessary conditions would suffice.

7. The basic reasoning behind the first argument and hints of the second can be found in Kitcher (1984). Elements of them can also be found in Hull (1974), Wimsatt (1976), and perhaps Maull (1977), and Darden and Maull (1977). A third argument can be constructed on the basis of Beatty's point that molecular biology will never completely explain CMG because it will never be able to reduce the evolutionary explanation of Mendelian principles (see 1983). Beatty has developed an important point about the limits of molecular reductionism in biology and it would be decisive if I was arguing that all of biology can be reduced to a science of proximate causes. But my interest concerns the question of whether the proximate theory of Mendelian genetics will be reduced by the proximate theory of molecular biology. Evolutionary questions about Mendelian phenomena will not go away upon achievement of this reduction; they will simply be reduced to evolutionary questions about molecular phenomena.

8. Kitcher (1984), for example, suggests something along this line when he writes that "molecular genetics on its own, cannot deliver the goods" (p. 366) and that "it would be folly to suggest that the [explanatory] extension is provided by molecular genetics alone" (p. 368).

The Antireductionist Consensus Won't Survive

9. Kitcher's (1984, p. 361) defense of this characterization is enmeshed within his distinctive account of the structure of scientific theories. I have challenged the central motivation for his radical departure from the traditional view that theories contain law-like claims (Waters 1989 and forthcoming). If I'm correct, the principle of gene replication should be viewed as a law of CMG.

10. *Drosophila* researchers have shifted their attention to genes that play significant roles in developmental processes. So, the search is mainly for genes with developmental significance.

11. The depth of Kitcher's (1984) account of this case stems from the fact that he has taken into account these underlying philosophical issues. But I believe the denial of the unconnectability objection, a more explicit account of molecular biology, and different philosophical views on structure of theories and the nature of explanation will lead to a different and more illuminating picture of the situation.

12. The theses defended in this article are further developed in Waters (forthcoming).

REFERENCES

Beatty, J. 1983. "The Insights and Oversights of Molecular Genetics: The Place of the Evolutionary Perspective." In P. Asquith and T. Nickles (eds.), *PSA 1982*, Volume 2, East Lansing, MI: Philosophy of Science Association.

Bhattacharyya, M. K., Smith, A. M., Ellis, T. H. N., Hedley, C., and Martin, C. (1990). "The Wrinkled-Seed Character of Pea Described by Mendel Is Caused by a Transposon-like Insertion in a Gene Encoding Starch-Branching Enzyme." *Cell* 60 : 115–22.

Carlson, E. A. 1989. *The Gene, A Critical History*. Ames, Iowa: Iowa State University Press.

Churchland, P. S. 1986. *Neurophilosophy*. Cambridge, Mass.: Bradford/MIT Press.

Darden, L., and N. Maull. 1977. "Interfield Theories." *Philosophy of Science* 44 : 43–64.

Holliday, R. 1964. "A Mechanism for Gene Conversion in Fungi." *Genetics Research* 5 : 282–303.

Hooker, C. A. 1981. "Towards a General Theory of Reduction, Part I: Historical and Scientific Setting, Part II: Identity in Reduction, Part III: Cross-categorical Reduction.", *Dialogue* 20 : 38–59, 201–36, 496–529.

Hull, D. 1972. "Reduction in Genetics—Biology or Philosophy?" *Philosophy of Science* 39 : 491–99.

———. 1974. *Philosophy of Biological Science*. Englewood Cliffs, N.J.: Prentice-Hall.

Kitcher, P. 1984. "1953 and All That. A Tale of Two Sciences." *Philosophical Review* 93 : 335–73.

Low, K. B. 1988. "Genetic Recombination: A Brief Overview." In K. B. Low (ed.), *The Recombination of the Genetic Material*. San Diego: Academic Press, pp. 1–23.

Maull, N. 1977. "Unifying Science without Reduction." *Studies in the History and Philosophy of Science* 8 : 143–71.

Morgan, T. H. 1926. *The Theory of the Gene*. New Haven: Yale University Press.

Nagel, E. 1961. *The Structure of Science*. London: Routledge and Kegan Paul.

Potter, H., and Dressler, D. 1979. "Branch Migration in Recombination." *Cold Spring Harbor Symposium of Quantitative Biology* 43 : 957–90.

———. 1988. "Genetic Recombination: Molecular Biology, Biochemistry, and Evolution." In K. Brooks Low (ed.), *The Recombination of the Genetic Material*. San Diego: Academic Press, pp. 218–283.

Richardson, R. C. 1979. "Functionalism and Reductionism." *Philosophy of Science* 48:533–58.

———. 1982. "Discussion: How Not to Reduce a Functional Psychology." *Philosophy of Science* 49:125–37.

Rosenberg, A. 1985. *The Structure of Biological Science*. Cambridge: Cambridge University Press.

Ruse, M. 1969. "Reduction in Genetics." In R. S. Cohen et al. (eds.), *PSA 1974*. Boston: D. Reidel, pp. 653–70.

Schaffner, K. 1969. "The Watson-Crick Model and Reductionism." *British Journal for the Philosophy of Science* 20, 325–48.

———. 1976. "Reductionism in Biology: Prospects and Problems." In R. S. Cohen et al. (eds.), *PSA 1974*. Boston: D. Reidel, pp. 613–32.

Suzuki, D. T. et al. 1986. *Introduction to Genetic Analysis*. New York: W. H. Freeman and Company.

Waters, C. K. 1989. "The Universal Laws of Biology." Presented to the 1989 Meeting of the Eastern Division of the American Philosophical Association, abstract in *Proceedings and Addresses of the American Philosophical Association* 62:74.

———. 1990. "Rosenberg's Rebellion." *Biology and Philosophy* 5:225–39.

———. Forthcoming. "Laws, Kinds and Generalities in Biology." Unpublished manuscript.

———. Forthcoming. "Genes Mude Molecular." *Philosophy of Science*.

X Ethics and Sociobiology

20 Moral Philosophy as Applied Science

Michael Ruse and Edward O. Wilson

For much of this century, moral philosophy has been constrained by the supposed absolute gap between *is* and *ought*, and the consequent belief that the facts of life cannot of themselves yield an ethical blueprint for future action. For this reason, ethics has sustained an eerie existence largely apart from science. Its most respected interpreters still believe that reasoning about right and wrong can be successful without a knowledge of the brain, the human organ where all the decisions about right and wrong are made. Ethical premises are typically treated in the manner of mathematical propositions: directives supposedly independent of human evolution, with a claim to ideal, eternal truth.

While many substantial gains have been made in our understanding of the nature of moral thought and action, insufficient use has been made of knowledge of the brain and its evolution. Beliefs in extrasomatic moral truths and in an absolute is/ought barrier are wrong. Moral premises relate only to our physical nature and are the result of an idiosyncratic genetic history—a history which is nevertheless powerful and general enough within the human species to form working codes. The time has come to turn moral philosophy into an applied science because, as the geneticist Hermann J. Muller urged in 1959, 100 years without Darwin are enough.[1]

The naturalistic approach to ethics, dating back through Darwin to earlier preevolutionary thinkers, has gained strength with each new advance in biology and the brain sciences. Its contemporary version can be expressed as follows:

Everything human, including the mind and culture, has a material base and originated during the evolution of the human genetic constitution and its interaction with the environment. To say this much is not to deny the great creative power of culture, or to minimize the fact that most causes of human thought and behavior are still poorly understood. The important point is that modern biology can account for many of the unique properties of the species. Research on the subject is accelerating, quickly enough to lend plausibility to

From *Philosophy*, 1986, 61:173–192.

the belief that the human condition can eventually be understood to its foundations, including the sources of moral reasoning.

This accumulating empirical knowledge has profound consequences for moral philosophy. It renders increasingly less tenable the hypothesis that ethical truths are extrasomatic, in other words divinely placed within the brain or else outside the brain awaiting revelation. Of equal importance, there is no evidence to support the view—and a great deal to contravene it—that premises can be identified as global optima favoring the survival of any civilized species, in whatever form or on whatever planet it might appear. Hence external goals are unlikely to be articulated in this more pragmatic sense.

Yet biology shows that internal moral premises do exist and can be defined more precisely. They are immanent in the unique programs of the brain that originated during evolution. Human mental development has proved to be far richer and more structured and idiosyncratic than previously suspected. The constraints on this development are the sources of our strongest feelings of right and wrong, and they are powerful enough to serve as a foundation for ethical codes. But the articulation of enduring codes will depend upon a more detailed knowledge of the mind and human evolution than we now possess. We suggest that it will prove possible to proceed from a knowledge of the material basis of moral feeling to generally accepted rules of conduct. To do so will be to escape—not a minute too soon—from the debilitating absolute distinction between *is* and *ought*.

All populations of organisms evolve through a law-bound causal process, as first described by Charles Darwin in his *Origin of Species*. The modern explanation of this process, known as natural selection, can be briefly summarized as follows. The members of each population vary hereditarily in virtually all traits of anatomy, physiology, and behavior. Individuals possessing certain combinations of traits survive and reproduce better than those with other combinations. As a consequence, the units that specify physical traits—genes and chromosomes—increase in relative frequency within such populations, from one generation to the next.

This change in different traits, which occurs at the level of the entire population, is the essential process of evolution. Although the agents of natural selection act directly on the outward traits and only rarely on the underlying genes and chromosomes, the shifts they cause in the latter have the most important lasting effects. New variation across each population arises through changes in the chemistry of the genes and their relative positions on the chromosomes. Nevertheless, these changes (broadly referred to as mutations) provide only the raw material of evolution. Natural selection, composed of the sum of differential survival and reproduction, for the most part determines the rate and direction of evolution.[2]

Although natural selection implies competition in an abstract sense between different forms of genes occupying the same chromosome positions or between different gene arrangements, pure competition, sometimes caricatured as "nature red in tooth and claw," is but one of several means by which natural selection can operate on the outer traits. In fact, a few species are known whose members do not compete among themselves at all. Depending on circumstances, survival and reproduction can be promoted equally well through the avoidance of predators, more efficient breeding, and improved cooperation with others.[3]

In recent years there have been several much-publicized controversies over the pace of evolution and the universal occurrence of adaptation.[4] These uncertainties should not obscure the key facts about organic evolution: that it occurs as a universal process among all kinds of organisms thus far carefully examined, that the dominant driving force is natural selection, and that the observed major patterns of change are consistent with the known principles of molecular biology and genetics. Such is the view held by the vast majority of the biologists who actually work on heredity and evolution.[5] To say that not all the facts have been explained, to point out that forces and patterns may yet be found that are inconsistent with the central theory—healthy doubts present in any scientific discipline—is by no means to call into question the prevailing explanation of evolution. Only a demonstration of fundamental inconsistency can accomplish that much, and nothing short of a rival explanation can bring the existing theory into full disarray.

There are no such crises. Even Motoo Kimura, the principal architect of the "neutralist" theory of genetic diversity—which proposes that most evolution at the molecular level happens through random factors—allows that "classical evolution theory has demonstrated beyond any doubt that the basic mechanism for adaptive evolution is natural selection acting on variations produced by changes in chromosomes and genes. Such considerations as population size and structure, availability of ecological opportunities, change of environment, life-cycle 'strategies', interaction with other species, and in some situations kin or possibly group selection play a large role in our understanding of the process."[6]

Human evolution appears to conform entirely to the modern synthesis of evolutionary theory as just stated. We know now that human ancestors broke from a common line with the great apes as recently as six or seven million years ago, and that at the biochemical level we are today closer relatives of the chimpanzees than the chimpanzees are of gorillas.[7] Furthermore, all that we know about human fossil history, as well as variation in genes and chromosomes among individuals and the key events in the embryonic assembly of the nervous system, is consistent with the prevailing view that natural selection has served as the principal agent in the origin of humanity.

It is true that until recently information on the brain and human evolution was sparse. But knowledge is accelerating, at least as swiftly as the remainder of natural science, about a doubling every ten to fifteen years. Several key developments, made principally during the past twenty years, will prove important to our overall argument for a naturalistic ethic developed as an applied science.

The number of human genes identified by biochemical assay or pedigree analysis is at the time of writing 3,577, with approximately 600 placed to one or the other of the twenty-three pairs of chromosomes.[8] Because the rate at which this number has been accelerating (up from 1,200 in 1977), most of the entire complement of 100,000 or so structural genes may be characterized to some degree within three or four decades.

Hundreds of the known genes affect behavior. The great majority do so simply by their effect on general processes of tissue development and metabolism, but a few have been implicated in more focused behavioral traits. For example, a single allele (a variant of one gene), prescribes the rare Lesch-Nyhan syndrome, in which people curse uncontrollably, strike out at others with no provocation, and tear at their own lips and fingers. Another allele at a different chromosome position reduces the ability to perform on certain standard spatial tests but not on the majority of such tests.[9] Still another allele, located tentatively on chromosome 15, induces a specific learning disability.[10]

These various alterations are of course strong and deviant enough to be considered pathological. But they are also precisely the kind usually discovered in the early stages of behavioral genetic analysis for any species. *Drosophila* genetics, for example, first passed through a wave of anatomical and physiological studies directed principally at chromosome structure and mechanics. As in present-day human genetics, the first behavioral mutants discovered were broadly acting and conspicuous, in other words, those easiest to detect and characterize. When behavioral and biochemical studies grew more sophisticated, the cellular basis of gene action was elucidated in the case of a few behaviors, and the new field of *Drosophila* neurogenetics was born. The hereditary bases of subtle behaviors such as orientation to light and learning were discovered somewhat later.[11]

We can expect human behavioral genetics to travel along approximately the same course. Although the links between genes and behavior in human beings are more numerous and the processes involving cognition and decision making far more complex, the whole is nevertheless conducted by cellular machinery precisely assembled under the direction of the human genome (that is, genes considered collectively as a unit). The techniques of gene identification, applied point by point along each of the twenty-three pairs of chromosomes, is beginning to make genetic dissection of human behavior a reality.

Yet to speak of genetic dissection, a strongly reductionist procedure, is not to suggest that the whole of any trait is under the control of a single gene, nor does it deny substantial flexibility in the final product. Individual alleles (gene-variants) can of course affect a trait in striking ways. To take a humble example, the possession of a single allele rather than another on a certain point on one of the chromosome pairs causes the development of an attached earlobe as opposed to a pendulous earlobe. However, it is equally true that a great many alleles at different chromosome positions must work together to assemble the entire earlobe. In parallel fashion, one allele can shift the likelihood that one form of behavior will develop as opposed to another, but many alleles are required to prescribe the ensemble of nerve cells, neurotransmitters, and muscle fibers that orchestrate the behavior in the first place. Hence classical genetic analysis cannot by itself explain all of the underpinnings of human behavior, especially those that involve complex forms of cognition and decision making. For this reason behavioral development viewed as the interaction of genes and environment should also occupy center stage in the discussion of human behavior. The most important advances at this level are being made in the still relatively young field of cognitive psychology.[12]

With this background, let us move at once to the central focus of our discussion: morality. Human beings, all human beings, have a sense of right and wrong, good and bad. Often, although not always, this "moral awareness" is bound up with beliefs about deities, spirits, and other supersensible beings. What is distinctive about moral claims is that they are prescriptive; they lay upon us certain obligations to help and to cooperate with others in various ways. Furthermore, morality is taken to transcend mere personal wishes or desires. "Killing is wrong" conveys more than merely "I don't like killing." For this reason, moral statements are thought to have an objective referent, whether the Will of a Supreme Being or eternal verities perceptible through intuition.

Darwinian biology is often taken as the antithesis of true morality. Something that begins with conflict and ends with personal reproduction seems to have little to do with right and wrong. But to reason along such lines is to ignore a great deal of the content of modern evolutionary biology. A number of causal mechanisms—already well confirmed in the animal world—can yield the kind of cooperation associated with moral behavior. One is so-called kin selection. Genes prescribing cooperation spread through the populations when self-sacrificing acts are directed at relatives, so that they (not the cooperators) are benefited, and the genes they share with the cooperators by common descent are increased in later generations. Another such cooperation-causing mechanism is "reciprocal altruism." As its name implies, this involves transactions (which can occur between nonrelatives) in which aid given is offset by the expectation of aid received. Such mutual assistance can be extended to a whole group, whose individual members contribute to a general pool and (as needed) draw from the pool.[13]

Sociobiologists (evolutionists concerned with social behavior) speak of acts mediated by such mechanisms as "altruistic." It must be recognized that this is now a technical biological term, and does not necessarily imply conscious free giving and receiving. Nevertheless, the empirical evidence suggests that cooperation between human beings was brought about by the same evolutionary mechanisms as those just cited. To include conscious, reflective beings is to go beyond the biological sense of altruism into the realm of genuine nonmetaphorical altruism. We do not claim that people are either unthinking genetic robots or that they cooperate only when the expected genetic returns can be calculated in advance. Rather, human beings function better if they are deceived by their genes into thinking that there is a disinterested objective morality binding upon them, which all should obey. We help others because it is "right" to help them and because we know that they are inwardly compelled to reciprocate in equal measure. What Darwinian evolutionary theory shows is that this sense of "right" and the corresponding sense of "wrong," feelings we take to be above individual desire and in some fashion outside biology, are in fact brought about by ultimately biological processes.

Such are the empirical claims. How exactly is biology supposed to exert its will on conscious, free beings? At one extreme, it is possible to conceive of a moral code produced entirely by the accidents of history. Cognition and moral sensitivity might evolve somewhere in some imaginary species in a wholly unbiased manner, creating the organic equivalent of an all-purpose computer. In such a blank-slate species, moral rules were contrived some time in the past, and the exact historical origin might now be lost in the mists of time. If proto-humans evolved in this manner, individuals that thought up and followed rules ensuring an ideal level of cooperation then survived and reproduced, and all others fell by the wayside.

However, before we consider the evidence, it is important to realize that any such even-handed device must also be completely gene based and tightly controlled, because an exact genetic prescription is needed to produce perfect openness to any moral rule, whether successful or not. The human thinking organ must be indifferently open to a belief such as "killing is wrong" or "killing is right," as well as to any consequences arising from conformity or deviation. Both a very specialized prescription and an elaborate cellular machinery are needed to achieve this remarkable result. In fact, the blank-slate brain might require a cranial space many times that actually possessed by human beings. Even then a slight deviation in the many feedback loops and hierarchical controls would shift cognition and preference back into a biased state. In short, there appears to be no escape from the biological foundation of mind.

It can be stated with equal confidence that nothing like all-purpose cognition occurred during human evolution. The evidence from both genetic and cognitive studies demonstrates decisively that the human brain is not a *tabula rasa*. Conversely, neither is the brain (and the consequent ability to think) genetically determined in the strict sense. No genotype is known that dictates a single behavior, precluding reflection and the capacity to choose from among alternative behaviors belonging to the same category. The human brain is something in between: a swift and directed learner that picks up certain bits of information quickly and easily, steers around others, and leans toward a surprisingly few choices out of the vast array that can be imagined.

This quality can be made more explicit by saying that human thinking is under the influence of "epigenetic rules," genetically based processes of development that predispose the individual to adopt one or a few forms of behaviors as opposed to others. The rules are rooted in the physiological processes leading from the genes to thought and action.[14] The empirical heart of our discussion is that we think morally because we are subject to appropriate epigenetic rules. These predispose us to think that certain courses of action are right and certain courses of action are wrong. The rules certainly do not lock people blindly into certain behaviors. But because they give the illusion of objectivity to morality, they lift us above immediate wants to actions which (unknown to us) ultimately serve our genetic best interests.

The full sequence in the origin of morality is therefore evidently the following: ensembles of genes have evolved through mutation and selection within a intensely social existence over tens of thousands of years; they prescribe epigenetic rules of mental development peculiar to the human species; under the influence of the rules certain choices are made from among those conceivable and available to the culture; and finally the choices are narrowed and hardened through contractual agreements and sanctification.

In a phrase, societies feel their way across the fields of culture with a rough biological map. Enduring codes are not created whole from absolute premises but inductively, in the manner of common law, with the aid of repeated experience, by emotion and consensus, through an expansion of knowledge and experience guided by the epigenetic rules of mental development, during which people sift the options and come to agree upon and to legitimate certain norms and directions.[15]

Only recently have the epigenetic rules of mental development and their adaptive roles become accepted research topics for evolutionary biology. It should therefore not be surprising that to date the best understood examples of epigenetic rules are of little immediate concern to moral philosophers. Yet what such examples achieve is to draw us from the realm of speculative philosophy into the center of ongoing scientific research. They provide the stepping stones to a more empirical basis of moral reasoning.

One of the most fully explored epigenetic rules concerns the constraint on color vision that affects the cultural evolution of color vocabularies. People see variation in the *intensity* of light (as opposed to color) the way one might intuitively expect to see it. That is, if the level of illumination is raised gradually, from dark to brightly lit, the transition is perceived as gradual. But if the *wavelength* is changed gradually, from a monochromatic purple all across the visible spectrum to a monochromatic red, the shift is not perceived as a continuum. Rather, the full range is thought to comprise four basic colors (blue, green, yellow, red), each persisting across a broad band of wavelengths and giving way through ambiguous intermediate color through narrow bands on either side. The physiological basis of this beautiful deception is partly known. There are three kinds of cones in the retina and four kinds of cells in the lateral geniculate nuclei of the visual pathways leading to the optical cortex. Although probably not wholly responsible, both sets of cells play a role in the coding of wavelength so that it is perceived in a discrete rather than continuous form. Also, some of the genetic basis of the cellular structure is known. Color-blindness alleles on two positions in the X-chromosome cause particular deviations in wavelength perception.

The following experiment demonstrated the effect of this biological constraint on the formation of color vocabularies. The native speakers of twenty languages from around the world were asked to place their color terms in a standard chart that displays the full visible color spectrum across varying shades of brightness. Despite the independent origins of many of the languages, which included Arabic, Ibidio, Thai, and Tzeltal, the terms placed together fall into four distinct clusters corresponding to the basic colors. Very few were located in the ambiguous intermediate zones.

A second experiment then revealed the force of the epigenetic rule governing this cultural convergence. Prior to European contact the Dani people of New Guinea possessed a very small color vocabulary. One group of volunteers was taught a newly invented Dani-like set of color terms placed variously on the four principal hue categories (blue, green, yellow, red). A second group was taught a similar vocabulary placed off-center, away from the main clusters formed by other languages. The first group of volunteers, those given the "natural" vocabulary, learned about twice as quickly as those given the off-center, less natural terms. Dani volunteers also selected these terms more readily when allowed to make a choice between the two sets.[16]

So far as we have been able to determine, all categories of cognition and behavior investigated to the present time show developmental biases. More precisely, whenever development has been investigated with reference to choice under conditions as free as possible of purely experimental influence, subjects automatically favored certain choices over others. Some of these epigenetic biases are moderate to very strong, as in the case of color vocabulary. Others are relatively weak. But all are sufficiently marked to exert a detectable influence on cultural evolution.

Examples of such deep biases included the optimum degree of redundancy in geometric design; facial expressions used to denote the basic emotions of fear, loathing, anger, surprise, and happiness; descending degrees of preference for sucrose, fructose, and other sugars; the particular facial expressions used to respond to various distasteful substances; and various fears, including the fear-of-strangers response in children. One of the most instructive cases is provided by the phobias. These intense reactions are most readily acquired against snakes, spiders, high places, running water, tight enclosures, and other ancient perils of mankind for which epigenetic rules can be expected to evolve through natural selection. In contrast, phobias very rarely appear in response to automobiles, guns, electric sockets, and other truly dangerous objects in modern life, for which the human species has not yet had time to adapt through genetic change.

Epigenetic rules have also been demonstrated in more complicated forms of mental development, including language acquisition, predication in logic, and the way in which objects are ordered and counted during the first steps in mathematical reasoning.[17]

We do not wish to exaggerate the current status of this area of cognitive science. The understanding of mental development is still rudimentary in comparison with that of most other aspects of human biology. But enough is known to see the broad outlines of complex processes. Moreover, new techniques are constantly being developed to explore the physical basis of mental

Michael Ruse and Edward O. Wilson

activity. For example, arousal can be measured by the degree of alpha wave blockage, allowing comparisons of the impact of different visual designs. Electroencephalograms of an advanced design are used to monitor moment-by-moment activity over the entire surface of the brain. In a wholly different procedure, radioactive isotopes and tomography are combined to locate sites of enhanced metabolic activity. Such probes have revealed the areas of the brain used in specific mental operations, including the recall of melodies, the visualization of notes on a musical staff, and silent reading and counting.[18] There seems to be no theoretical reason why such techniques cannot be improved eventually to address emotions, more complex reasoning, and decision making. There is similarly no reason why metabolic activity of the brain cannot be mapped in chimpanzees and other animals as they solve problems and initiate action, permitting the comparison of mental activity in human beings with that in lower species.

But what of morality? We have spoken of color perception, phobias, and other less value-laden forms of cognition. We argue that moral reasoning is likewise molded and constrained by epigenetic rules. Already biologists and behavioral scientists are moving directly into that area of human experience producing the dictates of right and wrong. Consider the avoidance of brother-sister incest, a negative choice made by the great majority of people around the world. By incest in this case is meant full sexual attraction and intercourse, and not merely exploratory play among children. When such rare matings do occur, lowered genetic fitness is the result. The level of homozygosity (a matching of like genes) in the children is much higher, and they suffer a correspondingly greater mortality and frequency of crippling syndromes due to the fact that some of the homozygous paris of genes are defective. Yet this biological cause and effect is not widely perceived in most societies, especially those with little or no scientific knowledge of heredity. What causes the avoidance instead is a sensitive period between birth and approximately six years. When children this age are exposed to each other under conditions of close proximity (both "use the same potty," as one anthropologist put it) they are unable to form strong sexual bonds during adolescence or later. The inhibition persists even when the pairs are biologically unrelated and encouraged to marry. Such a circumstance occurred, for example, when children from different families were raised together in Israeli kibbutzim and in Chinese households practicing minor marriages.[19]

A widely accepted interpretation of the chain of causation in the case of brother-sister incest avoidance is as follows. Lowered genetic fitness due to inbreeding led to the evolution of the juvenile sensitive period by means of natural selection; the inhibition experienced at sexual maturity led to prohibitions and cautionary myths against incest or (in many societies) merely a shared feeling that the practice is inappropriate. Formal incest taboos are the cultural reinforcement of the automatic inhibition, an example of the way culture is shaped by biology. But these various surface manifestations need not be consulted in order to formulate a more robust technique of moral reasoning. What matters in this case is the juvenile inhibition: the measures of its strength and universality, and a deeper understanding of why it came into being during the genetic evolution of the brain.

Sibling incest is one of several such cases showing that a tight and formal connection can be made between biological evolution and cultural change. Models of sociobiology have now been extended to include the full co-evolutionary circuit, from genes affecting the direction of cultural change to natural selection shifting the frequencies of these genes, and back again to open new channels for cultural evolution. The models also predict the pattern of cultural diversity resulting from a given genotype distributed uniformly through the human species. It has just been seen how the avoidance of brother-sister incest arises from a strong negative bias and a relative indifference to the preferences of others. The quantitative models incorporating these parameters yield a narrow range of cultural diversity, with a single peak at or near complete rejection on the part of the members of most societies. A rapidly declining percentage of societies possess higher rates of acceptance. If the bias is made less in the model than the developmental data indicate, the mode of this frequency curve (that is, the frequency of societies whose members display different percentages of acceptance) shifts from one end of the acceptance scale toward its center. If individuals are considerably more responsive to the preferences of others, the frequency curve breaks into two modes.[20]

Such simulations, employing the principles of population genetics as well as methods derived from statistical mechanics, are still necessarily crude and applicable only to the simplest forms of culture. But like behavioral genetics and the radionuclide-tomography mapping of brain activity, they give a fair idea of the kind of knowledge that is possible with increasing sophistication in theory and technique. The theory of the co-evolution of genes and culture can be used further to understand the origin and meaning of the epigenetic rules, including those that affect moral reasoning.

This completes the empirical case. To summarize, there is solid factual evidence for the existence of epigenetic rules—constraints rooted in our evolutionary biology that affect the way we think. The incest example shows that these rules, directly related to adaptive advantage, extend into the moral sphere. And the hypothesis of morality as a product of pure culture is refuted by the growing evidence of the co-evolution of genes and culture.

This perception of co-evolution is, of course, only a beginning. Prohibitions on intercourse with siblings hardly exhaust the human moral dimension. Philosophical reasoning based upon more empirical information is required to give a full evolutionary account of the phenomena of interest: philosophers' hands reaching down, as it were, to grasp the hands of biologists reaching up. Surely some of the moral premises articulated through ethical inquiry lie close to real epigenetic rules. For instance, the contractarians' emphasis on fairness and justice looks much like the result of rules brought about by reciprocal altruism, as indeed one distinguished supporter of that philosophy has already noted.[21]

We believe that implicit in the scientific interpretation of moral behavior is a conclusion of central importance to philosophy, namely, that there can be no genuinely objective external ethical premises. Everything that we know about the evolutionary process indicates that no such extrasomatic guides exist. Let us define ethics in the ordinary sense, as the area of thought and action governed by a sense of obligation—a feeling that there are certain standards one ought to live up to. In order not to prejudge the issue, let us also make no further assumptions about content. It follows from what we understand in the most general way about organic evolution that ethical premises are likely to differ from one intelligent species to another. The reason is that choices are made on the basis of emotion and reason directed to these ends, and the ethical premises composed of emotion and reason arise from the epigenetic rules of mental development. These rules are in turn the idiosyncratic products of the genetic history of the species and as such were shaped by particular regimes of natural selection. For many generations—more than enough for evolutionary change to occur—they favored the survival of individuals who practiced them. Feelings of happiness, which stem from positive reinforcers of the brain and other elements that compose the epigenetic rules, are the enabling devices that led to such right action.

It is easy to conceive of an alien intelligent species evolving rules its members consider highly moral but which are repugnant to human beings, such as cannibalism, incest, the love of darkness and decay, parricide, and the mutual eating of feces. Many animal species perform some or all of these things, with gusto and in order to survive. If human beings had evolved from a stock other than savanna-dwelling, bipedal, carnivorous man-apes we might do the same, feeling inwardly certain that such behaviors are natural and correct. In short, ethical premises are the peculiar products of genetic history, and they can be understood solely as mechanisms that are adaptive for the species that possess them. It follows that the ethical code of one species cannot be translated into that of another. No abstract moral principles exist outside the particular nature of individual species.

It is thus entirely correct to say that ethical laws can be changed, at the deepest level, by genetic evolution. This is obviously quite inconsistent with the notion of morality as a set of objective, eternal verities. Morality is rooted in contingent human nature, through and through.

Nor is it possible to uphold the true objectivity of morality by believing in the existence of an ultimate code, such that what is considered right corresponds to what is truly right—that the thoughts produced by the epigenetic rules parallel external premises.[22] The evolutionary explanation makes the objective morality redundant, for even if external ethical premises did not exist, we would go on thinking about right and wrong in the way that we do. And surely, redundancy is the last predicate that an objective morality can possess. Furthermore, what reason is there to presume that our present state of evolution puts us in correspondence with ultimate truths? If there are genuine external ethical premises, perhaps cannibalism is obligatory.

Thoughtful people often turn away from naturalistic ethics because of a belief that it takes the goodwill out of cooperation and reduces righteousness to a mechanical process. Biological "altruism" supposedly can never yield genuine altruism. This concern is based on a half truth. True morality, in other words behavior that most or all people can agree is moral, does consist in the readiness to do the "right" thing even at some personal cost. As pointed out, human beings do not calculate the ultimate effect of every given act on the survival of their own genes or those of close relatives. They are more than just gene replicators. They define each problem, weigh the options, and act in a manner conforming to a well-defined set of beliefs—with integrity, we like to say, and honor, and decency. People are willing to suppress their own desires for a while in order to behave correctly.

That much is true, but to treat such qualifications as objections to naturalistic ethics is to miss the entire force of the empirical argument. There is every reason to believe that most human behavior does protect the individual, as well as the family and the tribe and, ultimately, the genes common to all of these units. The advantage extends to acts generally considered to be moral and selfless. A person functions more efficiently in the social setting if he obeys the generally accepted moral code of his society than if he follows moment-by-moment egocentric calculations. This proposition has been well documented in the case of preliterate societies, of the kind in which human beings lived during evolutionary time. While far from perfect, the correlation is close enough to support the biological view that the epigenetic rules evolved by natural selection.[23]

It should not be forgotten that altruistic behavior is most often directed at close relatives, who possess many of the same genes as the altruist and perpetuate them through collateral descent. Beyond the circle of kinship, altruistic acts are typically reciprocal in nature, performed with the expectation of future reward either in this world or afterward. Note, however, that the expectation does not necessarily employ a crude demand for returns, which would be antithetical to true morality. Rather, I expect you (or God) to help me because it is right for you (or God) to help me, just as it was right for me to help you (or obey God). The reciprocation occurs in the name of morality. When people stop reciprocating, we tend to regard them as outside the moral framework. They are "sociopathic" or "no better than animals."

The very concept of morality—as opposed to mere moral decisions taken from time to time—imparts efficiency to the adaptively correct action. Moral feeling is the shortcut taken by the mind to make the best choices quickly. So we select a certain action and not another because we feel that it is "right," in other words, it satisfies the norms of our society or religion and thence, ultimately, the epigenetic rules and their prescribing genes. To recognize this linkage does not diminish the validity and robustness of the end result. Because moral consistency feeds mental coherence, it retains power even when understood to have a purely material basis.

For the same reason there is little to fear from moral relativism. A common argument raised against the materialist view of human nature is that if ethical premises are not objective and external to mankind, the individual is free to pick his own code of conduct regardless of the effect on others. Hence philosophy for the philosophers and religion for the rest, as in the Averrhoist doctrine. But our growing knowledge of evolution suggests that this is not at all the case. The epigenetic rules of mental development are relative only to the species. They are not relative to the individual. It is easy to imagine another form of intelligent life with nonhuman rules of mental development and therefore a radically different ethic. Human cultures, in contrast, tend to converge in their morality in the manner expected when a largely similar array of epigenetic rules meet a largely similar array of behavioral choices. This would not be the case if human beings differed greatly from one another in the genetic basis of their mental development.

Indeed, the materialist view of the origin of morality is probably less threatening to moral practice than a religious or otherwise nonmaterialistic view, for when moral beliefs are studied empirically, they are less likely to deceive. Bigotry declines because individuals cannot in any sense regard themselves as belonging to a chosen groups or as the sole bearers of revealed truth. The quest for scientific understanding replaces the hajj and the holy grail. Will it acquire a similar passion? That depends upon the value people place upon themselves, as opposed to their imagined rulers in the realms of the supernatural and the eternal.

Nevertheless, because ours is an empirical position, we do not exclude the possibility that some differences might exist between large groups in the epigenetic rules governing moral awareness. Already there is related work suggesting that the genes can cause broad social differences between groups—or, more precisely, that the frequency of genes affecting social behavior can shift across geographic regions.

An interesting example now being investigated is variation in alcohol consumption and the conventions of social drinking. Alcohol (ethanol) is broken down in two steps, first to acetaldehyde by the enzyme alcohol dehydrogenase and then to acetic acid by the enzyme acetaldehyde dehydrogenase. The reaction to alcohol depends substantially on the rate at which ethanol is converted into these two products. Acetaldehyde causes facial flushing, dizziness, slurring of words, and sometimes nausea. Hence the reaction to drinking depends substantially on the concentration of acetaldehyde in the blood, and this is determined by the efficiency of the two enzymes. The efficiency of the enzymes depends in turn on their chemical structure, which is prescribed by genes that vary within populations. In particular, two alleles (gene forms) are known for one of the loci (chromosome sites of the genes) encoding alcohol dehydrogenase, and two are known for a locus encoding acetaldehyde dehydrogenase. These various alleles produce enzymes that are either fast or slow in converting their target substances. Thus one combination of alleles causes a very slow conversion from ethanol to acetic acid, another the reverse, and so on through the four possibilities.

Independent evidence has suggested that the susceptibility to alcohol addiction is under partial genetic control. The tendency now appears to be substantially although not exclusively affected by the combination of genes determining the rates of ethanol and acetaldehyde conversion. Individuals who accumulate moderate levels of acetaldehyde are more likely to become addicted than those who sustain low levels. The propensity is especially marked in individuals who metabolize both ethanol and acetaldehyde rapidly and hence are more likely to consume large quantities to maintain a moderate acetaldehyde titer.

Differences among human populations also exist. Most caucasoids have slow ethanol and acetaldehyde conversion rates, and thus are able to sustain moderately high drinking levels while alone or in social gatherings. In contrast, most Chinese and Japanese convert ethanol rapidly and acetaldehyde slowly and thus build up acetaldehyde levels quickly. They reach intoxication levels with the consumption of a relatively small amount of alcohol.

Statistical differences in prevalent drinking habits are well known between the two cultures, with Europeans and North Americans favoring the consumption of relatively large amounts of alcohol during informal gatherings and eastern Asiatics favoring the consumption of smaller amounts on chiefly ceremonial occasions. The divergence would now seem not to be wholly a matter of historical accident but to stem from biological differences as well. Of course a great deal remains to be learned concerning the metabolism of alcohol and its effects on behavior, but enough is known to illustrate the potential of the interaction of varying genetic material and the environment to create cultural diversity.[24]

It is likely that such genetic variation accounts for only a minute fraction of cultural diversity. It can be shown that a large amount of the diversity can arise purely from the statistical scatter due to differing choices made by genetically identical individuals, creating patterns that are at least partially predictable from a knowledge of the underlying universal bias.[25] We wish only to establish that, contrary to prevailing opinion in social theory but in concert with the findings of evolutionary biology, cultural diversity can in some cases be enhanced by genetic diversity. It is wrong to exclude a priori the possibility that biology plays a causal role in the differences in moral attitude among different societies. Yet even this complication gives no warrant for extreme moral relativism. Morality functions within groups and now increasingly across groups, and the similarities between all human beings appear to be far greater then any differences.

The last barrier against naturalistic ethics may well be a lingering belief in the absolute distinction between *is* and *ought*. Note that we say "absolute." There can be no question that *is* and *ought* differ in meaning, but this distinction in no way invalidates the evolutionary approach. We started with Hume's own belief that morality rests ultimately on sentiments and feelings. But then we used the evolutionary argument to discount the possibility of an objective, external reference for morality. Moral codes are seen instead to be created by culture under the biasing influence of the epigenetic rules and legitimated by the illusion of objectivity. The more fully this process is understood, the sounder and more enduring can be the agreements.

Michael Ruse and Edward O. Wilson

Thus the explanation of a phenomenon such as biased color vision or altruistic feelings does not lead automatically to the prescription of the phenomenon as an ethical guide. But this explanation, the *is* statement, underlies the reasoning used to create moral codes. Whether a behavior is deeply ingrained in the epigenetic rules, whether it is adaptive or nonadaptive in modern societies, whether it is linked to other forms of behavior under the influence of separate developmental rules: all these qualities can enter the foundation of the moral codes. Of equal importance, the means by which the codes are created, entailing the estimation of consequences and the settling upon contractual arrangements, are cognitive processes and real events no less than the more elementary elements they examine.

No major subject is more important or relatively more neglected at the present time than moral philosophy. If viewed as a pure instrument of the humanities, it seems heavily worked, culminating a long and distinguished history. But if viewed as an applied science in addition to being a branch of philosophy, it is no better than rudimentary. This estimation is not meant to be derogatory. On the contrary, moral reasoning offers an exciting potential for empirical research and a new understanding of human behavior, providing biologists and psychologists join in its development. Diverse kinds of empirical information, best obtained through collaboration, are required to advance the subject significantly. As in twentieth-century science, the time of the solitary scholar pronouncing new systems in philosophy seems to have passed.

The very weakness of moral reasoning can be taken as a cause for optimism. By comparision with the financial support given other intellectual endeavors directly related to human welfare, moral philosophy is a starveling field. The current expenditure on health-related biology in the United States at the present time exceeds $3 billion. Support has been sustained at that level or close to it for over two decades, with the result that the fundamental processes of heredity and much of the molecular machinery of the cell have been elucidated. And yet a huge amount remains to be done: the cause of cancer is only partly understood, while the mechanisms by which cells differentiate and assemble into tissues and organs are still largely unknown. In contrast, the current support of research on subjects directly related to moral reasoning, including the key issues in neurobiology, cognitive development, and sociobiology, is probably less than 1 percent of that allocated to health-related biology. Given the complexities of the subject, it is not surprising that very little has been learned about the physical basis of morality—so little, in fact, that its entire validity can still be questioned by critics. We have argued that not only is the subject valid, but it offers what economists call increasing returns to scale. Small absolute increments in effort will yield large relative returns in concrete results. With this promise in mind, we will close with a brief characterization of several of the key problems of ethical studies as we see them.

First, only a few processes in mental development have been worked out in enough detail to measure the degree of bias in the epigenetic rules. The linkage from genes to cellular structure and thence to forms of social behavior is understood only partially. In addition, a curious disproportion exists: the human traits regarded as most positive, including altruism and creativity, have been among the least analyzed empirically. Perhaps they are protected by an unconscious taboo causing them to be regarded as matters of the "spirit" too sacred for material analysis.

Second, the interactive effects of cognition also remain largely unstudied. Among them are hierarchies in the expression of epigenetic rules. An extreme example is the suppression of preference in one cognitive category when another is activated. This is the equivalent to the phenomenon in heredity known as epistasis. We know in a very general way that certain desires and emotion-laden beliefs take precedence over others. Tribal loyalty can easily dominate other social bonds, especially when the group is threatened from the outside. Individual sacrifice becomes far more acceptable when it is believed to enhance future generations. The physical basis and relative quantitative strengths of such effects are almost entirely unknown.

Third, there is an equally enticing opportunity to create a comparative ethics, defined as the study of conceivable moral systems that might evolve in other intelligent species. Of course it is likely that even if such systems exist, we will never perceive them directly. But that is beside the point. Theoretical science, defined as the study of all conceivable worlds, imagines nonexistent phenomena in order to classify more precisely those that do exist. So long as we confine ourselves to one rather aberrant primate species (our own), we will find it difficult to identify the qualities of ethical premises that can vary and thus provide more than a narrow perspective in moral studies. The goal is to locate human beings within the space of all possible moral systems, in order to gauge our strengths and weaknesses with greater precision.

Fourth, there are pressing issues arising from the fact that moral reasoning is dependent upon the scale of time. The trouble is that evolution gave us abilities to deal principally with short-term moral problems. ("Save that child!" "Fight that enemy!") But, as we now know, short-term responses can easily lead to long-term catastrophes. What seems optional for the next ten years may be disastrous thereafter. Cutting forests and exhausting nonrenewable energy sources can produce a healthy, vibrant population for one generation—and starvation for the next ten. Perfect solutions probably do not exist for the full range of time in most categories of behavior. To choose what is best for the near future is relatively easy. To choose what is best for the distant future is also relatively easy, providing one is limited to broad generalities. But to choose what is best for both the near and distant futures is forbiddingly difficult, often drawing on internally contradictory sentiments. Only through study will we see how our short-term moral insights fail our long-term needs, and how correctives can be applied to formulate more enduring moral codes.

NOTES

1. H. J. Muller is quoted by G. G. Simpson in *This View of Life* (New York: Harcourt, Brace & World, 1964), 36.

2. See the following widely used textbooks: J. Roughgarden, *Theory of Population Genetics and Evolutionary Ecology: An Introduction* (New York: Macmillan, 1979); D. L. Hartl, *Principles of Population Genetics* (Sunderland, Mass.: Sinauer Associates, 1980); R. M. May (ed.), *Theoretical Ecology: Principles and Applications*, 2d ed. (Sunderland, Mass.: Sinauer Associates, 1981); J. R. Krebs and N. B. Davies (eds.) *Behavioral Ecology: An Evolutionary Approach*, 2d ed. (Sunderland, Mass.: Sinauer Associates, 1984).

3. Reviews of the various modes of selection, including forms that direct individuals away from competitive behavior, can be found in E. O. Wilson, *Sociobiology: The New Synthesis* (Cambridge, Mass.: Belknap Press of Harvard University Press, 1975); G. F. Oster and E. O. Wilson, *Caste and Ecology in the Social Insects* (Princeton: Princeton University Press, 1978); S. A. Boorman and P. R. Levitt, *The Genetics of Altruism* (New York: Academic Press, 1980); D. S. Wilson, *The Natural Selection of Populations and Communities* (Menlo Park, Calif.: Benjamin/Cummings, 1980).

4. For example, the debate over "punctuated equilibrium" versus "gradualism" among paleontologists and geneticists. For most biologists, the issue is not the mechanism of evolution but the conditions under which evolution sometimes proceeds rapidly and sometimes slows to a crawl. There is no difficulty in explaining the variation in rates. On the contrary, there is a surplus of plausible explanations, virtually all consistent with neo-Darwinian theory, but insufficient data to choose among them. See, for example, S. J. Gould and N. Eldredge, "Punctuated Equilibria: The Tempo and Mode of Evolution reconsidered," *Paleobiology* 3 (1977), 115–151; and J. R. G. Turner, "'The Hypothesis That Explains Mimetic Resemblance Explains Evolution': The Gradualist-Saltationist Schism," in M. Grene (ed.), *Dimensions of Darwinism* (Cambridge University Press, 1983), 129–169.

5. See footnote 2.

6. M. Kimura, *The Neutral Theory of Molecular Evolution* (Cambridge University Press, 1983).

7. C. G. Sibley and J. E. Ahlquist, "The Phylogeny of the Hominoid Primates, as Indicated by DNA-DNA Hybridization," *Journal of Molecular Evolution* 20 (1984), 2–15.

8. We are grateful to Victor A. McKusick for providing the counts of identified and inferred human genes up to 1984.

9. G. C. Ashton, J. J. Polovina, and S. G. Vandenberg, "Segregation Analysis of Family Data for 15 Tests of Cognitive Ability," *Behavior Genetics* 9 (1979), 329–347.

10. S. D. Smith, W. J. Kimberling, B. F. Pennington, and H. A. Lubs, "Specific Reading Disability: Identification of an Inherited Form through Linkage Analysis," *Science* 219 (1982), 1345–1347.

11. See J. C. Hall and R. J. Greenspan, "Genetic Analysis of *Drosophila* Neurobiology," *Annual Review of Genetics* 13 (1979), 127–195.

12. See, for example, the recent analysis by J. R. Anderson, *The Architecture of Cognition* (Cambridge, Mass.: Harvard University Press, 1983).

13. See footnote 3.

14. The evidence for biased epigenetic rules of mental development is summarized in C. J. Lumsden and E. O. Wilson, *Genes, Mind, and Culture* (Cambridge, Mass.: Harvard University Press, 1981) and *Promethean Fire: Reflections on the Origin of Mind* (Cambridge, Mass.: Harvard University Press, 1983).

15. A new discipline of decision making is being developed in cognitive psychology based upon the natural means—one can correctly say the epigenetic rules—by which people choose

among alternatives and reach agreements. See, for example, A. Tversky and D. Kahneman, "The Framing of Decisions and the Psychology of Choice," *Science* 211 (1981), 453–458; and R. Axelrod, *The Evolution of Cooperation* (New York: Basic Books, 1984).

16. E. Rosch, "Natural Categories," *Cognitive psychology* 4 (1973), 328–350.

17. The epigenetic rules of cognitive development analysed through the year 1980 are reviewed by C. J. Lumsden and E. O. Wilson, op. cit.

18. N. A. Lassen, D. H. Ingvar and E. Skinhøj, "Brain Function and Blood Flow," *Scientific American* 239 (1978), 62–71.

19. A. P. Wolf and C. S. Huang, *Marriage and Adoption in China, 1845–1945* (Stanford University Press, 1980); J. Shepher, *Incest: A Biosocial View* (New York: Academic Press, 1983); P. L. van den Berghe, "Human Inbreeding Avoidance: Culture in Nature," *Behavioral and Brain Sciences* 6 (1983), 91–123.

20. C. J. Lumsden and E. O. Wilson, op. cit. See also the précis of *Genes, Mind, and Culture* and commentaries on the book by twenty-three authors in *Behavioral and Brain Sciences* 5 (1982), 1–37.

21. J. Rawls, *A Theory of Justice* (Cambridge, Mass.: Harvard University Press, 1971), 502–503.

22. This is the argument proposed by R. Nozick in *Philosophical Explanations* (Cambridge, Mass.: Belknap Press of Harvard University Press, 1981) in order to escape the implications of sociobiology.

23. See footnote 16.

24. E. Jones and C. Aoki, "Genetic and Cultural Factors in Alcohol Use" (submitted to *Science*).

25. C. J. Lumsden and E. O. Wilson, op. cit., who show the way to predict cultural diversity caused by random choice patterns in different societies.

21 Four Ways of "Biologicizing" Ethics

Philip Kitcher

I

In 1975, E. O. Wilson invited his readers to consider "the possibility that the time has come for ethics to be removed temporarily from the hands of the philosophers and biologicized" (Wilson 1975:562). There should be no doubting Wilson's seriousness of purpose.[1] His writings from 1975 to the present demonstrate his conviction that nonscientific, humanistic approaches to moral questions are indecisive and uninformed, that these questions are too important for scholars to neglect, and that biology, particularly the branches of evolutionary theory and neuroscience that Wilson hopes to bring under a sociobiological umbrella, can provide much-needed guidance. Nevertheless, I believe that Wilson's discussions of ethics, those that he has ventured alone and those undertaken in collaboration first with the mathematical physicist Charles Lumsden and later with the philosopher Michael Ruse, are deeply confused through failure to distinguish a number of quite different projects. My aim in this chapter is to separate those projects, showing how Wilson and his co-workers slide from uncontroversial truisms to provocative falsehoods.

Ideas about "biologicizing" ethics are by no means new, nor are Wilson's suggestions the only proposals that attract contemporary attention.[2] By the same token, the distinctions that I shall offer are related to categories that many of those philosophers Wilson seeks to enlighten will find very familiar. Nonetheless, by developing the distinctions in the context of Wilson's discussions of ethics, I hope to formulate a map on which would-be sociobiological ethicists can locate themselves and to identify questions that they would do well to answer.

II

How do you "biologicize" ethics? There appear to be four possible endeavors:

From K. Bayertz (ed.), *Evolution und Ethik*. Reclam 1993.

1. Sociobiology has the task of explaining how people have come to acquire ethical concepts, to make ethical judgments about themselves and others, and to formulate systems of ethical principles.

2. Sociobiology can teach us facts about human beings that, in conjunction with moral principles that we already accept, can be used to derive normative principles that we had not yet appreciated.

3. Sociobiology can explain what ethics is all about and can settle traditional questions about the objectivity of ethics. In short, sociobiology is the key to metaethics.

4. Sociobiology can lead us to revise our system of ethical principles, not simply by leading us to accept new derivative statements—as in number 2 above—but by teaching us new fundamental normative principles. In short, sociobiology is not just a source of facts but a source of norms.

Wilson appears to accept all four projects, with his sense of urgency that ethics is too important to be left to the "merely wise" (1978:7) giving special prominence to endeavor 4. (Endeavors 2 and 4 have the most direct impact on human concerns, with endeavor 4 the more important because of its potential for fundamental changes in prevailing moral attitudes. The possibility of such changes seems to lie behind the closing sentences of Ruse and Wilson 1986.) With respect to some of these projects, the evolutionary parts of sociobiology appear most pertinent; in other instances, neurophysiological investigations, particularly the exploration of the limbic system, come to the fore.

Relatives of endeavors 1 and 2 have long been recognized as legitimate tasks. Human ethical practices have histories, and it is perfectly appropriate to inquire about the details of those histories. Presumably, if we could trace the history sufficiently far back into the past, we would discern the coevolution of genes and culture, the framing of social institutions, and the introduction of norms. It is quite possible, however, that evolutionary biology would play only a very limited role in the story. All that natural selection may have done is to equip us with the capacity for various social arrangemets and the capacity to understand and to formulate ethical rules. Recognizing that not every trait we care to focus on need have been the target of natural selection, we shall no longer be tempted to argue that any respectable history of our ethical behavior must identify some selective advantage for those beings who first adopted a system of ethical precepts. Perhaps the history of ethical think-ing instantiates one of those coevolutionary models that show cultural selec-tion's interfering with natural selection (Boyd and Richerson 1985). Perhaps what is selected is some very general capacity for learning and acting that is manifested in various aspects of human behavior (Kitcher 1990).

Nothing is wrong with endeavor 1, so long as it is not articulated in too simplistic a fashion and so long as it is not overinterpreted. The reminders of the last paragraph are intended to forestall the crudest forms of neo-Darwinian development of this endeavor. The dangers of overinterpretation, however, need more detailed charting. There is a recurrent tendency in Wilson's writings to draw unwarranted conclusions from the uncontroversial premise that our ability to make ethical judgments has a history, including, ultimately, an evolutionary history. After announcing that "everything human, including the mind and culture, has a material base and originated during the evolution of the human genetic constitution and its interaction with the environment" (Ruse and Wilson 1986:173), the authors assert that "accumulating empirical knowledge" of human evolution "has profound consequences for moral philosophy" (174). For that knowledge "renders increasingly less tenable the hypothesis that ethical truths are extrasomatic, in other words divinely placed within the brain or else outside the brain awaiting revelation" (174). Ruse and Wilson thus seem to conclude that the legitimacy of endeavor 1 dooms the idea of moral objectivity.

That this reasoning is fallacious is evident once we consider other systems of human belief. Plainly, we have capacities for making judgments in mathematics, physics, biology, and other areas of inquiry. These capacities, too, have historical explanations, including, ultimately, evolutionary components. Reasoning in parallel fashion to Ruse and Wilson, we could thus infer that objective truth in mathematics, physics, and biology is a delusion and that we cannot do *any* science without "knowledge of the brain, the human organ where all decisions ... are made" (173).

What motivates Wilson (and his collaborators Ruse and Lumsden) is, I think, a sense that ethics is different from arithmetic or statics. In the latter instances, we could think of history (including our evolutionary history) bequeathing to us a capacity to learn. That capacity is activated in our encounters with nature, and we arrive at objectively true beliefs about what nature is like. Since they do not see how a similar account could work in the case of moral belief, Wilson, Ruse, and Lumsden suppose that their argument does not generalize to a denunciation of the possibility of objective knowledge. This particular type of skepticism about the possibility of objectivity in ethics is revealed in the following passage: "But the philosophers and theologians have not yet shown us how the final ethical truths will be recognized as things apart from the idiosyncratic development of the human mind" (Lumsden and Wilson 1983:182–183).

There is an important challenge to those who maintain the objectivity of ethics, a challenge that begins by questioning how we obtain ethical knowledge. Evaluating that challenge is a complex matter I shall take up in connection with project 3. However, unless Wilson has independent arguments for resolving questions in metaethics, the simple move from the legitimacy of endeavor 1 to the "profound consequences for moral philosophy" is a blunder. The "profound consequences" result not from any novel information provided by recent evolutionary theory but from arguments that deny the possibility of assimilating moral beliefs to other kinds of judgments.

III

Like endeavor 1, endeavor 2 does not demand the removal of ethics from the hands of the philosophers. Ethicists have long appreciated the idea that facts about human beings, or about other parts of nature, might lead us to elaborate our fundamental ethical principles in previously unanticipated ways. Card-carrying Utilitarians who defend the view that morally correct actions are those that promote the greatest happiness of the greatest number, who suppose that those to be counted are presently existing human beings, and who identify happiness with states of physical and psychological well-being will derive concrete ethical precepts by learning how the maximization of happiness can actually be achieved. But sociobiology has no monopoly here. Numerous types of empirical investigations might provide relevant information and might contribute to a profitable division of labor between philosophers and others.

Consider, for example, a family of problems with which Wilson, quite rightly, has been much concerned. There are numerous instances in which members of small communities will be able to feed, clothe, house, and educate themselves and their children far more successfully if a practice of degrading the natural environment is permitted. Empirical information of a variety of types is required for responsible ethical judgment. What alternative opportunities are open to members of the community if the practice is banned? What economic consequences would ensue? What are the ecological implications of the practice? All these are questions that have to be answered. Yet while amassing answers is a prerequisite for moral decision, there are also issues that apparently have to be resolved by pondering fundamental ethical principles. How should we assess the different kinds of value (unspoiled environments, flourishing families) that figure in this situation? Whose interests, rights, or well-being deserve to be counted?

Endeavors like the second one are already being pursued, especially by workers in medical ethics and in environmental ethics. It might be suggested that sociobiology has a particularly important contribution to make to this general enterprise, because it can reveal to us our deepest and most entrenched desires. By recognizing those desires, we can obtain a fuller understanding of human happiness and thus apply our fundamental ethical principles in a more enlightened way. Perhaps. However, as I have argued at great length (Kitcher 1985), the most prominent sociobiological attempts to fathom the springs of human nature are deeply flawed, and remedying the deficiencies requires integrating evolutionary ideas with neuroscience, psychology, and various parts of social science (see Kitcher 1987a, 1987b, 1988, 1990). In any event, recognizing the legitimacy of endeavor 2 underscores the need to evaluate the different desires and interests of different people (and, possibly, of other organisms), and we have so far found no reason to think that sociobiology can discharge that quintessentially moral task.

IV

Wilson's claims about the status of ethical statements are extremely hard to understand. It is plain that he rejects the notion that moral principles are objective because they encapsulate the desires or commands of a deity (a metaethical theory whose credentials have been doubtful ever since Plato's *Euthyphro*). Much of the time he writes as though sociobiology settled the issue of the objectivity of ethics negatively. An early formulation suggests a simple form of emotivism:

Like everyone else, philosophers measure their personal emotional responses to various alternatives as though consulting a hidden oracle. That oracle resides within the deep emotional centers of the brain, most probably within the limbic system, a complex array of neurons and hormone-secreting cells located just below the "thinking" portion of the cerebral cortex. Human emotional responses and the more general ethical practices based on them have been programmed to a substantial degree by natural selection over thousands of generations. (1978:6)

Stripped of references to the neural machinery, the account Wilson adopts is a very simple one. The content of ethical statements is exhausted by reformulating them in terms of our emotional reactions. Those who assent to, "Killing innocent children is morally wrong," are doing no more than reporting on a feeling of repugnance, just as they might express gastronomic revulsion. The same type of metaethics is suggested in more recent passages, for example, in the denial that "ethical truths are extrasomatic" which I have already quoted.

Yet there are internal indications and explicit formulations that belie interpreting Wilson as a simple emotivist. Ruse and Wilson appear to support the claim that "'killing is wrong' conveys more than merely 'I don't like killing'" (1986:178). Moreover, shortly after denying that ethical truths are extrasomatic, they suggest that "our strongest feelings of right and wrong" will serve as "a foundation for ethical codes" (173), and their paper concludes with the visionary hope that study will enable us to see "how our short-term moral insights fail our long-term needs, and how correctives can be applied to formulate more enduring moral codes" (192). As I interpret them, they believe that some of our inclinations and disinclinations, and the moral judgments in which they are embodied, betray our deepest desires and needs and that the task of formulating an "objective" ("enduring," "corrected") morality is to identify these desires and needs, embracing principles that express them.

Even in Wilson's earlier writings, he sounds themes that clash with any simple emotivist metaethics. For example, he acknowledges his commitment to different sets of "moral standards" for different populations and different groups within the same population (1975:564). Population variation raises obvious difficulties for emotivism. On emotivist grounds, deviants who respond to the "limbic oracle" by wilfully torturing children must be seen as akin to those who have bizarre gastronomic preferences. The rest of us may be revolted, and our revulsion may even lead us to interfere. Yet if pressed to defend ourselves, emotivism forces us to concede that there is no standpoint from which our actions can be judged as objectively more worthy than the deeds we try to restrain. The deviants follow their hypothalamic imperative, and we follow ours.

I suspect that Wilson (as well as Lumsden and Ruse) is genuinely torn between two positions. One hews a hard line on ethical objectivity, drawing

Four Ways of "Biologicizing" Ethics

the "profound consequence" that there is no "extrasomatic" source of ethical truth and accepting an emotivist metaethics. Unfortunately, this position makes nonsense of Wilson's project of using biological insights to fashion an improved moral code and also leads to the unpalatable conclusion that there are no grounds for judging those whom we see as morally perverse. The second position gives priority to certain desires, which are to be uncovered through sociobiological investigation and are to be the foundation of improved moral codes, but it fails to explain what normative standard gives these desires priority or how that standard is grounded in biology. In my judgment, much of the confusion in Wilson's writings comes from oscillating between these two positions.

I shall close this section with a brief look at the line of argument that seems to lurk behind Wilson's emotivist leanings. The challenge for anyone who advocates the objectivity of ethics is to explain in what this objectivity consists. Skeptics can reason as follows: If ethical maxims are to be objective, then they must be objectively true or objectively false. If they are objectively true or objectively false, then they must be true or false in virtue of their correspondence with (or failure to correspond with) the moral order, a realm of abstract objects (values) that persists apart from the natural order. Not only is it highly doubtful that there is any such order, but, even if there were, it is utterly mysterious how we might ever come to recognize it. Apparently we would be forced to posit some ethical intuition by means of which we become aware of the fundamental moral facts. It would then be necessary to explain how this intuition works, and we would also be required to fit the moral order and the ethical intuition into a naturalistic picture of ourselves.

The denial of "extrasomatic" sources of moral truth rests, I think, on this type of skeptical argument, an argument that threatens to drive a wedge between the acquisition of our ethical beliefs and the acquisition of beliefs about physics or biology (see the discussion of endeavor 1 above). Interestingly, an exactly parallel argument can be developed to question the objectivity of mathematics. Since few philosophers are willing to sacrifice the idea of mathematical objectivity, the philosophy of mathematics contains a number of resources for responding to that skeptical parallel. Extreme Platonists accept the skeptic's suggestion that objectivity requires an abstract mathematical order, and they try to show directly how access to this order is possible, even on naturalistic grounds. Others assert the objectivity of mathematics without claiming that mathematical statements are objectively true or false. Yet others may develop an account of mathematical truth that does not presuppose the existence of abstract objects, and still others allow abstract objects but try to dispense with mathematical intuition.

Analogous moves are available in the ethical case. For example, we can sustain the idea that some statements are objectively justified without supposing that such statements are true. Or we can abandon the correspondence theory of truth for ethical statements in favor of the view that an ethical statement is true if it would be accepted by a rational being who proceeded in a particular way. Alternatively, it is possible to accept the thesis that there is a moral order but understand this moral order in naturalistic terms, proposing, for example, with the Utilitarians, that moral goodness is to be equated with the maximization of human happiness and that moral rightness consists in the promotion of the moral good. Yet another option is to claim that there are indeed nonnatural values but that these are accessible to us in a thoroughly familiar way—for example, through our perception of people and their actions. Finally, the defender of ethical objectivity may accept all the baggage that the skeptic assembles and try to give a naturalistic account of the phenomena that skeptics take to be incomprehensible.

I hope that even this brief outline of possibilities makes it clear how a quick argument for emotivist metaethics simply ignores a host of metaethical alternatives—indeed the main alternatives that the "merely wise" have canvassed in the history of ethical theory. Nothing in recent evolutionary biology or neuroscience forecloses these alternatives. Hence, if endeavor 3 rests on the idea that sociobiology yields a quick proof of emotivist metaethics, this project is utterly mistaken.

On the other hand, if Wilson and his co-workers intend to offer some rival metaethical theory, one that would accord with their suggestions that sociobiology might generate better ("more enduring") moral codes, then they must explain what this metaethical theory is and how it is supported by biological findings. In the absence of any such explanations, we should dismiss endeavor 3 as deeply confused.

V

In the search for new normative principles, project 4, it is not clear whether Wilson intends to promise or to deliver. His early writing sketches the improved morality that would emerge from biological analysis.

In the beginning the new ethicists will want to ponder the cardinal value of the survival of human genes in the form of a common pool over generations. Few persons realize the true consequences of the dissolving action of sexual reproduction and the corresponding unimportance of "lines" of descent. The DNA of an individual is made up of about equal contributions of all the ancestors in any given generation, and it will be divided about equally among all descendants at any future moment.... The individual is an evanescent combination of genes drawn from this pool, one whose hereditary material will soon be dissolved back into it. (1978:196–197)

I interpret Wilson as claiming that there is a fundamental ethical principle, which we can formulate as follows:

W: Human beings should do whatever is required to ensure the survival of a common gene pool for *Homo sapiens*.

He also maintains that this principle is not derived from any higher-level moral statement but is entirely justified by certain facts about sexual reproduction. Wilson has little time for the view that there is a fallacy in inferring values from facts (1980a:431; 1980b:68) or for the "absolute distinction between *is* and *ought"* (Ruse and Wilson 1986: 174). It appears, then, that there is supposed to be a good argument to W from a premise about the facts of sex:

S: The DNA of any individual human being is derived from many people in earlier generations and, if the person reproduces, will be distributed among many people in future generations.

I shall consider both the argument from S to W and the correctness of W.

Plainly, one cannot deduce W from S. Almost as obviously, no standard type of inductive or statistical argument will sanction this transition. As a last resort, one might propose that W provides the best explanation for S and is therefore acceptable on the grounds of S, but the momentary charm of this idea vanishes once we recognize that S is explained by genetics, not by ethical theory.

There are numerous ways to add ethical premises so as to license the transition from S to W, but making these additions only support the un-controversial enterprise 2, not the search for fundamental moral principles undertaken under the aegis of endeavor 4. Without the additions, the infer-ence is so blatantly fallacious that we can only wonder why Wilson thinks that he can transcend traditional criticisms of the practice of inferring values from facts.

The faults of Wilson's method are reflected in the character of the funda-mental moral principle he identifies. That principle, W, enjoins actions that appear morally suspect (to say the least). Imagine a stereotypical postholo-caust situation in which the survival of the human gene pool depends on copulation between two people. Suppose, for whatever reason, that one of the parties is unwilling to copulate with the other. (This might result from resent-ment at past cruel treatment, from recognition of the miserable lives that offspring would have to lead, from sickness, or whatever.) Under these cir-custances, W requires the willing party to coerce the unwilling person, using whatever extremes of force are necessary—perhaps even allowing for the murder of those who attempt to defend the reluctant one. There is an evident conflict between these consequences of W and other ethical principles, partic-ularly those that emphasize the rights and autonomy of individuals. More-over, the scenario can be developed so as to entail enormous misery for future descendants of the critical pair, thus flouting utilitarian standards of moral correctness. Faced with such difficulties for W, there is little consolation in the thought that our DNA was derived from many people and will be dispersed among many people in whatever future generations there may be. At stake are the relative values of the right to existence of future generations (possibly under dreadful conditions) and the right to self-determination of those now living. The biological facts of reproduction do not give us any information about that relationship.

In his more recent writings, Wilson has been less forthright about the principles of "scientific ethics." Biological investigations promise improved moral codes for the future: "Only by penetrating to the physical basis of moral thought and considering its evolutionary meaning will people have the power to control their own lives. They will then be in a better position to choose ethical precepts and the forms of social regulation needed to maintain the precepts" (Lumsden and Wilson 1983:183). Ruse and Wilson are surprisingly reticent in expressing substantive moral principles, apparently preferring to discuss general features of human evolution and results about the perception of colors. Their one example of an ethical maxim is not explicitly formulated, although since it has to do with incest avoidance, it could presumably be stated as, "Do not copulate with your siblings!" (see Ruse and Wilson 1986: 183–185; for discussion of human incest avoidance, see Kitcher 1990). If this is a genuine moral principle at all, it is hardly a central one and is certainly not fundamental.

I believe that the deepest problems with the sociobiological ethics recommended by Wilson, Lumsden, and Ruse can be identified by considering how the most fundamental and the most difficult normative questions would be treated. If we focus attention, on the one hand, on John Rawls's principles of justice (proposals about fundamental questions) or on specific claims about the permissibility of abortion (proposals about a very difficult moral question), we discover the need to evaluate the rights, interests, and responsibilities of different parties. Nothing in sociobiological ethics speaks to the issue of how these potentially conflicting sets of rights, interests, and responsibilities are to be weighed. Even if we were confident that sociobiology could expose the deepest human desires, thus showing how the enduring happiness *of a single individual* could be achieved, there would remain the fundamental task of evaluating the competing needs and plans of different people. Sociobiological ethics has a vast hole at its core—a hole that appears as soon as we reflect on the implications of doomsday scenarios for Wilson's principle (W). Nothing in the later writings of Wilson, Lumsden, and Ruse addresses the deficiency.

The gap could easily be plugged by retreating from project 4 to the uncontroversial project 2. Were Wilson a Utilitarian, he could address the question of evaluating competing claims by declaring that the moral good consists in maximizing total human happiness, conceding that this fundamental moral principle stands outside sociobiological ethics but contending that sociobiology, by revealing our evolved desires, shows us the nature of human happiness. As noted above in connection with project 2, there are grounds for wondering if sociobiology can deliver insights about our "deepest desires." In any case, the grafting of sociobiology onto utilitarianism hardly amounts to the fully naturalistic ethics proclaimed in Wilson's rhetoric.

If we try to develop what I take to be Wilson's strongest motivating idea, the appeal to some extrasociobiological principle is forced upon us. Contrasting our "short-term moral problems" with our "long-term needs," Ruse and Wilson hold out the hope that biological investigations, by providing a clearer picture of ourselves, may help us to reform our moral systems (1986:192). Such reforms would have to be carried out under the guidance of some principle that evaluated the satisfaction of different desires within the life of an individual. Why is the satisfaction of long-term needs preferable to the palliation of the desires of the moment? Standard philosophical answers to this question often presuppose that the correct course is to maximize the total life happiness of the individual, subject perhaps to some system of future discounting. Whether any of those answers is adequate or not, Wilson needs some principle that will play the same evaluative role if his vision of reforming morality is to make sense. Wilson's writings offer no reason for thinking of project 4 as anything other than a blunder, and Wilson's own program of moral reform presupposes the nonbiological ethics whose poverty he so frequently decries.

VI

Having surveyed four ways of "biologicizing" ethics, I shall conclude by posing some questions for the aspiring sociobiological ethicist. The first task for any sociobiological ethics is to be completely clear about which project (or projects) are to be undertaken. Genuine interchange between biology and moral philosophy will be achieved only when eminent biologists take pains to specify what they mean by the "biologicizations" of ethics, using the elementary categories I have delineated here.

Project 1 is relatively close to enterprises that are currently being pursued by biologists and anthropologists. Human capacities for moral reflection are phenotypic traits into whose histories we can reasonably inquire. However, those who seek to construct such histories would do well to ask themselves if they are employing the most sophisticated machinery for articulating coevolutionary processes and whether they are avoiding the adaptationist pitfalls of vulgar Darwinism.

Project 2 is continuous with much valuable work done in normative ethics over the last decades. Using empirical information, philosophers and collaborators from other disciplines have articulated various types of moral theory to address urgent concrete problems. If sociobiological ethicists intend to contribute to this enterprise, they must explicitly acknowledge the need to draw on extrabiological moral principles. They must also reflect on what ethical problems sociobiological information can help to illuminate and on whether human sociobiology is in any position to deliver such information. Although project 2 is a far more modest enterprise than that which Wilson and his collaborators envisage, I am very doubtful (for reasons given in Kitcher 1985, 1990) that human sociobiology is up to it.

Variants of the refrain that "there is no morality apart from biology" lead sociobiologists into the more ambitious project 3. Here it is necessary for the aspiring ethicists to ask themselves if they believe that some moral statements are true, others false. If they do believe in moral truth and falsity, they should be prepared to specify what grounds such truth and falsity. Those who think that moral statements simply record the momentary impulses of the person making the statement should explain how they cope with people who have deviant impulses. On the other hand, if it is supposed that morality consists in the expression of the "deepest" human desires, then it must be shown how, *without appeal to extrabiological moral principles*, certain desires of an individual are taken to be privileged and how the conflicting desires of different individuals are adjudicated.

Finally, those who undertake project 4, seeing biology as the source of fundamental normative principles, can best make their case by identifying such principles, by formulating the biological evidence for them, and by revealing clearly the character of the inferences from facts to values. In the absence of commitment to any specific moral principles, pleas that "the naturalistic fallacy has lost a great deal of its force in the last few years" (Wilson 1980a: 431) will ring hollow unless the type of argument leading from biology to morality is plainly identified. What kinds of premises will be used? What species of inference leads from those premises to the intended normative conclusion?

It would be folly for any philosopher to conclude that sociobiology can contribute nothing to ethics. The history of science is full of reminders that initially unpromising ideas sometimes pay off (but there are even more unpromising ideas that earn the right to oblivion). However, if success is to be won, criticisms must be addressed, not ignored. Those inspired by Wilson's vision of a moral code reformed by biology have a great deal of work to do.

NOTES

1. Some of Wilson's critics portray him as a frivolous defender of reactionary conservatism (see, for example, Lewontin, Rose, and Kamin 1984). While I agree with several of the substantive points that these critics make against Wilson's version of human sociobiology, I dissent from their assessment of Wilson's motives and commitments. I make the point explicit because some readers of my *Vaulting Ambition* (1985) have mistaken the sometimes scathing tone of that book for a questioning of Wilson's intellectual honesty or of his seriousness. As my title was intended to suggest, I view Wilson and other eminent scientists who have ventured into human sociobiology as treating important questions in a ham-fisted way because they lack crucial intellectual tools and because they desert the standards of rigor and clarity that are found in their more narrowly scientific work. The tone of my (1985) work stems from the fact that the issues are so important and the treatment of them often so bungled.

2. For historical discussion, see Richards (1986). Richard Alexander (1987) offers an alternative version of sociobiological ethics, while Michael Ruse (1986) develops a position that is closer to that espoused in Wilson's later writings (particularly in Ruse and Wilson 1986).

REFERENCES

Alexander, Richard. 1987. *The Morality of Biological Systems.* Chicago: Aldine.

Boyd, Robert, and Peter Richerson. 1985. *Culture and the Evolutionary Process.* Chicago: University of Chicago Press.

Kitcher, Philip. 1985. *Vaulting Ambition: Sociobiology and the Quest for Human Nature.* Cambridge: MIT Press.

———. 1987a. Precis of *Vaulting Ambition* and reply to twenty-two commentators ("Confessions of a Curmudgeon"). *Behavioral and Brain Sciences* 10:61–100.

———. 1987b. "The Transformation of Human Sociobiology." In A. Fine and P. Machamer (eds.), *PSA 1986*, Proceedings of the Philosophy of Science Association, 63–74.

————. 1988. "Imitating Selection." In Sidney Fox and Mae-Wan Ho (eds.), *Metaphors in the New Evolutionary Paradigm*. Chichester: John Wiley and Sons

————. 1990. "Developmental Decomposition and the Future of Human Behavioral Ecology." *Philosophy of Science* 57 : 96–117.

Lewontin, Richard, Stephen Rose, and Leon Kamin, 1984. *Not in Our Genes*. New York: Pantheon Books.

Lumsden, Charles, and Edward O. Wilson. 1983. *Promethean Fire*. Cambridge: Harvard University Press.

Richards, R. 1987: *Darwin and the Emergence of Evolutionary Theories of Mind and Behavior*. Chicago: University of Chicago Press.

Ruse, Michael. 1986. *Taking Darwin Seriously*. London: Routledge.

Ruse, Michael, and Edward O. Wilson. 1986. "Moral Philosophy as Applied Science." *Philosophy* 61 : 173–192.

Wilson, Edward O. 1975. *Sociobiology: The New Synthesis*. Cambridge: Harvard University Press.

————. 1978. *On Human Nature*. Cambridge: Harvard University Press.

————. 1980a. "The Relation of Science to Theology." *Zygon* 15 : 425–434.

————. 1980b. "Comparative Social Theory." *Tanner Lecture*, University of Michigan.

XI Cultural Evolution and Evolutionary Epistemology

22 Epistemology from an Evolutionary Point of View

Michael Bradie

Epistemology is the study of the foundations and nature of knowledge. The traditional approach, beginning with Plato and developed in its modern form by Descartes, is that epistemological questions have to be answered in ways that do not presuppose any particular knowledge. Such approaches might be termed "transcendental" insofar as the appeal to knowledge in order to answer these questions is rejected as question begging. The Darwinian revolution of the nineteenth century suggested an alternative approach, explored first by Dewey and the pragmatists. Human beings, as the products of evolutionary development, are natural beings. Their capacities for knowledge and belief are honed by evolutionary considerations. As such, there is reason to suspect that knowing, as a natural activity, can and should be treated and analyzed by the methods of science. On this view, there is no sharp division of labor between science and epistemology. In particular, the results of particular sciences such as evolutionary theory and cognitive psychology are deemed relevant to the solution of epistemological problems. Such approaches, in general, are called naturalistic epistemologies, whether they are directly motivated by evolutionary considerations or not. Those that are directly motivated by evolutionary considerations and argue that the growth of knowledge follows the pattern of evolution in biology are called evolutionary epistemologies.

Models, metaphors, and analogies play an important role in scientific reasoning. They have an obvious heuristic role as tools for suggesting new experiments, new theories, new ways to expand old theories, and new ways of explaining and understanding phenomena. The history of science is filled with instances of the metaphorical extension of theories from one domain to another. Evolutionary epistemology involves, in part, deploying models and metaphors drawn from evolutionary biology in the attempt to characterize and resolve issues arising in epistemology and conceptual change. As disciplines co-evolve, models are traded back and forth. Thus, evolutionary epistemology also involves attempts to understand how biological evolution proceeds by interpreting it through models drawn from our understanding of conceptual change and the development of theories (cf., Somenzi, Popper, Campbell, and Plotkin).

This essay appears here for the first time.

THREE DISTINCTIONS

There are two interrelated but distinct programs that go by the name evolutionary epistemology. One is the attempt to account for the characteristics of cognitive mechanisms in animals and humans by a straightforward extension of the biological theory of evolution to those aspects or traits of animals that are the biological substrates of cognitive activity: their brains, sensory systems, and motor systems, for example. The other program attempts to account for the evolution of ideas, scientific theories, and culture in general by using models and metaphors drawn from evolutionary biology. Both programs have their roots in nineteenth-century biology and social philosophy, in the work of Darwin, Spencer, James, and others. There have been a number of attempts in the intervening years to develop the programs in detail (see the bibliography and review in Campbell 1974a). Much of the contemporary work in evolutionary epistemology derives from the work of Konrad Lorenz (1977, 1982), Donald Campbell (1960, 1974a), Karl Popper (1968, 1972, 1976, 1978, 1984), and Stephen Toulmin (1967, 1972, 1974, 1981). I have labeled these two programs EEM, for the program that attempts to provide an evolutionary account of the development of cognitive structures, and EET, for the program that attempts to analyze the development of human knowledge and epistemological norms by appealing to relevant biological considerations (Bradie 1986). Some of these attempts involve analyzing the growth of human knowledge in terms of evolutionary (selectionist) models and metaphors (e.g., Popper 1968, 1972; Toulmin 1972; Hull 1988). Others (e.g., Ruse 1986; Rescher 1977) argue for a biological grounding of epistemological norms and methodologies but eschew selectionist models of the growth of human knowledge as such.

A second distinction concerns ontogeny versus phylogeny. Biological development involves both ontogenetic and phylogenetic considerations. Thus, the development of specific traits, such as the opposable thumb in humans, can be viewed from the point of view of the development of that trait in individual organisms (ontogeny) and the development of that trait in the human lineage (phylogeny). The development of knowledge and knowing mechanisms exhibits a parallel distinction. Thus, we can consider the growth of an individual's corpus of knowledge and epistemological norms or his or her brain (ontogeny) or the growth of human knowledge and establishment of epistemological norms across generations or the development of brains in the human lineage (phylogeny). The EEM/EET distinction cuts across this distinction since we may be concerned with either the ontogenetic or phylogenetic development of, for example, the brain or the ontogenetic or phylogenetic development of norms and knowledge corpora. One might expect that since current orthodoxy maintains that biological processes of ontogenesis proceed differently from the biological processes of phylogenesis, evolutionary epistemologies would reflect this difference. Curiously, however, they do not for the most part. An exception is the theory of neural Darwinism put forth by

Edelman (1987) and Changeaux (1985). They argue for a selectionist account of the ontogenetic development of the neural structures of the brain.

A third distinction concerns descriptive versus prescriptive approaches to epistemology and the growth of human knowledge. Many have argued that neither the EEM nor the EET programs have anything at all to do with epistemology properly (traditionally) understood. The basis for this contention is that epistemology, properly understood, is a normative discipline, whereas the EEM and EET programs are concerned with the construction of causal and genetic (factual) models. No such models, it is alleged, can have anything important to contribute to normative epistemology.

The Two Programs

A clear statement of the EEM program can be found in Vollmer (1975:102): "Our cognitive apparatus is a result of evolution. The subjective cognitive structures are adapted to the world because they have evolved, in the course of evolution, in adaptation to that world. And they match (partially) the real structures because only such matching has made such survival possible" (quoted by Bunge 1983:8). Lorenz (1977) expresses a similar sentiment: "I consider human understanding in the same way as any other phylogenetically evolved function which serves the purpose of survival, that is, as a function of a natural physical interaction with a physical external world" (p. 4).

As early as 1941, Lorenz had endorsed the "biologizing of Kant." The a priori categorical structures that organisms use to form their cognitive pictures of reality are to be understood as the a posteriori evolutionary products of phylogenetic development. Thus, "One familiar with the innate modes of reaction of subhuman organisms can readily hypothesize that the *a priori* is due to hereditary differentiations of the central nervous system which have become characteristic of the species, producing hereditary dispositions to think in certain forms" (Lorenz 1982:122).

Sir Karl Popper advocates both programs in evolutionary epistemology. In his 1984 paper on the topic, Popper reduced his view on evolutionary epistemology to five theses. The first thesis is an endorsement of the EEM program: "The specifically human ability to know, and also the ability to produce scientific knowledge, are the results of natural selection. They are closely connected with the evolution of a specifically human language. The first thesis is almost trivial" (Popper 1984:239).

In his "Reply to My Critics," Popper notes some further consequences of evolutionary theory. From the fact that humans are animals and that animal senses have evolved from primitive beginnings, it follows, Popper thinks, that human knowledge is almost as fallible as animal knowledge and that human senses, like animal senses, are part of a "decoding mechanism" (Schilpp 1974: 1059). Popper then reverses the metaphor and claims that human sensory organs are "conjectures!" (ibid.: 111).

Donald Campbell, with whose views Popper has expressed "almost complete agreement" (Schilpp 1974:1059), has developed an evolutionary approach to epistemology in great detail. In his masterful 1974 survey of the subject, Campbell endorses a number of points that are characteristic of the EEM program. In particular, he notes, with approval, Lorenz's biologizing of Kant and the implication that the categories, are to be read "descriptively" and not "prescriptively." He also advocates the view that "evolution—even in its biological aspects—is a knowledge process, and ... the natural-selection paradigm for such knowledge increments can be generalized to other epistemic activities, such as learning, thought and science" (ibid.:412). Campbell consistently endorses the applicability of a "blind-selection-and-retention" model to explain not only the evolution of all biological structures (not merely cognitive structures) but also the growth of scientific knowledge that is more properly viewed as part of the complementary EET program.

The EET program addresses the relevance of biological considerations for understanding the growth of knowledge and the development of epistemological norms. The two programs are interrelated, and often the same authors argue for both. The general trend (exemplified by Lorenz, Campbell, Popper, Toulmin, and Hull, among others) is to attempt to develop selectionist models for the growth of scientific knowledge. Rescher (1977) and Ruse (1986) demur.

Lorenz (1977) endorses Campbell's extension of selectionist models to thinking, learning, and the development of science: "The method of the genome, perpetually making experiments, matching their results against reality, and retaining what is fittest, differs from that adopted by man in his scientific quest for knowledge in only one respect, and that not a vital one, namely that the genome learns only from its successes, whereas man learns also from his failures" (p. 24).

The second thesis of Popper (1984) holds that "the evolution of scientific knowledge is, in the main, the evolution of better theories. This is, again, a Darwinian process. The theories become better adapted through natural selection: they give us better and better information about reality (they get nearer and nearer to the truth). All organisms are problem solvers: problems arise together with life" (p. 239).

Campbell endorses Popper's treatment of the succession of theories in science as due to a selective elimination process analogous to the eliminative role of natural selection in biological evolution. In addition, trial and error learning by animals, including humans, brings the evolutionary model to the ontogenesis of knowledge. (Schilpp 1974:415f.)

The core thesis of Stephen Toulmin's *Human Understanding* is a commitment to what Toulmin considers a form of epistemological Darwinism: "Darwin's populational theory of 'variation and natural selection' is one illustration of a more general form of historical explanation; and ... this same pattern is applicable also, on appropriate conditions, to historical entities and populations of other kinds" (Toulmin, 1972:135). Science, according to Toulmin, develops in a two-step process analogous to biological evolution. At each stage in the historical development of science, a pool of competing intellectual variants exists along with a selection process that determines which variants survive and which die out (Toulmin 1967:465).

David Hull, in a series of papers and a recent book (1973, 1975, 1978, 1982, 1983, 1988), has been working on a Toulminesque project. On Hull's view, neither biological evolution nor the growth of knowledge serves as the primary model in terms of which we are to understand the other. Hull prefers to develop a general analysis of "evolution through selection processes which applies equally to biological, social and cultural evolution" (Hull 1982: 275, 1988). Hull's rationale for treating both biological evolution and conceptual evolution as exemplifications of some common general selectionist model is to undercut objections to selectionist accounts of conceptual change that emphasize the disanalogies between biological and conceptual change (Hull 1988:418). Although the specific mechanisms of change are not the same in the two cases (ibid.:431) and there is no clear evidence that there is any "significant correlation between genetic and conceptual inclusive fitness" (pp. 282f.), Hull argues that both processes can be profitably analyzed in terms of interaction, selection, and differential replication.

Michael Ruse, although a fervent critic of evolutionary epistemologies that promote selectionist models of conceptual change, nevertheless sees an important role for Darwinian insights into the development of scientific methods and traditional epistemological problems in general (Ruse 1986). Ruse, in urging us to take Darwin seriously, argues against the bold attempts by "evolutionary epistemologists" to lift the model of variation and selection that characterizes evolution by natural selection and use it to model scientific methodology. Ruse contends that if we are to take Darwin seriously, we must stick to fundamentals and eschew the facile application of selectionist models drawn from evolutionary biology. Scientific reasoning is a specimen of culture, and if we are to give a Darwinian account of it, that account must rely on the building blocks of culture that modern Darwinian thought has bequeathed to us. Those building blocks, according to Ruse, are the "epigenetic rules" that shape the ways in which our minds work. The term "epigenetic rule" was coined by Charles Lumsden and E. O. Wilson (1981) as a label for a developmental regularity. It derives from the biological process of epigenesis, which is the result of the interaction of genes and the environment. From a Wilsonian point of view, where genes hold culture on a leash, we would expect such development to result in broad regularities in intraspecific mental and cultural patterns. Thus,

Most of the mental idiosyncrasies of the human species have not been translated by psychologists into a precise calculus of cultural choice. But the evidence accumulated so far indicates clearly that the genes have not freed the mind in the special sense of conferring pure cultural transmission. They hold thought and culture in an intermediate degree of dependency, and that, we felt, might contain the secret of the accelerated evolution of the human brain.... The exact configurations of the brain cells and the manner in which the hormones affect them are the chief determinants of the epigenetic rules. The rules then shape the outcome of culture. (Lumsden and Wilson 1983:72).

For Ruse, taking Darwin seriously in questions epistemological involves determining what the epigenetic rules of scientific reasoning are. To ascertain what those rules are, we need to examine the practice of scientific method and infer from that what the rules must be in order to produce the practice that we observe. The model of science that Ruse adopts is an admittedly noncontroversial version of the standard model, which includes elements of inductive and deductive reasoning: logic, mathematics, reasoning by analogy, the generation of laws, the attribution of causes, and appeals to simplicity and consilience (Ruse 1986: 156ff.)—a very proper nineteenth-century view, indeed. These principles and methods are "rooted in our biology" and justified by their adaptive value to us or our proto-human ancestors (ibid.: 155)

Taking Darwin seriously involves explaining how we come to use these rules and methods. We do so, Ruse argues, because the methods and principles of scientific reason mirror or mimic general intellectual tendencies that we would expect would be of selective advantage to those of our ancestors who happened to have the good fortune to act on them. The net effect is that although there is a general selectionist explanation of why human beings have evolved the scientific methodology that they have, the intellectual processes that result in the growth of knowledge are not themselves to be construed in terms of a selectionist model.

This brief survey should serve to illustrate that there are, in fact, two quite distinct programs parading under the name "evolutionary epistemology" and that they are, nevertheless, interrelated. They are not, however, identical. There is a sense in which some version of the EEM program must be true if our current understanding of evolutionary processes is anywhere near correct. What remains to be seen is what useful insights, if any, will be forthcoming from it about the evolution of the cognitive mechanisms of organisms. A further question is what, if anything, any such results have to do with epistemology, whether narrowly or broadly conceived. The success of the EET programs is much more problematic. Even if we could demonstrate that human brains and cognitive apparatuses in general have evolved under selection because of their cognizing abilities, it is by no means a straightforward extension from this to the conclusion that the methodological norms we have come to adopt we have done so because of selective advantage. Nor does it straightforwardly follow that a selectionist model of conceptual change is either a correct or even fruitful way of thinking about conceptual change. Ruse's view, although not endorsing a selectionist model of conceptual change, has an air of post hoc reconstructionism about it. Given that we endorse certain cognitive and epistemological methods, it is easy enough to claim that we do so for selective reasons but not so easy to see what we learn by so doing.

Ontogeny and Phylogeny

The evolutionary considerations addressed in the previous section were directed toward phylogenetic change. Here I briefly discuss some of the applications of selectionist models to ontogenetic processes. These come in two varieties, depending on whether one focuses on the ontogenetic development of individual brains (an EEM project) or the ontogenesis of knowledge or norms in the individual (an EET project). We turn first to selectionist accounts of neural development.

The theory of neural Darwinism, as presented by Edelman, Changeaux, and others, applies populational and variational models to the ontogenetic development of neuronal networks in the brains of individual organisms (Edelman 1985, 1987; Changeaux 1985; cf. Cain and Darden 1989). The basic idea is that the neurons of the brain do not develop according to a program "hardwired" in the genes. Rather, the specific interconnections and topology of neural networks form in accordance with selection pressures of various sorts. Edelman's version, which he calls the theory of neuronal group selection, rests on three premises. First, it is assumed that "groups of neurons are formed in nuclei or laminae of the brain" (Edelman 1985:10). These groups from into "populations." As brain development proceeds, "certain connections are selected over others, and the others disappear." These groups are called "primary repertoires" (Edelman 1987:5). Once in place, the geometry of the structure does not change as the organism interacts with the environment, but pathways are differentially reinforced, which leads to redundancy or "degeneracy." This set of alternative pathways for responding to similar stimuli forms the basis for the system to be able to generalize. These groups are called "secondary repertoires" (ibid.). Finally, it is assumed that there exist in the brain "reentrant circuits," which serve to correlate the independently sampled sensory modalities of the polymorphic (disjunctively defined) sets that make up the world of stimuli (Edelman 1985:11–15).

The theory has two empirical requirements in order to be a plausible candidate for explaining the development of circuitry in the brain. First, there must be some appropriate mechanism for generating the necessary diversity in neuronal circuitry: the CAM (cell adhesion molecule) system does this. The CAMs attached to cells are assumed to switch on and off in response to the local environment, which in turn leads to changes in the geometry of the resulting networks. The resulting "maps" are unique to individual organisms (Edelman 1985:20f.). Second, some evidence for the presence of group selection and the existence of reentrant circuits is needed. Some such evidence exists based on monkey studies (ibid.:22).

Edelman's theories are still somewhat controversial, and, as he himself admits, much work, both theoretical and practical, needs to be done before they can become part of mainstream neurophysiology.

What of the ontogenesis of knowledge in an individual? Campbell, as was noted earlier, endorses Popper's "extension" of the trial-and-error model to include learning processes by individual organisms. In "Of Clouds and Clocks," Popper claimed that organisms as well as phyla were "problem solvers." In fact, he puts forward there the curious analogy that as the actions of an individual organism are tentative solutions to problems faced by it in its environment, so individual organisms are tentative solutions to problems faced by the phylum of which they are members (Popper 1972:243). Elsewhere he draws the connection between the two in the following way. From an ontogenetic point of view, scientific explanations rest on the expectations of the newborn child. Children allegedly grow into critical adulthood by the well-known Popperian process of conjecture and refutation, with the initial conjectures formed on the basis of innate expectations. These innate expectations are the result of phylogenetic development, so from a phylogenetic point of view, today's science rests on the "expectations" of ancestral unicellular organisms. This is epitomized by the quip, "There is, as it were, only one step from the [ancestral] amoeba to Einstein" (ibid.:347). Campbell (1960), in his pre-Popperian days, expresses a similar sentiment. This way of putting the point blurs the difference between the two programs. The phylogenetic development of innate expectations in organisms in a lineage is a question appropriate to EEM. The ontogenetic development of knowledge in an individual (as opposed to the ontogenetic development of the biological structures necessary for an individual to become a competent adult discussed above in connection with neural Darwinism) is a question in the EET program. The corresponding phylogenetic EET question concerns, for example, the historical evolution of science from, say, Aristotle to Einstein. Toulmin characterizes the distinction within the EET program in a clearer way:

We ... face questions about the social, cultural, and intellectual changes that are responsible for the historical evolution of our various modes of life and thought—our institutions, our concepts, and our other practical procedures. (These questions correspond to questions about *phylogeny* in evolutionary biology.) Individually speaking, we ... face questions about the manner in which maturation and experience, socialization and enculturation shape the young child's capacities for rational thought and action—how the child comes to participate in his native society and culture. These questions correspond to the questions about *ontogeny* in developmental biology.) (Toulmin 1981:26)

Finally, although behaviorism is no longer in vogue, it should be noted that B. F. Skinner explicitly notes the parallels between his theory of operant conditioning, Thorndike's "Law of Effect," and Darwin's theory of natural selection (Skinner 1953:64, 1969: esp. chap. 7, 1981; cf. Edelman 1987:12).

Descriptive versus Prescriptive Epistemology

For Donald Campbell, evolutionary epistemology minimally takes cognizance of and is compatible with "man's status as a product of biological and social evolution" (Campbell 1974a:413). Campbell characterizes his approach to epistemology as "descriptive" rather than "analytic" or prescriptive. A descriptive epistemology is "descriptive of man as knower." Descriptive epistemologies, minimally, put constraints on prescriptive epistemologies. So, Campbell argues, an evolutionary picture of human development rules out (1) the view that truth is divinely revealed to humans; (2) direct realism, which assumes that human beings have viridical perception of the world; and (3) epistemologies based on ordinary language analysis. Descriptive epistemology is a branch of science rather than traditional philosophy (Campbell 1974b:141). As such, it is both hypothetical and contingent (Brewer and Collins 1981:12).

Dretske (1971:586) agrees that evolutionary considerations might be construed as telling us that "more often than not, or perhaps in the vast majority of cases, the normal members of a well-adapted species (such as you and I) are *right* in their perceptual judgments about their surroundings." However, Dretske argues, this has no particular epistemological significance, just as the fact that well-trained rats invariably find their way through mazes does not entail that the rats know their way through the maze. Human beings, unlike rats, make judgments (cf. Dretske 1985). Evolutionary theory and, by implication, evolutionary epistemology fail to address the central issues of "true" epistemology, such as what counts as the right to be sure, what counts as adequate evidence, what counts as a good or the best explanation, and how to distinguish between conclusive and inconclusive reasons (Dretske 1971:586).

Campbell does not disagree with this assessment. Descriptive epistemology, in his view, is trying to do something different from traditional epistemology (Campbell 1974b:140).

There are three possible configurations of the relationship between descriptive and traditional epistemology:

1. Descriptive epistemology is a competitor to traditional epistemology. On this view, both are trying to address the same concerns and offering competing solutions. Riedl (1984) defends this position. Dretske argues that descriptive epistemology in this sense fails to touch the traditional questions and thereby is epistemologically irrelevant. The force of this argument is tempered by the extent to which one rejects the "tradition" as irrelevant or uninteresting.

2. Descriptive epistemology might be seen as a successor discipline to traditional epistemology. On this reading, descriptive epistemology does not address the questions of traditional epistemology because it deems them irrelevant, unanswerable, or uninteresting. Many defenders of naturalized epistemologies fall into this camp (Quine 1960, 1969; Davidson 1973; Dennett 1978; Harman 1982, Kornblith 1985. Cf. Bartley 1976, 1987a, 1987b; Munz 1985; Dewey 1910).

3. Descriptive epistemology might be seen as complementary to traditional epistemology. This appears to be Campbell's view.

Campbell admits that descriptive epistemology, in his sense, does beg the traditional epistemological question of how knowledge is possible. It attempts to explore the problem of knowledge from "within the framework of contingent knowledge, and by assuming such knowledge" (Campbell 1974b:141). Brewer and Collins characterize Campbell's descriptive epistemology as "hypothetically normative." As they see it, it seeks to explain why science works to produce "valid knowledge" if and when it does work. It seeks to explain why science fails, if and when it does. Finally, it seeks to determine how to go about acquiring "valid knowledge." It does all this given some presumptive truths about the world (Brewer and Collins 1981:12). Among those presumptive truths is an ontological position that has come to be known as "hypothetical realism."

With respect to the traditional epistemological question (How is knowledge possible?), descriptive epistemology must, according to Campbell, be what he calls an "epistemology of the other one." Such a perspective abandons the task of justifying first-person knowledge and works instead "on the problem of how people in general, or other organisms, come to know" (Campbell 1974b: 141).

Dretske doubts the ability of an "evolutionary view of man's perceptual powers" to satisfy the skeptic (Dretske 1971: 588). Stroud, in criticizing Quine's naturalized epistemology, agrees that, in effect, all we get is an "epistemology of the other." But, he claims, the question of how knowledge is possible at all goes unanswered (Stroud 1981: 463–466). In response to Stroud's objections to attacking epistemological questions by "projecting ourselves into the other's place," Quine argues that "this projection must be seen not transcendentally but as a routine matter of analogies and causal hypotheses within our scientific theories" (Quine 1981: 474; cf. Olding 1983: 2). In such a way, Quine thinks, we get from an epistemology of the other to traditional epistemology.

Quine's move may appear to just beg the question against the skeptic again. But one can defuse, in part, the sting of traditional skepticism by dividing skeptics into two groups: those who will not accept anything and those who argue for skepticism on the basis of arguments from illusion or the fallibility of science or the like. To the former, we can say nothing and must leave them at the crossroads. Life is too short to take such objections seriously. To the latter, Quine's point is more telling, for it is, in effect, a rejection of the skeptic's move to "transcendentalize" objections, which, after all, were derived from intersubjective comparisons and errors in the first place. On this reading, both the skeptic and the epistemologist of the other start from the same intersubjective considerations, but the skeptic is the one who gives the argument a transcendental turn and then complains that appeals to intersubjective experience are question begging. The epistemologist of the other need only block the initial turn to thwart this line of argument. The wrong move would be to accept the problem as posed by the skeptic in its transcendental form and then try to argue back to intersubjectivity. This latter strategy runs afoul of all the "veil of illusions" objections that have plagued post-Cartesian epistemology. The solution is to avoid being seduced behind the veil in the first place.

Hull (1982) agrees "with the critics that a purely descriptive epistemology is 'epistemology' in name only" (Hull 1982: 273). He does not, however, take this as a significant criticism of Campbell's EET program but rather as an indication that such efforts, his own included, should be seen not as attempts to understand epistemology from an evolutionary point of view but rather to produce a "scientific theory of socio-cultural evolution." Hull (1988) develops a selectionist model of scientific change in some detail and includes a case study based on the development of cladism in evolutionary systematics, which, Hull argues, illustrates and supports his model.

THE EVOLUTIONARY METAPHOR

Campbell (1960:381) argues that underlying both trial-and-error problem solving and natural selection in evolution is a general model for "inductive gains." He sets three conditions for such models. They must incorporate (1) a means of producing variation, (2) a mechanism of selection, and (3) a mechanism for preservation and reproduction. The general consensus is that any biologically derived evolutionary model of conceptual change must exhibit at least these three features. It is left open whether such models are required to find specific analogies for all the key concepts in evolutionary biology: "gene," "genotype/phenotype," "organism," "species," and the like. (See Bradie 1986 for an extended discussion of recent attempts to construct such models.)

Is the analogy between trial-and-error problem solving and natural selection superficial (as argued, for example, by Thagard 1980: 187, 1988: chap. 6; Kary 1982; Lewontin 1982; Bunge 1983: 58) or deep (as argued by Toulmin 1967, 1972, 1981; Popper 1972, 1977; Hull 1988, with caveats)? In arguing for the significance of any alleged similarity between evolution in biology and the evolution of science or knowledge, one must either identify appropriate epistemological analogues for key biological concepts or argue for the irrelevance of any apparent disanalogies. There are any number of potentially devastating disanalogies, as a survey of the literature will show (see Bradie 1986 for details). Here I focus on one major difficulty: the problem of progress.

Almost all critics and defenders alike agree that in one important respect conceptual evolution differs from biological evolution. In science, it is claimed, there is progress toward a goal; in biological evolution, there are no goals. Kuhn is one of the few who sees a virtue where others see a flaw:

The analogy that relates the evolution of organisms to the evolution of scientific ideas can easily be pushed too far. But with respect to ... [the idea of progress through scientific revolutions] ... it is very nearly perfect.... The resolution of revolutions is the selection by conflict within the scientific community of the fittest way to practice future science. The net result of a sequence of such revolutionary selections, separated by periods of normal research, is the wonderfully adapted set of instruments we call modern scientific knowledge. Successive stages in that developmental process are marked by an increase in articulation and specialization. And the entire process may have occurred, as we now suppose biological evolution did, without benefit of a set goal, a permanent fixed scientific truth, of which each stage in the development of scientific knowledge is a better exemplar. (Kuhn 1962: 172f.)

Under pressure to clarify the sense in which his position did or did not avoid relativism, Kuhn restated his view as follows:

I believe it would be easy to design a set of criteria—including maximum accuracy to predictions, degree of specializations, number (but not scope) of concrete problem solutions—which would enable any observer involved with either theory to tell which was the older which the descendant. For me, therefore, scientific development is, like biological evolution, unidirectional and irreversible. One scientific theory is not as good as another for doing what scientists normally do. In that sense I am not a relativist. (Lakatos and Musgrave 1970:264)

Toulmin (1972:322ff.) objected to this reformulation as illegitimately attributing unidirectionality to biological evolution and as a throwback to a providential view of evolutionary theory rather than a populational view, which, he argued, is characteristic of Darwinism. In distancing his own view from this later Kuhnian position, Toulmin argues that his own view is evolutionary only insofar as "changes from one temporal cross-section to the next involve the selective perpetuation of conceptual variants" (ibid.:323). No unidirectionality is implied. Indeed, Toulmin argues for a "local" or "ecological" concept of contextual rationality. The recognition of the populational nature of concept change leads us to the conclusion that there are no universal criteria for rationality and no "global" selection criteria (ibid.:317; Toulmin 1981:31; cf. Blackwell 1973 and Hull 1973).

That the progress of science implies some such global criteria whereas natural selection does not has, in fact, been taken as a strong point of disanalogy between conceptual and biological evolution (Elster 1979; Thagard 1980; Blackwell 1973). Hull (1982) and Bechtel (1984) have argued that scientific progress need not involve a commitment to global criteria if it is recognized that the regularities of nature and the "laws of nature" exert a transcontextual constraint on the development of scientific theories.

The progress issue is a problematic test for the adequacy of evolutionary models of scientific development. If the views of Kuhn and Toulmin are correct, then our notion of progress needs to be reevaluated. For all we know, progress in science may turn out to be as chimerical as we now take immanent purpose in nature to be. Leading intellectuals 300 years ago may have disagreed. An appeal to realism, in and of itself, does not seem to be sufficient to guarantee that science can be progressive. Popperian realism may provide such guarantees, but that approach leads to well-o'-the-wisp chases after measures of progress such as "verisimilitude." The fact of the matter is that judgments of progress in science are always "local." Even if "global" criteria exist, we are never in a position to know what they are except by a local presumption or fiat. The history of science and the history of ideas should give us pause when we contemplate reifying a contemporary standard as an inviolable canon.

Evolutionary epistemology is not merely the attempt to understand knowing through the use of metaphors drawn from evolutionary biology but also the attempt to understand biological processes in terms of metaphors drawn from epistemology (Campbell 1960). Somenzi (1980) credits Samuel Butler with first introducing the metaphor of organisms as problem solvers. Peirce argued that

[in science we] ... proceed by experimentation. That is to say, we guess out the laws bit by bit. We ask, What if we were to vary our procedure a little? Would the result be the same? We try it. If we are on the wrong track, an emphatic negative soon gets put upon the guess, and so our conceptions get nearer and nearer right. The improvements of our inventions are made in the same manner. *The theory of natural selection in nature proceeds by similar experimentation* to adapt a stock of animals or plants precisely to its environment, and to keep it in adaptation to a slowly changing environment. (Skagestad 1979:94; the quotation is from *The Collected Papers of Charles Peirce*, 2:86)

Plotkin (1982) endorses the idea that biological adaptations are, in some sense, knowledge. For Plotkin, "information or knowledge, in biological terms, describes a *relationship* between the order of the world, whatever that order is, and the answering and reciprocal organization of an organism" (Plotkin 1982:6). The "reciprocal transfer" of models and metaphors from biology to knowledge or knowledge to biology is not necessarily detrimental to understanding either. It is, in fact, common practice in science (Bradie 1984a, 1984b).

Lewontin (1982) and Toulmin (1967, 1972, 1981) have noted that contemporary evolutionary models of the growth of knowledge based on "Darwinian" models do not represent the introduction of biological considerations into epistemology but rather a shift from one underlying metaphor to another. Lewontin sees two central growth metaphors permeating biology and the social sciences: unfolding (a transformational model characteristic of embryology, classically understood) and trial and error (a variational model characteristic of Darwinian evolution). The distinction between transformational models and variational models of development corresponds roughly to Toulmin's distinction between providential and populational processes. Lewontin explicitly (and Toulmin, implicitly) argues that "unfolding" is as much a metaphor for embryology and "trial-and-error adaptation" is as much a metaphor for biological evolution as they are allegedly metaphors for the evolution of knowledge or culture. Lewontin goes on to argue that both metaphors are bad metaphors for embryology and organic evolution, so, he thinks, "little wonder that they have failed to resolve the contradictions in the theory of cognition" (Lewontin 1982:155). Embryological development is not a "mere" unfolding of potentialities in the genome because of the fact that the phenotypes are produced through an interactive process between genotypes and environments ($P = G \times E$). The argument against treating organic evolution as a process of trial-and-error adaptations follows the familiar line, which includes pointing out the limitation of viewing organisms as optimality-seeking problem solvers, a reminder that frequency-dependent selection means natural selection does not always maximize or even increase the fitness of organisms, and reminders about the effects of pleiotropy, "hitch-hiking" effects, random drift, and the like (ibid.:157–159).

I am not persuaded by the argument that since these metaphors are imperfect models of organic development they cannot be expected to cast any significant light on the evolution of culture or knowledge. Indeed, their very imperfection in this respect may be a hidden blessing in disguise, for whenever biological data suggest the basic metaphor must be modified, the needed modifications can serve as a guide to rethinking how knowledge develops. Thus, the fact that organisms may not be optimizing problem solvers does not, in itself, undermine the view that they are problem solvers using, in some sense, a trial-and-error approach. They may be "satisficing" trial-and-error problem solvers. Similarly, even if we admit that not all organic evolution is adaptational, surely some is, and we are driven to consider how to formulate and solve corresponding problems in cultural and epistemic evolution. In any case, Lewontin is not arguing for a metaphor-free biology or social science; he just wants more appropriate metaphors. Lewontin sees both unfolding and trial-and-error adaptation metaphors as resting on a deeper metaphor that pervades all of contemporary Western thought and ideology—the alienated mechanistic world of Cartesian dualism. In its stead, Lewontin proposes a metaphor of "dialectical interdependence," which involves the "interpenetration of organism and environment" (Lewontin 1982:159–163). On this view, the metaphor of "adaptation" itself becomes suspect, relying as it does, according to Lewontin, on the view of "preexisting niches" waiting to be occupied by organisms capable of exploiting them. The Cartesian vision in evolutionary theory is of "an environment causally prior to, and ontologically independent of organisms.... The world is divided into causes and effects, the external and the internal, environments and the organisms they 'contain'" (ibid.:159).

In place of this, Lewontin suggests that "organisms and environments are dialectically related." Niches exist and can be identified only in virtue of the organisms that occupy them. Organisms and environments are constantly shaping and reshaping each other. The separation of organism from its environment can be, one is led to think, only conceptual and not real (cf. also Lewontin 1978).

The implications of this metaphor for epistemology are potentially quite profound. If niches are "created" by organisms and do not exist independently of them, perhaps problems may not properly exist independently of their tentative solutions. Taking Lewontin's dialectical metaphor seriously suggests that the theory-problem distinction may turn out to be as problematic as the theory-observation distinction. More radical consequences follow. If the organism-environment coupled pair is the analogue of the knower-known pair, then the idea of a constant regular nature—that is, many forms of realism—is in danger of going under. I suspect that Lewontin does not want to push the metaphor to this extreme. But he does conclude by suggesting that "the fundamental error of evolutionary epistemologies as they now exist is their failure to understand how much of what is 'out there' is the product of what is 'in here'" (Lewontin 1982:169; cf. Fleck 1979).

REALISM

Campbell argues that descriptive epistemology, which studies how organisms acquire and process knowledge, leads to the "debunking of the value of 'hard facts'" (Campbell 1979:195f.). Nevertheless, he attests, the ideology of "stubborn facts" has a functional truth. Evolutionary epistemology, in turn, abandons "literal truth" while retaining the "goal of truth." For Campbell, the central epistemological issues concern how organisms interact with their environments to produce knowledge. Evolutionary and processing considerations suggest that knowledge is analogous to a mosaic mural, "a compromise of vehicular characteristics and of referent attributes" (ibid.: 183). We find ourselves committed to an "epistemological relativity." Amoebas know what they know, frogs know other things, and humans still more. Each kind comes to know what it does through processing by cognitive structures that are the product of evolutionary development. Humans may know more things than other creatures or different things, but each kind of organism constructs an image of reality based on its own needs and capacities. However, Campbell, with Popper, is unwilling to push this epistemological relativism to an ontological relativism, although, he says, "the language of science is subjective, provincial, approximative, and metaphoric, never the language of reality itself" (Campbell 1975a: 1120).

That reality itself has a "language" is embodied in Campbell's presumptive ontological view, which he calls "hypothetical realism." The basic postulate of hypothetical realism is that there is an objective world of objects and relations that exists independently of any knowing and perceiving organisms. The organisms that inhabit and interact with this world, however, have only indirect, fallible knowledge, which is "edited" by the "objective referent" (Brewer and Collins 1981: 2f.).

Lorenz is also committed to a version of hypothetical realism, which he takes to be the view that "the categories and modes of perception of man's cognitive apparatus are the natural products of phylogeny and thus adapted to the parameters of external reality in the same way, and for the same reasons, as the horse's hooves are adapted to the prairie, or the fish's fins to the water" (Lorenz 1977: 37; cf. Lorenz 1982: 124f.).

Lorenz argues that the Kantian a priori categories and forms of intuition can be reinterpreted in the light of evolutionary theory. What Kant took to be a priori forms and categories are, in fact, the phylogenetic heritage of biological evolution. Each species has its own version of the forms and categories. Lorenz argues that since the "a priori" cognitive capacities of organisms are capable of evolving, "the boundaries of the transcendental begin to shift [and] ... the boundary separating the experienceable from the transcendental must vary for each individual type of organism" (Lorenz 1982: 123). The transcendental world, on this reading, is a reality that lies just beyond the grasp of cognizing organisms, who, as their lineage evolves, may converge upon it. Ruse (1986) sees this Lorenzian argument as a failed attempt to block Humean skepticism. It certainly takes liberties with the Kantian canon, but if we start from evolutionary theory and its implications, then the idea that as lineages evolve, organisms with more or less specific innate categorical structures come into being is not implausible.

Hypothetical realism, then, is the view that there is a real world that scientific theories only approximately "fit." As science "progresses," it converges on a "true description" of that world. The notion of "approximate fit" captures the sense in which the view of Campbell and Lorenz is a "transcendental" as well as a "hypothetical" realism. As such it controverts certain forms of instrumentalism, pragmatism, and relativism. Still, some form of ontological realism is possible. The extrapolation from historical consensuses to ultimate consensus through convergence requires additional assumptions (Skagestad 1981:88). That our theories change and may even converge locally does not entail the Campbellian conclusion (via Peirce and Popper) that an ultimate consensus at the end of scientific inquiry is a legitimate goal.

Following Peirce, with caveats, Skagestad argues that realism explains the fact of error elimination and the drift toward consensus (Skagestad 1981:87f.). The caveats are that stability of belief is no guarantee of truth and that the drift toward consensus may be due to the "invisible hand" of common influences and shared biases (cf. Brewer and Collins 1981:2). These cautions aside, convergent realism, according to Skagestad, explains why (1) "well established beliefs in science tend to remain stable," and (2) "investigators independently addressing the same question tend ultimately to come up with the same answer" (Skagestad 1981:89f.).

But weaker forms of nonconvergent realism can explain these facts as well. Well-established beliefs in science do tend to remain stable, as long as methods and practices remain more or less stable, but otherwise not. Similarly, the stability of community practices, such as they are, tend to rule out investigators who do not come up with the "same answer" when addressing the "same question." Much, of course, depends on how questions, answers, and beliefs are individuated. The radical work of the social constructionists in science is grist for this mill. (See, e.g., Latour 1987; Woolgar 1988; Pickering 1984.)

An evolutionary argument for a convergence-free version of realism can be constructed. Biological evolution is opportunistic. Organisms exploit their local environments, and the longevity of lineages is contingent not only on issues of selective advantage but also upon historical accidents. (See, e.g., Gould 1989.) Imagine a laboratory of many worlds, each ecological clones of the others. We would not expect, on evolutionary grounds, that either the organisms that appear or the lineages that evolve on one world would be identical or even similar to those organisms and lineages that evolve on another. One must allow for some similarity given the fact of developmental constraints and general environmental similarities. Still, the point stands. There is, in the biological world as far as we currently understand, no convergence to an ultimately perfect form. Most evolutionary epistemologists (Toulmin excepted) are unwilling to accept the implications of this for cognitive or conceptual evolution. They will admit that the local, intellectual, social, and cultural background out of which new ideas emerge is a relevant factor in forming the selective forces that determine which of those ideas will survive and which will not, but they are unwilling to accept the radical implication that the direction of conceptual evolution and change need not be leading anywhere in particular. They are unwilling to abandon the eschatological vision of a terminal consensus "in the long run" (cf. Hull 1988:466f.).

To be sure, there are good philosophical motives for their recalcitrance having to do with the virtues of "objectivity" and "truth." But there is no biological rationale for convergence to consensus, at least insofar as it draws upon Darwinism. Hypothetical realism neither suggests nor guarantees convergence. There is a long inferential leap from general coping with the environment, which all successful lineages must develop, to consensus of structure, content, and function of knowledge. The appeal to laws or regularities of nature to force convergence assumes that the constraints "reality" imposes on how and what we think are sufficient, in the long run, to wash out the social and cultural differences, which, for all we know, operate to induce divergence. Even if a global or galactic community eventually reached consensus, shared bias and mutual reinforcement could not be ruled out as major contributory causes. Accepting a nonconvergent form of realism means that the difficult issues of how to construe rationality and objectivity remain to be dealt with. But convergent forms of realism face these problems as well.

THE QUESTION OF NORMS

Traditional epistemology is, in large part, a normative discipline. Evolutionary epistemologies, in particular, and naturalistic epistemologies, in general, insofar as they construe epistemology as continuous with science, would seem to be purely descriptive. Can such enterprises deal with normative questions? How one approaches this question depends on how one construes the relationship of evolutionary epistemology to traditional epistemology. If evolutionary epistemology is seen as a successor discipline to traditional epistemology, then one may be well prepared to write off many of the traditional questions that epistemology set for itself to answer. Other naturalized theories of knowledge, such as Laudan's, involve reinterpreting normative claims in terms of empirical hypotheticals (Laudan 1984, 1990). Quine (1990) also argues that naturalizing epistemology does not lead to a rejection of norms; "normative epistemology gets naturalized into a chapter of engineering: the technology of anticipating sensory stimulation" (Quine 1990:19).

Campbell thinks that evolutionary epistemology can or should deal with normative issues but at the price of begging some of the traditional questions of epistemology, such as how knowledge is possible. (See Shimony 1981:99.) Ruse (1983:144–146) points out what he takes to be deficiencies in both Popper's and Toulmin's models of evolutionary epistemology. As Ruse sees it, Toulmin's model does not pretend to be normative but only to be descriptive of the evolution of knowledge. Insofar as it is merely descriptive, Ruse argues, it fails to address any central epistemological issues since these are, for the most part, normative. Popper's evolutionary epistemology, on the other hand, claims to be normative and so can be said to be addressing epistemological issues. However, Darwin's theory is a descriptive, not a normative, theory, and therefore Popper's program does not qualify as truly "evolutionary" if by that one means "Darwinian" in nature.

Deriving epistemological norms from the facts of human knowledge acquisition would seem to commit some version of the naturalistic fallacy. It is open to evolutionary epistemologists to challenge the contention that the naturalistic fallacy (or at least any attempt to base norms on facts) is a fallacy. An argument of this form can be found in Richards (1988) with respect to the naturalistic fallacy in ethics. Richards argues that not all derivations of moral norms from facts are fallacious. The general idea is that norms need to be calibrated against intuitively clear cases. If some evolutionary account is forthcoming as to why we are inclined to argue in certain ways with respect to morality (or logic or epistemology), then we can ground our normative principles in these evolutionary considerations. Richards's argument is complex and perhaps not completely successful, but it is not unpromising (cf. Bradie 1990). A similar argument should be constructible for epistemological norms (Vollmer 1987; Munz 1985; Bartley 1987a; Bradie 1989). In any case, even if some evolutionary account of the emergence of epistemological norms is forthcoming, it is very unlikely that specific norms are going to be derivable or justifiable from biological or genetic considerations alone. The specific form of the epistemological norms that are accepted by communities of cognizers will most certainly reflect local cultural contingencies in much the same way that specific moral codes reflect local cultural contingencies.

THE PROSPECTS FOR EVOLUTIONARY EPISTEMOLOGY

What are the prospects for evolutionary approaches to epistemology? Many of the key questions remain to be resolved, but it is time, as Ruse has argued, to take Darwin seriously. For epistemology, that means taking seriously the fact that human beings and other cognizing organisms are members (or parts) of evolving lineages. As such, we should expect that their cognitive apparatuses are subject to various phylogenetic and ontogenetic constraints. It behooves us to direct our attention toward the delimitation of the nature and scope of human knowledge from that perspective.

What I have called EEM programs are saddled with the typical uncertainties of phylogenetic reconstructions (is this or that organ an adaptation?) plus uncertainties from the necessarily sparse fossil record of brain and sensory organ development. The EET programs are even more problematic. Although it is plausible to think that the evolutionary imprint on our organs of thought influences what and how we do think, it is not at all clear that the influence is direct, significant, or detectible. Selectionist epistemologies that endorse a trial-and-error methodology as an appropriate model for understanding scientific change are not analytic consequences of accepting that the brain and other ancillary organs are adaptations that have evolved primarily under the influence of natural selection. The viability of such selectionist models is an empirical question resting on the development of adequate models. Hull's (1988) is, as he admits, but the first step in that direction. Much hard empirical work needs to be done to sustain this line of research (Giere 1988:17). Non-selectionist evolutionary epistemologies, along the lines of Ruse (1986), face a different range of difficulties. It remains to be shown that any biological considerations are sufficiently restrictive to narrow the range of potential methodologies in any meaningful way. It is one thing to show, as Ruse does, that the scientific method as advanced by Whewell, Mill, and others in the nineteenth century is consistent with what we would expect given certain evolutionary assumptions and quite another to show that these assumptions are capable of narrowing the field of effective or acceptable methods to a chosen few. More work needs to be done along these lines if we are to take the evolutionary constraints on patterns of thinking seriously.

Despite the difficulties facing the implementation of substantial Darwinian programs of various sorts in epistemology, there is always the methodological turn to consider. Taking Darwin seriously in epistemology means at least that we reconsider what it means to be human and a knower in the light of all the evidence that suggests that we are not the privileged observers of a divinely created universe as we once thought. The challenge of Darwin and of evolutionary theory is to reconsider, in the light of that evidence, what it is to be a human being (or biological organism) who can know and what it is to know that such creatures can do it (cf. Dewey 1910).

REFERENCES

Bartley, W. W. 1976. "The Philosophy of Karl Popper: Part I: Biology and Evolutionary Epistemology." *Philosophia* 6:463–494.

———. 1987a. "Philosophy of Biology versus Philosophy of Physics." In G. Radnitzky and W. W. Bartley III (eds.), *Evolutionary Epistemology: Theory of Rationality and the Sociology of Knowledge*, 7–45. LaSalle, Ill.: Open Court.

———. 1987b. "Theories of Rationality." In G. Radnitzky and W. W. Bartley III (eds.), *Evolutionary Epistemology: Theory of Rationality and the Sociology of Knowledge*, 205–214. LaSalle, Ill.: Open Court.

Bechtel, W. 1984. "The Evolution of Our Understanding of the Cell: A Study in the Dynamics of Scientific Progress," *Studies in the History and Philosophy of Science* 15:309–356.

Blackwell, R. J. 1973. "The Adaptation Theory of Science," *International Philosophical Quarterly*, 319–334.

Bradie, M. 1984a. "The Metaphorical Character of Science." *Philosophia Naturalis.* 21:229–243.

———. 1984b. "Metaphors in Science." Presented at the International Conference on System Research Informatics and Cybernetics, Baden-Baden, West Germany, August 1–5.

———. 1986. "Assessing Evolutionary Epistemology." *Biology and Philosophy* 1:401–459.

———. 1989. "Evolutionary Epistemology as Naturalized Epistemology." In K. Hahlweg and C. A. Hooker (eds.), *Issues in Evolutionary Epistemology*, 393–412. Albany: State University of New York Press.

———. 1990. "Should Epistemologists Take Darwin Seriously?" In N. Rescher (ed.), *Evolution, Cognition, and Realism*, 33–39. Lanham, Md.: University Press of America.

———. 1992. "Darwin's Legacy." *Biology and Philosophy* 7:111–126.

Brewer, M. B., and Collins, B. E. 1981. *Scientific Inquiry and the Social Sciences: A Volume in Honor of Donald T. Campbell.* San Francisco: Jossey-Bass.

Epistemology from an Evolutionary Point of View

Bunge, Mario. 1983. *Treatise on Basic Philosophy. vol. 5: Epistemology and Methodology I: Exploring the World*. Dordrecht: Reidel.

Cain J. and L. Darden 1989. "Selection Type Theories." *Philosophy of Science* 56:106–129.

Campbell, D. T. 1960. "Blind Variation and Selective Retention in Creative Thought As in Other Knowledge Processes." *Psychological Review* 67:380–400.

———. 1974a. "Evolutionary Epistemology." In P. A. Schilpp (ed.), *The Philosophy of Karl Popper, I*. 413–463. LaSalle, Ill.: Open Court.

———. 1974b. "Unjustified Variation and Selective Retention in Scientific Discovery." In F. J. Ayala and T. Dobzhansky (eds.), *Studies in the Philosophy of Biology*, 139–161. London: Macmillan.

———. 1975a. "On the Conflicts between Biological and Social Evolution and between Psychology and Moral Tradition." *American Psychologist* 30:1103–1126.

———. 1975b. "Reintroducing Konrad Lorenz to Psychology." In R. I. Evans (ed.), *Konrad Lorenz: The Man and His Ideas*. New York: Harcourt Brace Jovanovich.

———. 1977. "Comment on 'The Natural Selection Model of Conceptual Evolution.'" *Philosophy of Science* 44:502–507.

———. 1979. "A Tribal Model of the Social System Vehicle Carrying Scientific Knowledge." *Knowledge: Creation. Diffusion. Utilization* I, 2:181–201.

Changeux, Jean-Pierre. 1985. *Neuronal Man*. New York: Pantheon.

Darden, L., and Cain, J. A. 1989. "Selection Type Theories." *Philosophy of Science* 56:106–129.

Davidson, D. 1973. "On the Very Idea of a Conceptual Scheme." *Proceedings of the American Philosophical Association* 47:5–20.

Dennett, D. 1978. *Brainstorms*. Cambridge: MIT Press.

Dewey, J. 1910. *The Influence of Darwinism on Philosophy and Other Essays in Contemporary Thought*. New York: Henry Holt & Co.

Dretske, F. 1971. "Perception from an Epistemological Point of View." *Journal of Philosophy*. 68:584–591.

———. 1985. "Machines and the Mental." *Proceedings and Addresses of the American Philosophical Association* 59, 1:23–33.

Edelman, G.M. 1985: "Neural Darwinism: Population Thinking and Higher Brain Function." In M. Shafto (ed.), *How We Know: The Inner Frontier of Cognitive Science*. San Franciso: Harper & Row.

Edelman, G. M. 1987. *Neural Darwinism: The Theory of Neuronal Group Selection*. New York: Basic Books.

Elster, J. 1979. *Ulysses and the Sirens: Studies in Rationality and Irrationality*. Cambridge: Cambridge University Press.

Fleck L. 1979. *Genesis and Development of a Scientific Fact*. Chicago: University of Chicago Press.

Giere, R. 1988. *Explaining Science*. Chicago: University of Chicago Press.

Gould, S. J. 1989. *Wonderful Life: The Burgess Shale and the Nature of History*. New York: W. W. Norton & Company.

Harman, G. 1982. "Metaphysical Realism and Moral Relativism." *Journal of Philosophy* 79:568–575.

Hull, D. L. 1973. "A Populational Approach to Scientific Change." *Science* 182:1121–1124.

———. 1975. "Central Subjects and Historical Narratives." *History and Theory* 14:253–274.

———. 1978. "Altruism in Science: A Sociobiological Model of Co-operative Behavior among Scientists." *Animal Behavior* 26:685–697.

———. 1982. "The Naked Meme." In H. C. Plotkin. (ed.), *Learning, Development, and Culture.* New York: John Wiley & Sons.

———. 1983. "Exemplars and Scientific Change." In *PSA 1982.* 2:479–503. Edited by P. D. Asquith and T. Nickles. East Lansing, Mich.: Philosophy of Science Association.

———. 1988. *Science as a Process: An Evolutionary Account of the Social and Conceptual Development of Science.* Chicago: University of Chicago Press.

Kary, C. 1982. "Can Darwinian Inheritance Be Extended from Biology to Epistemology?" in *PSA 1982*, 1:356–369. Edited by P. D. Asquith and T. Nickles. East Lansing, Mich.: Philosophy of Science Association.

Kornblith, H. (ed.). 1985. *Naturalizing Epistemology.* Cambridge: MIT Press.

Kuhn, T. 1962. *The Structure of Scientific Revolutions.* Chicago: University of Chicago Press.

Lakatos, I., and Musgrave, A. (eds.). 1970. *Criticism and the Growth of Knowledge.* Cambridge: Cambridge University Press.

Latour, B. 1987. *Science in Action.* Cambridge: Harvard University Press.

Laudan, L. 1984. *Science and Values.* Berkeley: University of California Press.

———. 1990. "Normative Naturalism." *Philosophy of Science* 57:44–59.

Lewontin, R. C. 1982. "Organism and Environment." In *Learning, Development, and Culture,* 151–170. Edited by H. C. Plotkin. New York: John Wiley and Sone.

Lorenz, K. 1974. "Analogy as a Source of Knowledge." *Science* 185:229–234.

———. 1977. *Behind the Mirror.* London: Methuen.

———. 1982. "Kant's Doctrine of the A Priori in the Light of Contemporary Biology." In *Learning, Development, and Culture,* 121–143. Edited by H. C. Plotkin. New York: John Wiley & Sons.

Lumsden, C. J., and Wilson, E. O. 1981. *Genes, Mind and Culture.* Cambridge: Harvard University Press.

———. 1983. *Promethean Fire.* Cambridge: Harvard University Press.

Munz, P. 1985. *Our Knowledge of the Growth of Knowledge: Popper or Wittgenstein?* London: Routledge & Kegan Paul.

Olding, A. 1983. "Biology and Knowledge." *Theoria* 49:1–22.

Pickering. A. 1984. *Constructing Quarks.* Chicago: University of Chicago Press.

Plotkin, H. C. (ed.). 1982. *Learning, Development, and Culture: Essays in Evolutionary Epistemology.* New York: John Wiley & Sons.

Popper, K. R. 1968. *The Logic of Scientific Discovery.* New York: Harper.

———. 1972. *Objective Knowledge: An Evolutionary Approach.* Oxford: Clarendon Press.

———. 1976. "Darwinism as a Metaphysical Research Programme." *Methodology and Science* 9:103–119.

———. 1978. "Natural Selection and the Emergence of Mind." *Dialectica* 32:339–355.

———. 1984. "Evolutionary Epistemology." In *Evolutionary Theory: Paths into the Future.* Edited by J. W. Pollard. London: John Wiley & Sons.

Quine, W. V. O. 1960. *Word and Object*. Cambridge: MIT Press.

———. 1969. *Ontological Relativity and Other Essays*. New York: Columbia University Press.

———. 1981. "Reply to Stroud." In *Midwest Studies in Philosophy VI*. 473–475. Edited by P. A. French, T. E. Uehling, Jr., and H. K. Wettstein. Minneapolis: University of Minnesota Press.

Rescher, N. 1977. *Methodological Pragmatism*. Oxford: Basil Blackwell.

———. (ed.) 1990. *Evolution, Cognition, and Realism*. Lanham, Md.: University Press of America.

Richards, R. 1987. *Darwin and the Emergence of Evolutionary Theories of Mind and Behavior*. Chicago: University of Chicago Press.

Riedl, R. 1984. *Biology of Knowledge: The Evolutionary Basis of Reason*. Chichester: John Wiley & Sons.

Rosenfield, I. 1988. *The Invention of Memory: A New View of the Brain*. New York: Basic Books.

Ruse, M. 1983. "Darwin and Philosophy Today." In D. Oldroyd and I. Langham (eds.), *The Wider Domain of Evolutionary Thought*, 73–158. Dordrecht: Reidel.

———. 1986. *Taking Darwin Seriously: A Naturalistic Approach to Philosophy*. Oxford: Blackwell.

Schilpp, P. A. (ed.). 1974. *The Philosophy of Karl Popper*. LaSalle, Ill.: Open Court.

Shafto, M. (ed.). 1985. *How We Know*. San Francisco: Harper & Row.

Shimony, A. 1981. "Integral Epistemology." In M. B. Brewer and B. E. Collins (eds.), *Scientific Inquiry and the Social Sciences: A Volume in Honor of Donald T. Campbell*, 98–123. San Francisco: Jossey-Bass.

Skagestad, P. 1979. "C. S. Peirce on Biological Evolution and Scientific Progress." *Synthese* 41:85–114.

———. 1981. "Hypothetical Realism." In M. B. Brewer and B. E. Collins. (eds.), *Scientific Inquiry and the Social Sciences: A Volume in Honor of Donald T. Campbell* 77–97. San Francisco: Jossey-Bass.

Skinner, B. F. 1953. *Science and Human Behavior*. New York: Macmillan.

———. 1969. *Contingencies of Reinforcement: A Theoretical Analysis*. New York: Appleton-Century-Crofts.

———. 1981. "Selection by Consequences." *Science* 213:501–504.

Somenzi, V. 1980. "Scientific Discovery from the Viewpoint of Evolutionary Epistemology." In M. D. Grmk, R. S. Cohen, and G. Cimino (eds.), *On Scientific Discovery*, 167–177. Dordrecht: Reidel.

Stroud, B. 1981. "Evolution and the Necessity of Thought." In *Pragmatism and Purpose: Essays Presented to Thomas A. Goudge*, 236–247. Edited by L. W. Sumner, John G. Slater, and Fred Wilson. Toronto: University of Toronto Press.

Thagard, P. 1980. "Against Evolutionary Epistemology." In *PSA 1980* 1:187–196. Edited by P. D. Asquith and R. N. Giere. E. Lansing, Michigan: Philosophy of Science Association.

———. 1988. *Computational Philosophy of Science*. Cambridge: MIT Press.

Toulmin, S. 1967. "The Evolutionary Development of Natural Science." *American Scientist* 55:4.

————. 1972. *Human Understanding: The Collective Use and Evolution of Concepts*. Princeton N.J.: Princeton University Press.

————. 1974. "Rationality and Scientific Discovery." In K. Schaffner and R. Cohen (eds.), *Boston Studies in the Philosophy of Science XX*. Dordrecht: Reidel.

————. 1981. "Evolution, Adaptation, and Human Understanding." In M. B. Brewer and B. E. Collins (eds.), *Scientific Inquiry and the Social Sciences: A Volume in Honor of Donald T. Campbell*, 18–36. San Francisco: Jossey-Bass.

Vollmer, G. 1975. *Evolutionare Erkenntnistheorie*, Frankfurt: S. Hirzel.

————. 1987. "On Supposed Circularities in an Empirically Oriented Epistemology." In G. Radnitzky and W. W. Bartley III (eds.), *Evolutionary Epistemology. Theory of Rationality and the Sociology of Knowledge*, 163–200. LaSalle, Ill: Open Court.

Woolgar, S. 1988. *Science the Very Idea*. Chichester: Ellis Horwood.

23 Models of Cultural Evolution

Elliott Sober

As least since the time of Darwin, there has been a tradition of borrowing between evolutionary theory and the social sciences. Darwin himself owed a debt to the Scottish economists who showed him how order can be produced without conscious design. Adam Smith thought that socially beneficial characteristics can emerge in a society as if by an "invisible hand"; though each individual acts only in his or her narrow self-interest, the result, Smith thought, would be a society of order, harmony, and prosperity. The kind of theory Darwin aimed at—in which fitness improves in a population without any conscious guidance—found a suggestive precedent in the social sciences.

The use of game theory by Maynard Smith[1] and others provides a contemporary example in which an idea invented in the social sciences finds application in evolutionary theory. Economists and mathematicians were the first to investigate the payoffs that would accrue to players following different strategies in games of a given structure. Biologists were also to see that game theory does not require that the players be rational or even that they have minds. The behavior of organisms exhibits regularities; this is enough for us to talk of them as pursuing strategies. The payoffs of the behaviors that result from these strategies can be measured in the currency of fitness—i.e., in terms of their consequences for survival and reproduction. This means that the idea of payoffs within games allows us to describe evolution by natural selection. Here again is a case in which a social scientific idea has broader scope than its initial social science applications might have suggested.

At present, there is considerable interest and controversy surrounding borrowings that go in the opposite direction. Rather than apply social science ideas to biological phenomena, sociobiology and related research programs aim to apply evolutionary ideas to problems that have traditionally been thought to be part of the subject matter of the social sciences. Sociobiology is the best known of these enterprises. It has been criticized on a variety of fronts. Although I think that these criticisms differ in their force, I don't want to review them here. My interest is in a somewhat lesser-known movement

From P. Griffiths (ed.), *Trees of Life: Essays in the Philosophy of Biology* (Australasian Studies in the History and Philosophy of Science), Kluwer, 1991, 17–38.

within biology, one that strives to extend evolutionary ideas to social scientific phenomena. I want to discuss the models of cultural evolution put forward by Cavalli-Sforza and Feldman[2] and by Boyd and Richerson.[3] These authors have distanced themselves from the mistakes they see attaching to sociobiology. In particular, they wish to describe how cultural traits can evolve for reasons that have nothing to do with the consequences the traits have for reproductive fitness. In a very real sense, their models describe how it is possible for mind and culture to play an irreducible and autonomous role in cultural change. For this reason, there is at least one standard criticism of sociobiology that does not apply to these models of cultural evolution. They deserve a separate hearing.

In order to clarify how these models differ from some of the ideas put forward in sociobiology, it will be useful to describe some simple ways in which models of natural selection can differ. I focus here on natural selection, even though there is more to evolutionary theory than the theory of natural selection, and in spite of the fact that the two books I am considering sometimes exploit these nonselectionist ideas. Although there are nonselectionist ideas in these two books, the bulk of their models assigns a preeminent role to natural selection and its cultural analogs. So a taxonomy of selection models will help us see how models of cultural evolution are related to arguments put forward in sociobiology.

There are two crucial ingredients in a selection process. Given a set of objects that exhibit variation, what will it take for that ensemble to evolve by natural selection? By evolution, I mean that the frequency of some characteristic in the population changes. The first requirement is that the objects differ with respect to some characteristic that makes a difference in their abilities to survive and reproduce. Second, there must be some way to ensure that offspring resemble their parents. The first of these ingredients is called *differential fitness*; the second is *heritability*.

In standard formulations of the genetical theory of natural selection, different genes or gene complexes in a population encode different phenotypes. The phenotypes confer different capacities to survive and reproduce on the organisms that possess them. As a result, some genes are more successful in finding their way into the next generation than others. In consequence, the frequency of the phenotype in question changes. This is evolution by natural selection with a genetic mode of transmission. Note that traits differ in fitness because some organisms have more babies than others. It may seem odd to say that "having babies"[4] is one way to measure fitness, as if there could be others. My reason for saying this will become clearer later on.

The phenotype treated in such a selection model might be virtually any piece of morphology, physiology, or behavior. Biologists have developed different applications of this Darwinian pattern to characteristics of all three sorts in a variety of species. One way—the most straightforward way—to apply biology to the human sciences is to claim that some psychological or cultural characteristic became common in our species by a selection process of this sort. This is essentially the pattern of explanation that Wilson was using when he talked about aggression, xenophobia, and behavioral differences between the sexes. An ancestral population is postulated in which phenotypic differences have a genetic basis; then a claim is made about the consequences of those phenotypes for survival and reproduction. This is used to explain why the population changed to the configuration we now observe.

The second form that a selection process can take retains the idea that fitness is measured by how many babies an organism produces, but drops the idea that the relevant phenotypes are genetically transmitted. Strictly speaking, evolution by natural selection does not require genes. It simply requires that offspring resemble their parents. For example, if characteristics were transmitted by parents teaching their children, a selection process could occur without the mediation of genes.

A hypothetical example of how this might happen is afforded by that favorite subject of sociobiological speculation—the incest taboo. Suppose that incest avoidance is advantageous because individuals with the trait have more viable offspring than individuals without it. The reason is that outbreeding diminishes the chance that children will have deleterious recessive genes in double dose. If offspring learn whether to be incest avoiders from their parents, the frequency of the trait in the population may evolve. And this may occur without there being any genetic differences between those who avoid incest and those who do not. Indeed, incest avoidance could evolve in this way in a population of genetically identical individuals, provided that the environmental determinant of the behavior runs in families.[5]

In this second kind of selection model, mind and culture displace one but not the other of the ingredients found in models of the first type. In the first sort of model, a genetic mode of transmission works side by side with a concept of fitness defined in terms of reproductive output—what I have called "having babies." In the second, reproductive output is retained as the measure of fitness, but the genetic mode of transmission is replaced by a psychological one. Teaching can provide the requisite heritability just as much as genes.

The third pattern for applying the idea of natural selection abandons both of the ingredients present in the first. Genes are abandoned as the mode of transmission. And fitness is not measured by how many babies an organism has. Individuals acquire their ideas because they are exposed to the ideas of their parents, of their peers, and of their parents' generation. So the transmission patterns may be vertical, horizontal, and oblique. An individual exposed to a mix of ideas drawn from these different sources need not give them all equal credence. Some may be more attractive than others. If so, the frequency of ideas in the population may evolve over time. Notice that there is no need for organisms to differ in terms of their survivorship or degree of reproductive success in this case. Some ideas catch on while others become passé. In this third sort of selection model, ideas spread the way a contagion spreads.

It is evident that this way of modeling cultural change is tied to the genetical theory of natural selection no more than it is tied to epidemiology. Rumors and diseases exhibit a similar dynamic. The spread of a novel characteristic in a population by natural selection, like the spread of an infection or an idea, is a diffusion process.

This third type of selection model has a history that predates sociobiology and the models of cultural evolution that I eventually want to discuss. Consider the economic theory of the firm.[6] Suppose one wishes to explain why businesses of a certain sort in an economy behave as profit maximizers. One hypothesis might be that individual managers are rational and economically well informed; they adjust their behavior so as to cope with market conditions. Call this the learning hypothesis. An alternative hypothesis is that managers are not especially rational or well informed. Rather, firms that are not efficient profit maximizers go bankrupt and thereby disappear from the market. This second hypothesis posits a selection process.

Note that the selection hypothesis involved here is of type III. Individual firms stick to the same market strategies, or convert to new ones, by some process other than genetic transmission. In addition, the biological kind of survival and reproduction (what I have called "having babies") does not play a role. Firms survive differentially, but this does not require any individual organism to die or reproduce.

A different example of type III models, which will be familiar to philosophers of science, is involved in some versions of evolutionary epistemology. Karl Popper suggested that scientific theories compete with each other in a struggle for existence.[7] Better theories spread through the population of inquirers; inferior ones exit from the scene. Popper highlighted the nonbiological definition of fitness used in this view of the scientific process when he said that "our theories die in our stead."[8]

The three possible forms that a selection model can take are summarized in table 23.1. By "learning," I don't want to require anything that is especially cognitive; imitation is a kind of learning. In addition, "having students" should be interpreted broadly, as any sort of successful influence mediated by learning.[9]

The parallelism between type I and type III models is instructive. In the type I case, individuals produce different numbers of babies in virtue of their phenotypic differences (which are transmitted genetically). In the type III case, individuals produce different numbers of students in virtue of their phenotypic differences (which are transmitted by learning).

Selection models of cultural characteristics that are of either pattern I or pattern II can properly be said to provide a "biological" treatment of the

Table 23.1. Three types of selection models

	Heritability	Fitness
I	Genes	Having babies
II	Learning	Having babies
III	Learning	Having students

Note: The description of Type III models, in which fitness is measured by "having students," is due to Peter Richerson.

characteristic in question. Models of type III, on the other hand, do not really propose biological explanations at all. A selectional theory of the firm, or a diffusion model that describes the spread in popularity of an idiom in a language, are no more "biological" than their competitors. In type III models, the mode of transmission and the reason for differential survival and replication may have an entirely autonomous cultural basis. Genes and having babies are notable by their absence; the biological concept of natural selection plays the role of a suggestive metaphor, and nothing more.

It is important to recognize that this threefold taxonomy describes the process of natural selection, not the product that process may yield. For example, once a type I process of natural selection has run its course, it is an open question whether the variation that remains is genetic or nongenetic. Consider the work in sociobiology by Richard Alexander.[10] He believes that human beings behave so as to maximize their inclusive fitness. This means that there is an evolutionary explanation for the fact that people in one culture behave differently from those in another. But Alexander does not think that this is due to there being genetic differences between the two cultures. Rather, his idea is that the human genome has evolved so that a person will select the fittest behavior, given the environment he or she occupies. The fact that people behave differently is due to the fact that they occupy different environments. So, in terms of the current variation that we observe, Alexander is, in fact, a radical environmentalist. This is worth contemplating if you think that sociobiology stands or falls with the thesis of genetic determinism.

Matters change when we consider not the present situation, but the evolutionary past that generated it. The genome that Alexander postulates, which gives current humans their ability to modify behavior in the light of ecological conditions, evolved because it was fitter than the alternatives against which it competed. That is, the process of natural selection that led to the present configuration is one in which genetic differences account for differences in behavior.

So Alexander sees genetic differences as being crucial to the process of evolution, but environmental differences as characterizing the product of that evolution. He is a type I theorist, since these types pertain to the process of natural selection, not its product.

The distinction between process and product is perhaps a bit harder to grasp when we think of the evolution of some behavioral or psychological trait, but it really applies to any evolutionary event. For the fact of the matter is that evolution driven by a type I selection process feeds on (additive) genetic variation, and uses it up. A morphological character can display the same double aspect. The opposable thumb evolved because there was a genetic difference between those with the thumb and those without it. But once that trait has finished evolving, the difference between those with and those without a thumb may owe more to industrial accidents and harmful drugs taken prenatally than to genetic oddities.

This threefold division among selection models is of course consistent with there being models that combine two or more of these sorts of process. My taxonomy describes "pure types," so to speak, whereas it is often interesting to consider models in which various pure types are mixed. This is frequently the case in the examples worked out by Cavalli-Sforza and Feldman and by Boyd and Richerson. I want to describe one example from each of these books. The point is to discern the way in which quite different selection processes interact.

In the nineteenth century, Western societies exhibited an interesting demographic change, one that had three stages. First, oscillations in death rates due to epidemics and famines became both less frequent and less extreme. Second, overall mortality steadily declined. This latter change had a multiplicity of causes; improved nutrition, sanitation, and (if the more recent past is also considered) medical advances played a role. The third part of this demographic transition was a dramatic decline in birth rates. Typically, there was a time lag; birth rates began to decline only after death rates were already on the way down. Cavalli-Sforza and Feldman (p. 181) give the somewhat idealized rendition of this pattern shown in figure 23.1.

Cavalli-Sforza and Feldman consider the question of how fertility could have declined in Europe. From the point of view of a narrowly Darwinian outlook, this change is puzzling. A characteristic that increases the number of viable and fertile offspring will spread under natural selection, at least when that process is conceptualized from the point of view of a type I model. Cavalli-Sforza and Feldman are not tempted to appeal to the theory of optimal clutch size due to Lack, according to which a parent can sometimes augment the number of offspring surviving to adulthood by having fewer babies.[11] Presumably, this Darwinian option is not even worth exploring, because women in nineteenth-century Europe easily could have had more viable fertile offspring than they in fact did. People were not caught in the bind that Lack attributed to his birds.

The trait that increased in the modern demographic transition was one of reduced biological fitness. The trait spread *in spite of* its biological fitness, not *because of* it. In Italy, women changed from having about five children on

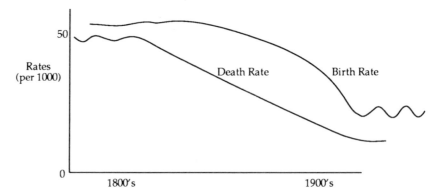

Figure 23.1 Cavalli-Sforza and Feldman's (p. 181) idealized representation of the demographic time lag in Europe. Mortality rates decline; then, after a time lag, the birth rate declines also. (Reprinted by permission of Princeton University Press.)

average to having about two. The trait of having two children, therefore, has a biological fitness of 2/5, when compared with the older trait it displaced.

Cavalli-Sforza and Feldman focus on the problem of explaining how the new custom spread. One possible explanation is that women in all social strata gradually and simultaneously reduced their fertilities. A second possibility is that two dramatically different traits were in play and that the displacement of one by another cascaded from one social class down to the next. The first hypothesis, which posits a gradual spread of innovation, says that fertilities declined from 5 to 4.8 to 4.5 and so on, with this process occurring simultaneously across all classes. The second hypothesis says that having five children competed with having two, and that the novel character was well on its way to displacing the more traditional one among educated people before the same process began among less educated people. This second hypothesis is illustrated in figure 23.2. There is some statistical evidence that the second pattern is more correct, at least in some parts of Europe.

Cavalli-Sforza and Feldman emphasize that this demographic change could not have taken place if traits were passed down solely from mothers to daughters. The Darwinian disadvantage of reduced fertility is so great that purely vertical transmission is not enough to offset it. This point holds true whether fertility is genetically transmitted or learned. A woman with the new trait will pass it along to fewer offspring than a woman with the old pattern, if a daughter is influenced only by her mother.

What is required for the process is some mixture of horizontal and oblique transmission. That is, a woman's reproductive behavior must be influenced by her peers and by her mother's contemporaries. However, it will not do for a woman to adopt the behavior that she finds represented on average in the group that influences her. What is required is that a woman find small family size more attractive than large family size even when very few of her peers

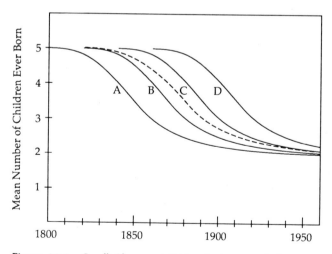

Figure 23.2 Cavalli-Sforza and Feldman's (p. 185) idealized picture of the demographic transition in Italy. A is the most educated class. B, C and D are progressively less educated. (Reprinted by permission of Princeton University Press.)

Models of Cultural Evolution

possess the novel characteristic. There must be a "transmission bias" in favor of the new trait.

Having a small family was more attractive than having a large one, even though the former trait had a lower Darwinian fitness than the latter. Cavalli-Sforza and Feldman show how the greater attractiveness of small family size can be modeled by using ideas drawn from population genetics. However, when these genetic ideas are transposed into a cultural setting, one is talking about cultural fitness, not biological fitness. So the model they end up with for the demographic transition combines two selection processes. When fitness is defined in terms of having babies, having a small family is selected against. When fitness is defined in terms of the attractiveness of an idea ("having students"), there is selection favoring a reduction in family size. Cavalli-Sforza and Feldman show how the cultural process can overwhelm the biological one; given that the trait is sufficiently attractive (and their models have the virtue of giving this idea quantitative meaning), the trait can evolve in spite of its Darwinian disutility.

The example I want to describe from Boyd and Richerson's book is developed in a chapter that begins with a discussion of Japanese kamikaze pilots during World War II. Self-sacrificial behavior—altruism—has been an important problem for recent evolutionary theory. Indeed, Wilson called it "the central problem of sociobiology."[12] Although some apparently altruistic behaviors can be unmasked—shown to be predicated on the selfish expectation of reciprocity, for example—Boyd and Richerson are not inclined to say this about the kamikazes. They died for their country. Nor can one explain their self-sacrifice by saying that it was coerced by leaders; kamikaze pilots volunteered. Nor is it arguable that the pilots volunteered in ignorance of the consequences; suicide missions were common knowledge in the Japanese air force.

So why did kamikaze pilots volunteer? Boyd and Richerson (pp. 204–5) refer to one historian who "argues that the complex of beliefs that gave rise to the kamikaze tactic can be traced back to the Samurai military code of feudal Japan which called for heroic self-sacrifice and put death before dishonour. When the Japanese military modernized in the nineteenth century, the officer corps was drawn from the Samurai class. These men brought their values and transmitted them to subsequent generations of officers who in turn inculcated these values in their men."

Boyd and Richerson (pp. 204–5) say that this historical explanation is "unsatisfactory for two reasons. First, it is incomplete. It tells us why a particular generation of Japanese came to believe in heroic self-sacrifice for the common good; it does not tell us how these beliefs came to predominate in the warrior class of feudal Japan. Second, it is not general enough. The beliefs that led the kamikazes to die for their country are just an especially stark example of a much more general tendency of humans to behave altruistically toward members of various groups of which they are members." They then impose two conditions of adequacy on any proposed explanation: (1) it must show how the "tendency to acquire self-sacrificial beliefs and values could have evolved"; (2) it must show "why altruistic cooperation is directed toward some individuals and not others" (p. 205).

In answer to these requirements, Boyd and Richerson then construct a group selection model that incorporates a certain form of learning. Altruists and selfish individuals exist in each of several groups. Within each group, altruists do less well than selfish people. However, groups of altruists go extinct less often and found more colonies than groups of selfish individuals. These ideas are standard fare in the models of group selection that evolutionary biologists have considered.[13] A type I selection model of the evolution of altruism will require a between-group process favoring altruism that offsets the within-group process that acts to eliminate the trait.

The new wrinkle introduced by the idea of cultural transmission is as follows. Boyd and Richerson postulate that cultural transmission favors common characteristics and works against rare ones. Within a group, individuals are especially biased toward adopting altruism if most individuals are altruists and toward becoming selfish if most people are selfish. What I mean by "especially" biased is illustrated in figure 23.3. In all cases of cultural transmission, the state that a naive individual acquires is influenced by the frequency of traits in the population. Boyd and Richerson impose a more extreme demand. They require that the probability of acquiring a common trait be higher than its population frequency; this is what they call "frequency-dependent biased transmission" (depicted in figure 23.3c).

The process of cultural transmission can work within the time frame of a single biological generation. The effect is to augment the amount of variation there is among groups. Whereas traditional genetic models of group selection allow for a continuum of local frequencies of altruism, the result of this biased transmission rule is to push each local population toward 100 percent altruism or 100 percent selfishness. This has the effect of raising the probability that altruism will evolve and be maintained.

Boyd and Richerson also raise the question of how this biased "conformist" transmission rule could have evolved in the first place. They speculate that if

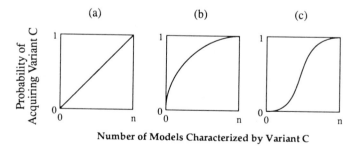

Figure 23.3 Boyd and Richerson's (p. 207) characterization of three patterns of cultural transmission. In all three cases, the probability that a naive individual will acquire a trait depends on the frequency of the trait among the individual's models. (a) represents unbiased transmission, (b) directly based transmission, and (c) frequency-dependent biased transmission. (Reprinted by permission of University of Chicago Press.)

Models of Cultural Evolution

a species is composed of a set of local populations, and if these populations inhabit qualitatively different micro habitats, an individual moving into a new habitat may do best by imitating the traits that are common there. Their proposal is a Darwinian explanation for acting Roman in Rome, so to speak. Once this transmission bias has evolved, it may have various spin-off consequences that have the effect of harming organisms rather than helping them. If you find yourself living with altruists, the transmission bias will lead you to become altruistic yourself, even though you would be better off remaining selfish. Boyd and Richerson admit that there is little or no psychological evidence that people deploy the extreme form of transmission bias that their model postulates.

Just as in the example discussed from Cavalli-Sforza and Feldman, this model of Boyd and Richerson's mixes together the concepts of biological and cultural fitness. Altruism is deleterious to individuals, when fitness is calibrated in terms of the survival and reproduction of organisms. But common characteristics are more contagious than rare ones, when the individuals use a conformist transmission rule. This means that when altruism is common, it is more catching than selfishness. In such cases, the cultural fitness of altruism is greater than the cultural fitness of selfishness, when one considers a group in which altruism is common. The net result is that the special cultural transmission rule can allow a characteristic to evolve that could not evolve without it. Within purely biological models, altruism is eliminated in a large range of parameter values. The prospects for altruism to evolve are enhanced when culture is included in the model. Just as in Cavalli-Sforza and Feldman's discussion of the demographic transition, assumptions about cultural transmission lead to predictions that would not be true if a purely biological and non-cultural process were postulated.

The two examples I have described are typical of the models discussed in the two books. The models aim to show how different patterns of cultural transmission make a difference for how a psychological or social characteristic will evolve. Although most of the emphasis is placed on identifying cultural analogs of natural selection, the authors do develop nonselective models of cultural change. For example, population geneticists have described how genes of nearly identical fitness can change frequency in a population by doing a random walk. The models developed for random genetic drift, as it is called, can be used to describe the process by which family names disappear. This helps explain why the descendants of the *Bounty* mutineers have come to share the same surname. A reduction in variation is the expected consequence of random walks, both genetic and cultural.[14]

What are we to make of the research program embodied in these books? Biologists interested in culture are often struck by the absence of viable general theories in the social sciences. All of biology is united by the theory of biological evolution. Perhaps progress in the social sciences is impeded because there is no general theory of cultural evolution. The analogies between cultural and genetic change are palpable. And at least some of the disanalogies can be taken into account when the biological models are transposed.

For example, the Weismann doctrine tells us that variation is "undirected"; mutations do not occur because they would be beneficial. But ideas are not invented at random. Individuals often create new ideas—in science, for example—precisely because they would be useful.[15] Another and related disanalogy concerns the genotype/phenotype distinction. An organism's genotype is a cause of the phenotype it develops; that same genotype also causally contributes to the genotype of the organism's offspring. But there is no further pathway by which a parental phenotype can causally shape the genotype of its offspring. This is one way of describing the idea that there is no "inheritance of acquired characteristics." No such constraints seems to apply to the learning that occurs in cultural transmission.

These disanalogies between genetic and cultural change do not show that it is pointless or impossible to write models of cultural evolution that draw on the mathematical resources provided by evolutionary theory. In a sense, it is precisely because of such differences that there is a point to seeing the consequences of models that take these differences into account. These structural differences between genetic and cultural evolution do not undermine the idea that models of cultural evolution have a point.

Another reservation that has been voiced about models of cultural evolution is that they atomize cultural characteristics. Having two children rather than five, or being a kamikaze pilot, are characteristics that are abstracted from a rich and interconnected network of traits. The worry is that by singling out these traits for treatment, we are losing sight of the context that gives them cultural meaning.

It is worth mentioning that precisely the same question has been raised about various models in genetic evolution itself. If you wish to understand the population frequency of sickle cell anemia, for example, you cannot ignore the fact that the trait is correlated with resistance to malaria. In both cultural and genetic evolution it is a mistake to think that each trait evolved independently of all the others. Of course, the lesson to be drawn from this is not that one should not atomize characteristics, but rather that the atoms one identifies should be understood in terms of their relationship to other atoms.

In fact, this emphasis on context is one of the virtues that Boyd and Richerson think their approach has over the approach taken by sociobiology. According to the models under review, genetic selection has given our species the ability to engage in social learning. Once in place, this cultural transmission system allows characteristics to evolve that could not have evolved without it. In other words, it is only because the traits in question evolve in the context of a cultural transmission system that they are able to evolve at all.

We need to recognize that the descriptors singled out for treatment in science always abstract from complexities. If there is an objection to the descriptors used in models of cultural evolution, it must concern the details of how these models are constructed, not the mere fact that they impose a descriptive framework of some sort or other.[16]

Although the criticisms I have reviewed so far do not seem very powerful, there is a rather simple fact about these models that does suggest that they may be of limited utility in the social sciences. Insofar as these models describe culture, they describe systems of cultural transmission and the evolutionary consequences of such systems. Given that the idea of having two children was more attractive than the idea of having five, and given the horizontal and oblique transmission systems thought to be in place, we can see why the demographic transition took place. But as Cavalli-Sforza and Feldman recognize, their model does not begin to describe why educated women in nineteenth-century Italy came to prefer having smaller families, or why patterns adopted in higher classes cascaded down to lower ones. The model describes the consequences of an idea's being attractive, not the causes of its being attractive.

This distinction between the consequences of fitness differences and the causes of fitness differences also applies to theories of biological evolution.[17] A population geneticist can tell you what the evolutionary consequences for a population will be, if the genes in the population bear various fitness relationships to each other. It is a separate question to say why a given gene in fact is fitter than the alternatives. For example, consider the simplest of one-locus two-allele models for a diploid population. There are three genotypes possible at the locus in question, which we might label AA, Aa, and aa. If the heterozygote genotype is fitter than the two homozygote forms, the population will evolve to a stable polymorphism. Neither allele will be eliminated by the selection process. This is a simple algebraic fact, one having nothing to do with the biological details of any living population. Models such as this one can be thought of as intellectual resources that biologists interested in some particular population might find reason to use.

When human geneticists apply this model to the sickle cell system, they say that Aa is the fittest genotype because heterozygotes at the locus in question have enhanced resistance to malaria and little or no anemia. The two homozygotes have lesser fitnesses because they are either anemic or lack the malaria resistance. These specific remarks about the locus in the relevant human population describe the sources of fitness differences. Alternatively, a fruitfly geneticist may take the same population genetics model and apply it to a locus in some Drosophila population by saying that the heterozygote has enhanced temperature tolerance over the two homozygotes. The population consequences of heterozygote superiority are the same in the two cases; a stable polymorphism evolves. It is the sources of the fitness differences that distinguish the human application from the application to fruitflies.

This, I think, is the main shortcoming of the models of cultural evolution I am considering. The illumination they offer of culture concerns the consequences of cultural transmission systems. But there is far more to culture than the consequences of the rules that describe who learns what from whom. Social scientists have not wholly ignored the way that patterns of influence are structured in specific cases. A historian of nineteenth-century Italy might attempt to explain why some traits found among educated people were transmitted to lower social strata, while others were not. Again, it is the sources of the transmission system that will interest the social scientist. The social scientist will take it for granted that the consequences of this influence will be that ideas cascade from one class to another.

Models of transmission systems describe the quantitative consequences of systems of cultural influence. Social scientists inevitably make qualitative assumptions about the consequences of these systems. If it could be shown that these qualitative assumptions were wrong in important cases, and that these mistakes actually undermine the plausibility of various historical explanations, that would be a reason or social scientists to take a greater interest in these models of cultural evolution. But if the qualitative assumptions turn out to be correct, it is perhaps understandable that historians should not accord much importance to these investigations.[18]

Population genetics really is a unifying framework within evolutionary theory. Fruitflies and human beings differ in many ways, but if a one-locus system exhibits heterozygote superiority, the population consequences will be the same, regardless of whether we are talking about people or Drosophila. Evolutionary theory is much less unified when we consider what it has to say about the sources of fitness differences. There are many, many models that treat a multiplicity of life-history characteristics and ecological relationships. Evolutionary theory achieves its greatest generality when it ignores sources and focuses on consequences.

The transposition of evolutionary models to the social sciences is a transposition of the most unified and complete part of evolutionary theory, one that leaves behind less unified theoretical ideas. This is not a criticism of the models of cultural evolution that result, but a fact about the price one pays for very general theorizing of this type. Cultural learning is a cultural universal. And patterns of cultural learning conveniently divide into vertical, horizontal, and oblique subcases. When ideas differ in their attractiveness, the system of transmission will determine the rate of change and the end-state that the population achieves. Only because they develop theories *within* this narrow compass do these models *of* cultural evolution have the generality they do.

Many of the examples discussed in the two books I have been considering describe evolution within a culture, not the evolution of the cultural transmission system itself. However, Boyd and Richerson, especially, also concern themselves with the way a system of cultural learning could have evolved by straightforward Darwinian means. Here the authors are not giving a model of how human cultures work, once they exist, but are trying to show how cultural learning became a possibility in the first place. This project obviously is a very important one, but not one that applies to many social scientific research programs. A correct genetic explanation of this important feature of the human phenotype would not provide a unifying framework within which social scientists would then do their work. They would not use this theory at all. It is one thing to explain the demographic transition in nineteenth-century Italy, something else to explain why human beings are able to learn from individuals who are not their biological parents.

In spite of these shortcomings, there is a basic achievement of these models of cultural evolution that deserves emphasis. A persistent theme in debates about sociobiology, about the nature/nurture controversy, and in other contexts as well is the relative "importance" that should be accorded to biology and culture. I place the term "importance" in quotation marks to indicate that it is a vague idea crying out for explication. Nonetheless, it has been a fundamental problem in these controversies to assess the relative "strength" or "power" of biological and cultural influences.

One virtue of these models of cultural evolution is that they place culture and biology into a common framework, so that the relative contributions to an outcome are rendered commensurable. What becomes clear in these models is that in assessing their relative importance of biology and culture, *time is of the essence*. Culture is often a more powerful determiner of change than biological evolution because cultural changes occur faster. When biological fitness is calibrated in terms of having babies, its basic temporal unit is the span of a human generation. Think how many replication events can occur in that temporal interval when the reproducing entities are ideas that jump from head to head. Ideas spread so fast that they can swamp the slower (and hence weaker) impact of biological natural selection.

There is a vague idea about the relation of biology and culture that these models help lay to rest. This is the idea that biology is "deeper" than the social sciences, not just in the sense that it has developed further, but in the sense that it investigates more fundamental causes. A social scientist will explain incest avoidance by describing the spread of a custom; the evolutionary biologist goes deeper by showing us why the behavior evolved. The mind-set expressed here is predisposed to think that culture is always a weak influence when it opposes biology. The works described here deserve credit for showing why this common opinion rests on a confusion.

In spite of this achievement, I doubt that these models of cultural evolution provide a general framework within which social scientific investigations may proceed. My main reason for skepticism is that these models concern themselves with the consequences of transmission systems and fitness differences, not with their sources. Social scientists interested in cultural change generally focus on sources and make do with intuitive and qualitative assessments of what the consequences will be. It isn't that the biologists and the social scientists are in conflict; rather, they are talking past each other.

Dobzhansky is famous for having said that "nothing in biology can be understood except in the light of evolution." His idea was not the modest one that evolution is necessary for full understanding; that would be true even if evolution's contribution were minor, though ineliminable. Rather, Dobzhansky had in mind the stronger claim that evolutionary considerations should be assigned pride of place in our understanding of the living world. A transposition of Dobzhansky's slogan to the topic of this chapter would say that "nothing in the social sciences can be understood except in the light of models

Elliott Sober

of cultural evolution." My suspicion is that only the weaker reading of this pronouncement is defensible.

NOTES

I worked on this chapter while a William Evans Fellow at the University of Otago during parts of July and August 1990; my thanks to the university and to the members of the Philosophy Department for inviting me and for making my stay such an enjoyable one. This work expands upon a talk I gave in December 1985 at the University of Palma de Mallorca, "Natural Selection and the Social Sciences." I'm grateful to Robert Boyd, Dan Hausman, Peter Richerson, and David S. Wilson for comments on an earlier draft.

1. John Maynard Smith (1982), *Evolution and the Theory of Games*, Cambridge University Press.

2. L. Cavalli-Sforza and M. Feldman (1981), *Cultural Transmission and Evolution: A Quantitative Approach*, Princeton University Press.

3. R. Boyd and P. Richerson (1985), *Culture and the Evolutionary Process*, University of Chicago Press.

4. "Having babies" should be interpreted broadly, so as to include "having grandbabies," "having greatgrandbabies," and so forth. In some selection models (e.g., Fisher's sex ratio argument), fitness differences require that one consider expected numbers of descendants beyond the first generation.

5. See R. Colwell and M. King (1983), "Disentangling Genetic and Cultural Influences on Human Behavior: Problems and Prospects," in D. Rajecki (ed.), *Comparing Behavior: Studying Man Studying Animals*, Lawrence Erlbaum Publishers.

6. These are reviewed in J. Hirshliefer (1977), "Economics from a Biological Viewpoint," *Journal of Law and Economics* 1:1–52.

7. See K. Popper (1973), *Objective Knowledge*, Oxford University Press.

8. A variety of "selective-retention" models of learning and of scientific change are reviewed in Donald Campbell (1974), "Evolutionary Epistemology," in P. Schilpp (ed.), *The Philosophy of Karl Popper*, Open Court Publishing. David Hull's *Science as a Process* (University of Chicago Press, 1988) develops some interesting ideas about how evolutionary ideas can be used to explain scientific change.

9. I do not claim that this taxonomy is exhaustive. For example, the spread of an infectious disease may be thought of as a selection process, in which the two states of an individual ("infected" and "not infected") differ in how catching they are. Clearly, this is not a type I process. Arguably, the concept of learning does not permit this process to be placed in type II. Perhaps the taxonomy would be exhaustive, if "learning" were replaced by "phenotypic resemblance not mediated by genetic resemblance."

10. See, for example, Richard Alexander (1979), *Darwinism and Human Affairs*, University of Washington Press.

11. See D. Lack (1954), *The Optimal Regulation of Animal Numbers*, Oxford University Press.

12. E. Wilson (1975), *Sociobiology: The New Synthesis*, Harvard University Press.

13. See E. Sober (1988), "What is Evolutionary Altruism?" New Essays on Philosophy and Biology (*Canadian Journal of Philosophy Supplementary* Volume 14), University of Calgary Press.

14. See Cavalli-Sforza and Feldman, pp. 255–66.

15. The difference between directed and undirected variation is conceptually different from the difference between biased and unbiased transmission. The former concerns the probability that a mutation will arise; the latter has to do with whether it will be passed along.

Directed variation (mutation) can be described as follows. Let u be the probability of mutating from A to a and v be the probability of mutating from a to A. Mutation is directed if (i) $u > v$ and (ii) $u > v$ because $w(a) > w(A)$, where $w(X)$ is the fitness of X.

16. See J. M. Smith (1989), *Did Darwin Get it Right?* Chapman and Hall.

17. See Elliott Sober (1984), *The Nature of Selection*, MIT Press.

18. So the question about the usefulness of these models of cultural evolution to the day-to-day research of social scientists comes to this: Are social scientists good at intuitive population thinking? If they are, then their explanations will not be undermined by precise models of cultural evolution. If they are not, then social scientists should correct their explanations (and the intuitions on which they rely) by studying these models.

Acknowledgments

1. Susan Mills and John Beatty, "The Propensity Interpretation of Fitness," *Philosophy of Science*, 1979, 46:263–286. Reprinted by permission of the publisher and the authors.

2. Larry Wright, "Functions," *Philosophical Review*, 1973, 82:139–168. Reprinted by permission of the publisher and the author.

3. Robert Cummins, "Functional Analysis," *Journal of Philosophy*, 1975, 72:741–764. Reprinted by permission of the publisher and the author.

4. Stephen Jay Gould and Richard Lewontin, "The Spandrels of San Marco and the Panglossian Paradigm: A Critique of the Adaptationist Programme," *Proc. R. Soc. London*, 1978, 205:581–598. Reprinted by permission of the publisher and the authors.

5. John Maynard Smith, "Optimization Theory in Evolution," *Annual Review of Ecology and Systematics*, 1978, 9:31–56. Reprinted by permission of the publisher and the author.

6. George C. Williams, excerpts from *Adaptation and Natural Selection*. Princeton University Press, 1966, 4–5, 16–19, 22–25, 92–101, 108–124, 208–212. © 1966 by Princeton University Press. Reprinted by permission of the publisher and the author.

7. David Sloan Wilson, "Levels of Selection: An Alternative to Individualism in Biology and the Human Sciences," *Social Networks*, 1989, 11:257–272. Reprinted by permission of the publisher and the author.

8. Ernst Mayr, "Typological versus Population Thinking," *Evolution and the Diversity of Life*, The Belknap Press of Harvard University Press, 1976, 26–29. Copyright by the President and Fellows of Harvard College. Reprinted by permission of the publisher and the author.

9. Elliott Sober, "Evolution, Population Thinking, and Essentialism," *Philosophy of Science*, 1980, 47:350–383. Reprinted by permission of the publisher and the author.

10. David Hull, "A Matter of Individuality," *Philosophy of Science*, 1978, 45: 335–360. Reprinted by permission of the publisher and the author.

11. Brent D. Mishler and Michael J. Donoghue, "Species Concepts: A Case for Pluralism," *Systematic Zoology*, 1982, 31:491–503. Reprinted by permission of the publisher and the authors.

12. Robert Sokal, "The Continuing Search for Order," *American Naturalist*, 1985, 126:729–749. Reprinted by permission of the author and publisher.

13. Willi Hennig, "Phylogenetic Systematics," *Annual Review of Entomology*, 1965, 10:97–116. Reprinted by permission of the publisher.

14. Ernst Mayr, "Biological Classification: Toward a Synthesis of Opposing Methodologies," *Science*, 1981, 214:510–516. © 1981 by the AAAS. Reprinted by permission of the author and publisher.

15. David L. Hull, "Contemporary Systematic Philosophies," *Annual Review of Ecology and Systematics*, 1970, 1:19–53. Reprinted by permission of the publisher and the author.

16. James Farris, "The Logical Basis of Phylogenetic Analysis," in N. Platnick and V. Funk (eds.), *Advances in Cladistics: Proceedings of the Second Meeting of the Willi Hennig Society*, Columbia University Press, 1982, 7–36. Reprinted by permission of the publisher and the author.

17. Joseph Felsenstein, "The Detection of Phylogeny," in D. Hawksworth (ed.), *Prospects in Systematics*. Systematics Association, Clarendon Press, 1988, 112–127. © The Systematics Association 1988. Reprinted by permission of Oxford University Press.

18. Philip Kitcher, "1953 and All That: A Tale of Two Sciences," *Philosophical Review*, 1984, 93:335–373. Reprinted by permission of the publisher and the author.

19. C. Kenneth Waters, "Why the Anti-reductionist Consensus Won't Survive the Case of Classical Mendelian Genetics," *PSA 1990*, Philosophy of Science Association, Volume 1, 125–139. Reprinted by permission of the publisher and the author.

20. Michael Ruse and Edward O. Wilson, "Moral Philosophy as Applied Science," *Philosophy*, 1986, 61:173–192. Reprinted by permission of Cambridge University Press and the authors.

21. Philip Kitcher, "Four Ways of 'Biologicizing' Ethics," First published in German as "Vier Arten, die Ethik zu 'biologisieren'" in K. Bayertz (ed.), *Evolution und Ethik*, Reclams Universal-Bibliothek Nr. 8857, 1993. © 1993 Philip Reclam Jun., Stuttgart. Reprinted by permission of Philipp Reclam Jun., Stuttgart and the author.

22. Michael Bradie, "Epistemology from an Evolutionary Point of View." This essay appears here for the first time.

23. Elliott Sober, "Models of Cultural Evolution," in P. Griffiths (ed.), *Trees of Life: Essays in the Philosophy of Biology* (Australasian Studies in the History and Philosophy of Science), Kluwer Academic Publishers, 1991, 17–38. Reprinted by permission of the publisher and the author.

Index

Coleman, W., 185n

Colless, D., 246, 303, 306, 308–309, 311, 325n, 352

Collins, B., 460–461, 467–468

Color perception, genetic basis of, 427–428

Comparative biology, 235

Compatibility methods. See Clique methods

Convergence. See Homoplasy

Coombs, E., 228n

Coon, C., 78

Copernicus, N., 158

Cosmides, L., 147, 150

Costa, R., 79

Cowan, S., 222

Cowe, J., 87

Cracraft, J., 217, 225, 239–240, 242

Crane, P., 225

Crick, F., 379, 383, 391, 398, 404

Cronquist, A., 225, 228n, 229n

Crovello, T., 219, 318–320, 322–324, 325n

Crow, J., 4, 6, 18, 93, 147, 150, 185n

Crozier, R., 113

Crumson, R., 205

Curio, E., 92, 97, 105–106, 114

Cuvier, G., 298, 307

Darden, L., 401, 415n, 459

Darwin, C., x, xii, 3, 5, 20n, 73, 76, 81, 82–84, 89, 91, 96, 98–99, 125, 128–129, 146, 157–159, 161, 166, 171, 175, 177, 185n–186n, 194, 204, 236, 277–279, 282, 285, 288, 290, 291n, 292n, 297–298, 323, 343, 421–422, 454, 457–458, 460, 470–471, 477

Darwin, F., 291n

Davidson, D., 461

Davis, D., 291n

Davis, P., 220, 227, 228n

Davitashvili, L., 78

Dawkins, R., xii, 143, 147, 150, 197–199, 399n

Dayhoff, M., 364

Debry, R., 366

Delbruck, M., 186n

Demographic transition, 482–484

de Montellano, B., 76

Dendograms, 258

Dennett, D., 461

Descartes, R., 340, 453, 466

Dewey, J., 453, 461, 471

Dickson, C., 124

Discontinuities, 242

in breeding, 223

in morphology, 223

in nature, 220–221

Distance matrix methods, 368. See also Phenetics

Dobzhansky, T., 5–6, 10, 17, 204, 229n, 295–297, 300–302, 304, 310, 313–316, 325n, 381, 490

Doner, M., 259

Dressler, D., 411

Dretske, F., 461–462

Duncan, T., 365

Dunn, L., 134

Dupraw, E., 239

Eck, R., 364

Edelman, G., 455, 459, 460

Edwards, A., 312, 345, 363–366, 369

Ehrenpreis, A., 152

Ehrlich, P., 221, 312, 323

Ehrlichs, A., 312

Einstein, A., 460

Eldredge, N., 167, 184, 196, 199, 202, 217, 225, 227, 239, 242, 282, 292n

Electrophoretic data, 346–347

Elster, J., 464

Emlen, J., 92

Emotivism, 445

Endler, J., 222

Environmental constraints, 98

Environmental variation, 109

Enzyme polymorphism, 97

Epigenetic rules, 426–433, 435–436, 437n, 457

Epistasis, 436

Epistemology

and Darwinism, 456

evolutionary vs. traditional, 460–462

norms in, 469–470

and relativism, 467

selectionist, 470

Essentialism, 161–163, 168, 175–176, 183–184, 186n, 314

bean-bag genetics and, 185n

chemical, 165–167

evolution and, 165

evolutionary theory and, 168, 177

explanation and, 164

Hull, D. and, 186n

individual variation and, 176

natural kinds and, 164

nonsaltative evolution and, 167

organismic perspective and, 185

population thinking and, xiii

Essentialism (cont.)
  the Sorites problem and, 166–167
  variation and, 162, 170–175, 178, 181,
    183–184
Estabrook, G., 240–241, 245, 335
Ethics. See Moral realism; Moral reasoning;
  Moral relativism; Sociobiology
Evolution, 83, 85, 91, 99, 101–102, 109–
    110, 113–114, 134, 138, 157–158, 162,
    193–194, 422–423, 478, 490
  biotic vs. organic, 127–129, 135–136
  brains and, 430, 458–459 (see also Neural
    Darwinism)
  competition and, 103
  conceptual vs. biological, 463–465
  constraints on, 73–76, 85–87, 96
  cultural, xvi, 478, 487–490, 492n
  as descent with modification, 223, 290, 488
  environment and, 101–103, 200
  essentialism and, 165
  ethics and, 421–422, 425–427, 431–432
  extinction and, 131, 135–137
  forces of, 133
  the fossil record and, 128
  function and, 41
  game theory and, 109–113, 150, 477 (see
    also Evolutionary stable strategy)
  historical factors in, 102
  human, 423–424, 447
  irreversibility of, 349–350, 353
  knowledge and, 456–458, 465
  macro-, 217, 227
  by natural selection, 161
  nonadaptive, 77
  parsimony and rate of, 367–368
  phenotypic, 95
  rate of, 352–354, 367, 372, 437n
  replication and, 197–198
  scientific theories and, 463, 480, 491n
  selection and, 197
  sexuality and, 181
  species and, 199
  stochastic models of, 339
  time and, 490
  units of, 195, 218, 225
Evolutionary classification, 277–279, 285,
    304
Evolutionary epistemology. See
  Epistemology
Evolutionary lag, 98
Evolutionary laws, 207–208
Evolutionary species concept, 228n
Evolutionary stable strategy, 100, 103, 110–

112, 150–151. See also Evolution, game
  theory and
Evolutionary taxonomy, xiv–xv, 284, 287–
    288, 290, 293n, 295–296, 300–301, 303,
    305–306, 319, 321–323. See also
    Cladistics; Phenetics; Systematics
  character analysis and, 283
  vs. phenetics, 307–308, 318
Evolutionary theory, 73, 80–81, 85, 89, 121,
    124, 143, 146, 159, 164, 172, 194, 196,
    207–208, 260, 262, 290, 295, 297, 298,
    304–305, 310, 325n, 453, 471, 478, 489
  Aristotle and, 170
  biological species concept and, 317
  circularity of, 3–7
  classification and, 236, 281
  essentialism and, 163, 168, 177
  function and, 60
  individuals vs. natural kinds in, 167–168
  natural selection and, 130
  phenetics and, 308
  population genetics and, 489
  population models and, 177
  population thinking and, 175, 185
  social science and, 477–493
  variability and, 184
Explanation, 20n
  covering law account of, 385
  deductive-nomological, 50, 54
  and explanatory power, 342–346, 348–
    349, 353
  essentialism and, 164
  function and, 43–45, 49–50, 53–57, 61–
    63, 65–67, 91, 100
Extinction, 131, 135–137
  in monophyletic groups, 265
  in paraphyletic groups, 265

Fahrenholz rule, 368
Falconer, D., 83
Falsification, 334–335. See also Popper, K.
Family resemblance, 209. See also Phenetics
Farris, J., 242, 246, 290, 292n, 293n, 335,
    340–342, 344–347, 349–351, 353, 364–
    365, 368–369, 372
Feldman, M., 478, 482–484, 486, 488
Felsenstein, J., 246, 338–341, 345, 349, 356–
    357, 366–368, 370–372
Fenner, F., 148
Feyerabend, P., 229n
Fiala, K., 247, 250, 252–253
Fink, W., 194n, 228n, 342
Fish, S., 228n

Quantitative characters, 370–371
Quetelet, A., 172–175, 186n
Quine, W., 161, 177, 462, 469

Rabel, G., 165
Ramsbottom, J., 165
Rapoport, A., 111
Ratcliffe, F., 148
Raven, P., 221–222, 228n, 236, 323
Rawls, J., 447
Realism, 464, 466–469
  causation and, 176–177, 183
  convergent, 468–469
  hypothetical, 467–469
  variability and, 178
Reductionism, xv, 143, 164, 177, 379–
    399, 401–416, 424. *See also* Classical
    genetics; Mendelian genetics; Molecular
    biology
Remane, A., 85, 236, 297
Rensch, B., 83, 104, 297
Replication, 392. *See also* Gene replication
  evolution and, 197–198
Reproductive communities, species and, 257
Reproductive isolation, 220–221, 223–225
Rescher, N., 454, 456
Richards, R., 449n, 470
Richerson, P., 153, 401, 440, 478, 482, 484–
    487, 489, 491n
Ricklefs, M., 217
Ridley, M., 211n
Riedl, R., 86, 236, 461
Riggins, R., 344
Rigid designation, 212n
Robertson, A., 92, 95
Robinet, J., 166
Robischon, G., 22n
Roger, J., 171
Rohlf, F., 243–246, 250, 252–253, 308,
    345–346
Romanes, G., 78, 81
Romberger, J., 293n
Rosado, T., 92, 95
Rosen, D., 224
Rosenberg, A., ix, 401, 405–406, 415n
Rowher, S., 99–100
Rudwick, M., 78
Ruse, M., ix, 3, 5, 20n, 36, 211n, 401, 404,
    439–441, 443, 446–447, 449n, 454, 456–
    458, 467, 470–471
Russell, E., 123

Sahlins, M., 76
Saltation, 186n, 196

Savage, R., 92, 94, 99, 107
Schaffner, K., 398n, 401, 404
Schilpp, P., 455–456
Schindewolf, O., 85, 292n
Schirmer, J., 253
Schoener, T., 94, 104–105
Schopf, J., 206
Schopf, T., 211n
Schuh, 345–346
Scientific change, xvi
Scientific theories
  as sets of statements, 380
  as species, 209
Scriven, M., 7
Segregation distortion, 135
Seilacher, A., 86–88
Selan, B., 211n
Selection, 89, 96–97, 110, 133, 194, 480,
    485. *See also* Units of selection
  between-group, 125, 148–149, 152
  between-individual vs. within-individual,
    150
  circularity of, 130
  differential, 222
  evolution and, 197
  functions and, 104
  genic, 123–124, 126–127, 131, 133–137,
    149, 196, 198, 487
  group, 95, 126, 131, 133–135, 138, 146–
    148, 150–151, 153, 187n, 198, 212n, 485
  individual, 124, 146–147, 151
  means of phenotype transmission and,
    479–481, 485–486
  phenotypic, 99, 123
  population, 198
  sexual, 102
Sex ratio, 93–95
Sexual reproduction, 124
Shao, K., 246, 252–253
Shea, B., 78
Sheppard, P., 18, 97, 124
Shimony, A., 469
Sibley, C., 292n
Sibly, R., 109
Similarity, 335. *See also* Phenetics
  individuation of species and, 203
  overall, 312
Similarity classes, 282
Simpson, G., 131–132, 137, 202–203, 218,
    273, 283, 285, 291n, 292n, 293n, 295–
    297, 300–301, 304, 314, 316, 322–324
Skagestad, P., 465, 468
Skepticism, 462
Skinner, B., 29, 460